W9-AVK-617

Short Courses

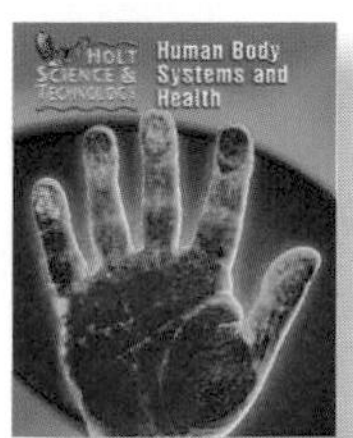

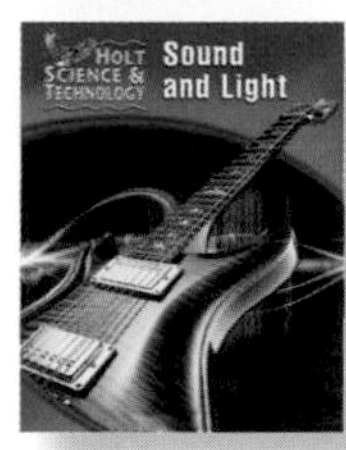

Teacher Edition WALK-THROUGH

Student Edition CONTENTS IN BRIEF

HOLT, RINEHART AND WINSTON
A Harcourt Education Company
Orlando • **Austin** • New York • San Diego • Toronto • London

Designed to meet the needs of all students

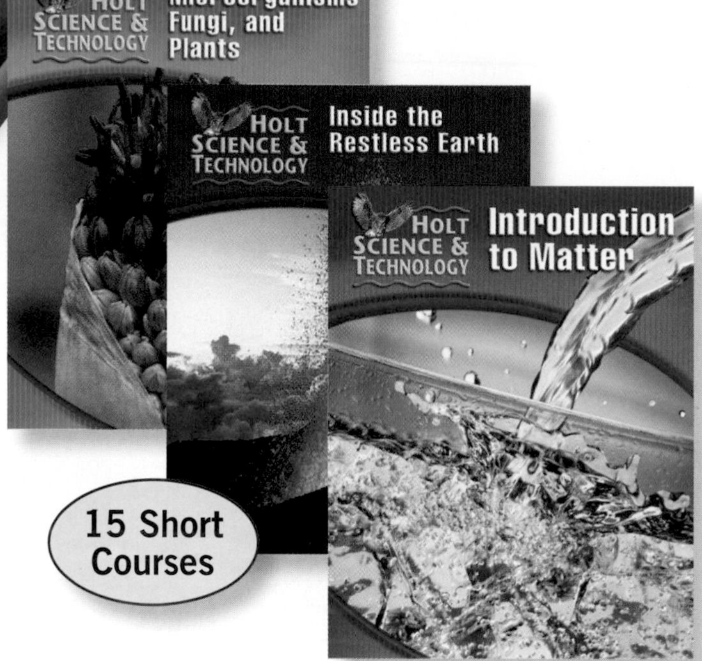

Holt Science & Technology: Short Course Series allows you to match your curriculum by choosing from 15 books covering life, earth, and physical sciences. The program reflects current curriculum developments and includes the strongest skills-development strand of any middle school science series. Students of all abilities will develop skills that they can use both in science as well as in other courses.

STUDENTS OF ALL ABILITIES RECEIVE THE READING HELP AND TAILORED INSTRUCTION THEY NEED.

- The *Student Edition* is accessible with a clean, easy-to-follow design and highlighted vocabulary words.
- Inclusion strategies and different learning styles help support all learners.
- Comprehensive **Section** and **Chapter Reviews** and **Standardized Test Preparation** allow students to practice their test-taking skills.
- **Reading Comprehension Guide** and **Guided Reading Audio CDs** help students better understand the content.

CROSS-DISCIPLINARY CONNECTIONS LET STUDENTS SEE HOW SCIENCE RELATES TO OTHER DISCIPLINES.

- **Mathematics, reading,** and **writing skills** are integrated throughout the program.
- Cross-discipline **Connection To** features show students how science relates to language arts, social studies, and other sciences.

A FLEXIBLE LABORATORY PROGRAM HELPS STUDENTS BUILD IMPORTANT INQUIRY AND CRITICAL-THINKING SKILLS.

- The laboratory program includes labs in each chapter, labs in the **LabBook** at the end of the text, six different lab books, and **Video Labs.**
- All labs are teacher-tested and rated by difficulty in the *Teacher Edition,* so you can be sure the labs will be appropriate for your students.
- A variety of labs, from **Inquiry Labs** to **Skills Practice Labs,** helps you meet the needs of your curriculum and work within the time constraints of your teaching schedule.

INTEGRATED TECHNOLOGY AND ONLINE RESOURCES EXPAND LEARNING BEYOND CLASSROOM WALLS.

- An **Enhanced Online Edition** or **CD-ROM Version** of the student text lightens your students' load.
- **SciLinks,** a Web service developed and maintained by the National Science Teachers Association (NSTA), contains current prescreened links directly related to the textbook.
- **Brain Food Video Quizzes** on videotape and DVD are game-show style quizzes that assess students' progress and motivate them to study.
- The **One-stop Planner® CD-ROM** with **Exam View® Test Generator** contains all of the resources you need including an *Interactive Teacher Edition,* worksheets, customizable lesson plans, **Holt Calendar Planner,** a powerful test generator, **Lab Materials QuickList Software,** and more.
- Spanish Resources include **Guided Reading Audio CD** in Spanish.

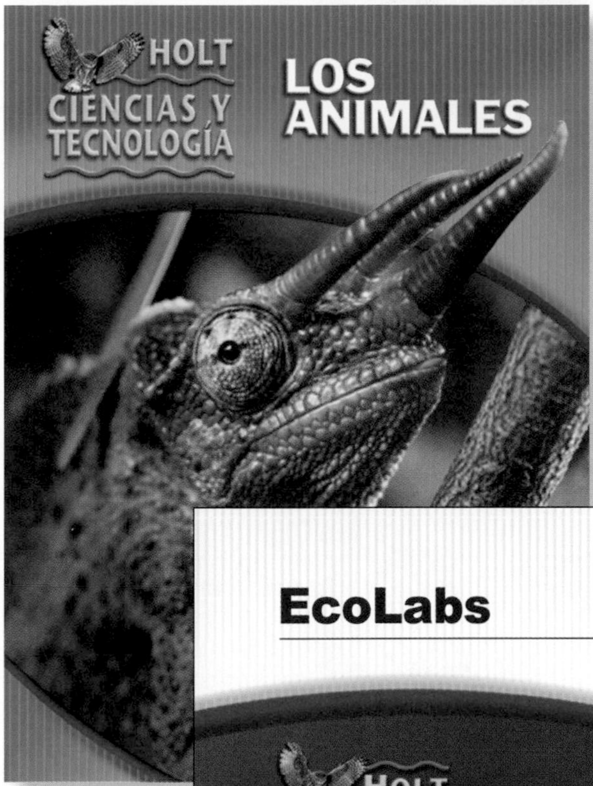

Life Science

	A MICROORGANISMS, FUNGI, AND PLANTS	B ANIMALS
CHAPTER 1	**It's Alive!! Or, Is It?** • Characteristics of living things • Homeostasis • Heredity and DNA • Producers, consumers, and decomposers • Biomolecules	**Animals and Behavior** • Characteristics of animals • Classification of animals • Animal behavior • Hibernation and estivation • The biological clock • Animal communication • Living in groups
CHAPTER 2	**Bacteria and Viruses** • Binary fission • Characteristics of bacteria • Nitrogen-fixing bacteria • Antibiotics • Pathogenic bacteria • Characteristics of viruses • Lytic cycle	**Invertebrates** • General characteristics of invertebrates • Types of symmetry • Characteristics of sponges, cnidarians, arthropods, and echinoderms • Flatworms versus roundworms • Types of circulatory systems
CHAPTER 3	**Protists and Fungi** • Characteristics of protists • Types of algae • Types of protozoa • Protist reproduction • Characteristics of fungi and lichens	**Fishes, Amphibians, and Reptiles** • Characteristics of vertebrates • Structure and kinds of fishes • Development of lungs • Structure and kinds of amphibians and reptiles • Function of the amniotic egg
CHAPTER 4	**Introduction to Plants** • Characteristics of plants and seeds • Reproduction and classification • Angiosperms versus gymnosperms • Monocots versus dicots • Structure and functions of roots, stems, leaves, and flowers	**Birds and Mammals** • Structure and kinds of birds • Types of feathers • Adaptations for flight • Structure and kinds of mammals • Function of the placenta
CHAPTER 5	**Plant Processes** • Pollination and fertilization • Dormancy • Photosynthesis • Plant tropisms • Seasonal responses of plants	
CHAPTER 6		
CHAPTER 7		

PROGRAM SCOPE AND SEQUENCE

Selecting the right books for your course is easy. Just review the topics presented in each book to determine the best match to your district curriculum.

C CELLS, HEREDITY, & CLASSIFICATION	D HUMAN BODY SYSTEMS & HEALTH	E ENVIRONMENTAL SCIENCE
Cells: The Basic Units of Life • Cells, tissues, and organs • Populations, communities, and ecosystems • Cell theory • Surface-to-volume ratio • Prokaryotic versus eukaryotic cells • Cell organelles	**Body Organization and Structure** • Homeostasis • Types of tissue • Organ systems • Structure and function of the skeletal system, muscular system, and integumentary system	**Interactions of Living Things** • Biotic versus abiotic parts of the environment • Producers, consumers, and decomposers • Food chains and food webs • Factors limiting population growth • Predator-prey relationships • Symbiosis and coevolution
The Cell in Action • Diffusion and osmosis • Passive versus active transport • Endocytosis versus exocytosis • Photosynthesis • Cellular respiration and fermentation • Cell cycle	**Circulation and Respiration** • Structure and function of the cardiovascular system, lymphatic system, and respiratory system • Respiratory disorders	**Cycles in Nature** • Water cycle • Carbon cycle • Nitrogen cycle • Ecological succession
Heredity • Dominant versus recessive traits • Genes and alleles • Genotype, phenotype, the Punnett square and probability • Meiosis • Determination of sex	**The Digestive and Urinary Systems** • Structure and function of the digestive system • Structure and function of the urinary system	**The Earth's Ecosystems** • Kinds of land and water biomes • Marine ecosystems • Freshwater ecosystems
Genes and Gene Technology • Structure of DNA • Protein synthesis • Mutations • Heredity disorders and genetic counseling	**Communication and Control** • Structure and function of the nervous system and endocrine system • The senses • Structure and function of the eye and ear	**Environmental Problems and Solutions** • Types of pollutants • Types of resources • Conservation practices • Species protection
The Evolution of Living Things • Adaptations and species • Evidence for evolution • Darwin's work and natural selection • Formation of new species	**Reproduction and Development** • Asexual versus sexual reproduction • Internal versus external fertilization • Structure and function of the human male and female reproductive systems • Fertilization, placental development, and embryo growth • Stages of human life	**Energy Resources** • Types of resources • Energy resources and pollution • Alternative energy resources
The History of Life on Earth • Geologic time scale and extinctions • Plate tectonics • Human evolution	**Body Defenses and Disease** • Types of diseases • Vaccines and immunity • Structure and function of the immune system • Autoimmune diseases, cancer, and AIDS	
Classification • Levels of classification • Cladistic diagrams • Dichotomous keys • Characteristics of the six kingdoms	**Staying Healthy** • Nutrition and reading food labels • Alcohol and drug effects on the body • Hygiene, exercise, and first aid	

Earth Science

	F INSIDE THE RESTLESS EARTH	G EARTH'S CHANGING SURFACE
CHAPTER 1	**Minerals of the Earth's Crust** • Mineral composition and structure • Types of minerals • Mineral identification • Mineral formation and mining	**Maps as Models of the Earth** • Structure of a map • Cardinal directions • Latitude, longitude, and the equator • Magnetic declination and true north • Types of projections • Aerial photographs • Remote sensing • Topographic maps
CHAPTER 2	**Rocks: Mineral Mixtures** • Rock cycle and types of rocks • Rock classification • Characteristics of igneous, sedimentary, and metamorphic rocks	**Weathering and Soil Formation** • Types of weathering • Factors affecting the rate of weathering • Composition of soil • Soil conservation and erosion prevention
CHAPTER 3	**The Rock and Fossil Record** • Uniformitarianism versus catastrophism • Superposition • The geologic column and unconformities • Absolute dating and radiometric dating • Characteristics and types of fossils • Geologic time scale	**Agents of Erosion and Deposition** • Shoreline erosion and deposition • Wind erosion and deposition • Erosion and deposition by ice • Gravity's effect on erosion and deposition
CHAPTER 4	**Plate Tectonics** • Structure of the Earth • Continental drifts and sea floor spreading • Plate tectonics theory • Types of boundaries • Types of crust deformities	
CHAPTER 5	**Earthquakes** • Seismology • Features of earthquakes • P and S waves • Gap hypothesis • Earthquake safety	
CHAPTER 6	**Volcanoes** • Types of volcanoes and eruptions • Types of lava and pyroclastic material • Craters versus calderas • Sites and conditions for volcano formation • Predicting eruptions	

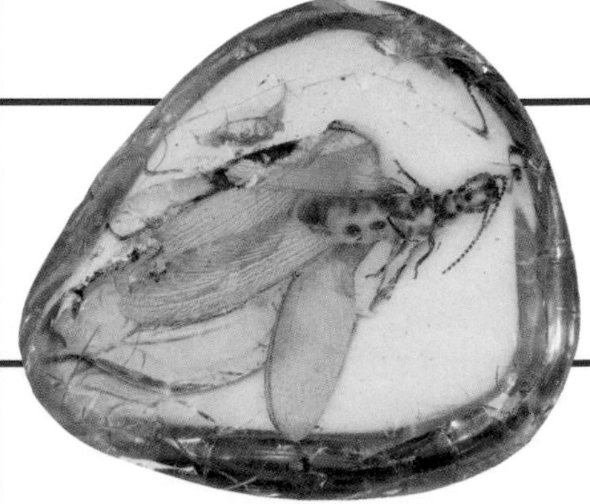

H WATER ON EARTH

The Flow of Fresh Water
- Water cycle
- River systems
- Stream erosion
- Life cycle of rivers
- Deposition
- Aquifers, springs, and wells
- Ground water
- Water treatment and pollution

Exploring the Oceans
- Properties and characteristics of the oceans
- Features of the ocean floor
- Ocean ecology
- Ocean resources and pollution

The Movement of Ocean Water
- Types of currents
- Characteristics of waves
- Types of ocean waves
- Tides

I WEATHER AND CLIMATE

The Atmosphere
- Structure of the atmosphere
- Air pressure
- Radiation, convection, and conduction
- Greenhouse effect and global warming
- Characteristics of winds
- Types of winds
- Air pollution

Understanding Weather
- Water cycle
- Humidity
- Types of clouds
- Types of precipitation
- Air masses and fronts
- Storms, tornadoes, and hurricanes
- Weather forecasting
- Weather maps

Climate
- Weather versus climate
- Seasons and latitude
- Prevailing winds
- Earth's biomes
- Earth's climate zones
- Ice ages
- Global warming
- Greenhouse effect

J ASTRONOMY

Studying Space
- Astronomy
- Keeping time
- Types of telescope
- Radioastronomy
- Mapping the stars
- Scales of the universe

Stars, Galaxies, and the Universe
- Composition of stars
- Classification of stars
- Star brightness, distance, and motions
- H-R diagram
- Life cycle of stars
- Types of galaxies
- Theories on the formation of the universe

Formation of the Solar System
- Birth of the solar system
- Structure of the sun
- Fusion
- Earth's structure and atmosphere
- Planetary motion
- Newton's Law of Universal Gravitation

A Family of Planets
- Properties and characteristics of the planets
- Properties and characteristics of moons
- Comets, asteroids, and meteoroids

Exploring Space
- Rocketry and artificial satellites
- Types of Earth orbit
- Space probes and space exploration

Physical Science

	 K INTRODUCTION TO MATTER	L INTERACTIONS OF MATTER
CHAPTER 1	**The Properties of Matter** • Definition of matter • Mass and weight • Physical and chemical properties • Physical and chemical change • Density	**Chemical Bonding** • Types of chemical bonds • Valence electrons • Ions versus molecules • Crystal lattice
CHAPTER 2	**States of Matter** • States of matter and their properties • Boyle's and Charles's laws • Changes of state	**Chemical Reactions** • Writing chemical formulas and equations • Law of conservation of mass • Types of reactions • Endothermic versus exothermic reactions • Law of conservation of energy • Activation energy • Catalysts and inhibitors
CHAPTER 3	**Elements, Compounds, and Mixtures** • Elements and compounds • Metals, nonmetals, and metalloids (semiconductors) • Properties of mixtures • Properties of solutions, suspensions, and colloids	**Chemical Compounds** • Ionic versus covalent compounds • Acids, bases, and salts • pH • Organic compounds • Biomolecules
CHAPTER 4	**Introduction to Atoms** • Atomic theory • Atomic model and structure • Isotopes • Atomic mass and mass number	**Atomic Energy** • Properties of radioactive substances • Types of decay • Half-life • Fission, fusion, and chain reactions
CHAPTER 5	**The Periodic Table** • Structure of the periodic table • Periodic law • Properties of alkali metals, alkaline-earth metals, halogens, and noble gases	
CHAPTER 6		

M FORCES, MOTION, AND ENERGY

Matter in Motion
- Speed, velocity, and acceleration
- Measuring force
- Friction
- Mass versus weight

Forces in Motion
- Terminal velocity and free fall
- Projectile motion
- Inertia
- Momentum

Forces in Fluids
- Properties in fluids
- Atmospheric pressure
- Density
- Pascal's principle
- Buoyant force
- Archimedes' principle
- Bernoulli's principle

Work and Machines
- Measuring work
- Measuring power
- Types of machines
- Mechanical advantage
- Mechanical efficiency

Energy and Energy Resources
- Forms of energy
- Energy conversions
- Law of conservation of energy
- Energy resources

Heat and Heat Technology
- Heat versus temperature
- Thermal expansion
- Absolute zero
- Conduction, convection, radiation
- Conductors versus insulators
- Specific heat capacity
- Changes of state
- Heat engines
- Thermal pollution

N ELECTRICITY AND MAGNETISM

Introduction to Electricity
- Law of electric charges
- Conduction versus induction
- Static electricity
- Potential difference
- Cells, batteries, and photocells
- Thermocouples
- Voltage, current, and resistance
- Electric power
- Types of circuits

Electromagnetism
- Properties of magnets
- Magnetic force
- Electromagnetism
- Solenoids and electric motors
- Electromagnetic induction
- Generators and transformers

Electronic Technology
- Properties of semiconductors
- Integrated circuits
- Diodes and transistors
- Analog versus digital signals
- Microprocessors
- Features of computers

O SOUND AND LIGHT

The Energy of Waves
- Properties of waves
- Types of waves
- Reflection and refraction
- Diffraction and interference
- Standing waves and resonance

The Nature of Sound
- Properties of sound waves
- Structure of the human ear
- Pitch and the Doppler effect
- Infrasonic versus ultrasonic sound
- Sound reflection and echolocation
- Sound barrier
- Interference, resonance, diffraction, and standing waves
- Sound quality of instruments

The Nature of Light
- Electromagnetic waves
- Electromagnetic spectrum
- Law of reflection
- Absorption and scattering
- Reflection and refraction
- Diffraction and interference

Light and Our World
- Luminosity
- Types of lighting
- Types of mirrors and lenses
- Focal point
- Structure of the human eye
- Lasers and holograms

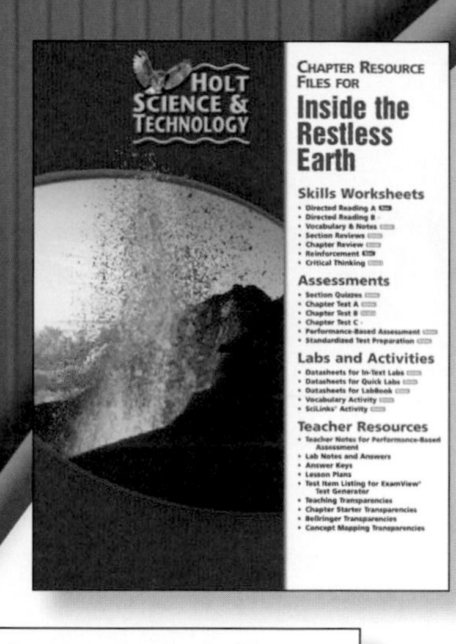

Program resources make teaching and learning easier.

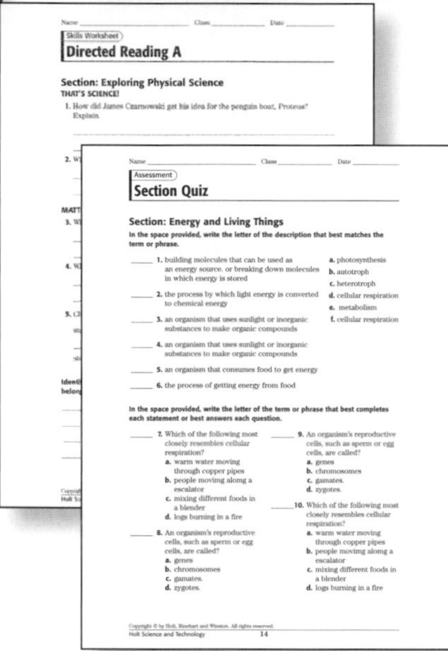

CHAPTER RESOURCES

A ***Chapter Resources book*** accompanies each of the 15 *Short Courses.* Here you'll find everything you need to make sure your students are getting the most out of learning science—all in one book.

Skills Worksheets

- Directed Reading A: Basic
- Directed Reading B: Special Needs
- Vocabulary and Chapter Summary
- Section Reviews
- Chapter Reviews
- Reinforcement
- Critical Thinking

Labs & Activities

- Datasheets for Chapter Labs
- Datasheets for Quick Labs
- Datasheets for LabBook
- Vocabulary Activity
- SciLinks® Activity

Assessments

- Section Quizzes
- Chapter Tests A: General
- Chapter Tests B: Advanced
- Chapter Tests C: Special Needs
- Performance-Based Assessments
- Standardized Test Preparation

Teacher Resources

- Lab Notes and Answers
- Teacher Notes for Performance-Based Assessment
- Answer Keys
- Lesson Plans
- Test Item Listing for ExamView® Test Generator
- Full-color **Teaching Transparencies,** plus section **Bellringers, Concept Mapping,** and **Chapter Starter Transparencies.**

SPANISH RESOURCES

Spanish materials are available for each *Short Course:*

- *Student Edition*
- ***Spanish Resources*** booklet contains worksheets and assessments translated into Spanish with an English **Answer Key.**
- **Guided Reading Audio CD Program**

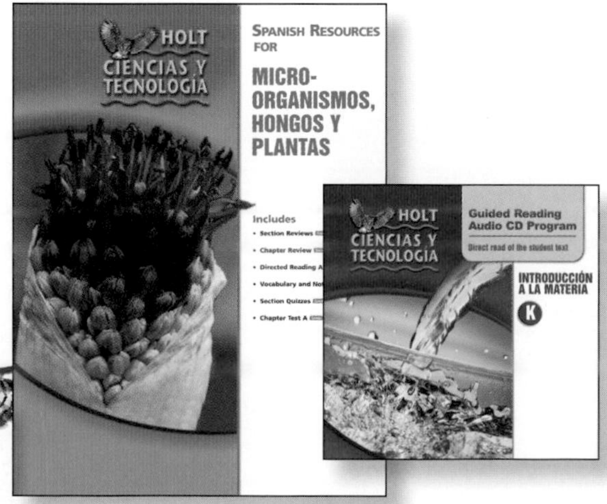

ONLINE RESOURCES

- *Enhanced Online Editions* engage students and assist teachers with a host of interactive features that are available anytime and anywhere you can connect to the Internet.
- **CNNStudentNews.com** provides award-winning news and information for both teachers and students.
- **SciLinks**—a Web service developed and maintained by the National Science Teachers Association—links you and your students to up-to-date online resources directly related to chapter topics.
- **go.hrw.com** links you and your students to online chapter activities and resources.
- **Current Science** articles relate to students' lives.

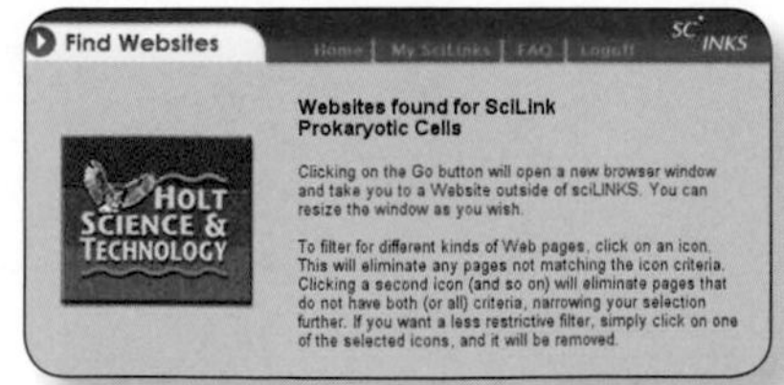

ADDITIONAL LAB AND SKILLS RESOURCES

- *Calculator-Based Labs* incorporates scientific instruments, offering students insight into modern scientific investigation.
- *EcoLabs & Field Activities* develops awareness of the natural world.
- *Holt Science Skills Workshop: Reading in the Content Area* contains exercises that target reading skills key.
- *Inquiry Labs* taps students' natural curiosity and creativity with a focus on the process of discovery.
- *Labs You Can Eat* safely incorporates edible items into the classroom.
- *Long-Term Projects & Research Ideas* extends and enriches lessons.
- *Math Skills for Science* provides additional explanations, examples, and math problems so students can develop their skills.
- *Science Skills Worksheets* helps your students hone important learning skills.
- *Whiz-Bang Demonstrations* gets your students' attention at the beginning of a lesson.

ADDITIONAL RESOURCES

- *Assessment Checklists & Rubrics* gives you guidelines for evaluating students' progress.
- *Holt Anthology of Science Fiction* sparks your students' imaginations with thought-provoking stories.
- *Holt Science Posters* visually reinforces scientific concepts and themes with seven colorful posters including **The Periodic Table of the Elements.**

- *Professional Reference for Teachers* contains professional articles that discuss a variety of topics, such as classroom management.
- *Program Introduction Resource File* explains the program and its features and provides several additional references, including lab safety, scoring rubrics, and more.
- *Science Fair Guide* gives teachers, students, and parents tips for planning and assisting in a science fair.
- *Science Puzzlers, Twisters & Teasers* activities challenge students to think about science concepts in different ways.

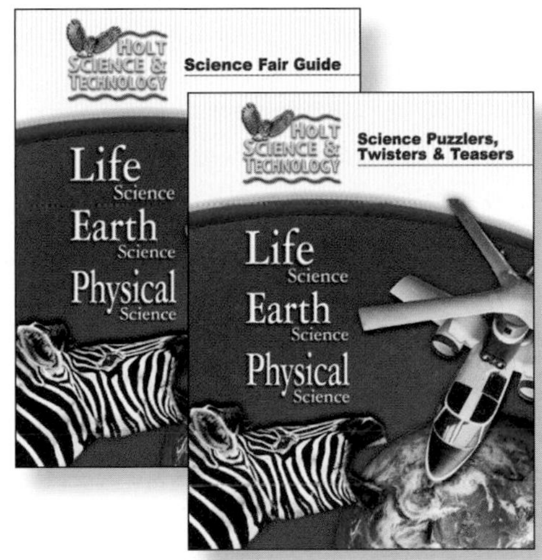

TECHNOLOGY RESOURCES

- *CNN Presents Science in the News: Video Library* helps students see the impact of science on their everyday lives with actual news video clips.
  - Multicultural Connections
  - Science, Technology & Society
  - Scientists in Action
  - Eye on the Environment
- *Guided Reading Audio CD Program*, available in English and Spanish, provides students with a direct read of each section.
- *HRW Earth Science Videotape* takes your students on a geology "field trip" with full-motion video.
- *Interactive Explorations CD-ROM Program* develops students' inquiry and decision-making skills as they investigate science phenomena in a virtual lab setting.

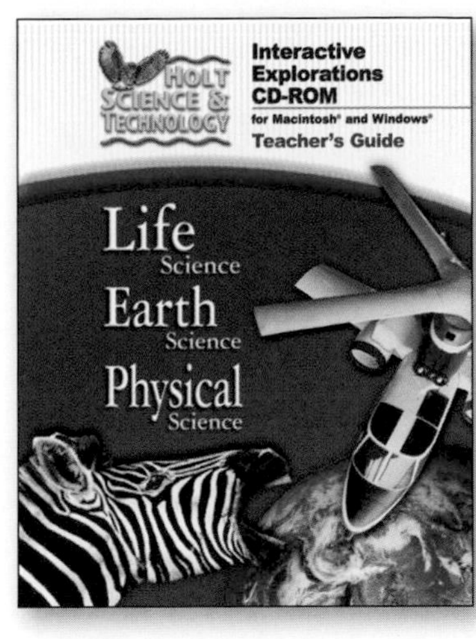

- *One-Stop Planner CD-ROM®* organizes everything you need on one disc, including printable worksheets, customizable lesson plans, a powerful test generator, **PowerPoint® LectureNotes, Lab Materials QuickList Software, Holt Calendar Planner, Interactive Teacher Edition,** and more.
- *Science Tutor CD-ROMs* help students practice what they learn with immediate feedback.
- *Lab Videos* make it easier to integrate more experiments into your lessons without the preparation time and costs. Available on DVD and VHS.
- Brain Food Video Quizzes are game-show style quizzes that assess students' progress. Available on DVD and VHS.
- *Visual Concepts CD-ROMs* include graphics, animations, and movie clips that demonstrate key chapter concepts.

Science and Math Worksheets

The ***Holt Science & Technology*** program helps you meet the needs of a wide variety of students, regardless of their skill level. The following pages provide examples of the worksheets available to improve your students' science and math skills whether they already have a strong science and math background or are weak in these areas. Samples of assessment checklists and rubrics are also provided.

In addition to the skills worksheets represented here, ***Holt Science & Technology*** provides a variety of worksheets that are correlated directly with each chapter of the program. Representations of these worksheets are found at the beginning of each chapter in this *Teacher Edition*.

Many worksheets are also available on the Holt Web site. The address is **go.hrw.com.**

Science Skills Worksheets: Thinking Skills

BEING FLEXIBLE

Worksheet 1 Thinking Skills *Being Flexible*

USING YOUR SENSES

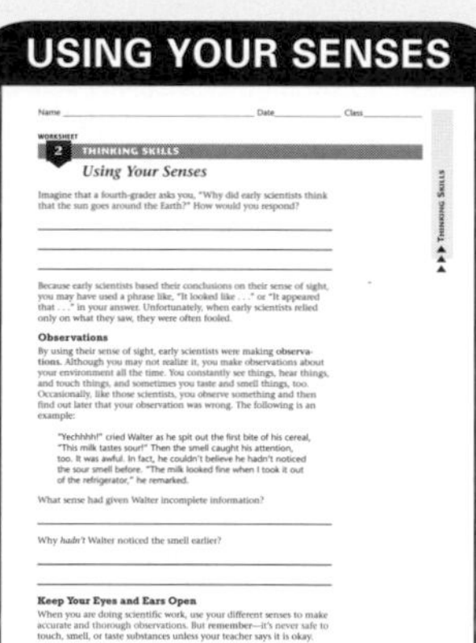

Worksheet 2 Thinking Skills *Using Your Senses*

THINKING OBJECTIVELY

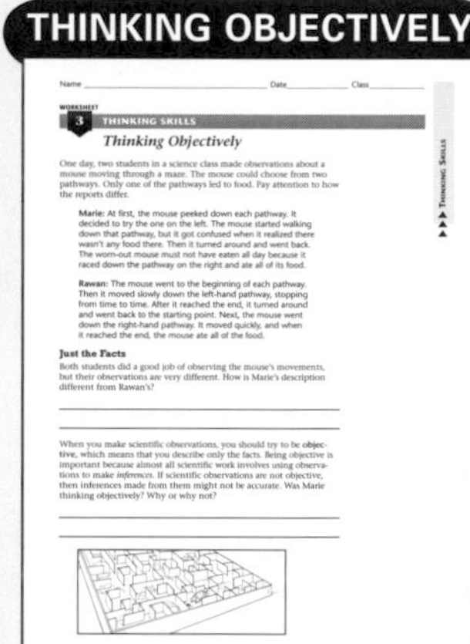

Worksheet 3 Thinking Skills *Thinking Objectively*

UNDERSTANDING BIAS

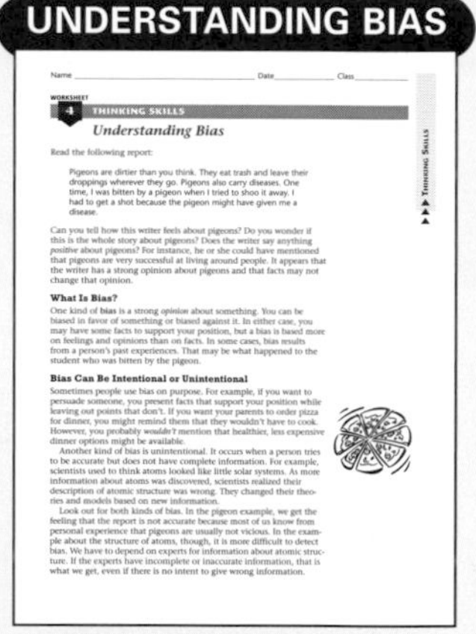

Worksheet 4 Thinking Skills *Understanding Bias*

USING LOGIC

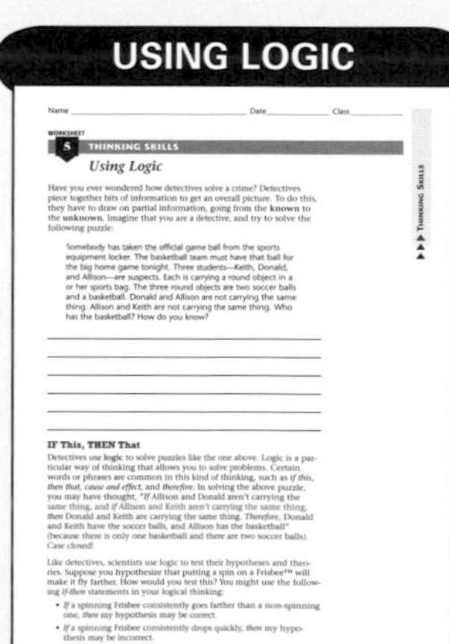

Worksheet 5 Thinking Skills *Using Logic*

BOOSTING YOUR MEMORY

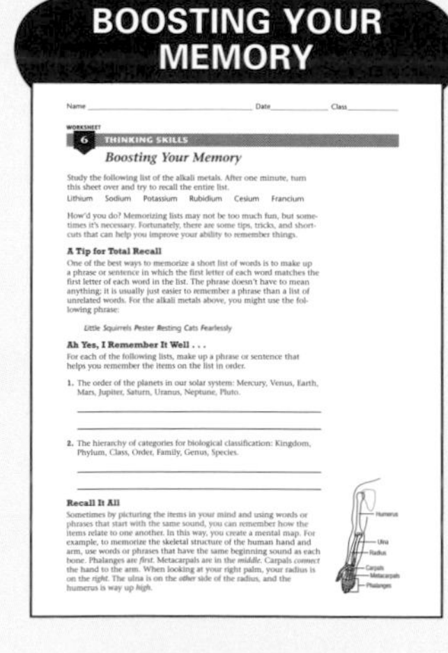

Worksheet 6 Thinking Skills *Boosting Your Memory*

IMPROVING YOUR STUDY HABITS

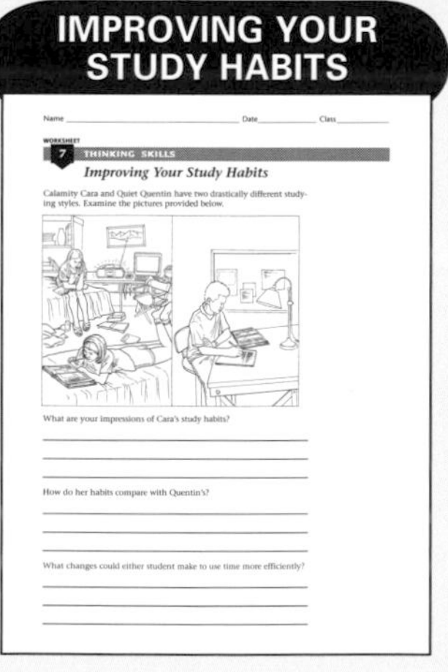

Worksheet 7 Thinking Skills *Improving Your Study Habits*

READING A SCIENCE TEXTBOOK

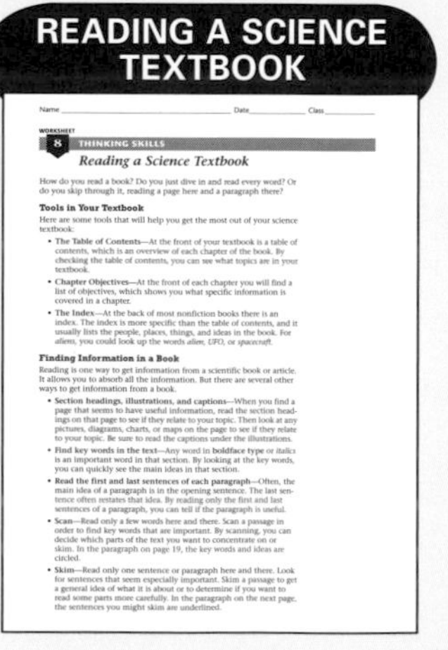

Worksheet 8 Thinking Skills *Reading a Science Textbook*

Science Skills Worksheets: Experimenting Skills

SAFETY RULES!

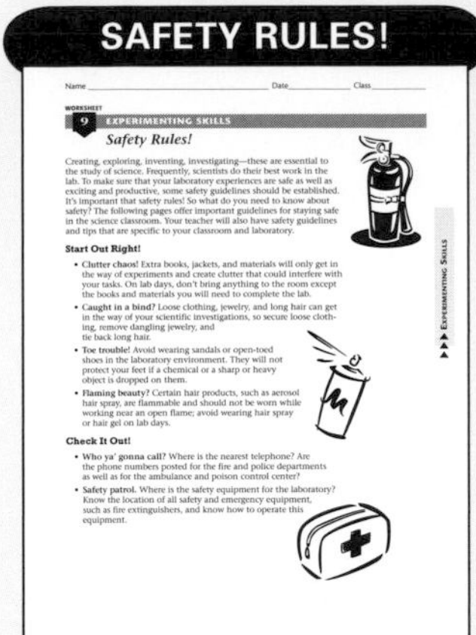
Worksheet 9 — Experimenting Skills — Safety Rules!

DOING A LAB WRITE-UP

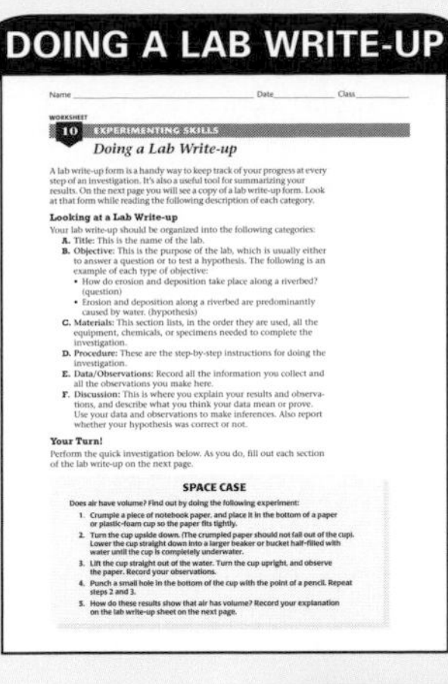
Worksheet 10 — Experimenting Skills — Doing a Lab Write-up

UNDERSTANDING VARIABLES

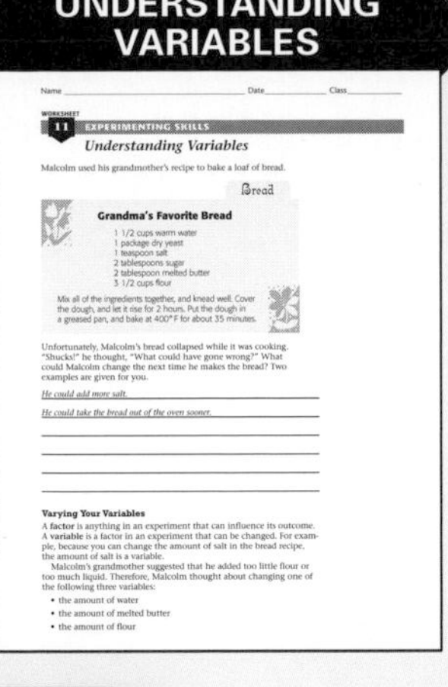
Worksheet 11 — Experimenting Skills — Understanding Variables

WORKING WITH HYPOTHESES

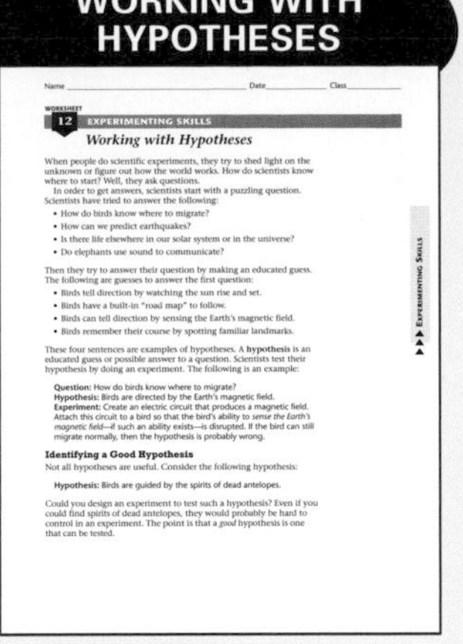
Worksheet 12 — Experimenting Skills — Working with Hypotheses

DESIGNING AN EXPERIMENT

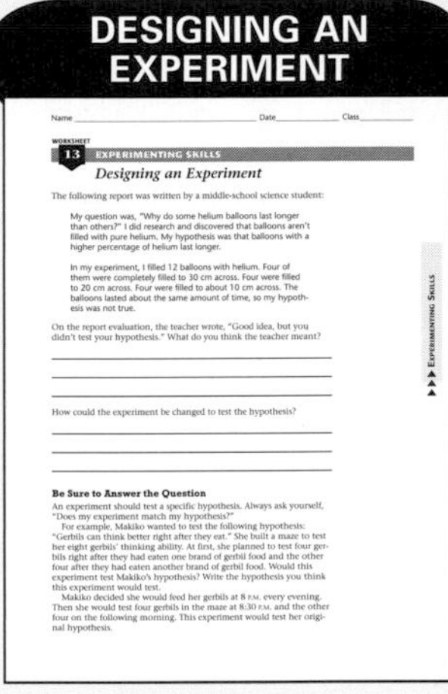
Worksheet 13 — Experimenting Skills — Designing an Experiment

USING THE INTERNATIONAL SYSTEM OF UNITS (SI)

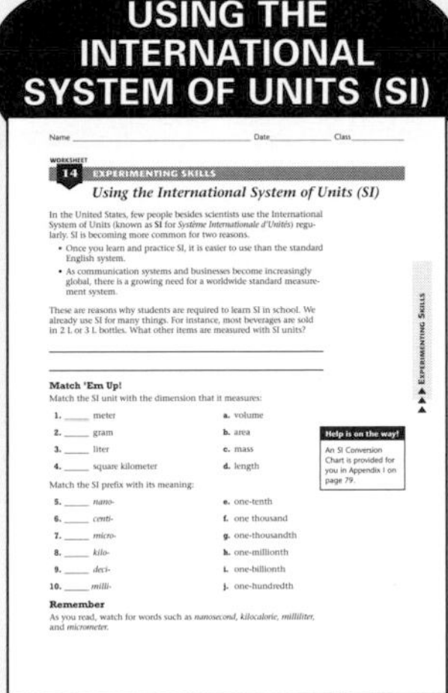
Worksheet 14 — Experimenting Skills — Using the International System of Units (SI)

MEASURING

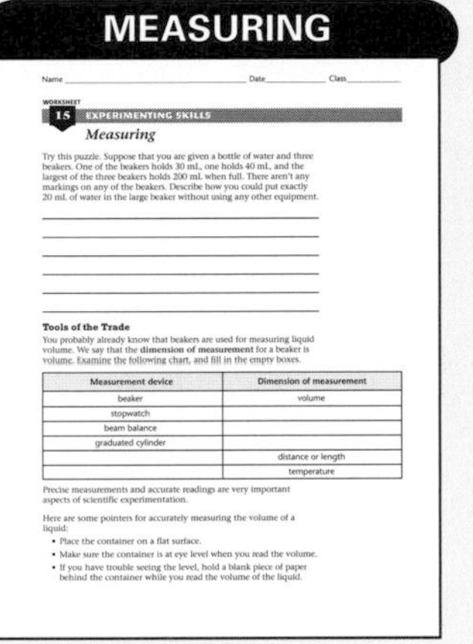
Worksheet 15 — Experimenting Skills — Measuring

Science Skills Worksheets: Researching Skills

CHOOSING YOUR TOPIC

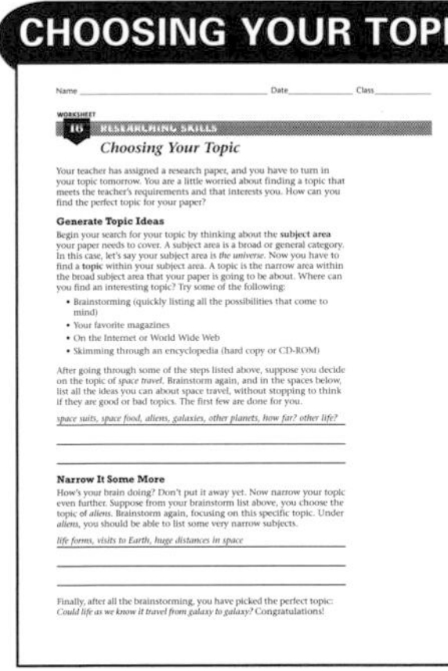
Worksheet 16 — Researching Skills — Choosing Your Topic

ORGANIZING YOUR RESEARCH

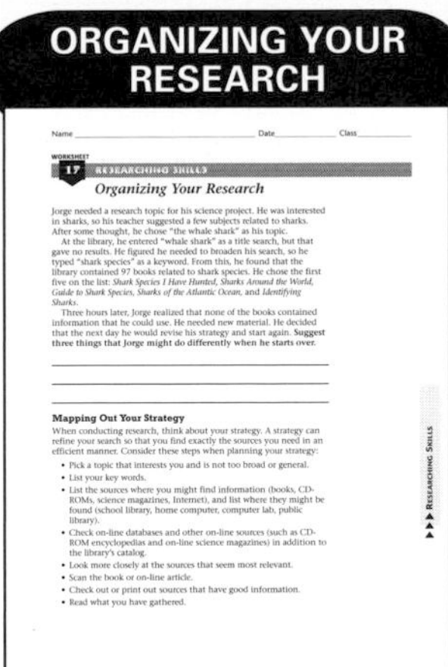
Worksheet 17 — Researching Skills — Organizing Your Research

FINDING USEFUL SOURCES

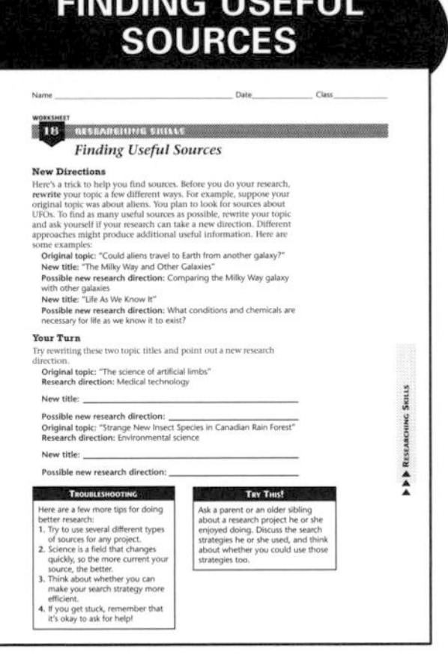
Worksheet 18 — Researching Skills — Finding Useful Sources

RESEARCHING ON THE WEB

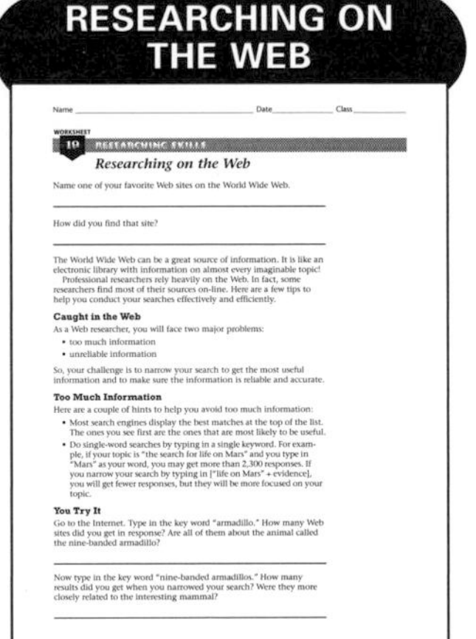
Worksheet 19 — Researching Skills — Researching on the Web

Science Skills Worksheets: Researching Skills (continued)

IDENTIFYING BIAS

Worksheet 20 — Researching Skills — *Identifying Bias*

TAKING NOTES

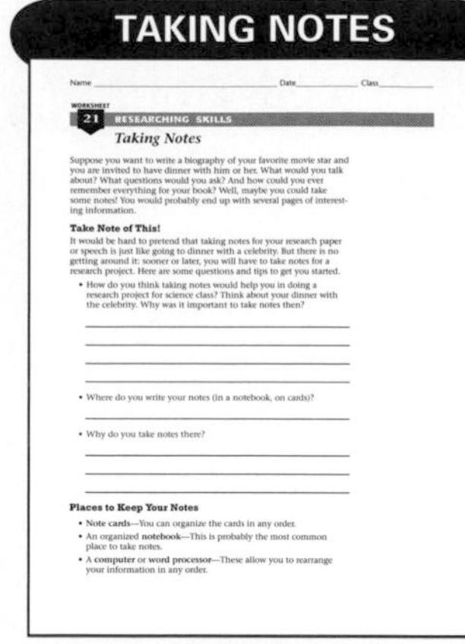
Worksheet 21 — Researching Skills — *Taking Notes*

Science Skills Worksheets: Communicating Skills

SCIENCE WRITING

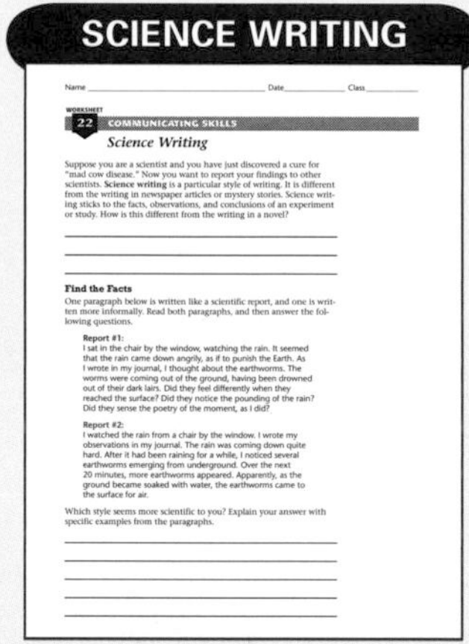
Worksheet 22 — Communicating Skills — *Science Writing*

SCIENCE DRAWING

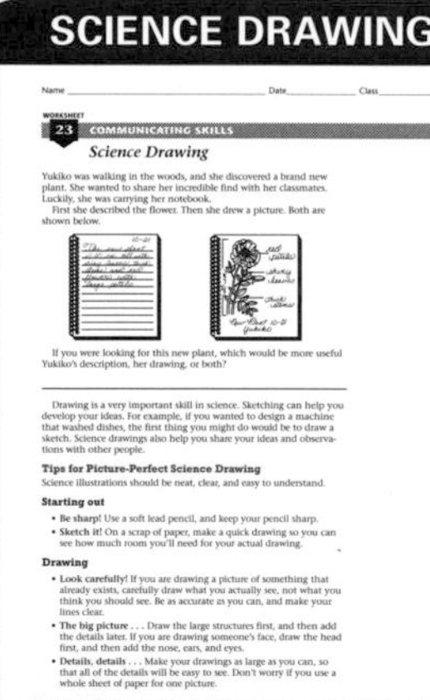
Worksheet 23 — Communicating Skills — *Science Drawing*

USING MODELS TO COMMUNICATE

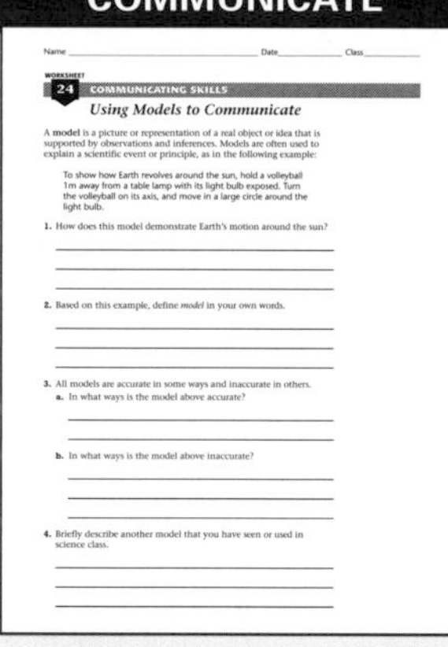
Worksheet 24 — Communicating Skills — *Using Models to Communicate*

INTRODUCTION TO GRAPHS

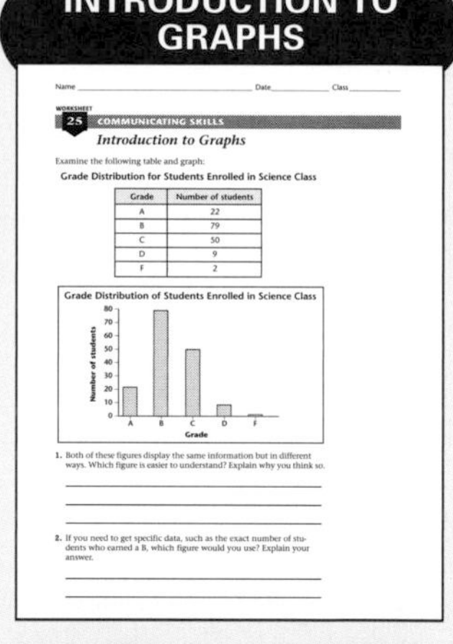
Worksheet 25 — Communicating Skills — *Introduction to Graphs*

GRASPING GRAPHING

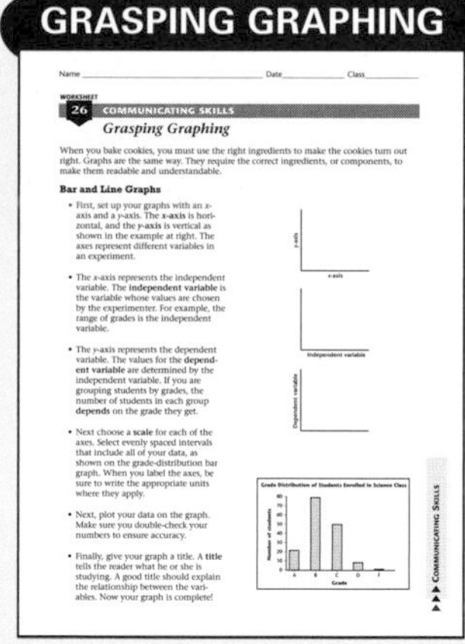
Worksheet 26 — Communicating Skills — *Grasping Graphing*

INTERPRETING YOUR DATA

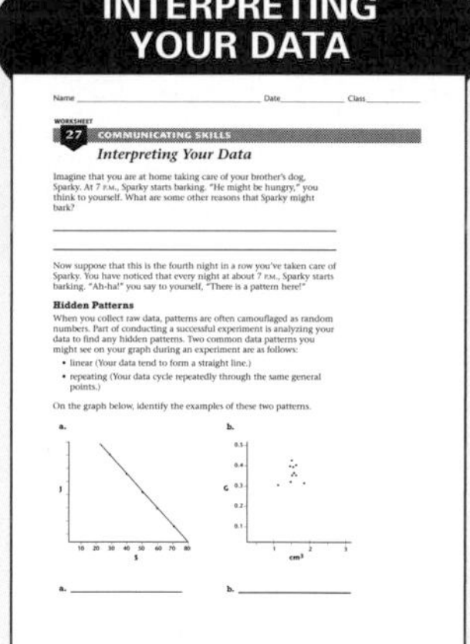
Worksheet 27 — Communicating Skills — *Interpreting Your Data*

RECOGNIZING BIAS IN GRAPHS

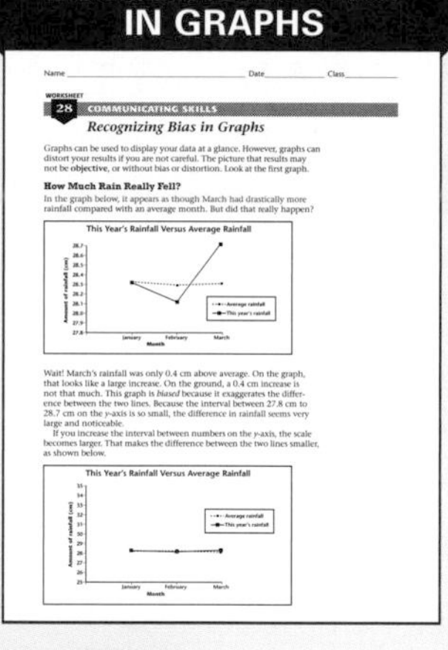
Worksheet 28 — Communicating Skills — *Recognizing Bias in Graphs*

MAKING DATA MEANINGFUL

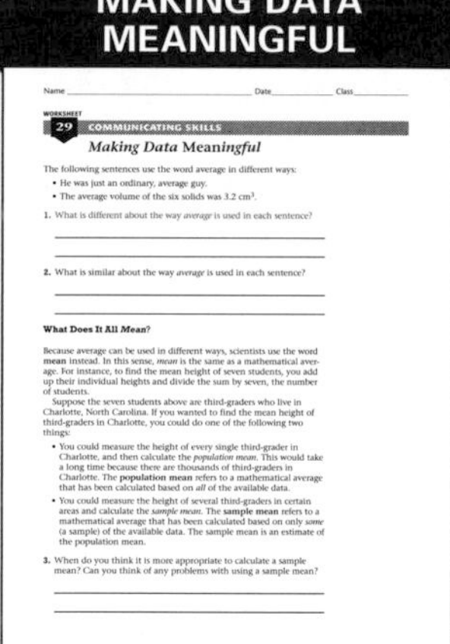
Worksheet 29 — Communicating Skills — *Making Data Meaningful*

HINTS FOR ORAL PRESENTATIONS

Worksheet 30 — Communicating Skills — *Hints for Oral Presentations*

Math Skills for Science

ADDITION AND SUBTRACTION

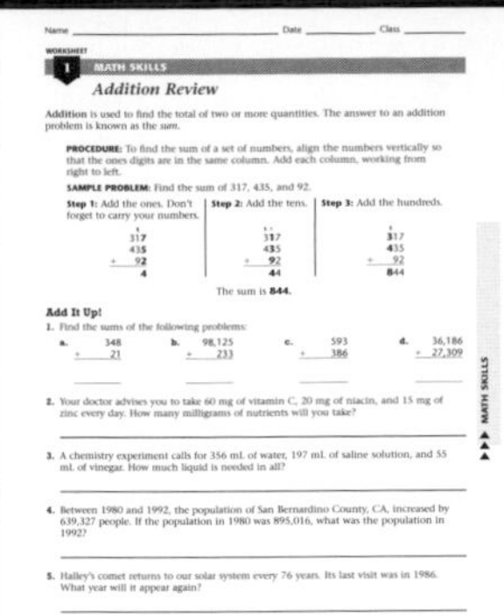

WORKSHEET 1 MATH SKILLS

Addition Review

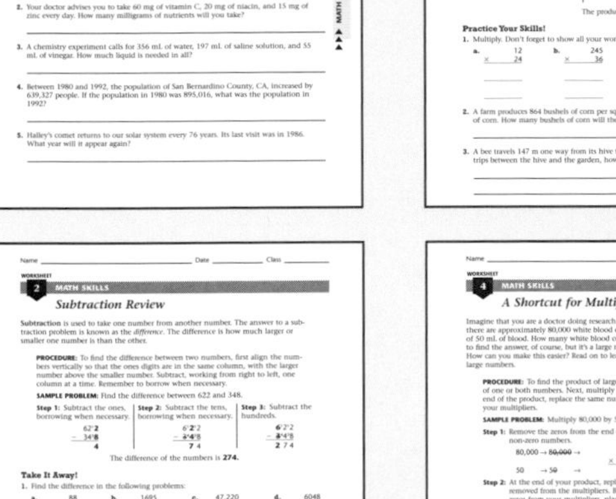

WORKSHEET 2 MATH SKILLS

Subtraction Review

MULTIPLICATION

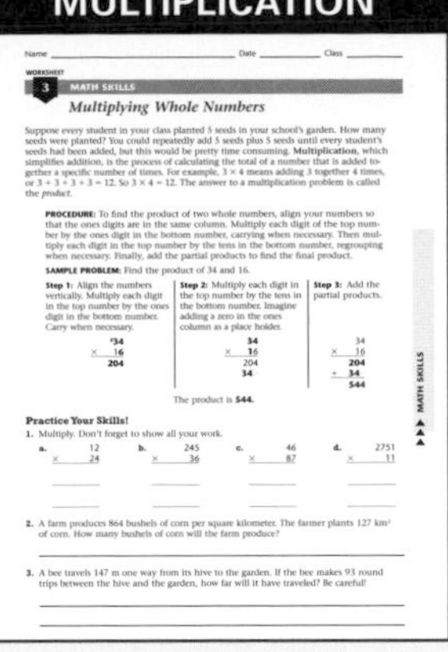

WORKSHEET 3 MATH SKILLS

Multiplying Whole Numbers

WORKSHEET 4 MATH SKILLS

A Shortcut for Multiplying Large Numbers

DIVISION

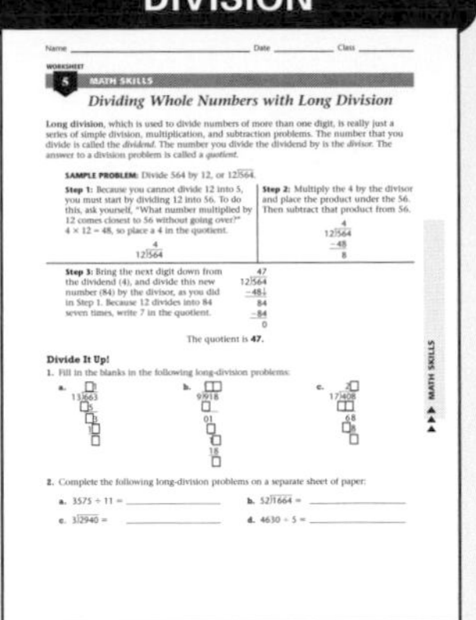

WORKSHEET 5 MATH SKILLS

Dividing Whole Numbers with Long Division

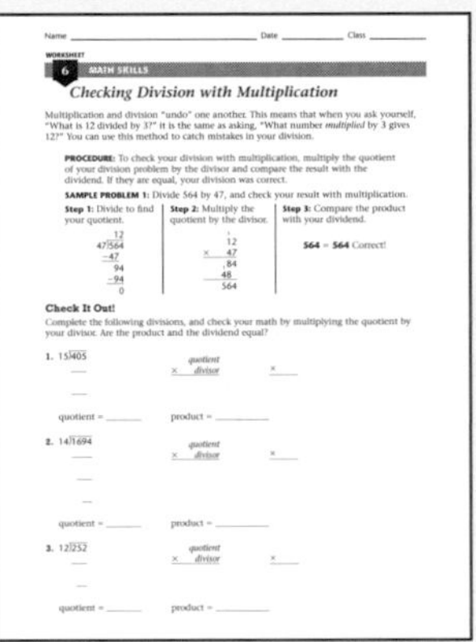

WORKSHEET 6 MATH SKILLS

Checking Division with Multiplication

AVERAGES

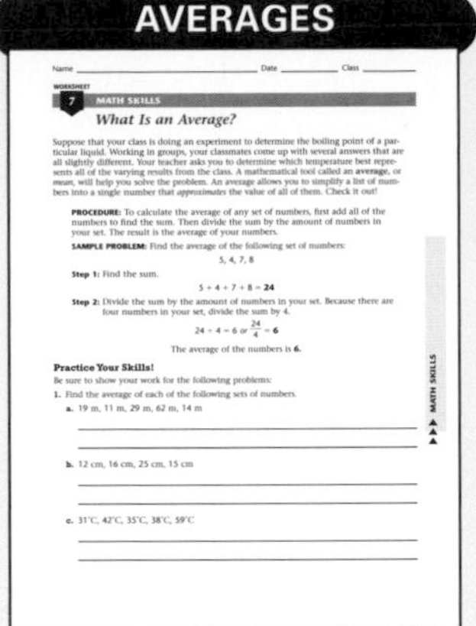

WORKSHEET 7 MATH SKILLS

What Is an Average?

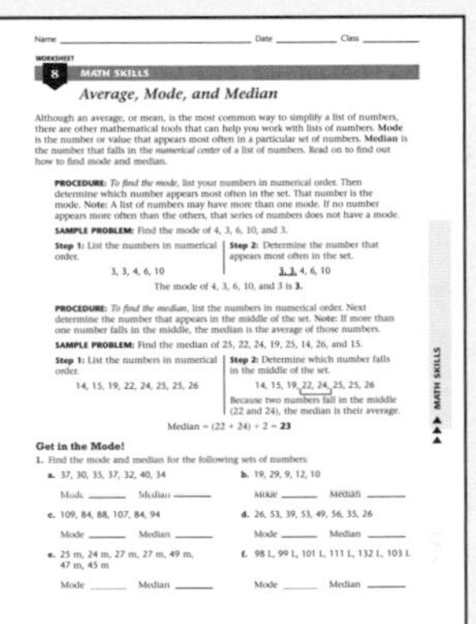

WORKSHEET 8 MATH SKILLS

Average, Mode, and Median

POSITIVE AND NEGATIVE NUMBERS

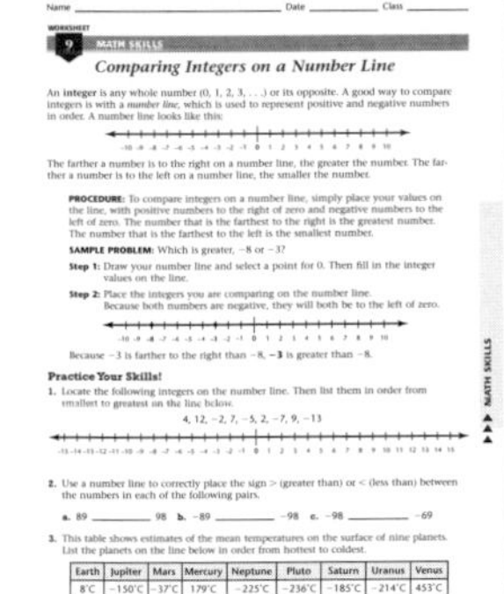

WORKSHEET 9 MATH SKILLS

Comparing Integers on a Number Line

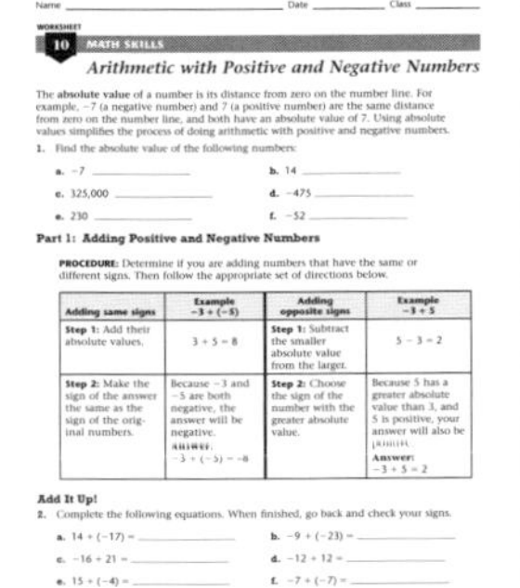

WORKSHEET 10 MATH SKILLS

Arithmetic with Positive and Negative Numbers

FRACTIONS

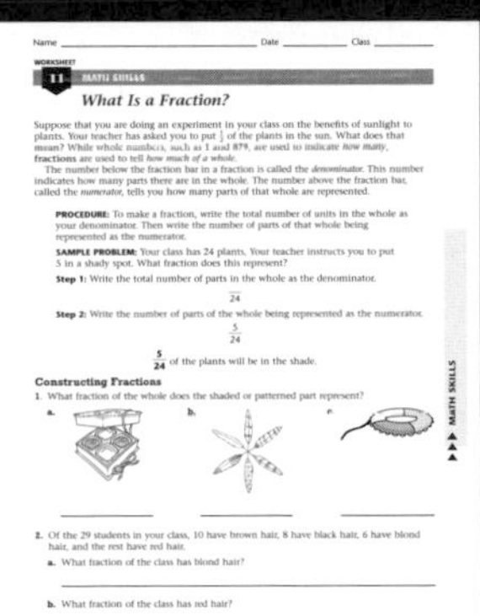

WORKSHEET 11 MATH SKILLS

What Is a Fraction?

WORKSHEET 12 MATH SKILLS

Reducing Fractions to Lowest Terms

WORKSHEET 13 MATH SKILLS

Improper Fractions and Mixed Numbers

WORKSHEET 14 MATH SKILLS

Adding and Subtracting Fractions

WORKSHEET 15 MATH SKILLS

Multiplying and Dividing Fractions

Math Skills for Science (continued)

RATIOS AND PROPORTIONS

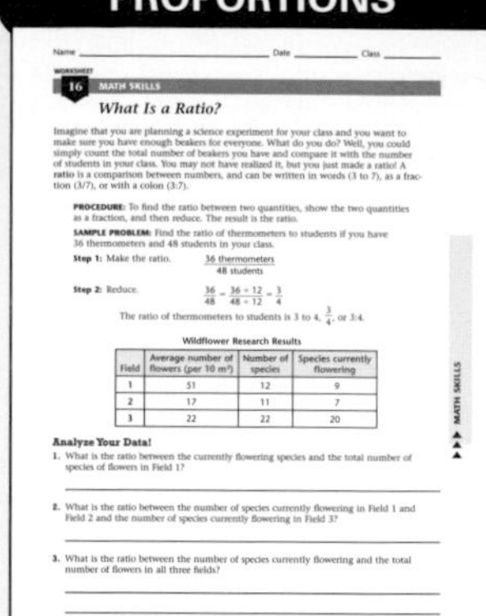
Worksheet 16 — Math Skills: What Is a Ratio?

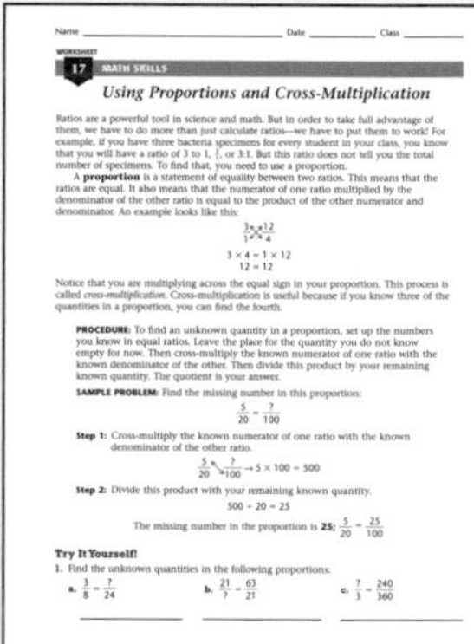
Worksheet 17 — Math Skills: Using Proportions and Cross-Multiplication

DECIMALS

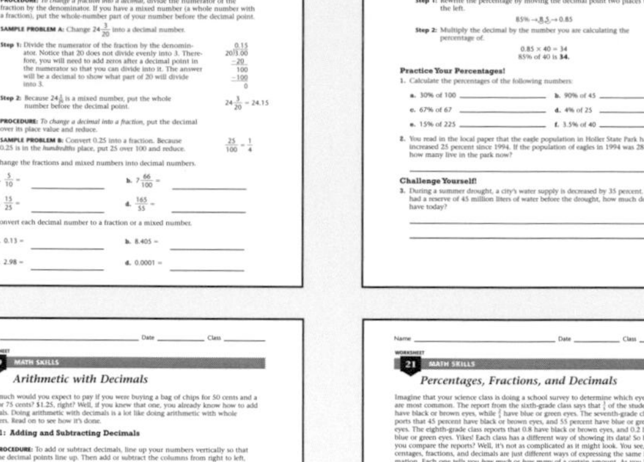
Worksheet 18 — Math Skills: Decimals and Fractions

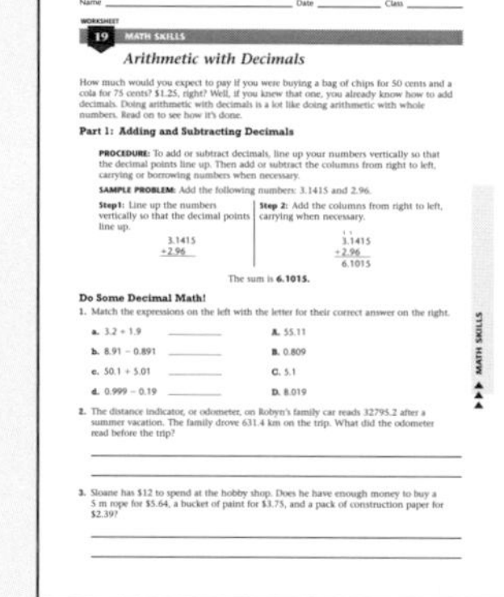
Worksheet 19 — Math Skills: Arithmetic with Decimals

PERCENTAGES

Worksheet 20 — Math Skills: Parts of 100: Calculating Percentages

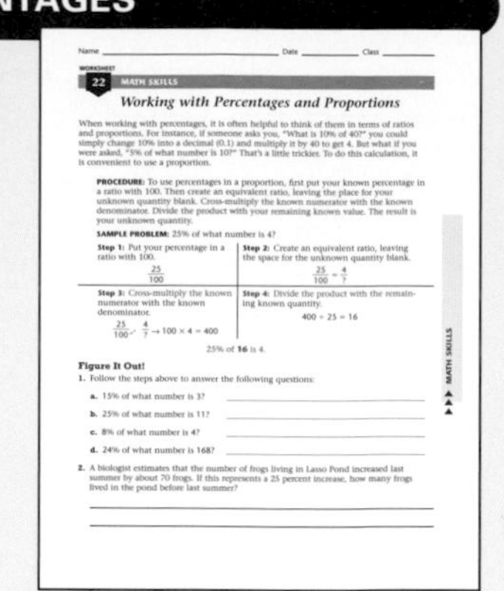
Worksheet 22 — Math Skills: Working with Percentages and Proportions

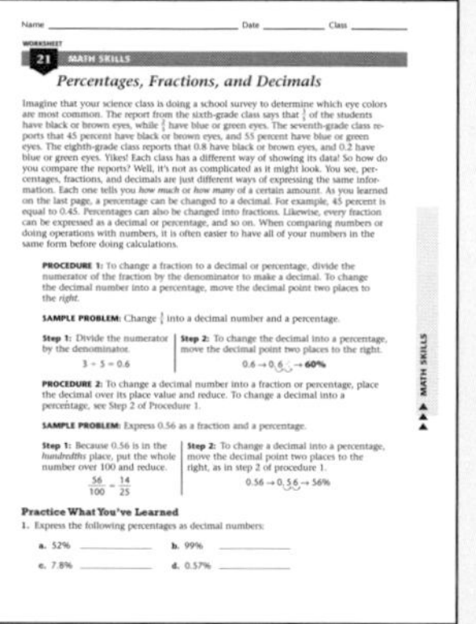
Worksheet 21 — Math Skills: Percentages, Fractions, and Decimals

POWERS OF 10

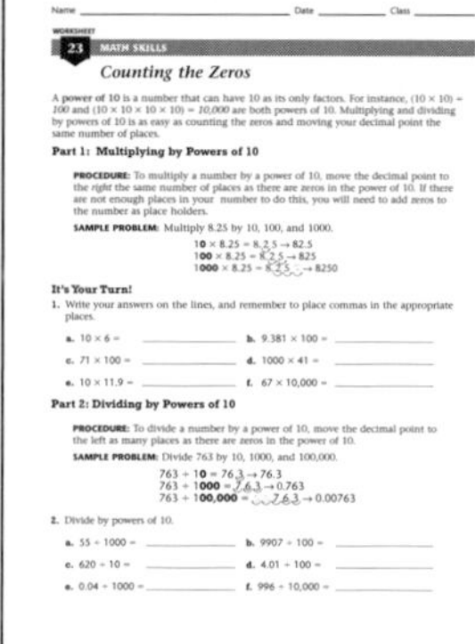
Worksheet 23 — Math Skills: Counting the Zeros

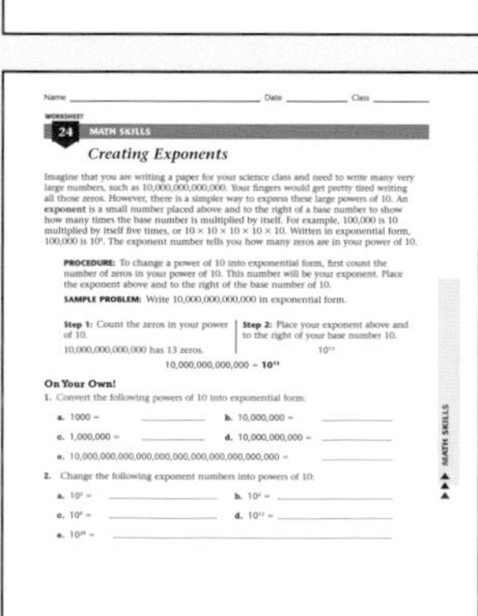
Worksheet 24 — Math Skills: Creating Exponents

SCIENTIFIC NOTATION

Worksheet 25 — Math Skills: What Is Scientific Notation?

Worksheet 26 — Math Skills: Multiplying and Dividing in Scientific Notation

SI MEASUREMENT AND CONVERSION

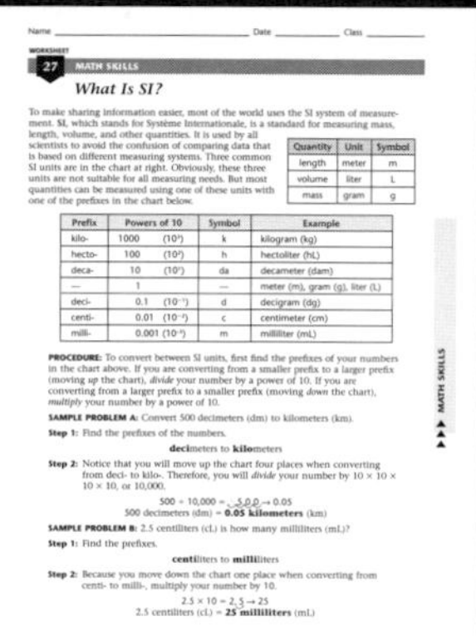
Worksheet 27 — Math Skills: What Is SI?

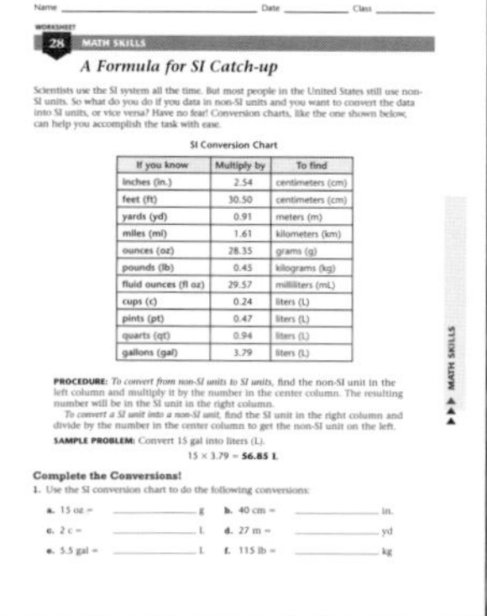
Worksheet 28 — Math Skills: A Formula for SI Catch-up

Math Skills for Science (continued)

GEOMETRY

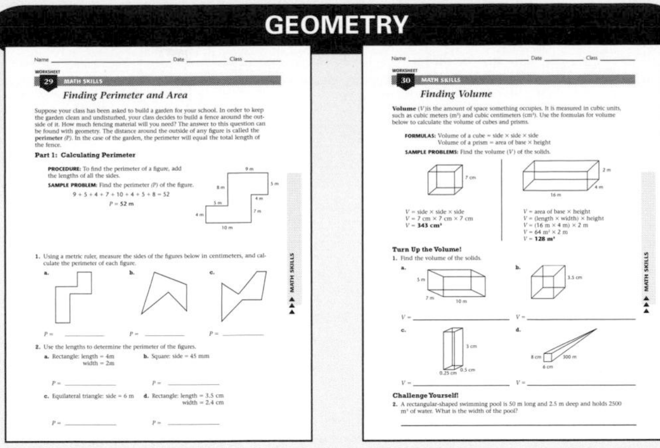

Worksheet 29 — Math Skills — Finding Perimeter and Area

Worksheet 30 — Math Skills — Finding Volume

THE UNIT FACTOR AND DIMENSIONAL ANALYSIS

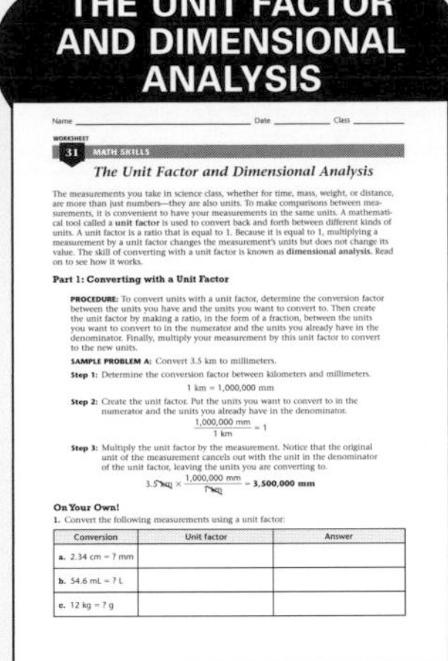

Worksheet 31 — Math Skills — The Unit Factor and Dimensional Analysis

MATH IN SCIENCE: INTEGRATED SCIENCE

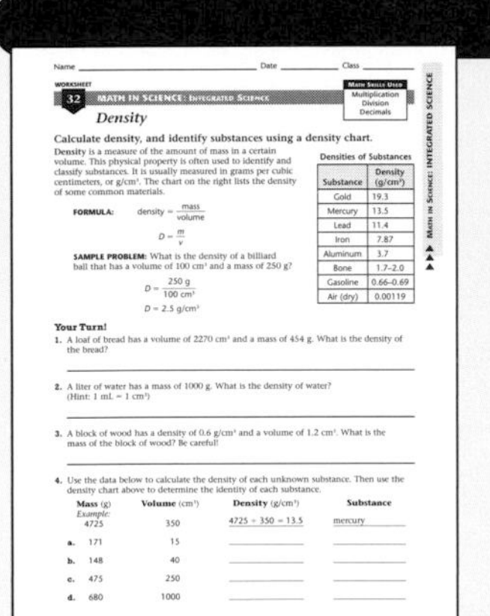

Worksheet 32 — Math in Science: Integrated Science — Density

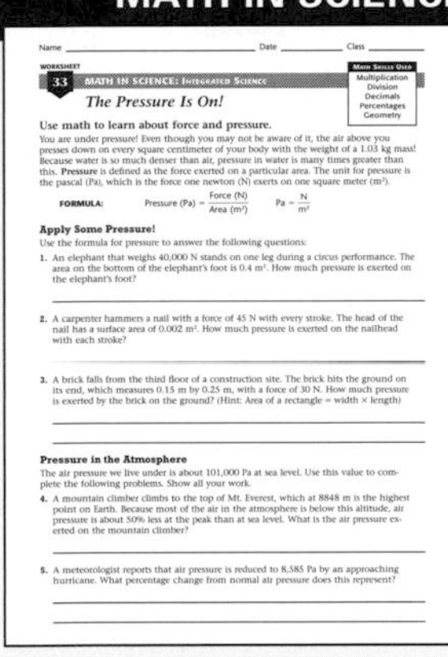

Worksheet 33 — Math in Science: Integrated Science — The Pressure Is On!

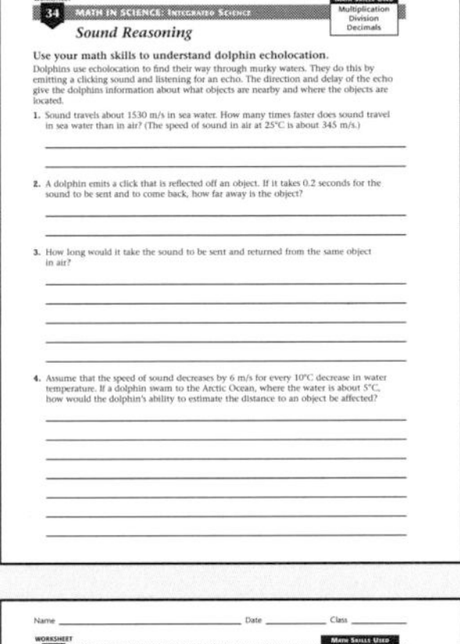

Worksheet 34 — Math in Science: Integrated Science — Sound Reasoning

Worksheet 35 — Math in Science: Integrated Science — Using Temperature Scales

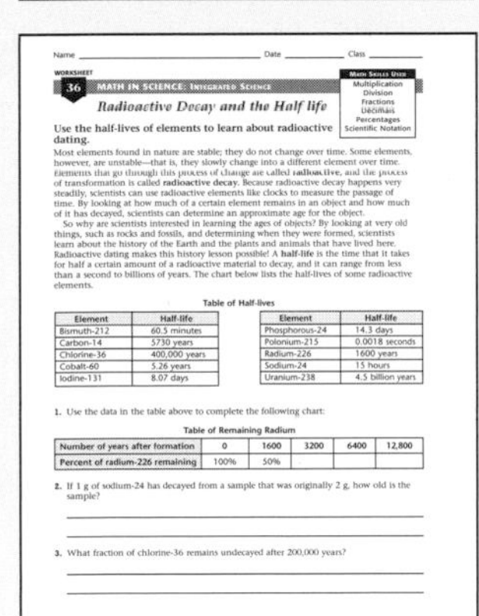

Worksheet 36 — Math in Science: Integrated Science — Radioactive Decay and the Half-life

Worksheet 37 — Math in Science: Integrated Science — Rain-Forest Math

Math Skills for Science (continued)

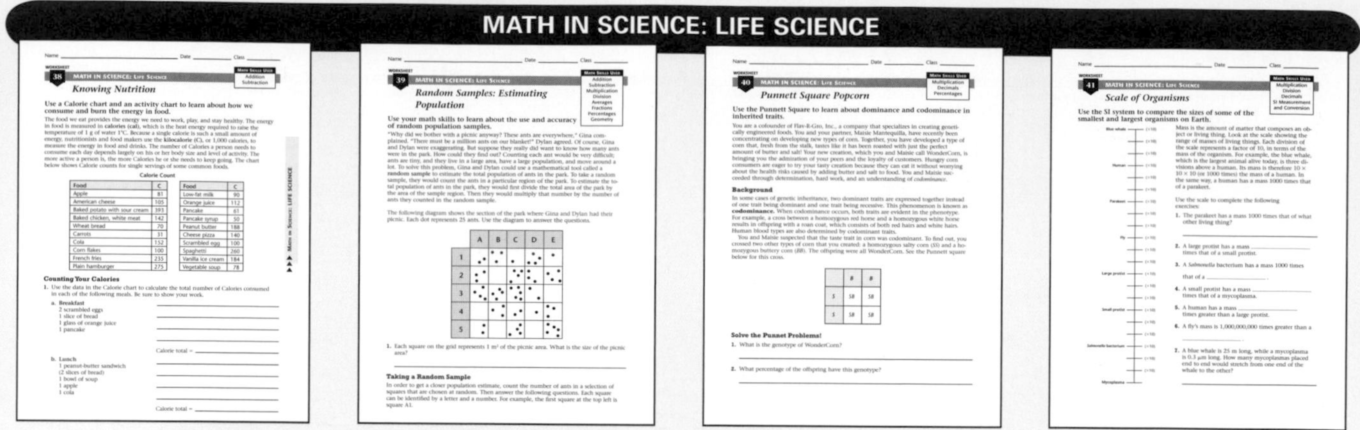

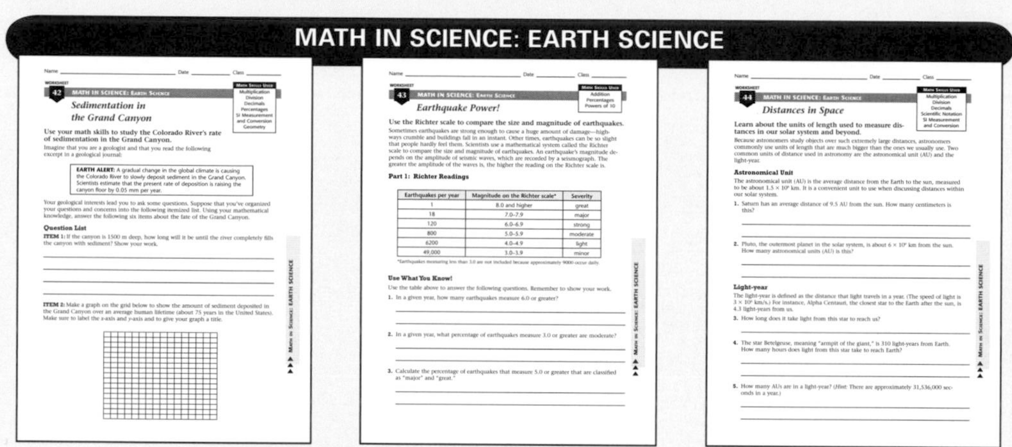

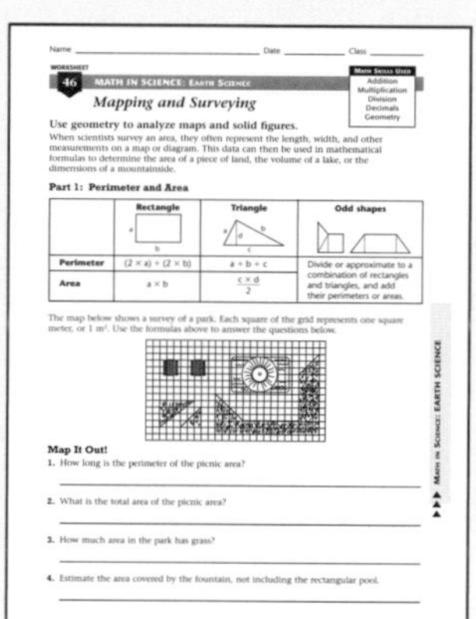

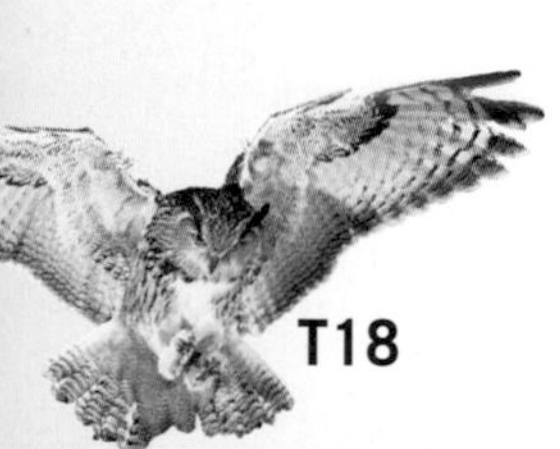

Math Skills for Science (continued)

MATH IN SCIENCE: PHYSICAL SCIENCE

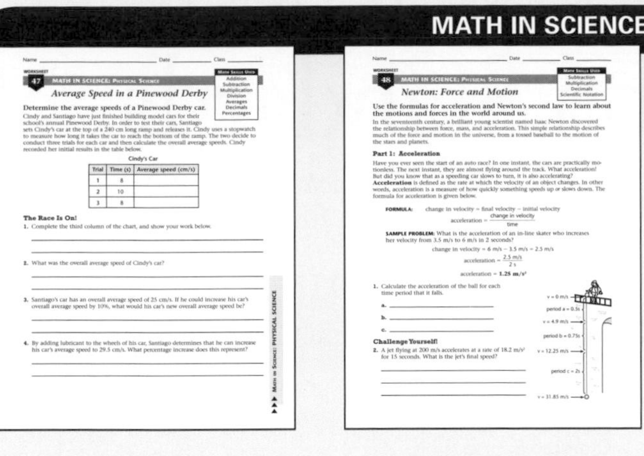
Worksheet 47 — Average Speed in a Pinewood Derby
Worksheet 48 — Newton: Force and Motion

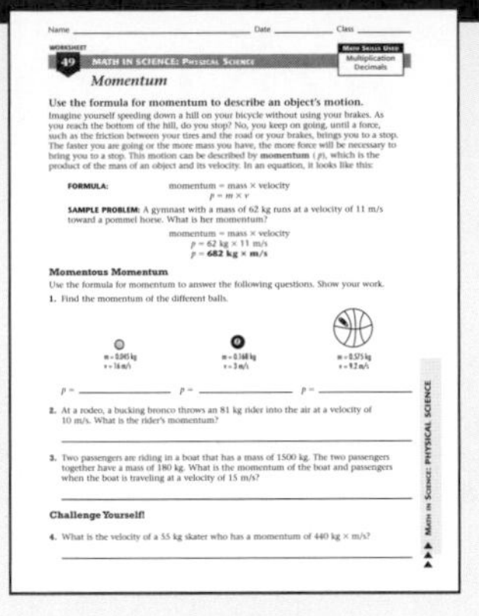
Worksheet 49 — Momentum

Worksheet 50 — Balancing Chemical Equations

Worksheet 51 — Work and Power

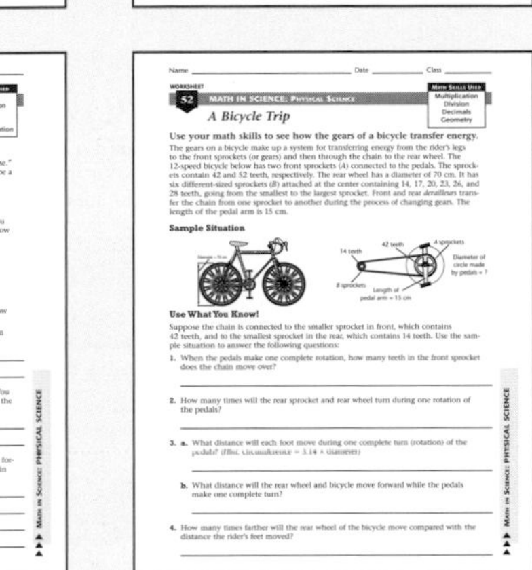
Worksheet 52 — A Bicycle Trip

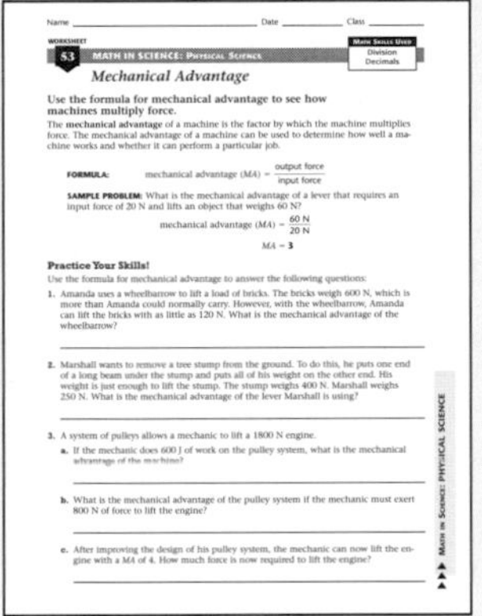
Worksheet 53 — Mechanical Advantage

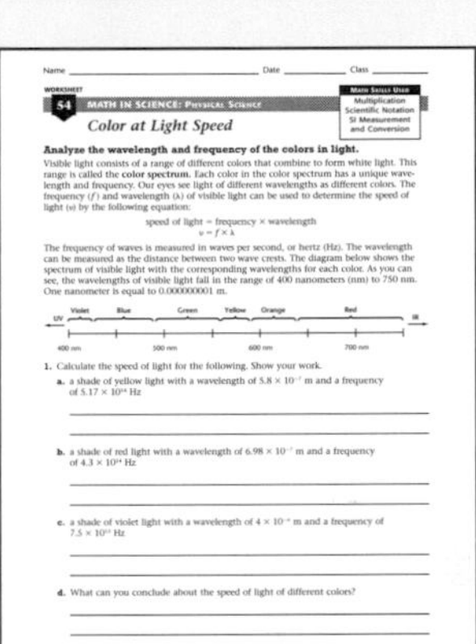
Worksheet 54 — Color at Light Speed

Assessment Checklist & Rubrics

The following is just a sample of over 50 checklists and rubrics contained in this booklet.

RUBRICS FOR WRITTEN WORK

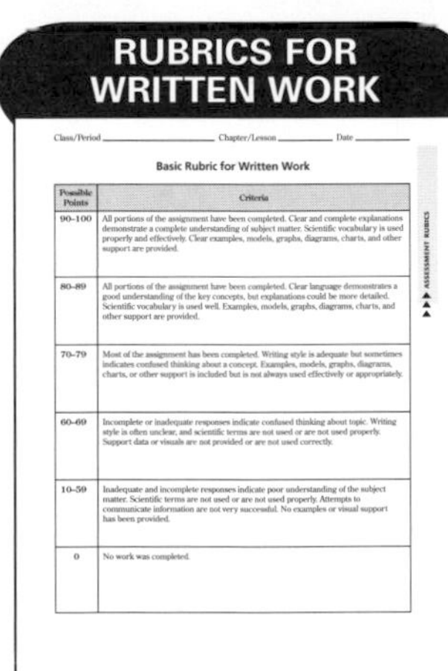
Basic Rubric for Written Work

RUBRIC FOR EXPERIMENTS

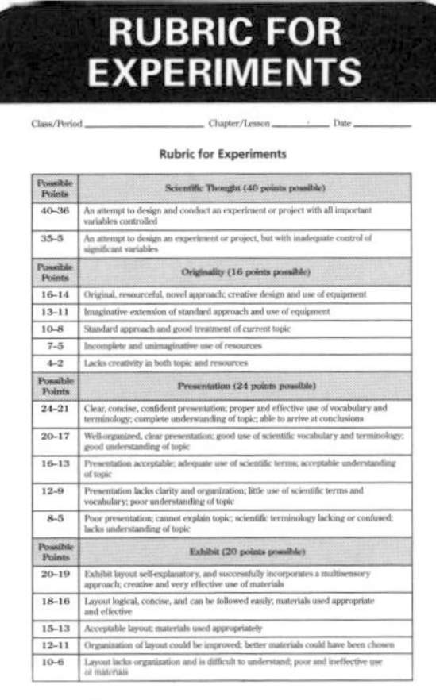
Rubric for Experiments

TEACHER EVALUATION OF COOPERATIVE LEARNING

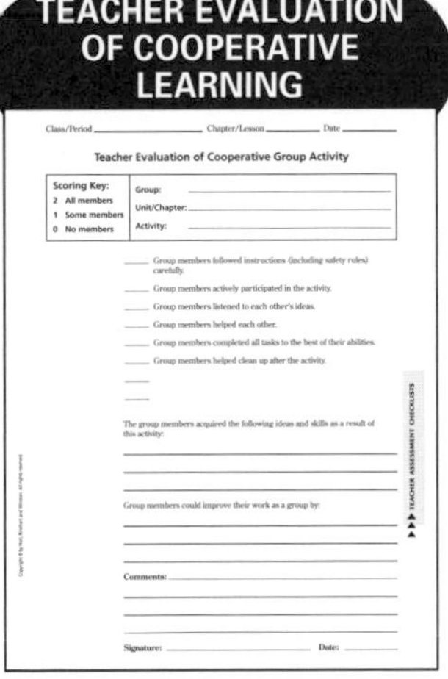
Teacher Evaluation of Cooperative Group Activity

TEACHER EVALUATION OF STUDENT PROGRESS

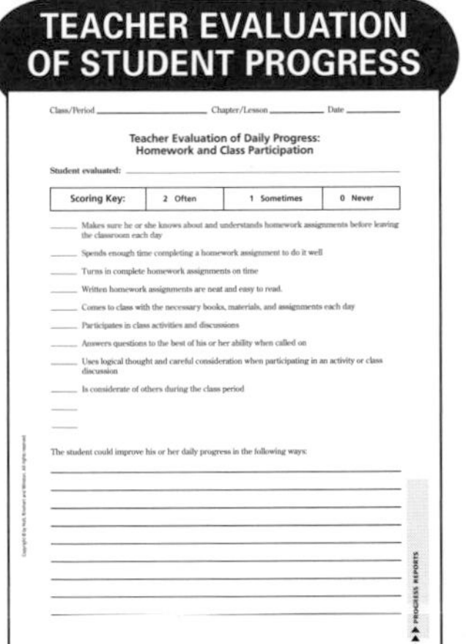
Teacher Evaluation of Daily Progress: Homework and Class Participation

National Science Education Standards

The following lists show the chapter correlation of ***Holt Science & Technology: Cells, Heredity, and Classification*** with the ***National Science Education Standards*** (grades 5–8).

Standard	Chapter Correlation
Unifying Concepts and Processes	
Systems, order, and organization Code: UCP 1	Chapter 1 1.2, 1.3; Chapter 2 2.1, 2.2; Chapter 3 3.1; Chapter 4 4.2; Chapter 5 5.2, 5.3; Chapter 6 6.1, 6.2; Chapter 7 7.1
Evidence, models, and explanation Code: UCP 2	Chapter 1 1.3; Chapter 2 2.1; Chapter 3 3.1, 3.2; Chapter 4 4.1; Chapter 5 5.1, 5.2; Chapter 6 6.1, 6.2, 6.3
Change, constancy, and measurement Code: UCP 3	Chapter 1 1.1; Chapter 2 2.1, 2.2; Chapter 3 3.2; Chapter 5 5.3; Chapter 6 6.2
Evolution and equilibrium Code: UCP 4	Chapter 1 1.2; Chapter 2 2.1, 2.2; Chapter 3 3.3; Chapter 4 4.2; Chapter 5 5.1, 5.2, 5.3; Chapter 6 6.1, 6.2, 6.3
Form and function Code: UCP 5	Chapter 1 1.1, 1.2, 1.3; Chapter 2 2.2; Chapter 3 3.3; Chapter 4 4.1, 4.2; Chapter 5 5.1, 5.2; Chapter 6 6.3; Chapter 7 7.2
Science as Inquiry	
Abilities necessary to do scientific inquiry Code: SAI 1	Chapter 1 1.1; Chapter 2 2.1, 2.2, 2.3; Chapter 3 3.1, 3.3; Chapter 4 4.1, 4.2; Chapter 5 5.2; Chapter 6 6.1, 6.2, 6.3; Chapter 7 7.2
Understandings about scientific inquiry Code: SAI 2	Chapter 1 1.1; Chapter 2 2.2; Chapter 3 3.1; Chapter 4 4.1, 4.2; Chapter 5 5.1, 5.2; Chapter 6 6.1, 6.3; Chapter 7 7.1
Science and Technology	
Abilities of technological design Code: ST 1	Chapter 4 4.1; Chapter 6 6.3
Understandings about science and technology Code: ST 2	Chapter 1 1.1; Chapter 2 2.2; Chapter 3 3.1; Chapter 4 4.1, 4.2; Chapter 6 6.3
Science in Personal Perspectives	
Personal health Code: SPSP 1	Chapter 2 2.2
Populations, resources, and environments Code: SPSP 2	Chapter 5 5.2
Risks and benefits Code: SPSP 4	Chapter 2 2.2; Chapter 4 4.2; Chapter 5 5.3
Science and technology in society Code: SPSP 5	Chapter 1 1.1; Chapter 2 2.2; Chapter 3 3.1, 3.3; Chapter 4 4.1, 4.2; Chapter 5 5.2, 5.3
History and Nature of Science	
Science as a human endeavor Code: HNS 1	Chapter 3 3.1; Chapter 4 4.1; Chapter 5 5.2; Chapter 6 6.1; Chapter 7 7.1, 7.2
Nature of science Code: HNS 2	Chapter 3 3.1, 3.3; Chapter 4 4.1; Chapter 5 5.1, 5.2; Chapter 6 6.1, 6.3; Chapter 7 7.1, 7.2
History of science Code: HNS 3	Chapter 3 3.1, 3.3; Chapter 4 4.1; Chapter 5 5.2; Chapter 6 6.1, 6.3; Chapter 7 7.1

Life Science Content Standards

Structure and Function in Living Systems

Standard	Chapter Correlation
Living systems at all levels of organization demonstrate the complementary nature of structure and function. Important levels of organization for structure and function include cells, organs, tissues, organ systems, whole organisms, and ecosystems. Code: LS 1a	**Chapter 1** 1.1, 1.2, 1.3 **Chapter 2** 2.1 **Chapter 4** 4.1 **Chapter 6** 6.1, 6.2, 6.3
All organisms are composed of cells—the fundamental unit of life. Most organisms are single cells; other organisms, including humans, are multicellular. Code: LS 1b	**Chapter 1** 1.1, 1.2 **Chapter 6** 6.2 **Chapter 7** 7.2
Cells carry on the many functions needed to sustain life. They grow and divide, thereby producing more cells. This requires that they take in nutrients, which they use to provide energy for the work that cells do and to make the materials that a cell or an organism needs. Code: LS 1c	**Chapter 1** 1.1, 1.2 **Chapter 2** 2.1, 2.2, 2.3 **Chapter 3** 3.3 **Chapter 4** 4.2
Specialized cells perform specialized functions in multicellular organisms. Groups of specialized cells cooperate to form a tissue, such as a muscle. Different tissues are in turn grouped together and form larger functional units, called organs. Each type of cell, tissue, and organ has a distinct structure and set of functions that serve the organism as a whole. Code: LS 1d	**Chapter 1** 1.3 **Chapter 3** 3.3
The human organism has systems for digestion, respiration, reproduction, circulation, excretion, movement, control and coordination, and protection from disease. These systems interact with one another. Code: LS 1e	**Chapter 1** 1.3 **Chapter 4** 4.2
Disease is the breakdown in structures or functions of an organism. Some diseases are the result of intrinsic failures of the system. Others are the result of damage by infection by other organisms. Code: LS 1f	**Chapter 4** 4.2 **Chapter 7** 7.2

Reproduction and Heredity

Standard	Chapter Correlation
Reproduction is a characteristic of all living systems; because no living organism lives forever, reproduction is essential to the continuation of every species. Some organisms reproduce asexually. Others reproduce sexually. Code: LS 2a	**Chapter 3** 3.2, 3.3 **Chapter 5** 5.2, 5.3 **Chapter 7** 7.2
In many species, including humans, females produce eggs and males produce sperm. Plants also reproduce sexually—the egg and sperm are produced in the flowers of flowering plants. An egg and sperm unite to begin development of a new individual. The individual receives genetic information from its mother (via the egg) and its father (via the sperm). Sexually produced offspring never are identical to either of their parents. Code: LS 2b	**Chapter 3** 3.1, 3.2, 3.3 **Chapter 4** 4.2 **Chapter 5** 5.2
Every organism requires a set of instructions for specifying its traits. Heredity is the passage of these instructions from one generation to another. Code: LS 2c	**Chapter 1** 1.1 **Chapter 3** 3.2, 3.3 **Chapter 4** 4.2
Hereditary information is contained in the genes, located in the chromosomes of each cell. Each gene carries a single unit of information. An inherited trait of an individual can be determined by one or by many genes, and a single gene can influence more than one trait. A human cell contains many thousands of different genes. Code: LS 2d	**Chapter 2** 2.3 **Chapter 3** 3.2, 3.3 **Chapter 4** 4.1, 4.2 **Chapter 5** 5.2
The characteristics of an organism can be described in terms of a combination of traits. Some traits are inherited and others result from interactions with the environment. Code: LS 2e	**Chapter 3** 3.1, 3.2 **Chapter 4** 4.2 **Chapter 5** 5.1, 5.2, 5.3

Standard	Chapter Correlation
Regulation and Behavior	
All organisms must be able to obtain and use resources, grow, reproduce, and maintain stable internal conditions while living in a constantly changing external environment. Code: LS 3a	**Chapter 1** 1.2 **Chapter 5** 5.1
Regulation of an organism's internal environment involves sensing the internal environment and changing physiological activities to keep conditions within the range required to survive. Code: LS 3b	**Chapter 1** 1.1
An organism's behavior evolves through adaptation to its environment. How a species moves, obtains food, reproduces, and responds to danger are based in the species' evolutionary history. Code: LS 3d	**Chapter 5** 5.1, 5.2, 5.3 **Chapter 6** 6.1, 6.2, 6.3
Populations and Ecosystems	
A population consists of all individuals of a species that occur together at a given place and time. All populations living together and the physical factors with which they interact compose an ecosystem. Code: LS 4a	**Chapter 5** 5.1
Populations of organisms can be categorized by the functions they serve in an ecosystem. Plants and some microorganisms are producers—they make their own food. All animals, including humans, are consumers, which obtain their food by eating other organisms. Decomposers, primarily bacteria and fungi, are consumers that use waste materials and dead organisms for food. Food webs identify the relationship among producers, consumers, and decomposers in an ecosystem. Code: LS 4b	**Chapter 7** 7.2
For ecosystems, the major source of energy is sunlight. Energy entering ecosystems as sunlight is transferred by producers into chemical energy through photosynthesis. That energy passes from organism to organism in food webs. Code: LS 4c	**Chapter 2** 2.2 **Chapter 7** 7.2
The number of organisms an ecosystem can support depends on the resources available and abiotic factors, such as the quantity of light and water, range of temperatures, and soil composition. Given adequate biotic and abiotic resources and no disease or predators, populations (including humans) increase at rapid rates. Lack of resources and other factors, such as predation and climate, limit the growth of populations in specific niches in the ecosystem. Code: LS 4d	**Chapter 5** 5.3
Diversity and Adaptations of Organisms	
Millions of species of animals, plants, and microorganisms are alive today. Although different species might look dissimilar, the unity among organisms becomes apparent from an analysis of internal structures, the similarity of their chemical processes, and the evidence of common ancestry. Code: LS 5a	**Chapter 1** 1.1, 1.2 **Chapter 4** 4.1 **Chapter 5** 5.1, 5.2 **Chapter 6** 6.2, 6.3 **Chapter 7** 7.1
Biological evolution accounts for the diversity of species developed through gradual processes over many generations. Species acquire many of their unique characteristics through biological adaptation, which involves the selection of naturally occurring variations in populations. Biological adaptations include changes in structures, behaviors, or physiology that enhance survival and reproductive success in a particular environment. Code: LS 5b	**Chapter 4** 4.2 **Chapter 5** 5.1, 5.2, 5.3 **Chapter 6** 6.1, 6.2, 6.3 **Chapter 7** 7.2
Extinction of a species occurs when the environment changes and the adaptive characteristics of a species are insufficient to allow its survival. Fossils indicate that many organisms that lived long ago are extinct. Extinction of species is common; most of the species that have lived on Earth no longer exist. Code: LS 5c	**Chapter 5** 5.1 **Chapter 6** 6.1, 6.2, 6.3

Cells, Heredity, and Classification

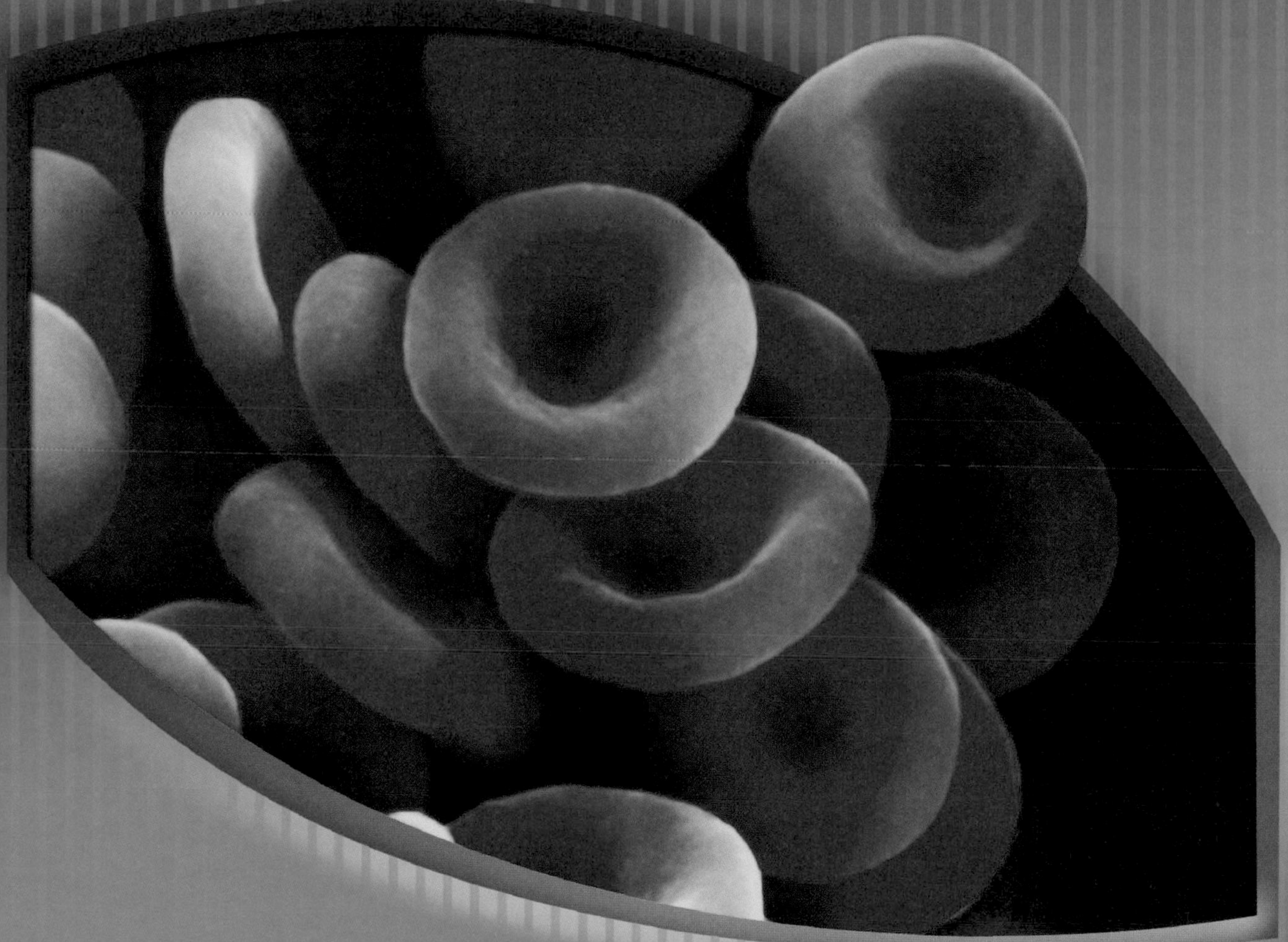

HOLT, RINEHART AND WINSTON
A Harcourt Education Company
Orlando • Austin • New York • San Diego • Toronto • London

Acknowledgments

Contributing Authors

Linda Ruth Berg, Ph.D.
Adjunct Professor
Natural Sciences
St. Petersburg College
St. Petersburg, Florida

Barbara Christopher
Science Writer and Editor
Austin, Texas

Mark F. Taylor, Ph.D.
Associate Professor of Biology
Biology Department
Baylor University
Waco, Texas

Inclusion Specialist

Ellen McPeek Glisan
Special Needs Consultant
San Antonio, Texas

Safety Reviewer

Jack Gerlovich, Ph.D.
Associate Professor
School of Education
Drake University
Des Moines, Iowa

Academic Reviewers

Glenn Adelson
Instructor
Biology Undergraduate Program
Harvard University
Cambridge, Massachusetts

Joe W. Crim, Ph.D.
Professor and Head of Cellular Biology
Department of Cellular Biology
University of Georgia
Athens, Georgia

Jim Denbow, Ph.D.
Associate Professor of Archaeology
Department of Anthropology and Archaeology
The University of Texas at Austin
Austin, Texas

David Haig, Ph.D.
Professor of Biology
Organismic and Evolutionary Biology
Harvard University
Cambridge, Massachusetts

Laurie Santos, Ph.D.
Assistant Professor
Department of Psychology
Yale University
New Haven, Connecticut

Patrick K. Schoff, Ph.D.
Research Associate
Natural Resources Research Institute
University of Minnesota—Duluth
Duluth, Minnesota

Richard P. Vari, Ph.D.
Research Scientist and Curator
Division of Fishes
National Museum of Natural History
Washington, D.C.

Teacher Reviewers

Diedre S. Adams
Physical Science Instructor
West Vigo Middle School
West Terre Haute, Indiana

Sarah Carver
Science Teacher
Jackson Creek Middle School
Bloomington, Indiana

Hilary Cochran
Science Teacher
Indian Crest Junior High School
Souderton, Pennsylvania

Karen Dietrich, S.S.J., Ph.D.
Principal and Biology Instructor
Mount Saint Joseph Academy
Flourtown, Pennsylvania

Debra S. Kogelman, MAed.
Science Teacher
University of Chicago Laboratory Schools
Chicago, Illinois

Elizabeth Rustad
Science Teacher
Higley School District
Gilbert, Arizona

Helen P. Schiller
Instructional Coach
The School District of Greenville County
Greenville, South Carolina

Stephanie Snowden
Science Teacher
Canyon Vista Middle School
Austin, Texas

Angie Williams
Teacher
Riversprings Middle School
Crawfordville, Florida

Lab Development

Diana Scheidle Bartos
Research Associate
School of Mines
Golden, Colorado

Carl Benson
General Science Teacher
Plains High School
Plains, Montana

Charlotte Blassingame
Technology Coordinator
White Station Middle School
Memphis, Tennessee

Marsha Carver
Science Teacher and Department Chair
McLean County High School
Calhoun, Kentucky

Printed in the United States of America

ISBN 0-03-025564-3

1 2 3 4 5 6 7 048 08 07 06 05 04

Kenneth E. Creese
Science Teacher
White Mountain Junior High School
Rock Springs, Wyoming

Linda Culp
Science Teacher and Department Chair
Thorndale High School
Thorndale, Texas

James Deaver
Science Teacher and Department Chair
West Point High School
West Point, Nebraska

Frank McKinney, Ph.D.
Professor of Geology
Appalachian State University
Boone, North Carolina

Alyson Mike
Science Teacher
East Valley Middle School
East Helena, Montana

C. Ford Morishita
Biology Teacher
Clackamas High School
Milwaukie, Oregon

Patricia D. Morrell, Ph.D.
Associate Professor
School of Education
University of Portland
Portland, Oregon

Hilary C. Olson, Ph.D.
Research Associate
Institute for Geophysics
The University of Texas at Austin
Austin, Texas

James B. Pulley
Science Editor and Former Science Teacher
North Kansas City, Missouri

Denice Lee Sandefur
Science Chairperson
Nucla High School
Nucla, Colorado

Patti Soderberg
Science Writer
The BioQUEST Curriculum Consortium
Biology Department
Beloit College
Beloit, Wisconsin

Phillip Vavala
Science Teacher and Department Chair
Salesianum School
Wilmington, Delaware

Albert C. Wartski, M.A.T.
Biology Teacher
Chapel Hill High School
Chapel Hill, North Carolina

Lynn Marie Wartski
Science Writer and Former Science Teacher
Hillsborough, North Carolina

Ivora D. Washington
Science Teacher and Department Chair
Hyattsville Middle School
Washington, D.C.

Lab Testing

Georgiann Delgadillo
Science Teacher
East Valley Continuous Curriculum School
Spokane, Washington

Susan Gorman
Science Teacher
North Ridge Middle School
North Richland Hills, Texas

Karma Houston-Hughes
Science Mentor
Kyrene Middle School
Tempe, Arizona

Kerry A. Johnson
Science Teacher
Isbell Middle School
Santa Paula, California

M. R. Penny Kisiah
Science Teacher and Department Chair
Fairview Middle School
Tallahassee, Florida

Kathy LaRoe
Science Teacher
East Valley Middle School
East Helena, Montana

Maurine O. Marchani
Science Teacher and Department Chair
Raymond Park Middle School
Indianapolis, Indiana

Terry J. Rakes
Science Teacher
Elmwood Junior High School
Rogers, Arkansas

Debra A. Sampson
Science Teacher
Booker T. Washington Middle School
Elgin, Texas

Feature Development

Hatim Belyamani
John A. Benner
David Bradford
Jennifer Childers
Mickey Coakley
Susan Feldkamp
Jane Gardner
Erik Hahn
Christopher Hess
Deena Kalai
Charlotte W. Luongo, MSc
Michael May
Persis Mehta, Ph.D.
Eileen Nehme, MPH
Catherine Podeszwa
Dennis Rathnaw
Daniel B. Sharp
April Smith West
John M. Stokes
Molly F. Wetterschneider

Answer Checking

Hatim Belyamani
Austin, Texas

C Cells, Heredity, and Classification

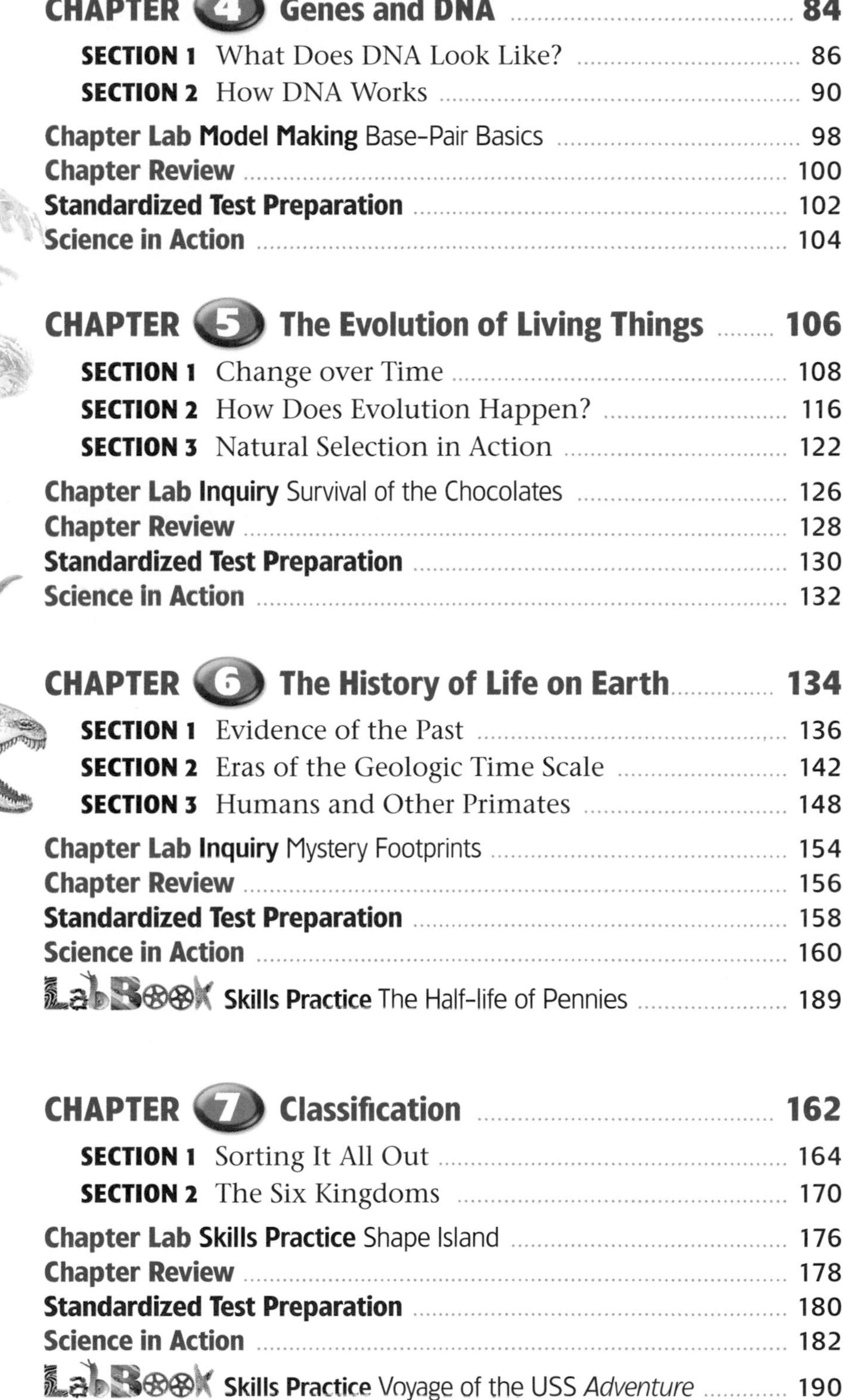

Skills Development

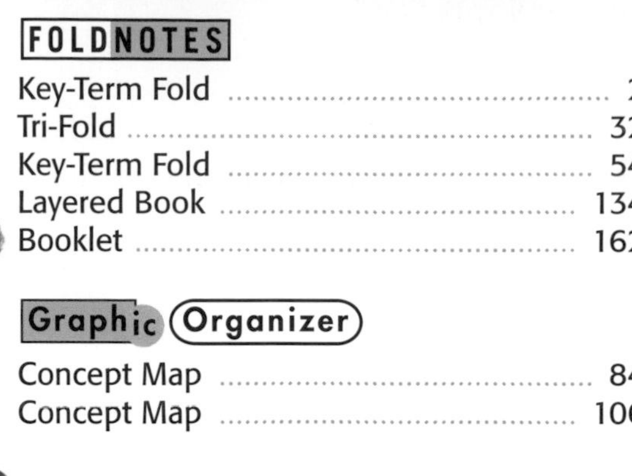

PRE-READING ACTIVITY

START-UP ACTIVITY

Quick Lab

School to Home

READING STRATEGY

INTERNET ACTIVITY

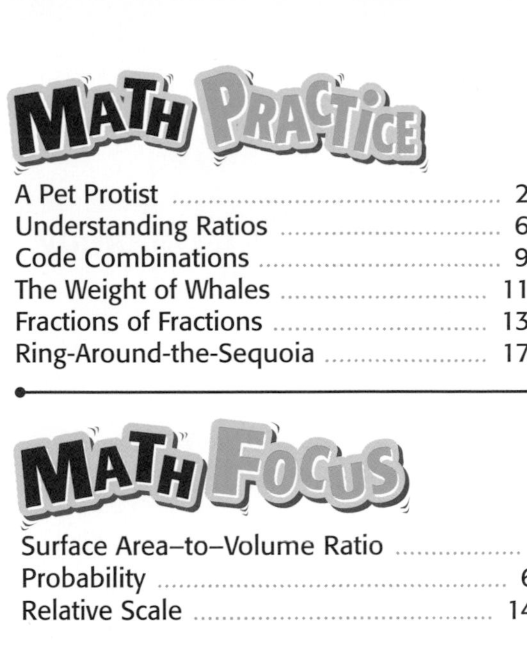

Connection to . . .

Science in Action

How to Use Your Textbook

Your Roadmap for Success with Holt Science and Technology

Reading Warm-Up

A Reading Warm-Up at the beginning of every section provides you with the section's objectives and key terms. The objectives tell you what you'll need to know after you finish reading the section.

Key terms are listed for each section. Learn the definitions of these terms because you will most likely be tested on them. Each key term is highlighted in the text and is defined at point of use and in the margin. You can also use the glossary to locate definitions quickly.

STUDY TIP Reread the objectives and the definitions to the key terms when studying for a test to be sure you know the material.

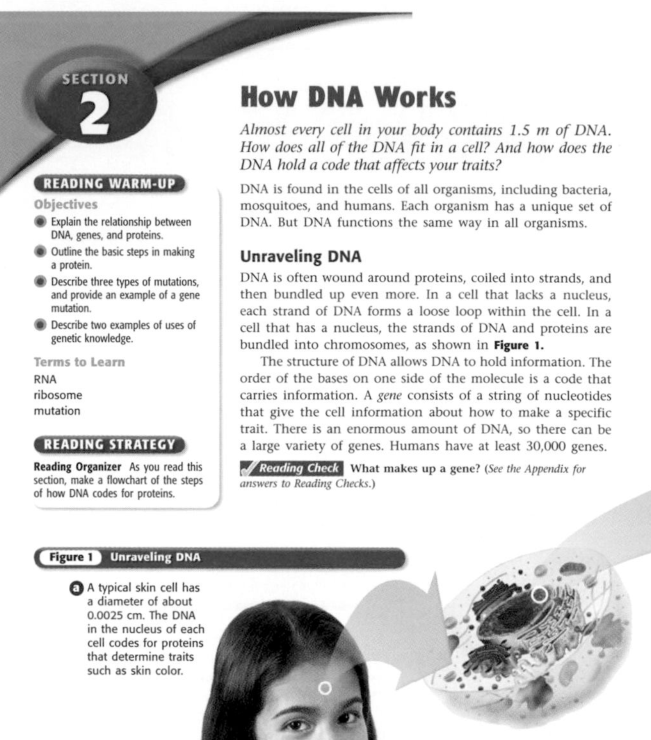
SECTION 2

How DNA Works

Almost every cell in your body contains 1.5 m of DNA. How does all of the DNA fit in a cell? And how does the DNA hold a code that affects your traits?

DNA is found in the cells of all organisms, including bacteria, mosquitoes, and humans. Each organism has a unique set of DNA. But DNA functions the same way in all organisms.

Unraveling DNA

DNA is often wound around proteins, coiled into strands, and then bundled up even more. In a cell that lacks a nucleus, each strand of DNA forms a loose loop within the cell. In a cell that has a nucleus, the strands of DNA and proteins are bundled into chromosomes, as shown in **Figure 1.**

The structure of DNA allows DNA to hold information. The order of the bases on one side of the molecule is a code that carries information. A *gene* consists of a string of nucleotides that give the cell information about how to make a specific trait. There is an enormous amount of DNA, so there can be a large variety of genes. Humans have at least 30,000 genes.

Reading Check **What makes up a gene?** (*See the Appendix for answers to Reading Checks.*)

READING WARM-UP

Objectives

- Explain the relationship between DNA, genes, and proteins.
- Outline the basic steps in making a protein.
- Describe three types of mutations, and provide an example of a gene mutation.
- Describe two examples of uses of genetic knowledge.

Terms to Learn

RNA
ribosome
mutation

READING STRATEGY

Reading Organizer As you read this section, make a flowchart of the steps of how DNA codes for proteins.

Figure 1 Unraveling DNA

a A typical skin cell has a diameter of about 0.0025 cm. The DNA in the nucleus of each cell codes for proteins that determine traits such as skin color.

b The DNA in the nucleus is part of a material called *chromatin*. Long strands of chromatin are usually bundled loosely within the nucleus.

148 Chapter 6

Get Organized

A Reading Strategy at the beginning of every section provides tips to help you organize and remember the information covered in the section. Keep a science notebook so that you are ready to take notes when your teacher reviews the material in class. Keep your assignments in this notebook so that you can review them when studying for the chapter test.

Be Resourceful—Use the Web

Internet Connect boxes in your textbook take you to resources that you can use for science projects, reports, and research papers. Go to scilinks.org, and type in the SciLinks code to get information on a topic.

Visit go.hrw.com Find worksheets, **Current Science**® magazine articles online, and other materials that go with your textbook at **go.hrw.com.** Click on the textbook icon and the table of contents to see all of the resources for each chapter.

An Example of a Substitution

A mutation, such as a substitution, can be harmful because it may cause a gene to produce the wrong protein. Consider the DNA sequence GAA. When copied as mRNA, this sequence gives the instructions to place the amino acid glutamic acid into the growing protein. If a mistake happens and the original DNA sequence is changed to GTA, the sequence will code for the amino acid valine instead.

This simple change in an amino acid can cause the disease *sickle cell anemia.* Sickle cell anemia affects red blood cells. When valine is substituted for glutamic acid in a blood protein, as shown in **Figure 4,** the red blood cells are changed into a sickle shape.

The sickle cells are not as good at carrying oxygen as normal red blood cells are. Sickle cells are also likely to get stuck in blood vessels and cause painful and dangerous clots.

Reading Check What causes sickle cell anemia?

School to Home

An Error in the Message

The sentence below is the result of an error similar to a DNA mutation. The original sentence was made up of three-letter words, but an error was made in this copy. Explain the idea of mutations to your parent. Then, work together to find the mutation, and write the sentence correctly.

THE IGB ADC ATA TET HEB IGR EDR AT.

Activity

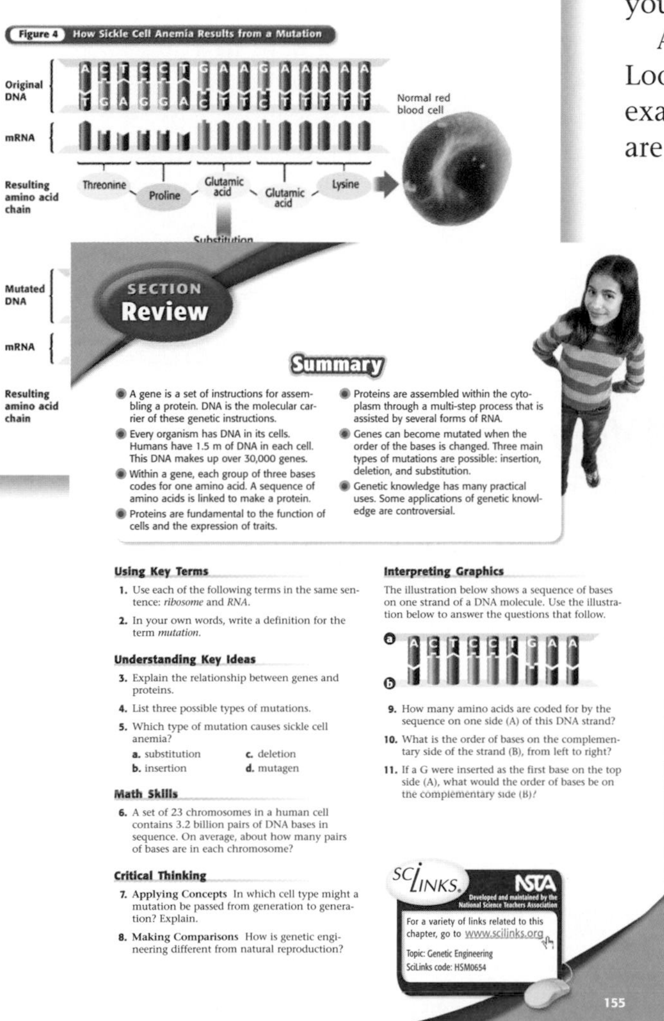

Use the Illustrations and Photos

Art shows complex ideas and processes. Learn to analyze the art so that you better understand the material you read in the text.

Tables and graphs display important information in an organized way to help you see relationships.

A picture is worth a thousand words. Look at the photographs to see relevant examples of science concepts that you are reading about.

Answer the Section Reviews

Section Reviews test your knowledge of the main points of the section. Critical Thinking items challenge you to think about the material in greater depth and to find connections that you infer from the text.

STUDY TIP When you can't answer a question, reread the section. The answer is usually there.

Do Your Homework

Your teacher may assign worksheets to help you understand and remember the material in the chapter.

STUDY TIP Don't try to answer the questions without reading the text and reviewing your class notes. A little preparation up front will make your homework assignments a lot easier. Answering the items in the Chapter Review will help prepare you for the chapter test.

Visit Holt Online Learning

If your teacher gives you a special password to log onto the Holt Online Learning site, you'll find your complete textbook on the Web. In addition, you'll find some great learning tools and practice quizzes. You'll be able to see how well you know the material from your textbook.

Visit CNN Student News

You'll find up-to-date events in science at **cnnstudentnews.com.**

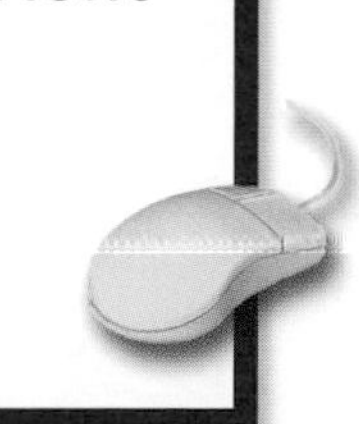

Exploring, inventing, and investigating are essential to the study of science. However, these activities can also be dangerous. To make sure that your experiments and explorations are safe, you must be aware of a variety of safety guidelines. You have probably heard of the saying, "It is better to be safe than sorry." This is particularly true in a science classroom where experiments and explorations are being performed. Being uninformed and careless can result in serious injuries. Don't take chances with your own safety or with anyone else's.

The following pages describe important guidelines for staying safe in the science classroom. Your teacher may also have safety guidelines and tips that are specific to your classroom and laboratory. Take the time to be safe.

Safety Rules!

Start Out Right

Always get your teacher's permission before attempting any laboratory exploration. Read the procedures carefully, and pay particular attention to safety information and caution statements. If you are unsure about what a safety symbol means, look it up or ask your teacher. You cannot be too careful when it comes to safety. If an accident does occur, inform your teacher immediately regardless of how minor you think the accident is.

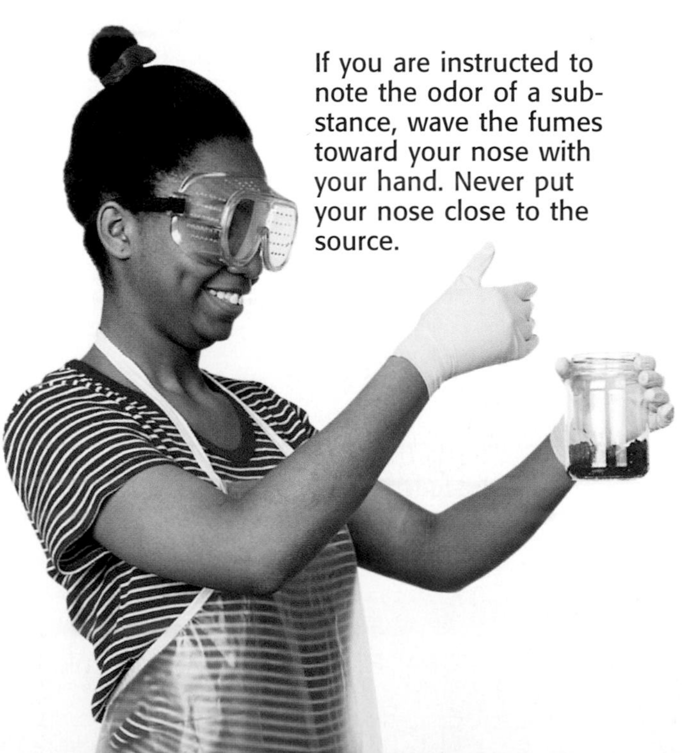

If you are instructed to note the odor of a substance, wave the fumes toward your nose with your hand. Never put your nose close to the source.

Safety Symbols

All of the experiments and investigations in this book and their related worksheets include important safety symbols to alert you to particular safety concerns. Become familiar with these symbols so that when you see them, you will know what they mean and what to do. It is important that you read this entire safety section to learn about specific dangers in the laboratory.

Eye protection

Clothing protection

Hand safety

Heating safety

Electric safety

Chemical safety

Animal safety

Sharp object

Plant safety

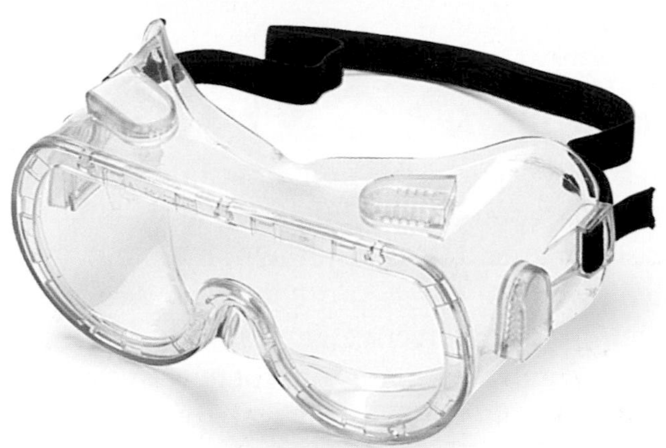

Eye Safety

Wear safety goggles when working around chemicals, acids, bases, or any type of flame or heating device. Wear safety goggles any time there is even the slightest chance that harm could come to your eyes. If any substance gets into your eyes, notify your teacher immediately and flush your eyes with running water for at least 15 minutes. Treat any unknown chemical as if it were a dangerous chemical. Never look directly into the sun. Doing so could cause permanent blindness.

Avoid wearing contact lenses in a laboratory situation. Even if you are wearing safety goggles, chemicals can get between the contact lenses and your eyes. If your doctor requires that you wear contact lenses instead of glasses, wear eye-cup safety goggles in the lab.

Safety Equipment

Know the locations of the nearest fire alarms and any other safety equipment, such as fire blankets and eyewash fountains, as identified by your teacher, and know the procedures for using the equipment.

Neatness

Keep your work area free of all unnecessary books and papers. Tie back long hair, and secure loose sleeves or other loose articles of clothing, such as ties and bows. Remove dangling jewelry. Don't wear open-toed shoes or sandals in the laboratory. Never eat, drink, or apply cosmetics in a laboratory setting. Food, drink, and cosmetics can easily become contaminated with dangerous materials.

Certain hair products (such as aerosol hair spray) are flammable and should not be worn while working near an open flame. Avoid wearing hair spray or hair gel on lab days.

Sharp/Pointed Objects 

Use knives and other sharp instruments with extreme care. Never cut objects while holding them in your hands. Place objects on a suitable work surface for cutting.

Be extra careful when using any glassware. When adding a heavy object to a graduated cylinder, tilt the cylinder so the object slides slowly to the bottom.

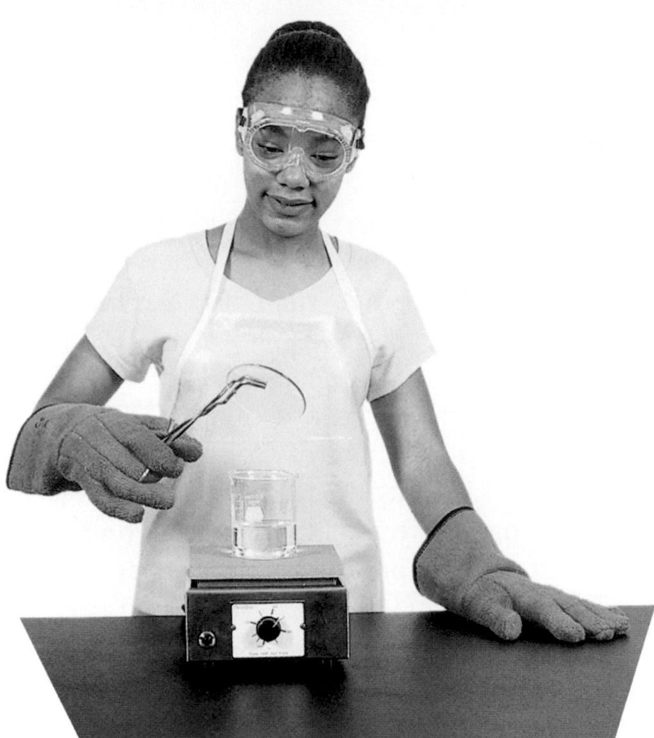

Chemicals 

Wear safety goggles when handling any potentially dangerous chemicals, acids, or bases. If a chemical is unknown, handle it as you would a dangerous chemical. Wear an apron and protective gloves when you work with acids or bases or whenever you are told to do so. If a spill gets on your skin or clothing, rinse it off immediately with water for at least 5 minutes while calling to your teacher.

Never mix chemicals unless your teacher tells you to do so. Never taste, touch, or smell chemicals unless you are specifically directed to do so. Before working with a flammable liquid or gas, check for the presence of any source of flame, spark, or heat.

Heat

Wear safety goggles when using a heating device or a flame. Whenever possible, use an electric hot plate as a heat source instead of using an open flame. When heating materials in a test tube, always angle the test tube away from yourself and others. To avoid burns, wear heat-resistant gloves whenever instructed to do so.

Electricity

Be careful with electrical cords. When using a microscope with a lamp, do not place the cord where it could trip someone. Do not let cords hang over a table edge in a way that could cause equipment to fall if the cord is accidentally pulled. Do not use equipment with damaged cords. Be sure that your hands are dry and that the electrical equipment is in the "off" position before plugging it in. Turn off and unplug electrical equipment when you are finished.

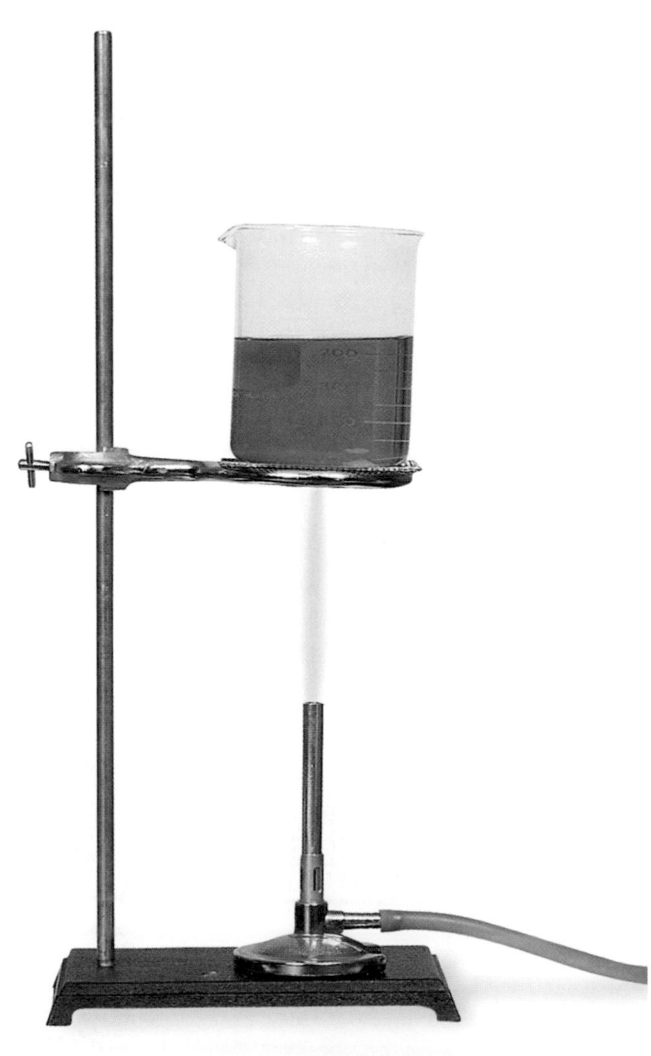

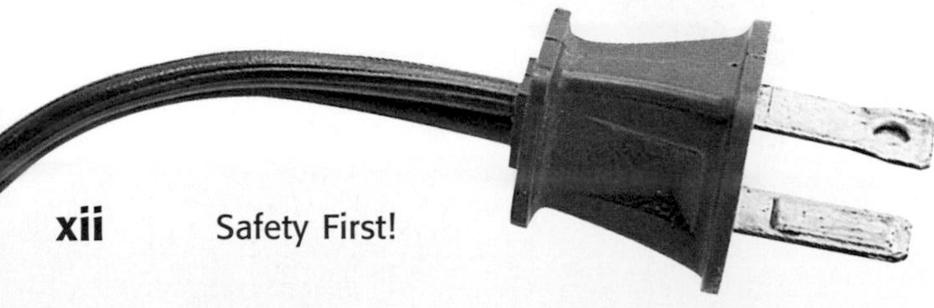

Animal Safety 

Always obtain your teacher's permission before bringing any animal into the school building. Handle animals only as your teacher directs. Always treat animals carefully and respectfully. Wash your hands thoroughly after handling any animal.

Plant Safety

Do not eat any part of a plant or plant seed used in the laboratory. Wash your hands thoroughly after handling any part of a plant. When in nature, do not pick any wild plants unless your teacher instructs you to do so.

Glassware

Examine all glassware before use. Be sure that glassware is clean and free of chips and cracks. Report damaged glassware to your teacher. Glass containers used for heating should be made of heat-resistant glass.

Cells: The Basic Units of Life
Chapter Planning Guide

Compression guide: To shorten instruction because of time limitations, omit the Chapter Lab.

OBJECTIVES	LABS, DEMONSTRATIONS, AND ACTIVITIES	TECHNOLOGY RESOURCES
PACING • 90 min pp. 2–11 **Chapter Opener**	**SE Start-up Activity,** p. 3 ◆ GENERAL	**OSP Parent Letter** ■ GENERAL **CD Student Edition on CD-ROM** **CD Guided Reading Audio CD** ■ **TR Chapter Starter Transparency*** **VID Brain Food Video Quiz**
Section 1 The Diversity of Cells • State the parts of the cell theory. • Explain why cells are so small. • Describe the parts of a cell. • Describe how eubacteria are different from archaebacteria. • Explain the difference between prokaryotic cells and eukaryotic cells.	**TE Activity** Modeling Cell Discovery, p. 4 ◆ GENERAL **SE Connection to Physics** Microscopes, p. 5 GENERAL **SE Quick Lab** Bacteria in Your Lunch?, p. 8 ◆ GENERAL **CRF Datasheet for Quick Lab*** **TE Group Activity** Archaebacteria, p. 8 ADVANCED **SE Connection to Social Studies** Where Do They Live?, p. 9 GENERAL **SE Skills Practice Labs** Elephant-Sized Amoebas?, p. 24 ◆ GENERAL **CRF Datasheet for Chapter Lab***	**CRF Lesson Plans*** **TR Bellringer Transparency*** **TR** Math Focus: Surface Area–to-Volume Ratio* **TR** A Typical Eukaryotic Cell* **VID Lab Videos for Life Science** **TE Internet Activity,** p. 10 GENERAL
PACING • 45 min pp. 12–19 **Section 2 Eukaryotic Cells** • Identify the different parts of a eukaryotic cell. • Explain the function of each part of a eukaryotic cell.	**TE Demonstration** Cell Walls and Cell Membranes, p. 13 BASIC **TE Activity** Cellular Sieve, p. 13 ◆ BASIC **TE Group Activity** Drawing Cells, p. 14 BASIC **TE Activity** Cell Models, p. 15 GENERAL **TE Activity** Vacuole Model, p. 18 ◆ BASIC **SE Skills Practice Lab** Cells Alive!, p. 184 ◆ GENERAL **CRF Datasheet for LabBook*** **LB Whiz-Bang Demonstrations** Grand Strand* ◆ GENERAL **LB Labs You Can Eat** The Incredible Edible Cell* ◆ ADVANCED **LB Long-Term Projects & Research Ideas** Ewe Again, Dolly?* ◆ ADVANCED	**CRF Lesson Plans*** **TR Bellringer Transparency*** **TR** *LINK TO PHYSICAL SCIENCE* Structural Formulas* **TR** Organelles and Their Functions* **CRF SciLinks Activity*** GENERAL
PACING • 45 min pp. 20–23 **Section 3 The Organization of Living Things** • List three advantages of being multicellular. • Describe the four levels of organization in living things. • Explain the relationship between the structure and function of a part of an organism.	**TE Activity** Concept Mapping, p. 20 ◆ GENERAL **TE Activity** Explain It to a Friend, p. 23 BASIC **SE Science in Action** Math, Social Studies, and Language Arts Activities, pp. 30–31 GENERAL	**CRF Lesson Plans*** **TR Bellringer Transparency*** **TR** Levels of Organization in the Cardiovascular System*

PACING • 90 min

CHAPTER REVIEW, ASSESSMENT, AND STANDARDIZED TEST PREPARATION

- **CRF Vocabulary Activity*** GENERAL
- **SE Chapter Review,** pp. 26–27 GENERAL
- **CRF Chapter Review*** ■ GENERAL
- **CRF Chapter Tests A*** ■ GENERAL, **B*** ADVANCED, **C*** SPECIAL NEEDS
- **SE Standardized Test Preparation,** pp. 28–29 GENERAL
- **CRF Standardized Test Preparation*** GENERAL
- **CRF Performance-Based Assessment*** GENERAL
- **OSP Test Generator** GENERAL
- **CRF Test Item Listing*** GENERAL

Online and Technology Resources

Visit **go.hrw.com** for a variety of free resources related to this textbook. Enter the keyword **HL5CEL.**

Students can access interactive problem-solving help and active visual concept development with the *Holt Science and Technology* Online Edition available at **www.hrw.com.**

Guided Reading Audio CD

A direct reading of each chapter using instructional visuals as guideposts. For auditory learners, reluctant readers, and Spanish-speaking students. Available in English and Spanish.

KEY

SE Student Edition **TE** Teacher Edition	**CRF** Chapter Resource File **OSP** One-Stop Planner **LB** Lab Bank **TR** Transparencies	**SS** Science Skills Worksheets **MS** Math Skills for Science Worksheets **CD** CD or CD-ROM **VID** Classroom Video/DVD	* Also on One-Stop Planner ◆ Requires advance prep ■ Also available in Spanish

SKILLS DEVELOPMENT RESOURCES	SECTION REVIEW AND ASSESSMENT	STANDARDS CORRELATIONS
SE Pre-Reading Activity, p. 2 GENERAL **OSP Science Puzzlers, Twisters & Teasers** GENERAL		National Science Education Standards UCP 4; HNS 3; LS 1b, 5b
CRF Directed Reading A* ■ BASIC, **B*** SPECIAL NEEDS **CRF Vocabulary and Section Summary*** ■ GENERAL **SE Reading Strategy** Reading Organizer, p. 4 GENERAL **TE Inclusion Strategies,** p. 5 **SE Math Focus** Surface Area-to-Volume Ratio, p. 6 GENERAL **TE Reading Strategy** Prediction Guide, p. 6 GENERAL **TE Reading Strategy** Prediction Guide, p. 7 GENERAL **TE Research** Be a Good Host, p. 9 GENERAL **MS Math Skills for Science** What Is a Ratio?* GENERAL **MS Math Skills for Science** Finding Perimeter and Area* GENERAL **MS Math Skills for Science** Finding Volume* GENERAL	**SE Reading Checks,** pp. 5, 6, 7, 9, 10 GENERAL **TE Reteaching,** p. 10 BASIC **TE Quiz,** p. 10 GENERAL **TE Alternative Assessment,** p. 10 GENERAL **SE Section Review,*** p. 11 ■ GENERAL **CRF Section Quiz*** ■ GENERAL	UCP 3, 5; SAI 1, 2; ST 2; SPSP 5; LS 1a, 1b, 1c, 2c, 3b, 5a; *Chapter Lab:* UCP 1, 2, 3; SAI 2; LS 1b, 1c, 3a, 3b
CRF Directed Reading A* ■ BASIC, **B*** SPECIAL NEEDS **CRF Vocabulary and Section Summary*** ■ GENERAL **SE Reading Strategy** Reading Organizer, p. 12 GENERAL **SE Connection to Language Arts** The Great Barrier, p. 13 GENERAL **TE Inclusion Strategies,** p. 14 **TE Reading Strategy** Prediction Guide, p. 16 GENERAL **CRF Critical Thinking** Cellular Construction* ADVANCED **CRF Reinforcement Worksheet** Building a Eukaryotic Cell* BASIC	**SE Reading Checks,** pp. 12, 13, 14, 16, 18 GENERAL **TE Homework,** p. 13 GENERAL **TE Homework,** p. 15 GENERAL **TE Homework,** p. 17 GENERAL **TE Reteaching,** p. 18 BASIC **TE Quiz,** p. 18 GENERAL **TE Alternative Assessment,** p. 18 GENERAL **SE Section Review,*** p. 19 ■ GENERAL **CRF Section Quiz*** ■ GENERAL	UCP 1, 4, 5; LS 1b, 1c, 3a, 5a, 5b; *LabBook:* UCP 1, 2, 5; SAI 1; ST 2; SPSP 5; HNS 1, 3; LS 1a, 1b, 1c, 1d, 2c, 3a, 5a
CRF Directed Reading A* ■ BASIC, **B*** SPECIAL NEEDS **CRF Vocabulary and Section Summary*** ■ GENERAL **SE Reading Strategy** Paired Summarizing, p. 20 GENERAL **SE Math Practice** A Pet Protist, p. 21 GENERAL	**SE Reading Checks,** pp. 20, 21, 22 GENERAL **TE Homework,** p. 21 GENERAL **TE Reteaching,** p. 22 BASIC **TE Quiz,** p. 22 GENERAL **TE Alternative Assessment,** p. 22 GENERAL **SE Section Review,*** p. 23 ■ GENERAL **CRF Section Quiz*** ■ GENERAL	UCP 1, 2, 5; LS 1a, 1b, 1d

This convenient CD-ROM includes:
- **Lab Materials QuickList Software**
- **Holt Calendar Planner**
- **Customizable Lesson Plans**
- **Printable Worksheets**
- **ExamView® Test Generator**

cnnstudentnews.com

Find the latest news, lesson plans, and activities related to important scientific events.

www.scilinks.org

Maintained by the **National Science Teachers Association.** See Chapter Enrichment pages for a complete list of topics.

Current Science®

Check out ***Current Science*** articles and activities by visiting the HRW Web site at **go.hrw.com.** Just type in the keyword **HL5CS03T.**

Classroom Videos
- **Lab Videos** demonstrate the chapter lab.
- **Brain Food Video Quizzes** help students review the chapter material.

1 Chapter Resources

Visual Resources

CHAPTER STARTER TRANSPARENCY

Cells: The Basic Units of Life — CHAPTER STARTER

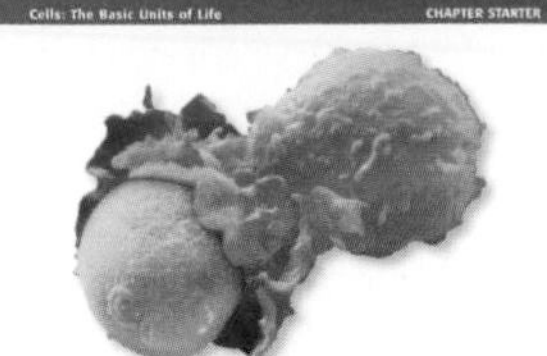

What If . . . ?

Imagine this scene from a horror film. A young man sits down to dinner to find that his mother has made asparagus again. The young man eats the dreaded asparagus stalks. Later, he finds out that instead of being digested, one of the stalks has taken up residence inside his body and is very much alive! Too horrifying to think about? What if the asparagus began to do wonderful things for the young man, such as giving him more energy than he ever dreamed possible? Lynn Margulis, a scientist, thinks that something similar may have happened to certain one-celled organisms that lived more than a billion years ago, giving rise to the kinds of cells that we are made of today.

According to Margulis's theory, about 1.2 billion years ago, some larger cells began eating smaller cells for dinner. Like the white blood cell on this page, these larger cells trapped the smaller cells with extensions of their cell body. But some of these smaller cells resisted being digested. In fact, they began to do very well in their new homes. The larger cells also benefited from their new guests. The smaller cells released large amounts of energy from food taken in by the larger cells. Other kinds of small cells used the energy in sunlight to make enough food to feed themselves and the larger cell. The energy-producing structures of most cells, including yours, are thought to have descended from these smaller cells. In this chapter, you will learn more about cells and their structures.

BELLRINGER TRANSPARENCIES

Cells: The Basic Units of Life — BELLRINGER TRANSPARENCY

Section: The Diversity of Cells

Why do you think cells weren't discovered until 1665? What invention do you think made their discovery possible? Do you think people can ever see cells with the naked eye? Explain your answer.

Write your responses in your **science journal.**

Section: Eukaryotic Cells

List three differences between *prokaryotic* and *eukaryotic* cells. Draw two diagrams illustrating the differences.

Write your responses in your **science journal.**

TEACHING TRANSPARENCIES

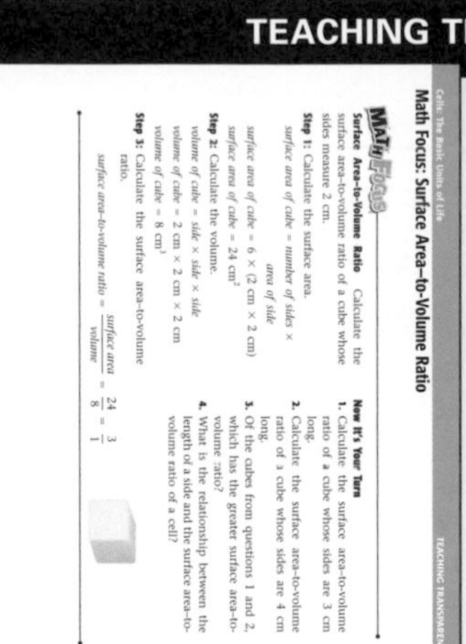

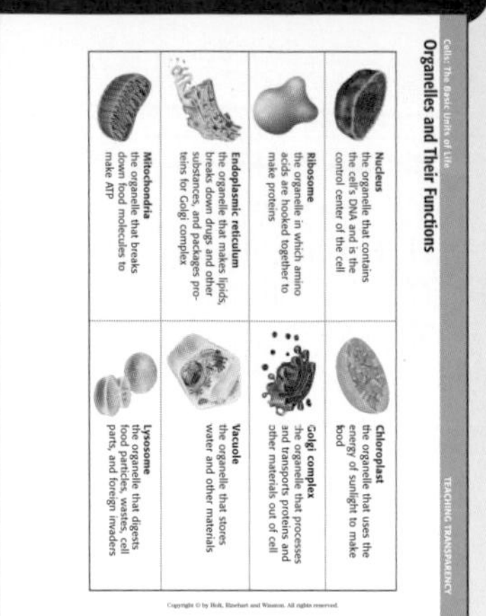

TEACHING TRANSPARENCIES

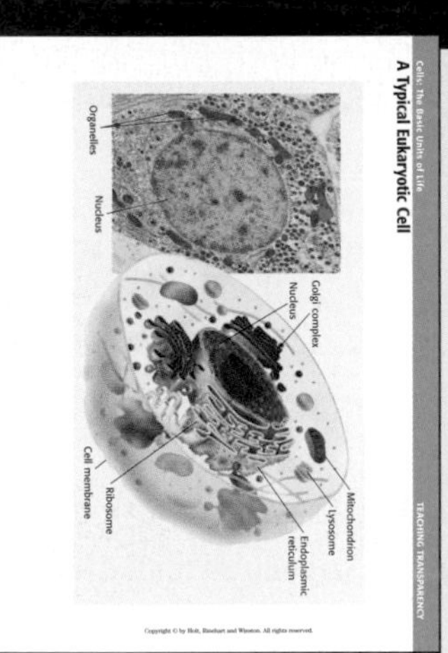

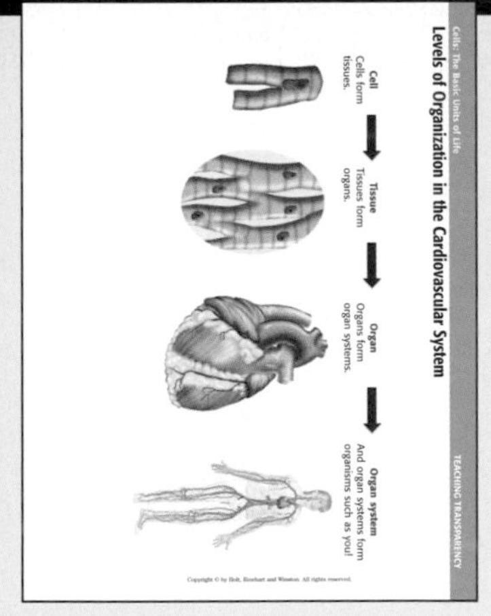

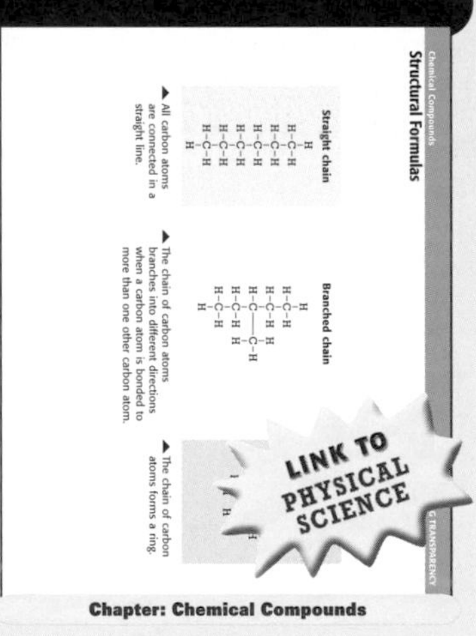

CONCEPT MAPPING TRANSPARENCY

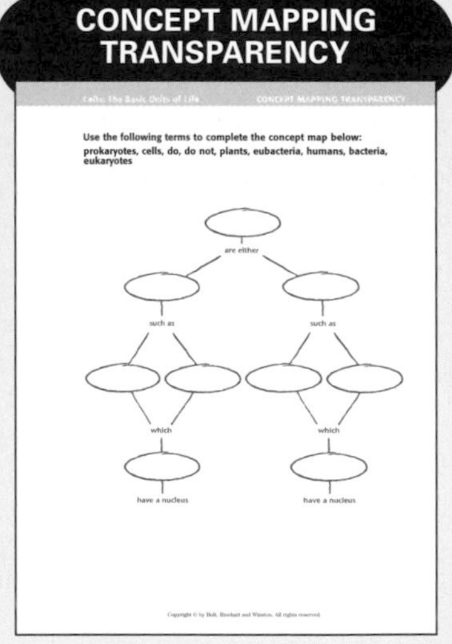

Planning Resources

LESSON PLANS

Lesson Plan SAMPLE

Section: Waves

Pacing

Regular Schedule: with lab(s):2 days without lab(s)2 days

Block Schedule: with lab(s):1 1/2 days without lab(s)1 day

Objectives

1. Relate the seven properties of life to a living organism.
2. Describe seven themes that can help you to organize what you learn about biology.
3. Identify the tiny structures that make up all living organisms.
4. Differentiate between reproduction and heredity and between metabolism and homeostasis.

National Science Education Standards Covered

LSInter6: Cells have particular structures that underlie their functions.

LSMat1: Most cell functions involve chemical reactions.

LSBeh1: Cells store and use information to guide their functions.

UCP1: Cell functions are regulated.

SI1: Cells can differentiate and form complete multicellular organisms.

PS1: Species evolve over time.

ESS1: The great diversity of organisms is the result of more than 3.5 billion years of evolution.

ESS2: Natural selection and its evolutionary consequences provide a scientific explanation for the fossil record of ancient life forms as well as for the striking molecular similarities observed among the diverse species of living organisms.

ST1: The millions of different species of plants, animals, and microorganisms that live on Earth today are related by descent from common ancestors.

ST2: The energy for life primarily comes from the sun.

SPSP1: The complexity and organization of organisms accommodates the need for obtaining, transforming, transporting, releasing, and eliminating the matter and energy used to sustain the organism.

SPSP6: As matter and energy flows through different levels of organization of living systems—cells, organs, communities—and between living systems and the physical environment, chemical elements are recombined in different ways.

HNS1: Organisms have behavioral responses to internal changes and to external stimuli.

PARENT LETTER

SAMPLE

Dear Parent,

Your son's or daughter's science class will soon begin exploring the chapter entitled "The World of Physical Science." In this chapter, students will learn about how the scientific method applies to the world of physical science and the role of physical science in the world. By the end of the chapter, students should demonstrate a clear understanding of the chapter's main ideas and be able to discuss the following topics:

1. physical science as the study of energy and matter (Section 1)
2. the role of physical science in the world around them (Section 1)
3. careers that rely on physical science (Section 1)
4. the steps used in the scientific method (Section 2)
5. examples of technology (Section 2)
6. how the scientific method is used to answer questions and solve problems (Section 2)
7. how our knowledge of science changes over time (Section 2)
8. how models represent real objects or systems (Section 3)
9. examples of different ways models are used in science (Section 3)
10. the importance of the International System of Units (Section 4)
11. the appropriate units to use for particular measurements (Section 4)
12. how area and density are derived quantities (Section 4)

Questions to Ask Along the Way

You can help your son or daughter learn about these topics by asking interesting questions such as the following:

- What are some surprising careers that use physical science?
- What is a characteristic of a good hypothesis?
- When is it a good idea to use a model?
- Why do Americans measure things in terms of inches and yards and meters ?

ALSO IN SPANISH

TEST ITEM LISTING

TEST ITEM LISTING

The World of Science SAMPLE

MULTIPLE CHOICE

1. A limitation of models is that
   a. they are large enough to see.
   b. they do not act exactly like the things that they model.
   c. they are smaller than the things that they model.
   d. they model unfamiliar things.
   Answer: B Difficulty: I Section: 3 Objective: 2
2. The length 10 m is equal to
   a. 100 cm. c. 10,000 mm.
   b. 1,000 cm. d. Both (b) and (c)
   Answer: B Difficulty: I Section: 3 Objective: 2
3. To be valid, a hypothesis must be
   a. testable. c. made into a law.
   b. supported by evidence. d. Both (a) and (b)
   Answer: B Difficulty: I Section: 3 Objective: 2 1
4. The statement "Sheila has a stain on her shirt" is an example of a(n)
   a. law. c. observation.
   b. hypothesis. d. prediction.
   Answer: B Difficulty: I Section: 3 Objective: 2
5. A hypothesis is often developed out of
   a. observations. c. laws.
   b. experiments. d. Both (a) and (b)
   Answer: B Difficulty: I Section: 3 Objective: 2
6. How many milliliters are in 3.5 kL?
   a. 3,500 mL c. 3,500, 000 mL
   b. 0.0035 mL d. 35,000 mL
   Answer: B Difficulty: I Section: 3 Objective: 2
7. A map of Seattle is an example of a
   a. law. c. model.
   b. theory. d. unit.
   Answer: B Difficulty: I Section: 3 Objective: 2
8. A lab has the safety icons shown below. These icons mean that you should wear
   a. only safety goggles. c. safety goggles and a lab apron.
   b. only a lab apron. d. safety goggles, a lab apron, and gloves.
   Answer: B Difficulty: I Section: 3 Objective: 2
9. The law of conservation of mass says the tot al mass before a chemical change is
   a. more than the total mass after the change.
   b. less than the total mass after the change.
   c. the same as the total mass after the change.
   d. not the same as the total mass after the change.
   Answer: B Difficulty: I Section: 3 Objective: 2
10. In which of the following areas might you find a geochemist at work?
    a. studying the chemistry of rocks c. studying fishes
    b. studying forestry d. studying the atmosphere
    Answer: B Difficulty: I Section: 3 Objective: 2

One-Stop Planner® CD-ROM

This CD-ROM includes all of the resources shown here and the following time-saving tools:

- *Lab Materials QuickList Software*
- *Customizable lesson plans*
- *Holt Calendar Planner*
- *The powerful ExamView® Test Generator*

For a preview of available worksheets covering math and science skills, see pages T12–T19. All of these resources are also on the One-Stop Planner®.

Meeting Individual Needs

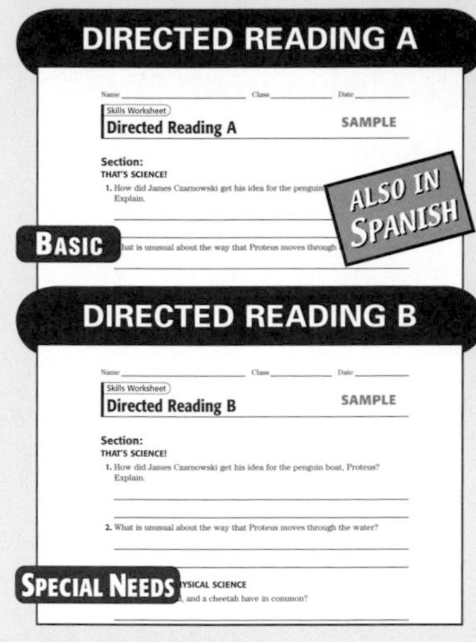

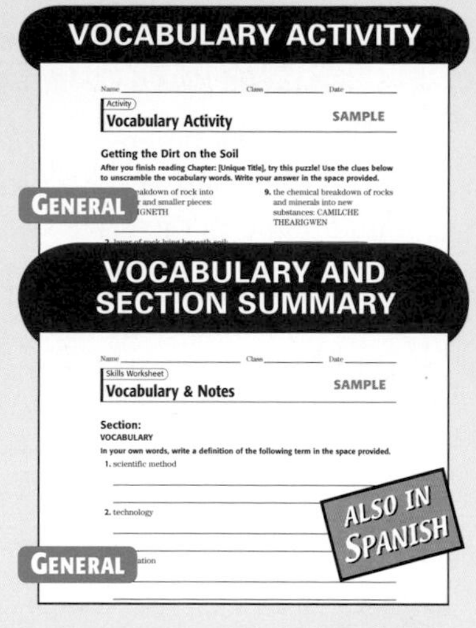

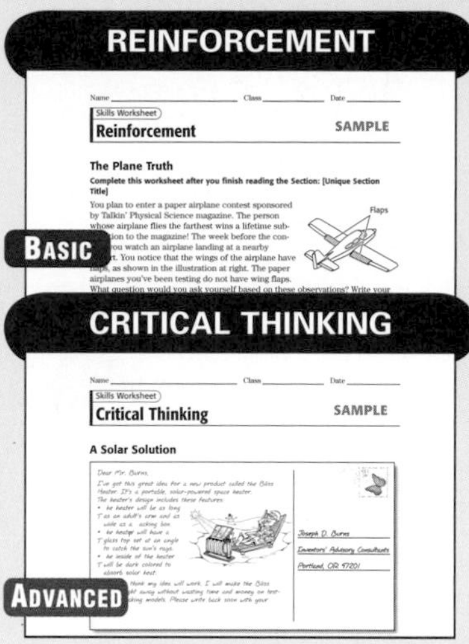

Labs and Activities

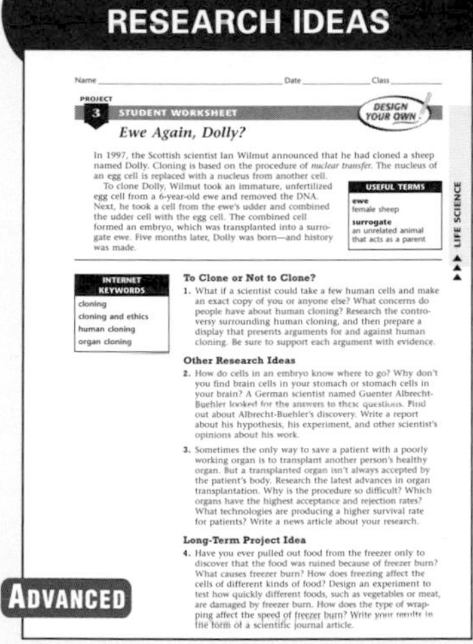

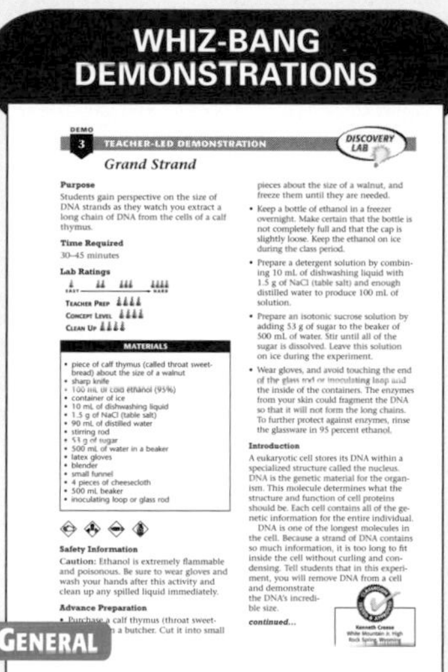

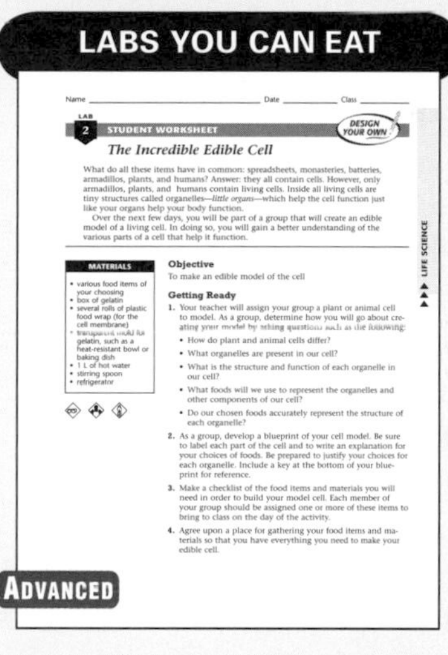

Review and Assessments

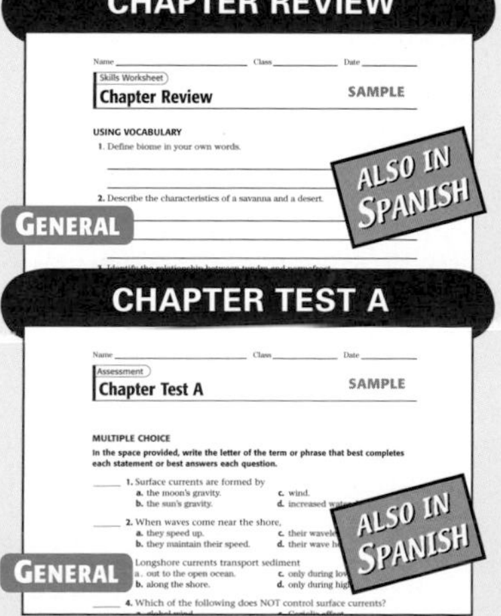

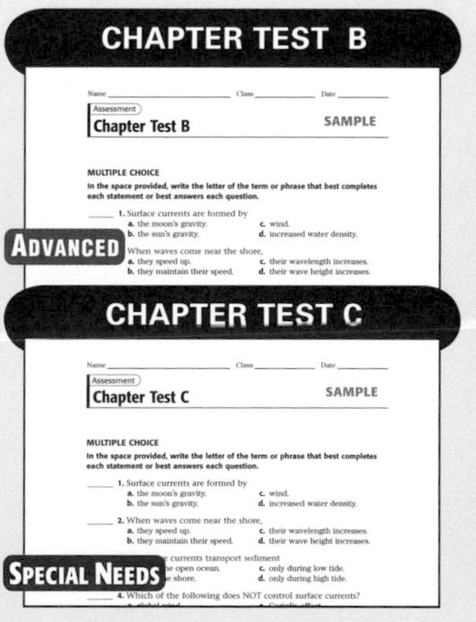

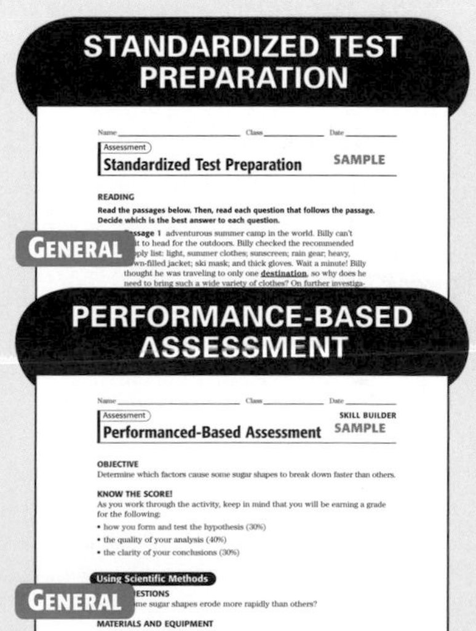

1 Chapter Enrichment

This Chapter Enrichment provides relevant and interesting information to expand and enhance your presentation of the chapter material.

Section 1

The Diversity of Cells

Microtomy

- The development of high-magnification microscopes required that the preparation of specimens for viewing also become more sophisticated. Microtomy once referred only to specimen cutting, because a microtome is the instrument used to slice tissue sections. Today, microtomy refers collectively to the art of preparing specimens by any number of techniques.
- When microscopic organisms are viewed as whole- mounts, they are preserved, stained, dried (alcohol removes the water), and made transparent with clove or cedar oil. Then, the organism is mounted in a drop of resin on a glass slide and covered with a piece of glass only 0.005 mm thick.

Physiology and the Cell Theory

- The development of the cell theory aided research in other fields. In the mid-1800s, French physiologist Claude Bernard proposed that plants and animals are composed of sets of control mechanisms that work to maintain the internal conditions necessary for life. He recognized that a mammal can sustain a constant body temperature regardless of the outside temperature. Today, we recognize the ability of organisms to regulate their physiological processes to maintain specific conditions as *homeostasis*. But at the time, no one knew what the "organized sets of control mechanisms" were. The discovery of cells and the way their many components function to sustain life in an organism gave credence to Bernard's position.

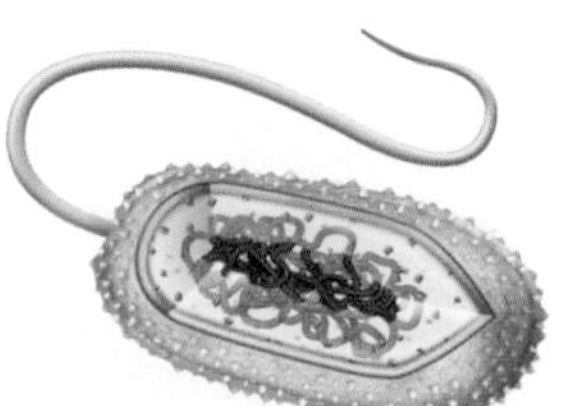

Is That a Fact!

- The Earth is 4.5 billion years old, and the oldest fossilized cells found so far are 3.5 billion years old!
- Aeolid nudibranchs are mollusks that eat hydroids, small polyps that have protective stinging cells. The nudibranch's digestive system carefully sorts out the hydroid's stinging cells and sends them to the protective tentacles on the nudibranch's own back.

Section 2

Eukaryotic Cells

"Protein" Therapy

- Decades of investigation into cell biology have produced what scientists call *gene therapy,* which refers to the use of genetic material to cure disease. It might be more appropriate to call this rapidly expanding field of science *protein therapy.*

- The gene can be thought of as a recipe for the proteins essential to life. For example, people with Duchenne muscular dystrophy lack dystrophin, an essential muscle protein that maintains the structure of muscle cells. Researchers have been able to remove the harmful genetic components of a virus and replace them with the gene for dystrophin. Their plan is to inject the dystrophin gene (the gene that codes for the dystrophin protein) directly into the muscles of Duchenne patients. If the process is successful, the dystrophin gene in the virus will compensate for patients' faulty dystrophin gene.

Tiny Scientists?

- Microbiologists study the characteristics of bacteria and other microorganisms to understand how they interact with other organisms. Virologists investigate viruses, which are active only inside a living host cell. Mycologists study fungi, which include molds and yeasts. Environmental microbiologists inspect the water in rivers and lakes. Microbiologists in agriculture are concerned with organisms that affect soil quality.

For background information about teaching strategies and issues, refer to the ***Professional Reference for Teachers.***

Is That a Fact!

- ◆ The oldest unquestionably eukaryotic fossil is about 2.1 billion years old.

Section 3

The Organization of Living Things

In a Heartbeat

- The heart will function properly only if the cells that form the connective tissue and muscle perform their jobs in coordination. Scientists can use an enzyme to dissolve an embryonic heart into its individual cells. When placed in a dish, these cells, called *myocytes,* will continue to beat, although they are out of sync with each other. After a couple of days, sheets of interconnected cells form, and the myocytes beat in unison. Why do these changes happen? Openings develop between cells that touch, and their cytoplasms connect, which allows the cells to communicate directly with each other.

Organs: Delicate Workhorses

- The most frequently transplanted organ is the kidney, followed by the liver, the heart, and the lung. Most transplants must be done within a few hours after the organ is removed from a donor because organs are too delicate to survive current long-term storage procedures.
- Cryobiologists, scientists who study how life systems tolerate low temperatures, are studying the possibility of storing organs and organ systems at subfreezing temperatures. They are investigating the fluids that keep insects alive during subfreezing temperatures. Cryobiologists hope that this knowledge can be applied to human organs.

Development

- In a multicellular organism, almost every cell has the same set of genes. (Some specialized cells delete or duplicate sections of their DNA.) Yet, different cell types are structurally distinct and perform widely different functions. Part of the reason is that each cell expresses some genes but not others. Sometimes, genes can be expressed in tissues where they should not be. Doctors have occasionally operated on people and removed tumors that had hair and teeth!

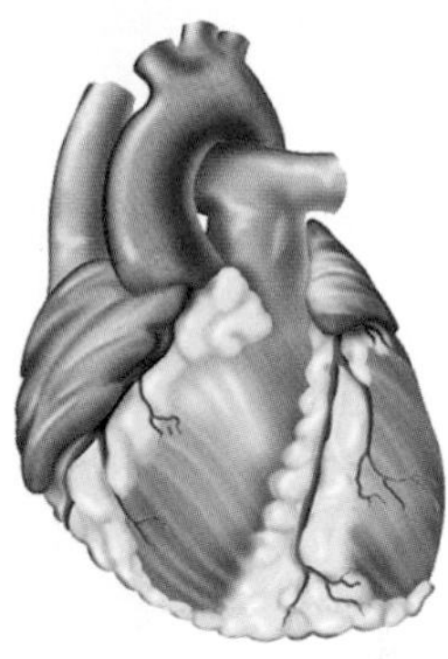

Is That a Fact!

- ◆ The oldest fossils of multicellular organisms are fossils of tiny algae approximately 1.2 billion years old.
- ◆ In 1931, a doctor removed a patient's parathyroid glands in error. These glands control the amount of calcium in the blood, which in turn regulates the heart. As a last-ditch effort to save the patient, a cow's parathyroid glands were ground up and injected into the patient. The patient recuperated and lived another 30 years with similar treatments.

SCI LINKS® NSTA Developed and maintained by the National Science Teachers Association

SciLinks is maintained by the National Science Teachers Association to provide you and your students with interesting, up-to-date links that will enrich your classroom presentation of the chapter.

Visit www.scilinks.org and enter the SciLinks code for more information about the topic listed.

Topic: Prokaryotic Cells
SciLinks code: HSM1225

Topic: Eukaryotic Cells
SciLinks code: HSM0541

Topic: Cell Structures
SciLinks code: HSM0240

Topic: Archaebacteria
SciLinks code: HSM0091

Topic: Organization of Life
SciLinks code: HSM1080

Topic: Body Systems
SciLinks code: HSM0184

Overview

This chapter will help students understand the great diversity of cells. The chapter will take students from the time when cells were unknown through the discovery of cells to the understanding of the tremendous diversity of cells. Students will learn about cell structures and will also learn how cells, tissues, and organs form organisms.

Assessing Prior Knowledge

Students should be familiar with the following topic:

- characteristics of a living thing

Identifying Misconceptions

Students may not understand that all cells and organisms have the same basic structures. Also, students may not have a sense of scale. When asked to draw a molecule, most students will draw something that resembles a cell. Instruction should emphasize the relationship between molecules and cells. For example, many students believe that proteins and molecules are bigger than cells.

1

Cells: The Basic Units of Life

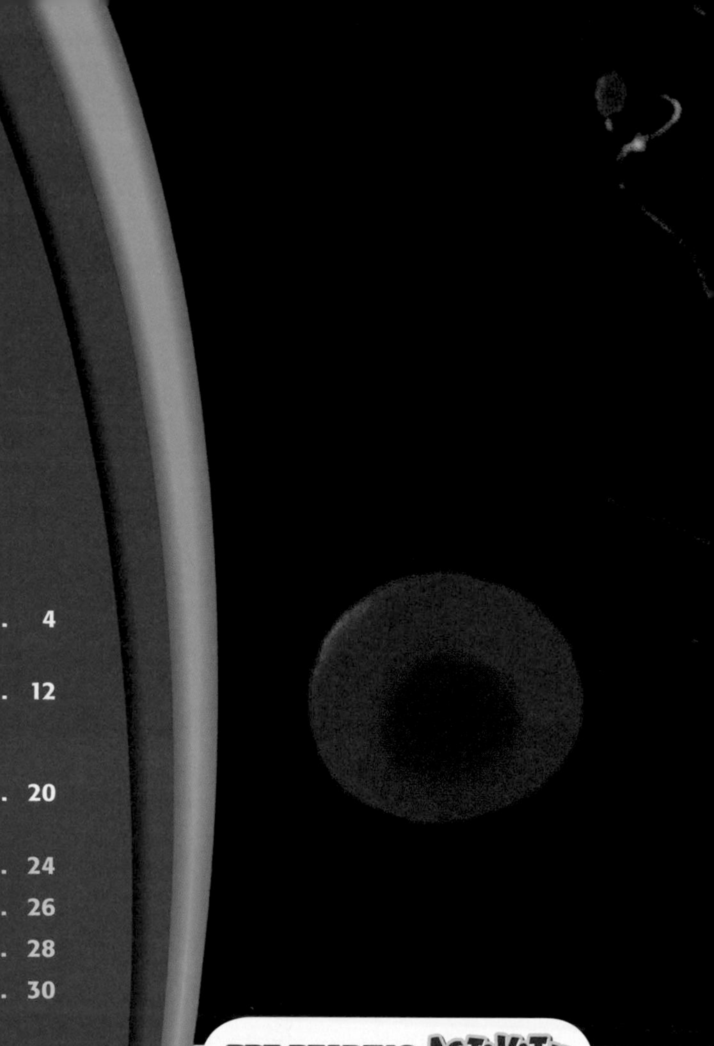

About the PHOTO

Harmful bacteria may invade your body and make you sick. But wait—your white blood cells come to the rescue! In this image, a white blood cell (the large, yellowish cell) reaches out its pseudopod to destroy bacteria (the purple cells). The red discs are red blood cells.

PRE-READING ACTIVITY

FOLDNOTES **Key-Term Fold** Before you read the chapter, create the FoldNote entitled "Key-Term Fold" described in the **Study Skills** section of the Appendix. Write a key term from the chapter on each tab of the key-term fold. Under each tab, write the definition of the key term.

Standards Correlations

National Science Education Standards

The following codes indicate the National Science Education Standards that correlate to this chapter. The full text of the standards is at the front of the book.

Chapter Opener
UCP 1; HNS 3; LS 1b, 5b

Section 1 The Diversity of Cells
UCP 4, 5; SAI 1, 2; ST 2; SPSP 5; LS 1a, 1b, 1c, 2c, 3b, 5a; *LabBook:* UCP 1, 2, 5; SAI 1; ST 2; SPSP 5; HNS 1, 3; LS 1a, 1b, 1c, 1d, 2c, 3a, 5a

Section 2 Eukaryotic Cells
UCP 1, 4, 5; LS 1b, 1c, 3a, 5a, 5b

Section 3 The Organization of Living Things
UCP 1, 2, 5; LS 1a, 1b, 1d

Chapter Lab
UCP 1, 2, 3; SAI 2; LS 1b, 1c, 3a, 3b

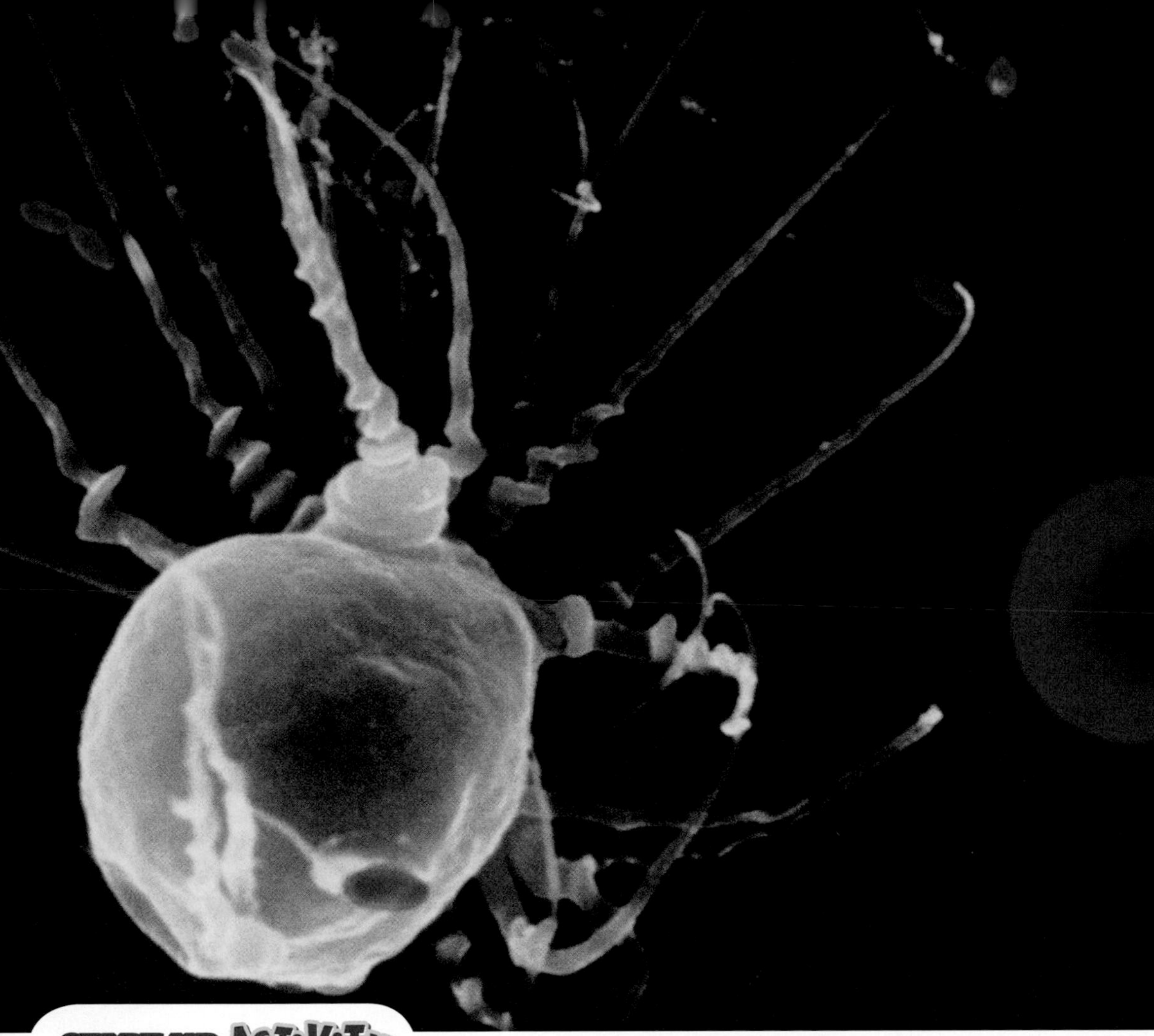

START-UP ACTIVITY

What Are Plants Made Of?

All living things, including plants, are made of cells. What do plant cells look like? Do this activity to find out.

Procedure

1. Tear off a **small leaf** from near the tip of an **Elodea sprig.**
2. Using **forceps,** place the whole leaf in a **drop of water** on a **microscope slide.**
3. Place a **coverslip** on top of the water drop by putting one edge of the coverslip on the slide near the water drop. Next, lower the coverslip slowly so that the coverslip does not trap air bubbles.
4. Place the slide on your **microscope.**
5. Using the lowest-powered lens first, find the plant cells. When you can see the cells under the lower-powered lens, switch to a higher-powered lens.
6. Draw a picture of what you see.

Analysis

1. Describe the shape of the *Elodea* cells. Are all of the cells in the *Elodea* the same?
2. Do you think human cells look like *Elodea* cells? How do you think they are different? How might they be similar?

START-UP ACTIVITY

MATERIALS

FOR EACH STUDENT

- coverslip, plastic
- Elodea, small leaf
- forceps
- microscope
- microscope slide, plastic
- water

Safety Caution: Remind students to review all safety cautions and icons before beginning this activity.

Answers

1. Students should be able to describe accurately the cells that they see. Students should observe that all of the cells share similar structures but the cells may not be exactly the same.
2. Accept all reasonable responses. Students may note that plant cells differ from human body cells but that plant and animal cells share many of the same structures.

Chapter Review
UCP 1; SAI 1; HNS 1; LS 1a, 1b, 1c, 1d, 3a, 3b

Science in Action
SAI 2; ST 2; SPSP 5; HNS 1; LS 3a, 3c

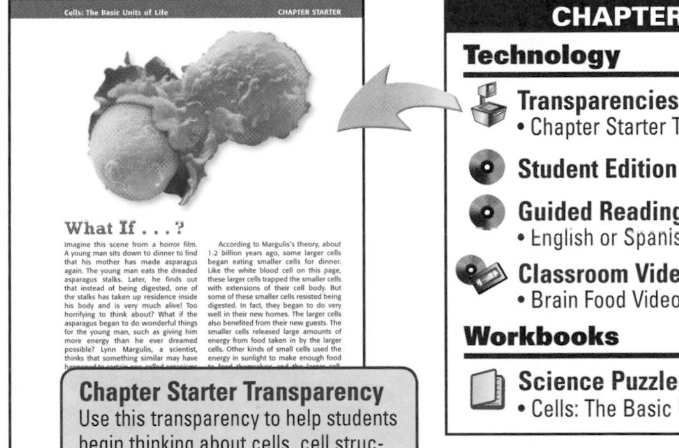

Cells: The Basic Units of Life — CHAPTER STARTER

What If . . . ?

Imagine this scene from a horror film. A young man sits down to dinner to find that his mother has made asparagus again. The young man eats the dreaded asparagus stalks. Later, he finds out that instead of being digested, one of the stalks has taken up residence inside his body and is very much alive! Too horrifying to think about? What if the asparagus began to do wonderful things for the young man, such as giving him more energy than he ever dreamed possible? Lynn Margulis, a scientist, thinks that something similar may have

According to Margulis's theory, about 1.2 billion years ago, some larger cells began eating smaller cells for dinner. Like the white blood cell on this page, these larger cells trapped the smaller cells with extensions of their cell body. But some of these smaller cells resisted being digested. In fact, they began to do very well in their new homes. The larger cells also benefited from their new guests. The smaller cells released large amounts of energy from food taken in by the larger cells. Other kinds of small cells used the energy in sunlight to make enough food

Chapter Starter Transparency
Use this transparency to help students begin thinking about cells, cell structures, and organisms.

CHAPTER RESOURCES

Technology

- **Transparencies** — READING SKILLS
  - Chapter Starter Transparency
- **Student Edition on CD-ROM**
- **Guided Reading Audio CD**
  - English or Spanish
- **Classroom Videos**
  - Brain Food Video Quiz

Workbooks

- **Science Puzzlers, Twisters & Teasers**
  - Cells: The Basic Units of Life — GENERAL

SECTION 1

Focus

Overview

This section introduces students to cells, their discovery, and their diversity. Students will learn about the parts of a cell and the reason that cells are so small. Finally, students will learn about eubacterial, archaebacterial, and eukaryotic cells.

Bellringer

Write the following questions on the board:

Why weren't cells discovered until 1665? What invention made their discovery possible? (Cells weren't discovered until 1665 because almost all cells are too small to be seen with the naked eye. The microscope is the invention that made their discovery possible.)

Motivate

ACTIVITY — GENERAL

Modeling Cell Discovery Before students begin this section, have them model Robert Hooke's discovery. Organize the class into small groups. Provide each group with a microscope and a prepared slide of cork cells. Have students describe and sketch their observations. LS **Visual**

SECTION 1

The Diversity of Cells

Most cells are so small they can't be seen by the naked eye. So how did scientists find cells? By accident, that's how! The first person to see cells wasn't even looking for them.

All living things are made of tiny structures called cells. A **cell** is the smallest unit that can perform all the processes necessary for life. Because of their size, cells weren't discovered until microscopes were invented in the mid-1600s.

READING WARM-UP

Objectives

- State the parts of the cell theory.
- Explain why cells are so small.
- Describe the parts of a cell.
- Describe how eubacteria are different from archaebacteria.
- Explain the difference between prokaryotic cells and eukaryotic cells.

Terms to Learn

cell	nucleus
cell membrane	prokaryote
organelle	eukaryote

READING STRATEGY

Reading Organizer As you read this section, create an outline of the section. Use the headings from the section in your outline.

Cells and the Cell Theory

Robert Hooke was the first person to describe cells. In 1665, he built a microscope to look at tiny objects. One day, he looked at a thin slice of cork. Cork is found in the bark of cork trees. The cork looked like it was made of little boxes. Hooke named these boxes *cells,* which means "little rooms" in Latin. Hooke's cells were really the outer layers of dead cork cells. Hooke's microscope and his drawing of the cork cells are shown in **Figure 1.**

Hooke also looked at thin slices of living plants. He saw that they too were made of cells. Some cells were even filled with "juice." The "juicy" cells were living cells.

Hooke also looked at feathers, fish scales, and the eyes of houseflies. But he spent most of his time looking at plants and fungi. The cells of plants and fungi have cell walls. This makes them easy to see. Animal cells do not have cell walls. This absence of cell walls makes it harder to see the outline of animal cells. Because Hooke couldn't see their cells, he thought that animals weren't made of cells.

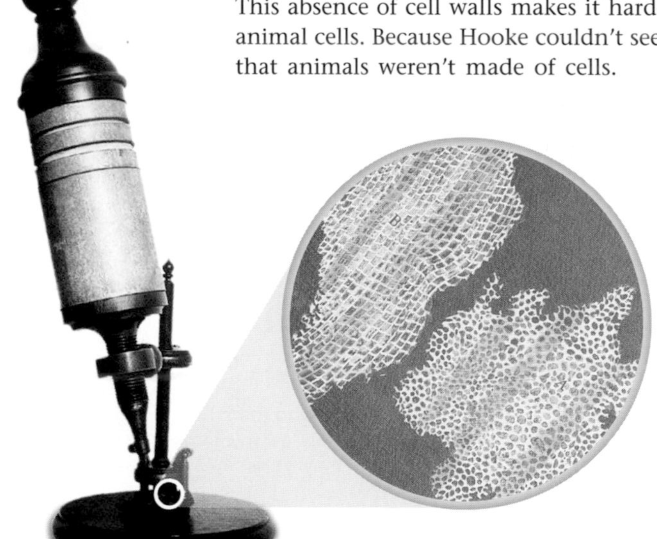

Figure 1 *Hooke discovered cells using this microscope. Hooke's drawing of cork cells is shown to the right of his microscope.*

CHAPTER RESOURCES

Chapter Resource File

- Lesson Plan
- Directed Reading A  BASIC
- Directed Reading B  SPECIAL NEEDS

Technology

 **Transparencies**
- Bellringer

Cultural Awareness GENERAL

Yeast Yeast is a fungus. Yeast used in baking is related to wild fungi living in the air around us. Strains of native yeasts vary regionally. For example, sourdough from San Francisco has its characteristic taste because bakers there use a yeast that is common in the air around that city. Not all breads require yeast. Many cultures have flat breads, such as tortillas from Mexico.

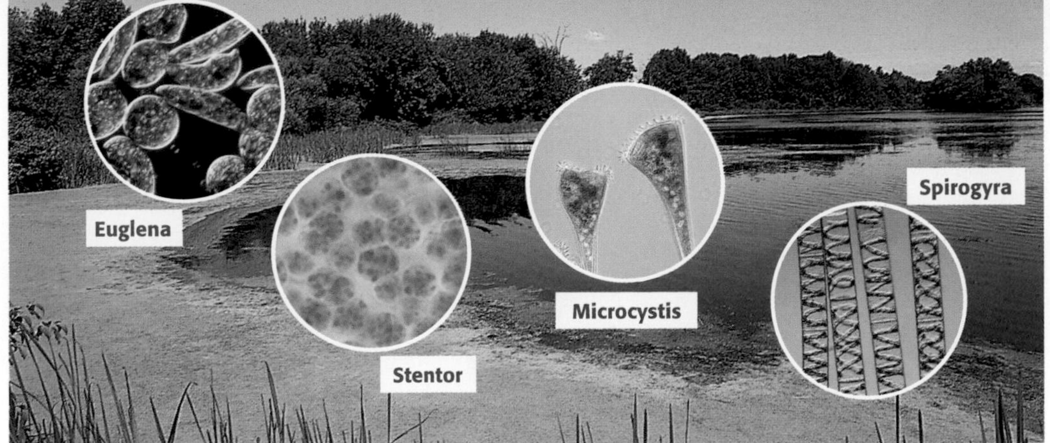

Figure 2 *The green area at the edge of the pond is a layer of pond scum. This pond scum contains organisms called* protists, *such as those shown above.*

Finding Cells in Other Organisms

In 1673, Anton van Leeuwenhoek (LAY vuhn HOOK), a Dutch merchant, made his own microscopes. Leeuwenhoek used one of his microscopes to look at pond scum. Leeuwenhoek saw small organisms in the water. He named these organisms *animalcules,* which means "little animals." Today, we call these single-celled organisms protists (PROH tists). Pond scum and some of the protists it contains are shown in **Figure 2.**

Leeuwenhoek also looked at animal blood. He saw differences in blood cells from different kinds of animals. For example, blood cells in fish, birds, and frogs are oval. Blood cells in humans and dogs are round and flat. Leeuwenhoek was also the first person to see bacteria. And he discovered that yeasts that make bread dough rise are single-celled organisms.

cell in biology, the smallest unit that can perform all life processes; cells are covered by a membrane and have DNA and cytoplasm

The Cell Theory

Almost 200 years passed before scientists concluded that cells are present in all living things. Scientist Matthias Schleiden (mah THEE uhs SHLIE duhn) studied plants. In 1838, he concluded that all plant parts were made of cells. Theodor Schwann (TAY oh dohr SHVAHN) studied animals. In 1839, Schwann concluded that all animal tissues were made of cells. Soon after that, Schwann wrote the first two parts of what is now known as the *cell theory.*

- All organisms are made of one or more cells.
- The cell is the basic unit of all living things.

Later, in 1858, Rudolf Virchow (ROO dawlf FIR koh), a doctor, stated that all cells could form only from other cells. Virchow then added the third part of the cell theory.

- All cells come from existing cells.

Reading Check **What are the three parts of the cell theory?**
(*See the Appendix for answers to Reading Checks.*)

CONNECTION TO Physics

Microscopes The microscope Hooke used to study cells was much different from microscopes today. Research different kinds of microscopes, such as light microscopes, scanning electron microscopes (SEMs), and transmission electron microscopes (TEMs). Select one type of microscope. Make a poster or other presentation to show to the class. Describe how the microscope works and how it is used. Be sure to include images.

ACTIVITY

Teach

INCLUSION *Strategies*

- ***Learning Disabled***
- ***Attention Deficit Disorder***
- ***Behavior Control Issues***

Give students a chance to associate dates with specific familiar reference points. Draw a timeline on the board. On opposite ends of the timeline, place "1492, Columbus sails to America" and "1879, Edison invents the electric light bulb." Tell students to add dates and information for five major cell discoveries that happened between 1492 and 1879. Ask students with behavior control issues to add their teams' information to the timeline. English Language Learners **LS Interpersonal**

CONNECTION to *Astronomy* GENERAL

Writing **Magnifiers** Both biologists and astronomers use magnifiers. Biologists use microscopes to see things that are too small to see with the unaided eye. Astronomers use telescopes to see planets, moons, and stars that are huge but too far away to view otherwise. Ask students to write a list of words that have the prefixes *micro-* and *tele-* and to use their examples to define those prefixes. English Language Learners **LS Verbal**

Answer to Reading Check

All organisms are made of one or more cells, the cell is the basic unit of all living things, and all cells come from existing cells.

Teach, continued

READING STRATEGY — GENERAL

Prediction Guide Before students read this page, ask them to choose one of the following reasons for why they think cells are so small:

1. There isn't enough microscopic food available for them.
2. There isn't enough room in a multicellular organism.
3. another reason (ask for suggestions)

Have students evaluate their answer after they read the page. LS **Logical**

MISCONCEPTION ALERT

Molecular Mix-Up The physical relationship between molecules and cells may be confusing to students. Molecules are not alive and are much smaller than cells. Cells and cell structures are made of molecules.

Answer to Reading Check

If a cell's volume gets too large, the cell's surface area will not be able to take in enough nutrients or get rid of wastes fast enough to keep the cell alive.

Cell Size

Most cells are too small to be seen without a microscope. It would take 50 human cells to cover the dot on this letter *i*.

A Few Large Cells

Most cells are small. A few, however, are big. A chicken egg, shown in **Figure 3,** is one big cell. The egg can be this large because it does not have to take in more nutrients.

Figure 3 *The white and yolk of this chicken egg are part of a single cell.*

Many Small Cells

There is a physical reason why most cells are so small. Cells take in food and get rid of wastes through their outer surface. As a cell gets larger, it needs more food and produces more waste. Therefore, more materials pass through its outer surface.

As the cell's volume increases, its surface area grows too. But the cell's volume grows faster than its surface area. If a cell gets too large, the cell's surface area will not be large enough to take in enough nutrients or pump out enough wastes. So, the area of a cell's surface—compared with the cell's volume—limits the cell's size. The ratio of the cell's outer surface area to the cell's volume is called the *surface area–to-volume ratio,* which can be calculated by using the following equation:

$$\textit{surface area–to-volume ratio} = \frac{\textit{surface area}}{\textit{volume}}$$

Reading Check Why are most cells small?

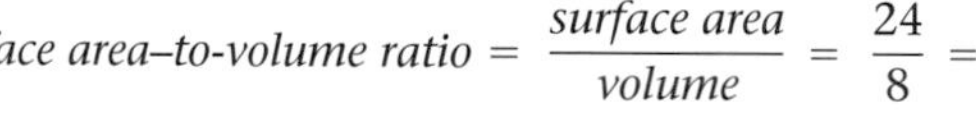

MATH FOCUS

Surface Area-to-Volume Ratio Calculate the surface area–to-volume ratio of a cube whose sides measure 2 cm.

Step 1: Calculate the surface area.

$\textit{surface area of cube} = \textit{number of sides} \times \textit{area of side}$

$\textit{surface area of cube} = 6 \times (2\text{ cm} \times 2\text{ cm})$

$\textit{surface area of cube} = 24\text{ cm}^2$

Step 2: Calculate the volume.

$\textit{volume of cube} = \textit{side} \times \textit{side} \times \textit{side}$

$\textit{volume of cube} = 2\text{ cm} \times 2\text{ cm} \times 2\text{ cm}$

$\textit{volume of cube} = 8\text{ cm}^3$

Step 3: Calculate the surface area–to-volume ratio.

$$\textit{surface area–to-volume ratio} = \frac{\textit{surface area}}{\textit{volume}} = \frac{24}{8} = \frac{3}{1}$$

Now It's Your Turn

1. Calculate the surface area–to-volume ratio of a cube whose sides are 3 cm long.
2. Calculate the surface area–to-volume ratio of a cube whose sides are 4 cm long.
3. Of the cubes from questions 1 and 2, which has the greater surface area–to-volume ratio?
4. What is the relationship between the length of a side and the surface area–to-volume ratio of a cell?

CHAPTER RESOURCES

Technology

Transparencies
- Math Focus: Surface Area–to-Volume Ratio

Answers to Math Focus

1. Surface area of cube (SA) = $(3\text{ cm} \times 3\text{ cm}) \times 6 = 54\text{ cm}^2$
   Volume of cube (V) = $3\text{ cm} \times 3\text{ cm} \times 3\text{ cm} = 27\text{ cm}^3$
   SA:V ratio = 54:27 or 2:1
2. SA = $(4\text{ cm} \times 4\text{ cm}) \times 6 = 96\text{ cm}^2$
   V = $4\text{ cm} \times 4\text{ cm} \times 4\text{ cm} = 64\text{ cm}^3$
   SA:V = 96:64 or 1.5:1
3. the cube whose sides are 3 cm long
4. The larger the cell is, the smaller the surface-to-volume ratio is.

Parts of a Cell

Cells come in many shapes and sizes. Cells have many different functions. But all cells have the following parts in common.

The Cell Membrane and Cytoplasm

All cells are surrounded by a cell membrane. The **cell membrane** is a protective layer that covers the cell's surface and acts as a barrier. It separates the cell's contents from its environment. The cell membrane also controls materials going into and out of the cell. Inside the cell is a fluid. This fluid and almost all of its contents are called the *cytoplasm* (SIET oh PLAZ uhm).

Organelles

Cells have organelles that carry out various life processes. **Organelles** are structures that perform specific functions within the cell. Different types of cells have different organelles. Most organelles are surrounded by membranes. For example, the algal cell in **Figure 4** has membrane-bound organelles. Some organelles float in the cytoplasm. Other organelles are attached to membranes or other organelles.

Reading Check What are organelles?

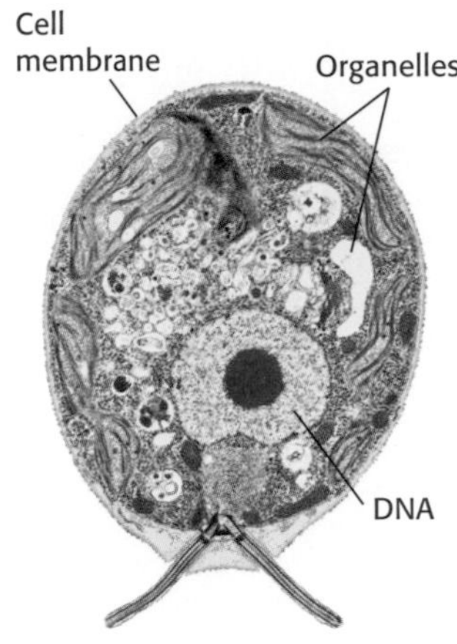

Figure 4 *This green alga has organelles. The organelles and the fluid surrounding them make up the cytoplasm.*

Genetic Material

All cells contain DNA (**d**eoxyribo**n**ucleic **a**cid) at some point in their life. *DNA* is the genetic material that carries information needed to make new cells and new organisms. DNA is passed on from parent cells to new cells and controls the activities of a cell. **Figure 5** shows the DNA of a bacterium.

In some cells, the DNA is enclosed inside an organelle called the **nucleus.** For example, your cells have a nucleus. In contrast, bacterial cells do not have a nucleus.

In humans, mature red blood cells lose their DNA. Red blood cells are made inside bones. When red blood cells are first made, they have a nucleus with DNA. But before they enter the bloodstream, red blood cells lose their nucleus and DNA. They survive with no new instructions from their DNA.

cell membrane a phospholipid layer that covers a cell's surface; acts as a barrier between the inside of a cell and the cell's environment

organelle one of the small bodies in a cell's cytoplasm that are specialized to perform a specific function

nucleus in a eukaryotic cell, a membrane-bound organelle that contains the cell's DNA and that has a role in processes such as growth, metabolism, and reproduction

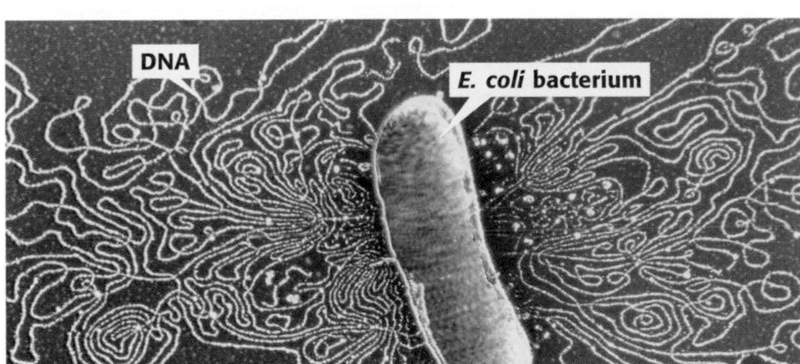

Figure 5 *This photo shows an* Escherichia coli *bacterium. The bacterium's cell membrane has been treated so that the cell's DNA is released.*

READING STRATEGY — GENERAL

Prediction Guide Before students read this page, ask them if the following statement is true or false: At some point, all cells contain DNA. (true; Even though some cells, such as red blood cells, lose their DNA when they mature, all cells have DNA at some point.)

Ask students to explain the reasons for their answer. Have students evaluate their answer after they read the page. **LS Verbal**

MISCONCEPTION ALERT

DNA and Complexity
Students may believe that larger or more-complex organisms have more DNA. This is not the case. In general, eukaryotes have more DNA than bacteria or viruses do. Among eukaryotes however, there is no strong correlation between body size or measures of complexity and DNA content. Although the fruit fly *Drosophila melanogaster* has about one-fourth as much DNA as a human, the protist *Amoeba dubia* has about 200 times more DNA than a human being does! Part of the reason for this apparent discrepancy is that some DNA does not code for any genes. Species with very large genomes have a lot of non-coding DNA.

CONNECTION to Language Arts — GENERAL

Writing **Smallest Living Thing** Is the smallest living thing an organism called a *Mycoplasma genitalium*? Or is it something called a "nanobacteria"? Have students conduct Internet or library research and write a report on the smallest living thing. (Scientists are not certain. "Nanobacteria" do not always show all the characteristics of living things. More research may settle the issue.) **LS Verbal/Logical**

Answer to Reading Check

Organelles are structures within a cell that perform specific functions for the cell.

Two Kinds of Cells

All cells have cell membranes, organelles, cytoplasm, and DNA in common. But there are two basic types of cells—cells without a nucleus and cells with a nucleus. Cells with no nucleus are *prokaryotic* (proh KAR ee AHT ik) *cells.* Cells that have a nucleus are *eukaryotic* (yoo KAR ee AHT ik) *cells.* Prokaryotic cells are further classified into two groups: *eubacteria* (yoo bak TIR ee uh) and *archaebacteria* (AHR kee bak TIR ee uh).

Prokaryotes: Eubacteria and Archaebacteria

Eubacteria and archaebacteria are prokaryotes (pro KAR ee OHTS). **Prokaryotes** are single-celled organisms that do not have a nucleus or membrane-bound organelles.

Eubacteria

The most common prokaryotes are eubacteria (or just *bacteria*). Bacteria are the world's smallest cells. These tiny organisms live almost everywhere. Bacteria do not have a nucleus, but they do have DNA. A bacteria's DNA is a long, circular molecule, shaped sort of like a rubber band. Bacteria have no membrane-covered organelles. But they do have ribosomes. *Ribosomes* are tiny, round organelles made of protein and other material.

Bacteria also have a strong, weblike exterior cell wall. This wall helps the cell retain its shape. A bacterium's cell membrane is just inside the cell wall. Together, the cell wall and cell membrane allow materials into and out of the cell.

Some bacteria live in the soil and water. Others live in, or on, other organisms. For example, you have bacteria living on your skin and teeth. You also have bacteria living in your digestive system. These bacteria help the process of digestion. A typical bacterial cell is shown in **Figure 6.**

prokaryote an organism that consists of a single cell that does not have a nucleus

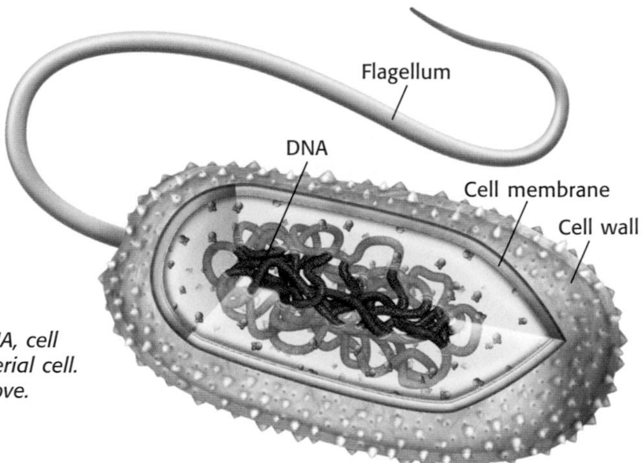

Figure 6 *This diagram shows the DNA, cell membrane, and cell wall of a eubacterial cell. The flagellum helps the bacterium move.*

Bacteria in Your Lunch?

Most of the time, you don't want bacteria in your food. Many bacteria make toxins that will make you sick. However, some foods—such as yogurt—are supposed to have bacteria in them! The bacteria in these foods are not dangerous.

In yogurt, masses of rod-shaped bacteria feed on the sugar (lactose) in milk. The bacteria convert the sugar into lactic acid. Lactic acid causes milk to thicken. This thickened milk makes yogurt.

1. Using a **cotton swab,** put a **small dot of yogurt** on a **microscope slide.**
2. Add a **drop of water.** Use the cotton swab to stir.
3. Add a **coverslip.**
4. Use a **microscope** to examine the slide. Draw what you observe.

Teach, *continued*

Quick Lab

MATERIALS

For Each Student

- cotton swab
- coverslip, plastic
- microscope
- microscope slide, plastic
- water
- yogurt with active culture

Answer

4. Drawings should depict rod-shaped bacteria.

CONNECTION to Earth Science — GENERAL

Writing **Subsurface Cells** Astronomers are interested in the work of scientists who investigate bacteria and other microscopic organisms in Earth's crust. Microbiologists have drilled deep into the crust and found microbes nearly 3 km below the surface, where the temperature is 75°C (167°F). Because other planets have surface conditions similar to the harsh environment within the Earth's crust, astronomers believe that microbes may live elsewhere in the solar system. Have students research and write a brief report on the conditions in Earth's crust, and have students learn about the organisms that live there. **LS Verbal**

Group Activity — ADVANCED

Archaebacteria Have students work in pairs to find out if archaebacteria are more similar to eubacteria or eukaryotes. What kinds of evidence do scientists use to answer this question? Have students make a poster or other visual presentation of their results. **LS Visual/Interpersonal**

WEIRD SCIENCE

In 1969, the *Apollo 12* crew retrieved a space probe from the moon that had been launched nearly 3 years earlier. NASA scientists found a stowaway in the probe's camera. The bacterium *Streptococcus mitis* had traveled to the moon and back. Despite the rigors of space travel, more than 2.5 years of radiation exposure, and freezing temperatures, the *Streptococcus mitis* was successfully reconstituted.

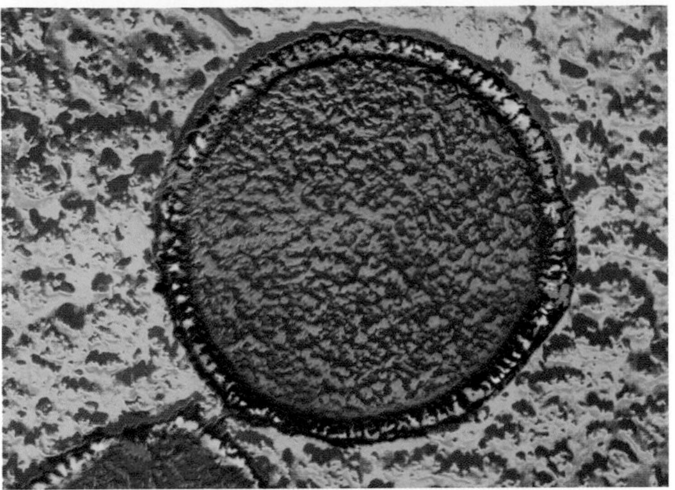

Figure 7 *This photograph, taken with an electron microscope, is of an archaebacterium that lives in the very high temperatures of deep-sea volcanic vents. The photograph has been colored so that the cell wall is green and the cell contents are pink.*

Archaebacteria

The second kind of prokaryote are the archaebacteria. These organisms are also called *archaea* (ahr KEE uh). Archaebacteria are not as common as bacteria, but they are similar to bacteria in some ways. For example, both are single-celled organisms. Both have ribosomes, a cell membrane, and circular DNA. And both lack a nucleus and membrane-bound organelles. But archaebacteria are different from bacteria. For example, archaebacterial ribosomes are different from eubacterial ribosomes.

Archaebacteria are similar to eukaryotic cells in some ways, too. For example, archaebacterial ribosomes are more like the ribosomes of eukaryotic cells. But archaebacteria also have some features that no other cells have. For example, the cell wall and cell membranes of archaebacteria are different from the cell walls of other organisms. And some archaebacteria live in places where no other organisms could live.

Three types of archaebacteria are *heat-loving, salt-loving,* and *methane-making.* Methane is a kind of gas frequently found in swamps. Heat-loving and salt-lovng archaebacteria are sometimes called extremophiles. *Extremophiles* live in places where conditions are extreme. They live in very hot water, such as in hot springs, or where the water is extremely salty. **Figure 7** shows one kind of methane-making archaebacteria that lives deep in the ocean near volcanic vents. The temperature of the water from those vents is extreme: it is above the boiling point of water at sea level.

Reading Check **What is one difference between eubacteria and archaebacteria?**

CONNECTION TO Social Studies

Where Do They Live? While most archaebacteria live in extreme environments, scientists have found that archaebacteria live almost everywhere. Do research about archaebacteria. Select one kind of archaebacteria. Create a poster showing the geographical location where the organism lives, describing its physical environment, and explaining how it survives in its environment.

ACTIVITY

CONNECTION to Earth Science — GENERAL

Writing **Discovering Ancient Earth** The work of paleontologists helps us understand the antiquity of unicellular and multicellular life on Earth. Some paleontologists specialize in ancient plant life, and some specialize in ancient climates. Have students conduct Internet or library research and write a report on different types of paleontologists. **LS Verbal**

MISCONCEPTION ALERT

Extreme Bacteria? Students may believe that all archaebacteria live in extreme environments because biology books often highlight the unusual and extreme environments in which some archaebacteria live. In these environments, other types of organisms cannot survive. In fact, many archaebacteria live in "normal" environments along with other eubacterial and eukaryotic species.

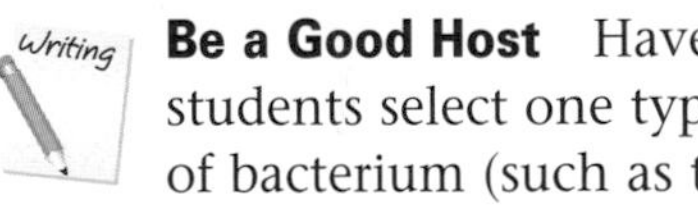

Research — GENERAL

Writing **Be a Good Host** Have students select one type of bacterium (such as the *Streptococcus mutans,* which causes tooth decay) that lives on or in the body, conduct Internet or library research on it, and then write and illustrate a report on the bacterium they have selected. **LS Verbal/Visual**

Is That a Fact!

The archaebacteria *Pyrodictium* prefers to live at 105°C. That temperature is 5°C hotter than the boiling temperature of water (at sea level)!

Answer to Reading Check

One difference between eubacteria and archaea is that bacterial ribosomes are different from archaebacterial ribosomes.

Close

Reteaching — BASIC

Drawing Cells Ask students to create a short picture book. Have them draw a picture of a typical prokaryotic cell on one page. Have them draw a picture of a typical eukaryotic cell on the next page. Students should label all the parts of both cells.
LS Visual

Quiz — GENERAL

1. When Robert Hooke saw "juice" in some cells, what was he looking at? (cytoplasm)
2. Why did Hooke think that cells existed only in plants and fungi and not in animals? (Plant and fungal cells have cell walls. Hooke's microscope wasn't strong enough to view the more delicate cell membranes of animal cells.)

Alternative Assessment — GENERAL

Writing **Vocabulary Game** Organize the students into groups, and assign two or three vocabulary words to each group. Ask students to write a descriptive statement about each word without using the vocabulary word in the sentence. Each group should challenge the other groups to guess the word described. For example, if "genetic material" is the definition, "What is DNA?" is the correct response. **LS Verbal**

eukaryote an organism made up of cells that have a nucleus enclosed by a membrane; eukaryotes include animals, plants, and fungi, but not archaebacteria or eubacteria

Eukaryotic Cells and Eukaryotes

Eukaryotic cells are the largest cells. Most eukaryotic cells are still microscopic, but they are about 10 times larger than most bacterial cells. A typical eukaryotic cell is shown in **Figure 8.**

Unlike bacteria and archaebacteria, eukaryotic cells have a nucleus. The nucleus is one kind of membrane-bound organelle. A cell's nucleus holds the cell's DNA. Eukaryotic cells have other membrane-bound organelles as well. Organelles are like the different organs in your body. Each kind of organelle has a specific job in the cell. Together, organelles, such as the ones shown in **Figure 8,** perform all the processes necessary for life.

All living things that are not bacteria or archaebacteria are made of one or more eukaryotic cells. Organisms made of eukaryotic cells are called **eukaryotes.** Many eukaryotes are multicellular. *Multicellular* means "many cells." Multicellular organisms are usually larger than single-cell organisms. So, most organisms you see with your naked eye are eukaryotes. There are many types of eukaryotes. Animals, including humans, are eukaryotes. So are plants. Some protists, such as amoebas, are single-celled eukaryotes. Other protists, including some types of green algae, are multicellular eukaryotes. Fungi are organisms such as mushrooms or yeasts. Mushrooms are multicellular eukaryotes. Yeasts are single-celled eukaryotes.

Reading Check How are eukaryotes different from prokaryotes?

Figure 8 Organelles in a Typical Eukaryotic Cell

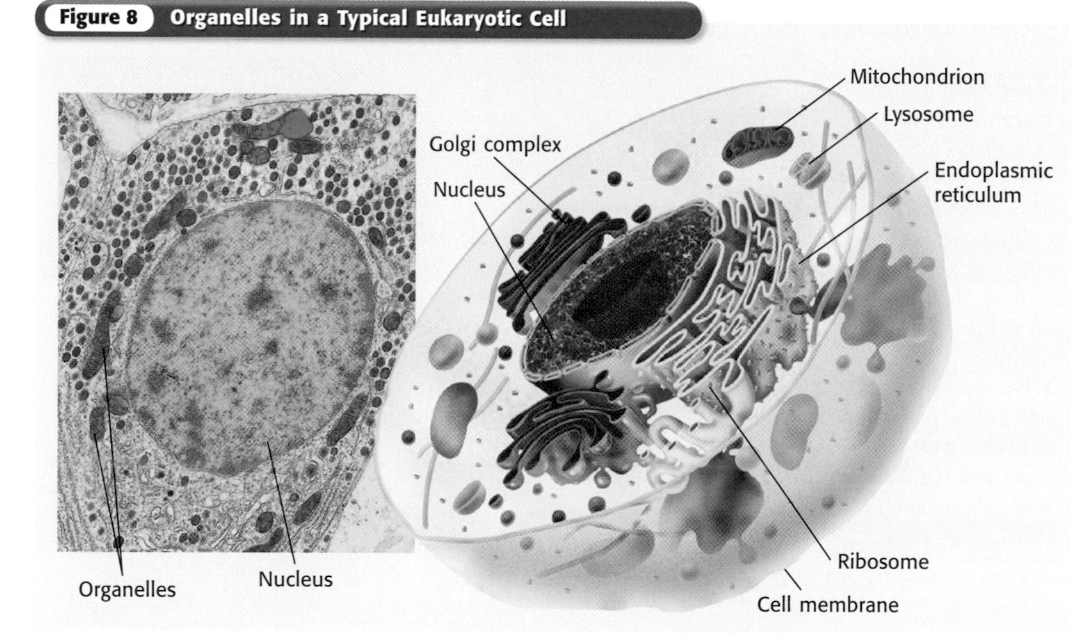

Answer to Reading Check

The main difference between prokaryotes and eukaryotes is that eukaryotic cells have a nucleus and membrane-bound organelles and prokaryotic cells do not.

Brochure — GENERAL

For an internet activity related to this chapter, have students go to **go.hrw.com** and type in the keyword **HL5CELW.**

SECTION Review

Summary

- Cells were not discovered until microscopes were invented in the 1600s.
- Cell theory states that all organisms are made of cells, the cell is the basic unit of all living things, and all cells come from other cells.
- All cells have a cell membrane, cytoplasm, and DNA.
- Most cells are too small to be seen with the naked eye. A cell's surface area–to-volume ratio limits the size of a cell.
- The two basic kinds of cells are prokaryotic cells and eukaryotic cells. Eukaryotic cells have a nucleus and membrane-bound organelles. Prokaryotic cells do not.
- Prokaryotes are classified as archaebacteria and eubacteria.
- Archaebacterial cell walls and ribosomes are different from the cell walls and ribosomes of other organisms.
- Eukaryotes can be single-celled or multicellular.

Using Key Terms

1. In your own words, write a definition for the term *organelle*.
2. Use the following terms in the same sentence: *prokaryotic, nucleus,* and *eukaryotic.*

Understanding Key Ideas

3. Cell size is limited by the
   a. thickness of the cell wall.
   b. size of the cell's nucleus.
   c. cell's surface area–to-volume ratio.
   d. amount of cytoplasm in the cell.
4. What are the three parts of the cell theory?
5. Name three structures that every cell has.
6. Give two ways in which archaebacteria are different from bacteria.

Critical Thinking

7. **Applying Concepts** You have discovered a new single-celled organism. It has a cell wall, ribosomes, and long, circular DNA. Is it a eukaryote or a prokaryote cell? Explain.
8. **Identifying Relationships** One of your students brings you a cell about the size of the period at the end of this sentence. It is a single cell, but it also forms chains. What characteristics would this cell have if the organism is a eukaryote? If it is a prokaryote? What would you look for first?

Interpreting Graphics

The picture below shows a particular organism. Use the picture to answer the questions that follow.

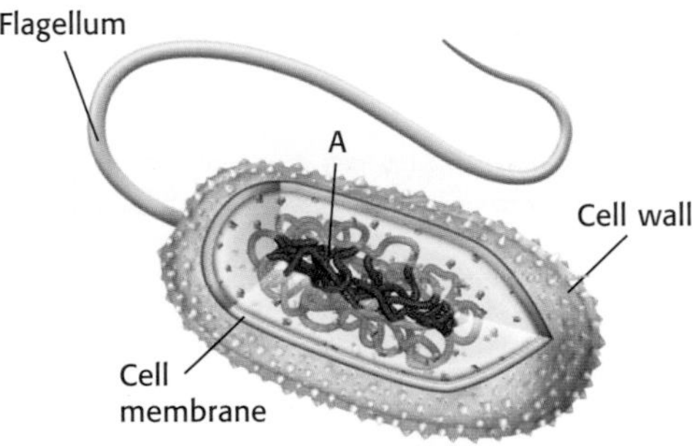

9. What type of organism does the picture represent? How do you know?
10. Which structure helps the organism move?
11. What part of the organism does the letter *A* represent?

Answers to Section Review

1. Sample answer: An organelle is a structure inside a cell that performs a specific function for the cell.
2. Sample answer: Eukaryotic cells have a nucleus, but prokaryotic cells do not.
3. c
4. All organisms are made of one or more cells, the cell is the basic unit of all living things, and all cells come from existing cells.
5. Every cell has a cell membrane, DNA, and cytoplasm.
6. Sample answer: The cell walls and the ribosomes of archaebacteria are different from those structures in eubacteria.
7. Sample answer: The cell is probably a prokaryote; The key is that it does not appear to have a nucleus and its DNA is long and circular.
8. Sample answer: eukaryote: whether the chains were a multicellular organism or a collection of individual cells, membrane-bound organelles, certain types of ribosomes and certain materials in the cell membranes, where the DNA is located, and a nucleus; prokaryote: other types of ribosomes and cell membrane materials, the structure of the DNA, and a nucleus; The first thing I would look for is a nucleus.
9. a typical eubacterial cell; It has no nucleus, and its DNA is long and circular.
10. the flagellum
11. the cell's DNA

CHAPTER RESOURCES

Chapter Resource File

- Section Quiz GENERAL
- Section Review GENERAL
- Vocabulary and Section Summary GENERAL
- Reinforcement Worksheet BASIC

Technology

Transparencies
- A Typical Eukaryotic Cell

SECTION 2

Focus

Overview

In this section, students will learn the names and functions of the cell structures, called *organelles,* in a eukaryotic cell.

Bellringer

On the board, write the following:

List three differences between prokaryotic and eukaryotic cells. (Prokaryotic cells have circular DNA, no nucleus, and no membrane-covered organelles. Eukaryotic cells have linear DNA, a nucleus, and membrane-covered organelles.)

Motivate

Discussion — GENERAL

Cellular Activity Ask students if they can feel the flurry of activity within their cells that keeps them alive. (no; But even though students can't feel activity in the cells, they can feel the heat produced by cellular activity. Students are not likely to know this.)

Ask students how they know their cells are working. (The students are alive: they can breathe, digest food, and move.) **LS Logical/Intrapersonal**

Answer to Reading Check

Plant, algae, and fungi cells have cell walls.

SECTION 2

Eukaryotic Cells

Most eukaryotic cells are small. For a long time after cells were discovered, scientists could not see what was going on inside cells. They did not know how complex cells are.

Now, scientists know a lot about eukaryotic cells. These cells have many parts that work together and keep the cell alive.

READING WARM-UP

Objectives

- Identify the different parts of a eukaryotic cell.
- Explain the function of each part of a eukaryotic cell.

Terms to Learn

cell wall	mitochondrion
ribosome	Golgi complex
endoplasmic reticulum	vesicle
	lysosome

READING STRATEGY

Reading Organizer As you read this section, make a table comparing plant cells and animal cells.

Cell Wall

Some eukaryotic cells have cell walls. A **cell wall** is a rigid structure that gives support to a cell. The cell wall is the outermost structure of a cell. Plants and algae have cell walls made of cellulose (SEL yoo LOHS). *Cellulose* is a complex sugar that most animals can't digest.

cell wall a rigid structure that surrounds the cell membrane and provides support to the cell

The cell walls of plant cells allow plants to stand upright. In some plants, the cells must take in water for the cell walls to keep their shape. When such plants lack water, the cell walls collapse and the plant droops. **Figure 1** shows a cross section of a plant cell and a close-up of the cell wall.

Fungi, including yeasts and mushrooms, also have cell walls. Some fungi have cell walls made of *chitin* (KIE tin). Other fungi have cell walls made from a chemical similar to chitin. Eubacteria and archaebacteria also have cell walls, but those walls are different from plant or fungal cell walls.

Reading Check What types of cells have cell walls? *(See the Appendix for answers to Reading Checks.)*

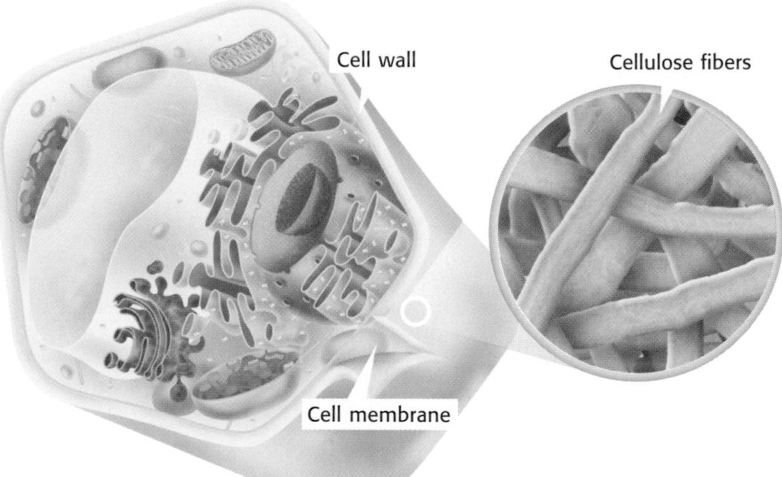

Figure 1 *The cell walls of plant cells help plants retain their shape. Plant cell walls are made of cellulose.*

CHAPTER RESOURCES

Chapter Resource File

- Lesson Plan
- Directed Reading A BASIC
- Directed Reading B SPECIAL NEEDS

Technology

Transparencies
- Bellringer

MISCONCEPTION ALERT

Cells Are Three-Dimensional Students often think of cells as flat. Looking at pictures and even viewing cells in a microscope can reinforce that misconception. Make sure that students understand that even though most cells are very small, they do have three dimensions, and they do take up space.

Cell Membrane

All cells have a cell membrane. The *cell membrane* is a protective barrier that encloses a cell. It separates the cell's contents from the cell's environment. The cell membrane is the outermost structure in cells that lack a cell wall. In cells that have a cell wall, the cell membrane lies just inside the cell wall.

The cell membrane contains proteins, lipids, and phospholipids. *Lipids,* which include fats and cholesterol, are a group of compounds that do not dissolve in water. The cell wall has two layers of phospholipids (FAHS fo LIP idz), shown in **Figure 2.** A *phospholipid* is a lipid that contains phosphorus. Lipids are "water fearing," or *hydrophobic.* Lipid ends of phospholipids form the inner part of the membrane. Phosphorus-containing ends of the phospholipids are "water loving," or *hydrophilic.* These ends form the outer part of the membrane.

Some of the proteins and lipids control the movement of materials into and out of the cell. Some of the proteins form passageways. Nutrients and water move into the cell, and wastes move out of the cell, through these protein passageways.

Reading Check What are two functions of a cell membrane?

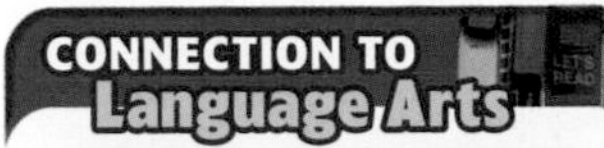

CONNECTION TO Language Arts

WRITING SKILL **The Great Barrier** In your **science journal,** write a science fiction story about tiny travelers inside a person's body. These little explorers need to find a way into or out of a cell to solve a problem. You may need to do research to find out more about how the cell membrane works. Illustrate your story.

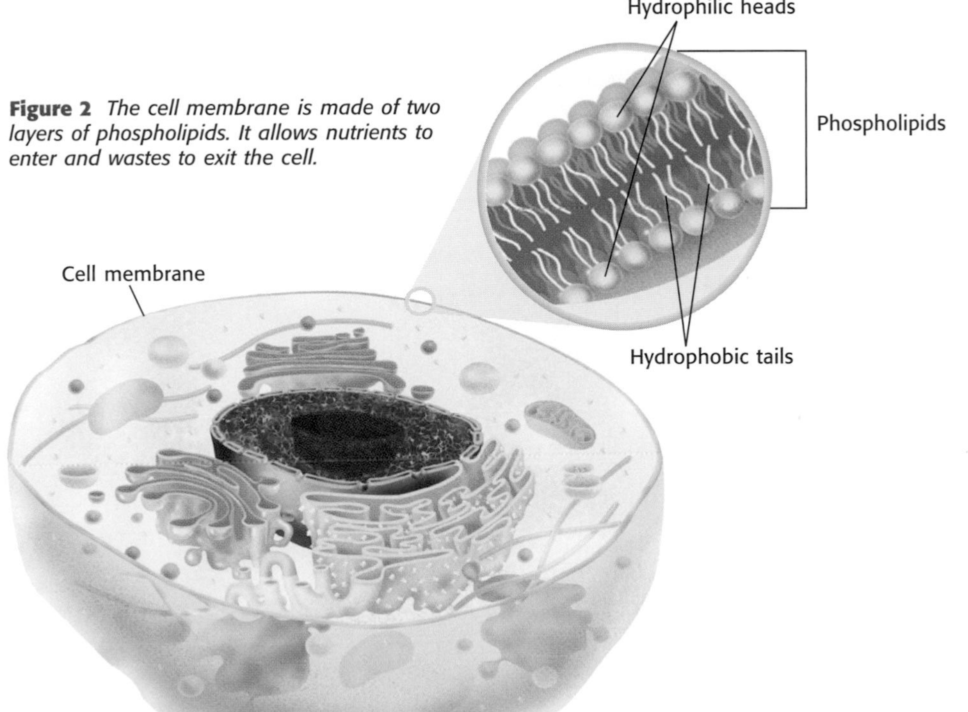

Figure 2 *The cell membrane is made of two layers of phospholipids. It allows nutrients to enter and wastes to exit the cell.*

Teach

Demonstration — BASIC

Cell Walls and Cell Membranes Using a stick and your own hand, you can illustrate the difference between a rigid cell wall, found in plant cells, and a flexible cell membrane, found in human skin cells. Bend the stick, and it will break. Make a fist, and your skin stretches to accommodate the flexing of muscles and bone joints. If we had rigid cell walls, we would find moving extremely difficult. English Language Learners
LS Visual

Activity — BASIC

MATERIALS

For Each Group

- food strainer, wire mesh
- gravel (such as gravel used to line aquaria), 250 mL
- marbles (or pebbles), 250 mL
- pan (to place under strainer)
- sand, 250 mL
- water, 250 mL

Cellular Sieve Have students place each material into the strainer, and have them observe and explain the results. Lead students to understand that a cell membrane functions somewhat like the strainer. The cell membrane lets some materials pass through but not others. Also, explain that the process works in both directions.

Teacher's Note: The effects demonstrated by this Activity also apply to the membranes of organelles within the cell.
LS Kinesthetic/Visual

Homework — GENERAL

PORTFOLIO **Poster Project** Have students investigate red blood cells and create a poster comparing red blood cells with other human cells. (RBCs are the only cells in the human body that do not have a nucleus or mitochondria when they are mature. Without a nucleus. RBCs cannot divide and reproduce. They live for only about 120 days, but new RBCs are made by bone marrow at the rate of up to 200 billion per day.) **LS Visual**

Answer to Reading Check

A cell membrane encloses the cell and separates and protects the cell's contents from the cell's environment. The cell membrane also controls movement of materials into and out of the cell.

Teach, *continued*

Group Activity—Basic

Drawing Cells Arrange students in pairs. Tell each pair to draw a plant or animal cell based on information presented in the text. Instruct students not to label the cell's parts. Then, have students exchange drawings with another pair. Students should put the proper labels on their classmates' picture. Finally, have each group of two pairs compare and discuss each other's work. LS **Visual/Interpersonal**

Answer to Reading Check

The cytoskeleton is a web of proteins in the cytoplasm. It gives the cell support and structure.

CONNECTION to *Physical Science*—General

Studying Cells Biophysics uses tools and techniques of physics to study the life processes of cells. Biophysicists are interested in the relationship between a molecule's structure and its function. Sophisticated techniques, such as electron microscopy, X-ray diffraction, magnetic resonance spectroscopy, and electrophoresis, allow biophysicists to study the structure of proteins, nucleic acids, and even parts of cells, such as ribosomes. Use the teaching transparency "Structural Formulas" to illustrate molecular structure.
LS **Visual**

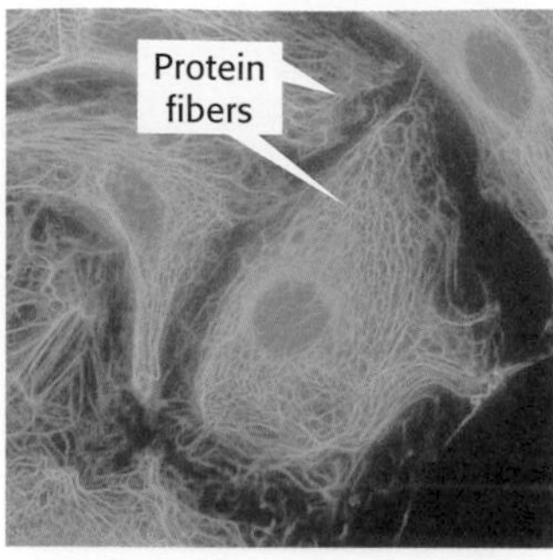

Figure 3 *The cytoskeleton, made of protein fibers, helps a cell retain its shape and move its organelles.*

Cytoskeleton

The *cytoskeleton* (SIET oh SKEL uh tuhn) is a web of proteins in the cytoplasm. The cytoskeleton, shown in **Figure 3,** acts as both a muscle and a skeleton. It keeps the cell's membranes from collapsing. The cytoskeleton also helps some cells move.

The cytoskeleton is made of three types of protein. One protein is a hollow tube. The other two are long, stringy fibers. One of the stringy proteins is also found in muscle cells.

Reading Check **What is the cytoskeleton?**

Nucleus

All eukaryotic cells have the same basic membrane-bound organelles, starting with the nucleus. The *nucleus* is a large organelle in a eukaryotic cell. It contains the cell's DNA, or genetic material. DNA contains the information on how to make a cell's proteins. Proteins control the chemical reactions in a cell. They also provide structural support for cells and tissues. But proteins are not made in the nucleus. Messages for how to make proteins are copied from the DNA. These messages are then sent out of the nucleus through the membranes.

The nucleus is covered by two membranes. Materials cross this double membrane by passing through pores. **Figure 4** shows a nucleus and nuclear pores. The nucleus of many cells has a dark area called the nucleolus (noo KLEE uh luhs). The *nucleolus* stores materials that will be used to make ribosomes.

Figure 4 *The nucleus contains the cell's DNA. Pores, shown in the diagram at right, allow materials to pass from the nucleus to the cytoplasm.*

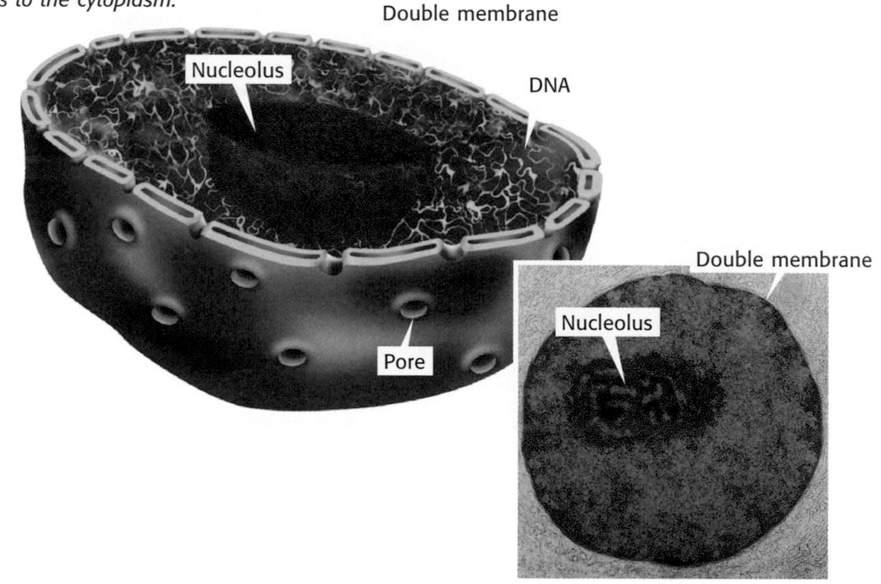

CHAPTER RESOURCES

Technology

Transparencies
- ***LINK TO PHYSICAL SCIENCE*** Structural Formulas

INCLUSION *Strategies*

- ***Developmentally Delayed***
- ***Attention Deficit Disorder***
- ***Behavior Control Issues***

Have six volunteers line up at the front of the room. Then, write "ABC order" on the board, and ask students to line up correctly. Next, add "by last letter of first name," and ask them to line up correctly. Discuss that having DNA in cells is like having complete, specific directions. LS **Kinesthetic** English Language Learners

Ribosomes

Organelles that make proteins are called **ribosomes.** Ribosomes are the smallest of all organelles. And there are more ribosomes in a cell than there are any other organelles. Some ribosomes float freely in the cytoplasm. Others are attached to membranes or the cytoskeleton. Unlike most organelles, ribosomes are not covered by a membrane.

Proteins are made within the ribosomes. Proteins are made of amino acids. An *amino acid* is any one of about 20 different organic molecules that are used to make proteins. All cells need proteins to live. All cells have ribosomes.

ribosome cell organelle composed of RNA and protein; the site of protein synthesis

endoplasmic reticulum a system of membranes that is found in a cell's cytoplasm and that assists in the production, processing, and transport of proteins and in the production of lipids

Endoplasmic Reticulum

Many chemical reactions take place in a cell. Many of these reactions happen on or in the endoplasmic reticulum (EN doh PLAZ mik ri TIK yuh luhm). The **endoplasmic reticulum,** or ER, is a system of folded membranes in which proteins, lipids, and other materials are made. The ER is shown in **Figure 5.**

The ER is the internal delivery system of the cell. Its folded membrane contains many tubes and passageways. Substances move through the ER to different places within the cell.

Endoplasmic reticulum is either rough ER or smooth ER. The part of the ER covered in ribosomes is rough ER. Rough ER is usually found near the nucleus. Ribosomes on rough ER make many of the cell's proteins. The ER delivers these proteins throughout the cell. ER that lacks ribosomes is smooth ER. The functions of smooth ER include making lipids and breaking down toxic materials that could damage the cell.

Figure 5 *The endoplasmic reticulum (ER) is a system of membranes. Rough ER is covered with ribosomes. Smooth ER does not have ribosomes.*

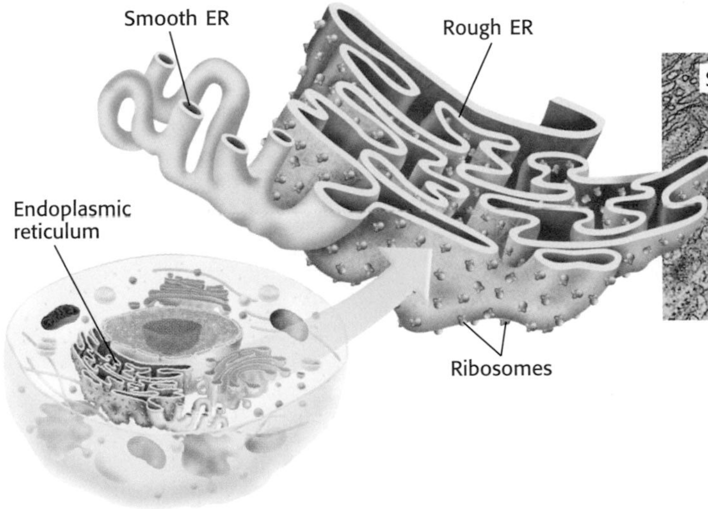

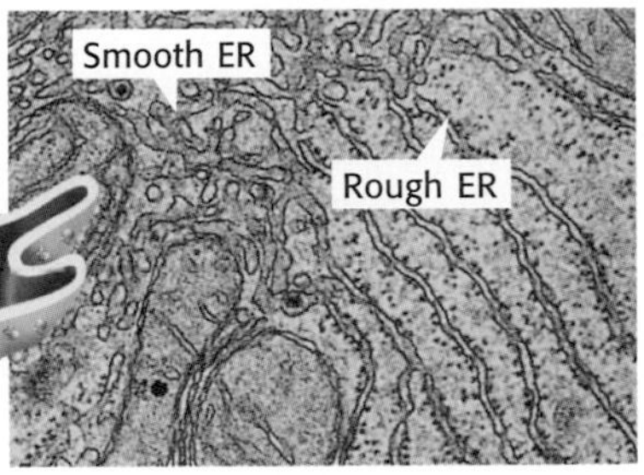

Homework — General

Cell Search Have students search the Internet for images (photomicrographs) of cells. Encourage students to compare images of cells from different types of organisms. Also, have students compare images of the same type of cell made by different microscopes, such as light microscopes, scanning electron microscopes, and transmission electron microscopes. Have students describe the cells that they find. **LS Logical/Visual**

Activity — General

Cell Models Students will be making edible models of cells. Students can bring edible items for the cell wall or cell membrane, such as crackers or pita bread. They can also bring items to represent organelles within the cell, such as small pieces of different kinds of candy (the hard candy shell on some candies may represent an organelle's membrane), olives, or other items. Have students explain the way they have represented the cell's structure in food. Pictures of the edible cells can be displayed in the classroom. **LS Kinesthetic** English Language Learners

Connection to Chemistry — Advanced

Ribosome Structure Ribosomes make proteins. The structure of ribosomes has been intensely studied. Scientists now know how ribosome structure relates to ribosome function. Have students research and report on how ribosomes work. Their report should include a diagram of ribosome structure and its relationship to ribosome function. **LS Verbal/Visual**

Connection to Language Arts — Advanced

No Energy? Mitochondrial diseases are a group of illnesses caused by malfunctioning mitochondria. These diseases can be caused by genes in the mitochondria or genes in the cell. Any activity or organ that requires energy is affected by these diseases. Have students conduct Internet or library research on mitochondrial diseases. Have them create a brochure or a pamphlet explaining one or more of these diseases. (Interested students may want to read *A Wind in the Door,* by Madeleine L'Engle, which is a story about a little boy with mitochondrial disease.) **LS Verbal**

READING STRATEGY — GENERAL

Prediction Guide Before students read this page, ask them if the following statement is true or false: Animal cells are completely different from plant cells. (false; Animal cells and plant cells have many features in common, such as membrane-covered organelles and a cell membrane. The main difference between animal and plant cells is that animal cells do not have a cell wall and they do not have chloroplasts and chlorophyll.)

Have students explain their answer. Then, have them evaluate their answer after they read the page. LS **Logical**

Answer to Reading Check

Most of a cell's ATP is made in the cell's mitochondria.

CONNECTION to *Language Arts* — ADVANCED

Writing **Far-Out Fiction** Have students write a story about an animal whose cells are invaded by chloroplasts. Students should describe how the animal's life processes at the cellular level would be affected. Students may also describe how the animal might use this chloroplast invasion to its advantage. Encourage students to write about an animal other than a mammal (corals might be an interesting subject). LS **Verbal/Logical**

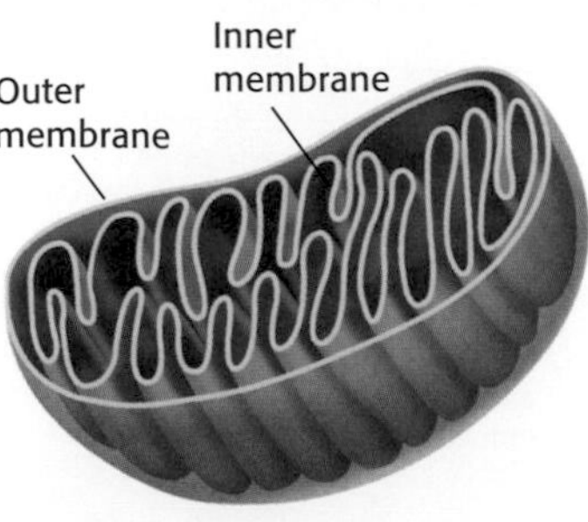

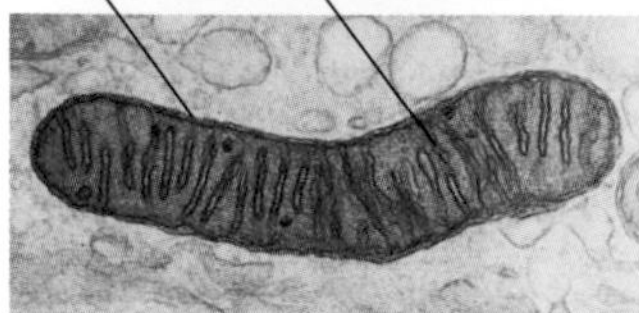

Figure 6 *Mitochondria break down sugar and make ATP. ATP is produced on the inner membrane.*

Mitochondria

A mitochondrion (MIET oh KAHN dree uhn) is the main power source of a cell. A **mitochondrion** is the organelle in which sugar is broken down to produce energy. Mitochondria are covered by two membranes, as shown in **Figure 6.** Energy released by mitochondria is stored in a substance called *ATP* (**a**denosine **t**ri**p**hosphate). The cell then uses ATP to do work. ATP can be made at several places in a cell. But most of a cell's ATP is made in the inner membrane of the cell's mitochondria.

Most eukaryotic cells have mitochondria. Mitochondria are the size of some bacteria. Like bacteria, mitochondria have their own DNA, and mitochondria can divide within a cell.

Reading Check Where is most of a cell's ATP made?

mitochondrion in eukaryotic cells, the cell organelle that is surrounded by two membranes and that is the site of cellular respiration

Chloroplasts

Animal cells cannot make their own food. Plants and algae are different. They have chloroplasts (KLAWR uh PLASTS) in some of their cells. *Chloroplasts* are organelles in plant and algae cells in which photosynthesis takes place. Like mitochondria, chloroplasts have two membranes and their own DNA. A chloroplast is shown in **Figure 7.** *Photosynthesis* is the process by which plants and algae use sunlight, carbon dioxide, and water to make sugar and oxygen.

Chloroplasts are green because they contain *chlorophyll,* a green pigment. Chlorophyll is found inside the inner membrane of a chloroplast. Chlorophyll traps the energy of sunlight, which is used to make sugar. The sugar produced by photosynthesis is then used by mitochondria to make ATP.

Figure 7 *Chloroplasts harness and use the energy of the sun to make sugar. A green pigment—chlorophyll—traps the sun's energy.*

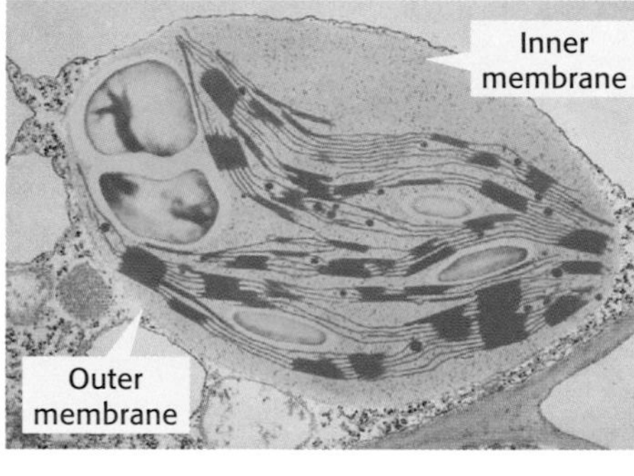

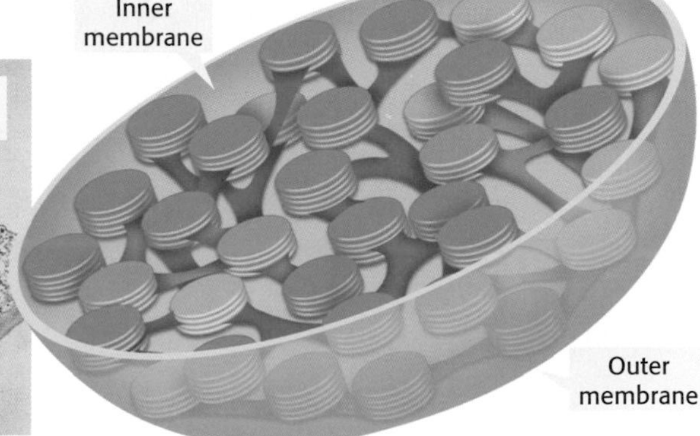

SCIENTISTS AT ODDS

Acquiring Genomes Dr. Lynn Margulis knew that mitochondria and chloroplasts have their own DNA and divide by binary fission. She proposed that these organelles were once bacteria that entered organisms and became parts of those cells. Other scientists disagreed, but research proved Dr. Margulis right. Now, Margulis proposes that all eukaryotes developed as a result of genetic mergers between different kinds of organisms. And other scientists disagree. Only more research will settle the debate.

Is That a Fact!

About 100 eukaryotic species do not have mitochondria. *Giardia* is a freshwater protist that lacks mitochondria. *Giardia* can make people sick if they drink water from an infected lake or stream.

Golgi Complex

The organelle that packages and distributes proteins is called the **Golgi complex** (GOHL jee KAHM PLEKS). It is named after Camillo Golgi, the Italian scientist who first identified the organelle.

The Golgi complex looks like ER, as shown in **Figure 8.** Lipids and proteins from the ER are delivered to the Golgi complex. There, the lipids and proteins may be modified to do different jobs. The final products are enclosed in a piece of the Golgi complex's membrane. This membrane pinches off to form a small bubble. The bubble transports its contents to other parts of the cell or out of the cell.

Golgi complex cell organelle that helps make and package materials to be transported out of the cell

vesicle a small cavity or sac that contains materials in a eukaryotic cell

Cell Compartments

The bubble that forms from the Golgi complex's membrane is a vesicle. A **vesicle** (VES i kuhl) is a small sac that surrounds material to be moved into or out of a cell. All eukaryotic cells have vesicles. Vesicles also move material within a cell. For example, vesicles carry new protein from the ER to the Golgi complex. Other vesicles distribute material from the Golgi complex to other parts of the cell. Some vesicles form when part of the cell membrane surrounds an object outside the cell.

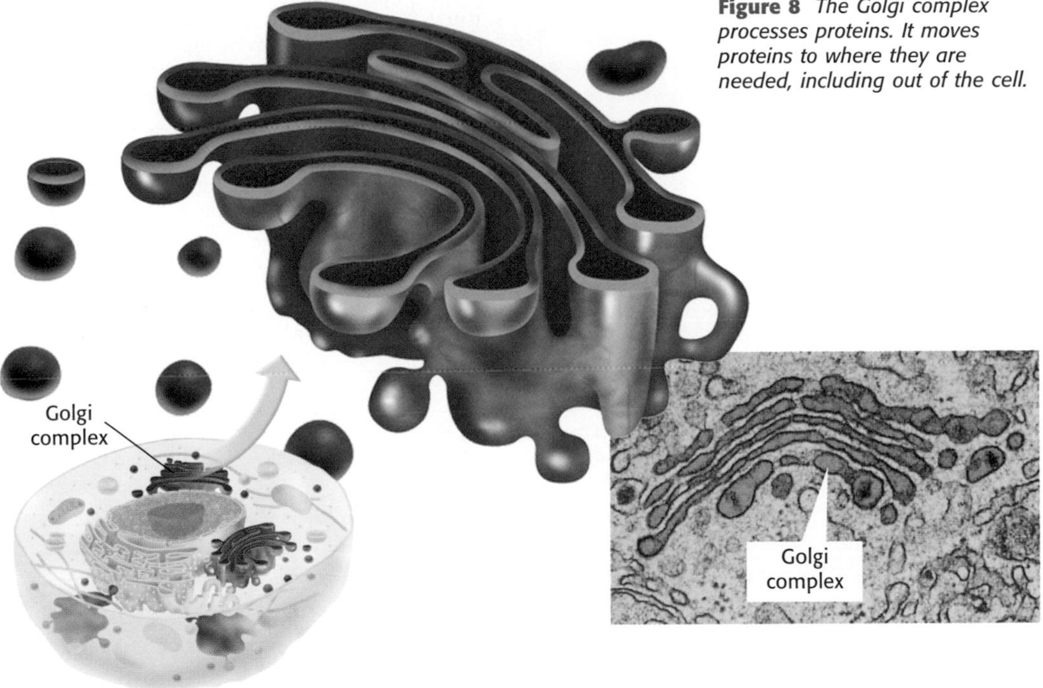

Figure 8 *The Golgi complex processes proteins. It moves proteins to where they are needed, including out of the cell.*

Cultural Awareness — GENERAL

Golgi and His Organelle
Camillo Golgi was born in 1843 in the small town Corteno, in the northern Italian province of Lombardy. He received a medical degree in 1865. Although he was a psychiatrist, he was interested in the microscopic nature of the nervous system. He discovered the "black reaction," which is a method of staining tissue using silver nitrate so that the tissue can be viewed clearly with a microscope. In 1897, using this method, Golgi saw the cellular structures we know today as the Golgi complex. For his work, Golgi shared the Nobel Prize for Medicine in 1906. Golgi died in 1926 at the age of 83. **LS Verbal**

Homework — GENERAL

Writing **Ornery Organelles** The organelles in a cell are rebelling against the nucleus. They all think that they work too hard, and they want to take a vacation. Have students write a dialogue between the nucleus and the other organelles. Tell students to help each organelle present a case for why it needs a rest, and then have the nucleus explain what would happen if even one of them took time off. **LS Verbal/Logical** PORTFOLIO

Scientists at Odds

Is It There or Not? Many scientists did not believe Golgi's claims about the organelle he observed and described. Those scientists thought that Golgi just saw tiny globs of the staining material. The existence of the organelle that was eventually named the *Golgi complex* was finally confirmed in the mid-1950s with the aid of the electron microscope.

Close

Reteaching BASIC

Organelles and Their Functions Give students a table similar to the table on this page, but leave the boxes blank. Have students work in pairs to fill in both the illustrations and the text of the table. LS **Interpersonal/Visual**

Quiz GENERAL

1. Why do scientists sometimes say that some plant cell vesicles are just large lysosomes? (because these vesicles store digestive enzymes and aid in cellular digestion)
2. What is the difference between the cytoskeleton and the cytoplasm? (The cytoplasm is the fluid—and almost all of its contents—inside a cell, and the cytoskeleton is a web of proteins in the cytoplasm that gives the cell shape and may help the cell move.)

Alternative Assessment GENERAL

Cell Model Have students create a cell using a shoe box for the cell wall. Any material readily available can be used for the cell parts. Parts can be hung on a string, glued to the box, or attached by any method the student chooses. The boxes can be displayed in the classroom. English Language Learners
LS **Kinesthetic**

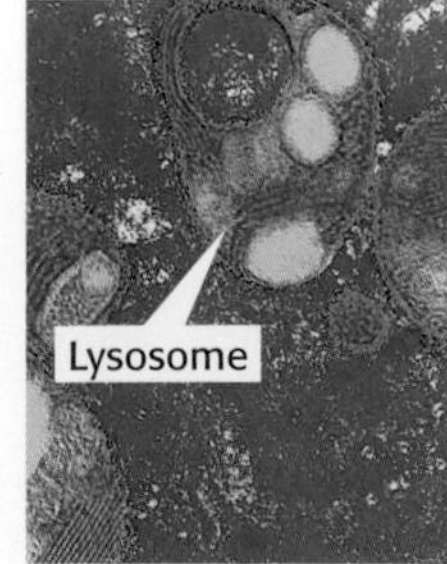

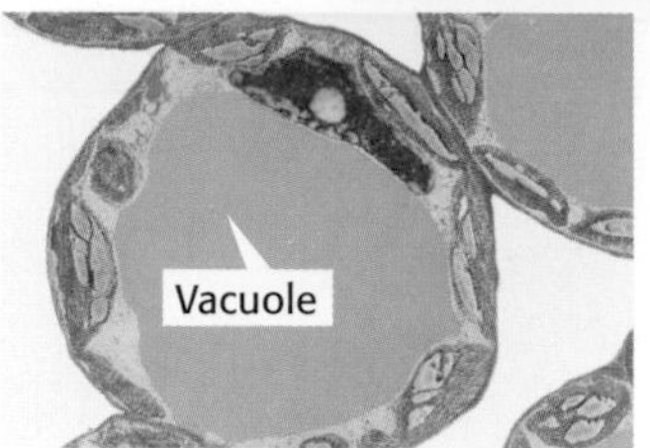

Figure 9 *Lysosomes digest materials inside a cell. In plant and fungal cells, vacuoles often perform the same function.*

Cellular Digestion

Lysosomes (LIE suh SOHMZ) are vesicles that are responsible for digestion inside a cell. **Lysosomes** are organelles that contain digestive enzymes. They destroy worn-out or damaged organelles, get rid of waste materials, and protect the cell from foreign invaders. Lysosomes, which come in a wide variety of sizes and shapes, are shown in **Figure 9.**

Lysosomes are found mainly in animal cells. When eukaryotic cells engulf particles, they enclose the particles in vesicles. Lysosomes bump into these vesicles and pour enzymes into them. These enzymes digest the particles in the vesicles.

Reading Check Why are lysosomes important?

Vacuoles

A *vacuole* (VAK yoo OHL) is a large vesicle. In plant and fungal cells, some vacuoles act like large lysosomes. They store digestive enzymes and aid in digestion within the cell. Other vacuoles in plant cells store water and other liquids. Vacuoles that are full of water, such as the one in **Figure 9,** help support the cell. Some plants wilt when their vacuoles lose water. **Table 1** shows some organelles and their functions.

lysosome a cell organelle that contains digestive enzymes

Table 1 Organelles and Their Functions

Organelle	Function	Organelle	Function
Nucleus	the organelle that contains the cell's DNA and is the control center of the cell	**Chloroplast**	the organelle that uses the energy of sunlight to make food
Ribosome	the organelle in which amino acids are hooked together to make proteins	**Golgi complex**	the organelle that processes and transports proteins and other materials out of cell
Endoplasmic reticulum	the organelle that makes lipids, breaks down drugs and other substances, and packages proteins for Golgi complex	**Vacuole**	the organelle that stores water and other materials
Mitochondria	the organelle that breaks down food molecules to make ATP	**Lysosome**	the organelle that digests food particles, wastes, cell parts, and foreign invaders

Answer to Reading Check

Lysosomes destroy worn-out organelles, attack foreign invaders, and get rid of waste material from inside the cell.

ACTIVITY BASIC

Vacuole Model You can make a demonstration model of a vacuole within a cell by filling a balloon with water or air and putting it inside a clear plastic storage bag. Show that the vacuole is part of the cell and is distinct within the cell. You can demonstrate that cells are full of motion and activity by moving the balloon around inside the bag. English Language Learners
LS **Visual**

SECTION Review

Summary

- Eukaryotic cells have organelles that perform functions that help cells remain alive.
- All cells have a cell membrane. Some cells have a cell wall. Some cells have a cytoskeleton.
- The nucleus of a eukaryotic cell contains the cell's genetic material, DNA.
- Ribosomes are the organelles that make proteins. Ribosomes are not covered by a membrane.
- The endoplasmic reticulum (ER) and the Golgi complex make and process proteins before the proteins are transported to other parts of the cell or out of the cell.
- Mitochondria and chloroplasts are energy-releasing organelles.
- Lysosomes are organelles responsible for digestion within a cell. In plant cells, organelles called *vacuoles* store cell materials and sometimes act like large lysosomes.

Using Key Terms

1. In your own words, write a definition for each of the following terms: *ribosome, lysosome,* and *cell wall.*

Understanding Key Ideas

2. Which of the following are found mainly in animal cells?
   - **a.** mitochondria
   - **b.** lysosomes
   - **c.** ribosomes
   - **d.** Golgi complexes
3. What is the function of a Golgi complex? What is the function of the endoplasmic reticulum?

Critical Thinking

4. **Making Comparisons** Describe three ways in which plant cells differ from animal cells.
5. **Applying Concepts** Every cell needs ribosomes. Explain why.
6. **Predicting Consequences** A certain virus attacks the mitochondria in cells. What would happen to a cell if all of its mitochondria were destroyed?
7. **Expressing Opinions** Do you think that having chloroplasts gives plant cells an advantage over animal cells? Support your opinion.

Interpreting Graphics

Use the diagram below to answer the questions that follow.

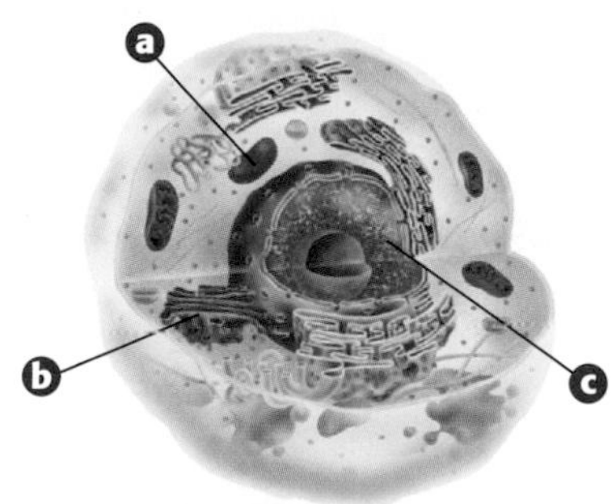

8. Is this a diagram of a plant cell or an animal cell? Explain how you know.
9. What organelle does the letter *b* refer to?

CHAPTER RESOURCES

Chapter Resource File

- **Section Quiz** GENERAL
- **Section Review** GENERAL
- **Vocabulary and Section Summary** GENERAL
- **SciLinks Activity** GENERAL

Technology

Transparencies
- Organelles and Their Functions

Answers to Section Review

1. Sample answer: Ribosomes are organelles where amino acids are joined together to make proteins. Lysosomes are organelles that carry out cellular digestion. The cell wall is the outermost structure in cells of plants, fungi, and algae.
2. b
3. Sample answer: Golgi complex: packages and distributes proteins within a cell; endoplasmic reticulum: a series of folded membranes on which lipids, proteins, and other materials are made, and through which those materials are delivered to other places in the cell
4. Sample answer: Plant cells have cell walls, but animal cells do not. Plant cells have chloroplasts, which animal cells do not have. Plant cells do not seem to have small lysosomes (they have large vacuoles instead), which animal cells do have.
5. Sample answer: Ribosomes are the organelles where proteins are made. All cells need protein in order to live.
6. Sample answer: Mitochondria are organelles that produce most of a cell's energy. If its mitochondria were destroyed, a cell would eventually die because it would have no source of energy.
7. Sample answer: I think plants have an advantage over animals because plants can make their own food just by using sunlight and other nutrients. Animals have to wait for plants to grow in order to get food.
8. This diagram is of an animal cell; the first clue is that the cell has no cell wall.
9. the Golgi complex

SECTION 3

Focus

Overview

In this section, students will learn that a cell is the smallest unit of life. In most multicellular organisms, groups of cells form tissues that compose organs. Two or more organs can interact to form an organ system.

Bellringer

Write the following questions on the board for students to answer:

Why can't you use your teeth to breathe? Why can't you use your arm muscles to digest food?

Motivate

ACTIVITY — GENERAL

Concept Mapping Organize the class into small groups. Provide each group with pictures of tissues, organs, and organ systems. Have students arrange the pictures into concept maps. Encourage students to notice similarities and differences between organs. For example, the stomach and the heart are very different organs, but both are made of muscle tissue, and both function by holding and moving substances through their cavities. LS **Intrapersonal/Visual** Co-op Learning

SECTION 3

The Organization of Living Things

In some ways, organisms are like machines. Some machines have just one part. But most machines have many parts. Some organisms exist as a single cell. Other organisms have many—even trillions—of cells.

READING WARM-UP

Objectives

- List three advantages of being multicellular.
- Describe the four levels of organization in living things.
- Explain the relationship between the structure and function of a part of an organism.

Terms to Learn

tissue	organism
organ	structure
organ system	function

READING STRATEGY

Paired Summarizing Read this section silently. In pairs, take turns summarizing the material. Stop to discuss ideas that seem confusing.

Most cells are smaller than the period that ends this sentence. Yet, every cell in every organism performs all the processes of life. So, are there any advantages to having many cells?

The Benefits of Being Multicellular

You are a *multicellular organism*. This means that you are made of many cells. Multicellular organisms grow by making more small cells, not by making their cells larger. For example, an elephant is bigger than you are, but its cells are about the same size as yours. An elephant just has more cells than you do. Some benefits of being multicellular are the following:

- **Larger Size** Many multicellular organisms are small. But they are usually larger than single-celled organisms. Larger organisms are prey for fewer predators. Larger predators can eat a wider variety of prey.
- **Longer Life** The life span of a multicellular organism is not limited to the life span of any single cell.
- **Specialization** Each type of cell has a particular job. Specialization makes the organism more efficient. For example, the cardiac muscle cell in **Figure 1** is a specialized muscle cell. Heart muscle cells contract and make the heart pump blood.

Reading Check List three advantages of being multicellular. *(See the Appendix for answers to Reading Checks.)*

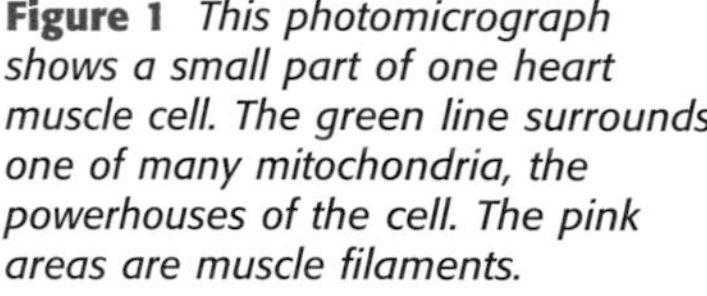

Figure 1 *This photomicrograph shows a small part of one heart muscle cell. The green line surrounds one of many mitochondria, the powerhouses of the cell. The pink areas are muscle filaments.*

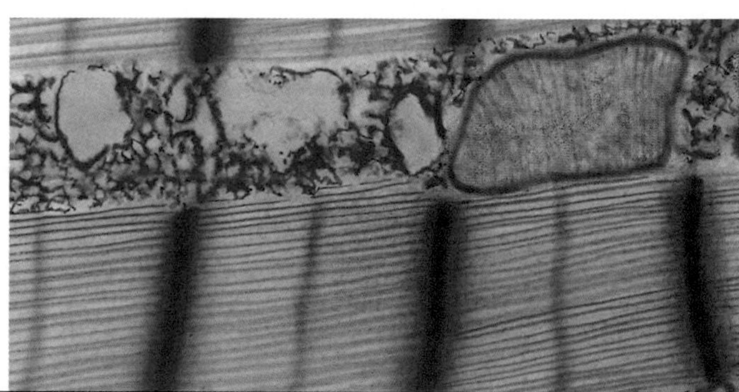

CHAPTER RESOURCES

Chapter Resource File

- Lesson Plan
- Directed Reading A BASIC
- Directed Reading B SPECIAL NEEDS

Technology

Transparencies
- Bellringer
- Levels of Organization in the Cardiovascular System

Answer to Reading Check

Sample answer: larger size, longer life, cell specialization

Figure 2 This photomicrograph shows cardiac muscle tissue. Cardiac muscle tissue is made up of many cardiac cells.

Cells Working Together

A **tissue** is a group of cells that work together to perform a specific job. The material around and between the cells is also part of the tissue. The cardiac muscle tissue, shown in **Figure 2,** is made of many cardiac muscle cells. Cardiac muscle tissue is just one type of tissue in a heart.

Animals have four basic types of tissues: nerve tissue, muscle tissue, connective tissue, and protective tissue. In contrast, plants have three types of tissues: transport tissue, protective tissue, and ground tissue. Transport tissue moves water and nutrients through a plant. Protective tissue covers the plant. It helps the plant retain water and protects the plant against damage. Photosynthesis takes place in ground tissue.

tissue a group of similar cells that perform a common function

organ a collection of tissues that carry out a specialized function of the body

Tissues Working Together

A structure that is made up of two or more tissues working together to perform a specific function is called an **organ.** For example, your heart is an organ. It is made mostly of cardiac muscle tissue. But your heart also has nerve tissue and tissues of the blood vessels that all work together to make your heart the powerful pump that it is.

Another organ is your stomach. It also has several kinds of tissue. In the stomach, muscle tissue makes food move in and through the stomach. Special tissues make chemicals that help digest your food. Connective tissue holds the stomach together, and nervous tissue carries messages back and forth between the stomach and the brain. Other organs include the intestines, brain, and lungs.

Plants also have different kinds of tissues that work together as organs. A leaf is a plant organ that contains tissue that traps light energy to make food. Other examples of plant organs are stems and roots.

Reading Check What is an organ?

A Pet Protist

Imagine that you have a tiny box-shaped protist for a pet. To care for your pet protist properly, you have to figure out how much to feed it. The dimensions of your protist are roughly 25 µ × 20 µ × 2 µ. If seven food particles per second can enter through each square micrometer of surface area, how many particles can your protist eat in 1 min?

Teach

Discussion — GENERAL

Muscles Ask students to list all of the ways they use their muscles. Responses will probably include walking, riding a bike, swimming, and throwing or kicking a ball. Lead students to understand that muscles are also involved in swallowing food (tongue and esophagus), digestion (stomach and intestines), and blinking eyes (eyelids). Also, help students understand that sometimes muscles act voluntarily (jumping, writing), and sometimes they act involuntarily (heart beating). **LS Auditory/Logical**

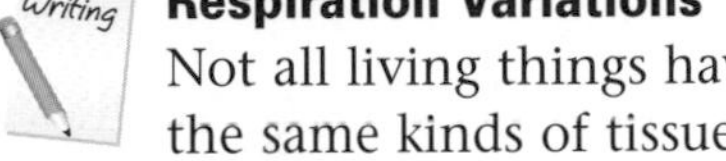

Writing **Respiration Variations** Not all living things have the same kinds of tissues and organs. Yet, all must perform similar life processes. Have students do research in order to compare the structures a fish uses to breathe with those that a human uses. Students' reports should also answer the question, What parts of the human and fish respiratory systems are similar? (Even though a fish has gills and a human has lungs, both have cells that exchange and transport oxygen and carbon dioxide.) **LS Logical**

Answer to Math Practice

The surface area of the protist is $[(25\ \mu \times 20\ \mu) + (25\ \mu \times 2\ \mu) + (20\ \mu \times 2\ \mu)] \times 2 = 1{,}180\ \mu^2$, so it can eat $1{,}180 \times 7 = 8{,}260$ particles of food every second, or $60 \times 8{,}260 = 495{,}600$ particles of food per minute.

Answer to Reading Check

An organ is a structure of two or more tissues working together to perform a specific function in the body.

Is That a Fact!

In your lifetime, your body will shed about 18 kg (almost 40 lb) of dead skin.

MISCONCEPTION ALERT

Dead Cells Students may think that hair is alive: advertisements for shampoo create the impression that hair is living tissue. Hair, and fingernails, too, are dead. Hair and fingernails grow out of specialized skin cells. They grow continuously, but both are composed of dead cells and a protein called *keratin*. If hair and fingernails were alive and contained nerve cells, as the deep skin layers do, haircuts and manicures would be quite painful.

Close

Reteaching — Basic

Levels of Organization Write the following headings on the board:

Cell, Tissue, Organ, Organ system, Organism

Have students write these headings on their paper and list at least two examples under each heading. English Language Learners
LS Verbal/Logical

Quiz — General

1. What is the relationship between your digestive system, stomach, and intestines? (The digestive system is an organ system. The stomach and intestines are organs that are parts of the digestive system.)
2. What is the main difference between a unicellular organism and a multicellular organism in the way life processes are carried out? (Sample answer: A unicellular organism must perform all life functions by itself. A multicellular organism may have specialized cells that work together to carry out each function.)

Alternative Assessment — General

Concept Mapping Have students choose an organ system and identify its component organs. Then, have students make a concept map describing the function of the organs and their relationship to one another. **LS Logical/Visual**

organ system a group of organs that work together to perform body functions

organism a living thing; anything that can carry out life processes independently

structure the arrangement of parts in an organism

function the special, normal, or proper activity of an organ or part

Organs Working Together

A group of organs working together to perform a particular function is called an **organ system.** Each organ system has a specific job to do in the body.

For example, the digestive system is made up of several organs, including the stomach and intestines. The digestive system's job is to break down food into small particles. Other parts of the body then use these small particles as fuel. In turn, the digestive system depends on the respiratory and cardiovascular systems for oxygen. The cardiovascular system, shown in **Figure 3,** includes organs and tissues such as the heart and blood vessels. Plants also have organ systems. They include leaf systems, root systems, and stem systems.

Reading Check List the levels of organization in living things.

Organisms

Anything that can perform life processes is an **organism.** An organism made of a single cell is called a *unicellular organism.* Bacteria, most protists, and some kinds of fungi are unicellular. Although some of these organisms live in colonies, the organisms are still unicellular. They are unicellular because all of the cells in the colony are the same. Each cell must carry out all life processes. In contrast, even the simplest multicellular organisms have specialized cells. Complex organisms are composed of cells, tissues, organs, and organ systems that perform specialized functions.

Figure 3 Levels of Organization in the Cardiovascular System

Cell Cells form tissues.

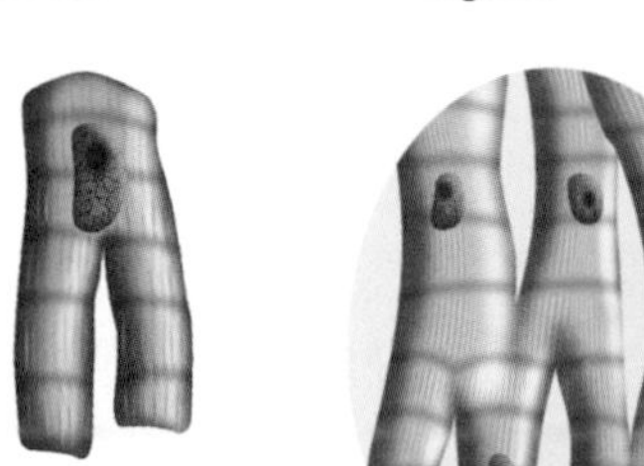

Tissue Tissues form organs.

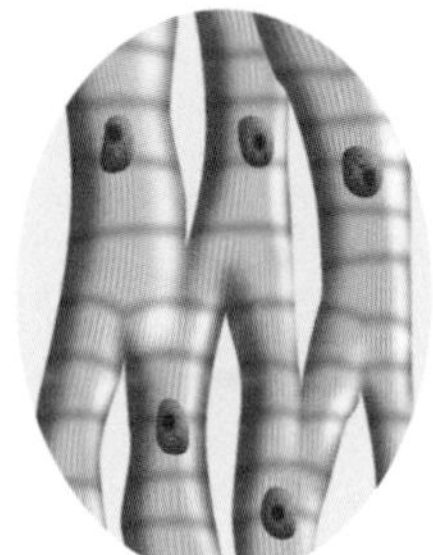

Organ Organs form organ systems.

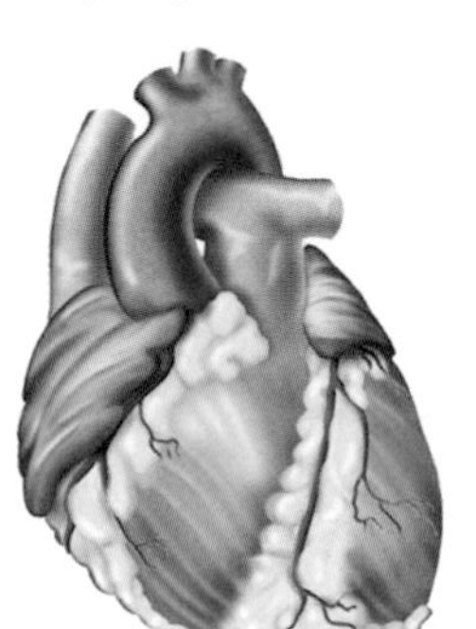

Organ system And organ systems form organisms such as you!

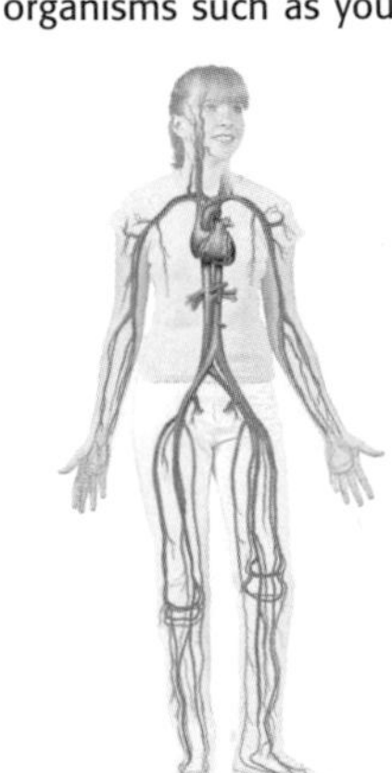

Is That a Fact!

An elephant's trunk is constructed of 135 kg (300 lb) of hair, skin, connective tissue, nerves, and muscles. The muscle tissue is composed of 150,000 tiny subunits of muscle, each of which is coordinated with the others to enable an elephant to greet its friends, breathe, grab, and drink.

Answer to Reading Check

cell, tissue, organ, organ system, organism

Structure and Function

In organisms, structure and function are related. **Structure** is the arrangement of parts in an organism. It includes the shape of a part and the material of which the part is made. **Function** is the job the part does. For example, the structure of the lungs is a large, spongy sac. In the lungs, there are millions of tiny air sacs called *alveoli*. Blood vessels wrap around the alveoli, as shown in **Figure 4.** Oxygen from air in the alveoli enters the blood. Blood then brings oxygen to body tissues. Also, in the alveoli, carbon dioxide leaves the blood and is exhaled.

The structures of alveoli and blood vessels enables them to perform a function. Together, they bring oxygen to the body and get rid of its carbon dioxide.

Figure 4 The Structure and Function of Alveoli

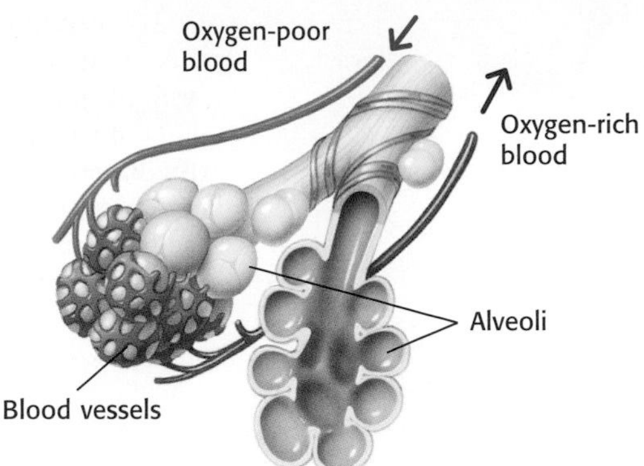

SECTION Review

Summary

- Advantages of being multicellular are larger size, longer life, and cell specialization.
- Four levels of organization are cell, tissue, organ, and organ system.
- A *tissue* is a group of cells working together. An *organ* is two or more tissues working together. An *organ system* is two or more organs working together.
- In organisms, a part's structure and function are related.

Using Key Terms

1. Use each of the following terms in a separate sentence: *tissue, organ,* and *function*.

Understanding Key Ideas

2. What are the four levels of organization in living things?
   a. cell, multicellular, organ, organ system
   b. single cell, multicellular, tissue, organ
   c. larger size, longer life, specialized cells, organs
   d. cell, tissue, organ, organ system

Math Skills

3. One multicellular organism is a cube. Each of its sides is 3 cm long. Each of its cells is 1 cm^3. How many cells does it have? If each side doubles in length, how many cells will it then have?

Critical Thinking

4. **Applying Concepts** Explain the relationship between structure and function. Use alveoli as an example. Be sure to include more than one level of organization.
5. **Making Inferences** Why can multicellular organisms be more complex than unicellular organisms? Use the three advantages of being multicellular to help explain your answer.

For a variety of links related to this chapter, go to www.scilinks.org

Topic: Organization of Life
SciLinks code: HSM1080

ACTIVITY — BASIC

Writing **Explain It to a Friend** A great way to learn something is to teach it to someone else. Have students write a letter to a friend explaining how cells, tissues, organs, and organ systems are related.

LS Logical/Verbal

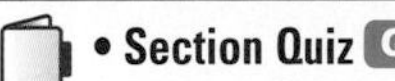

CHAPTER RESOURCES

Chapter Resource File

- Section Quiz GENERAL
- Section Review GENERAL
- Vocabulary and Section Summary GENERAL
- Reinforcement Worksheet BASIC
- Critical Thinking ADVANCED

Answers to Section Review

1. Sample answer: The body has several different kinds of tissue. I think that the most important organ in the body is the brain. Sometimes a part of the body with a certain structure performs more than one function.
2. d
3. $3 \text{ cm} \times 3 \text{ cm} \times 3 \text{ cm} = 27 \text{ cm}^3$
   $27 \text{ cm}^3 \div 1 \text{ cm}^3 = 27$ cells;
   If each side doubles in length, the organism will have 216 cells ($6 \times 6 \times 6 = 216$).
4. Sample answer: Alveoli are tiny sacs whose function is to contain and exchange gases such as oxygen and carbon dioxide. The structure of alveoli, as tiny sacs surrounded by tiny blood vessels, includes the cells that make up the tissue of the alveoli and the tissue that joins the alveoli to the bronchioles, which are part of the lung.The lungs are made of several kinds of tissue, such as the bronchi, bronchioles, and alveoli.
5. Sample answer: The main reason that multicellular organisms can be more complex than unicellular organisms is that multicellular organisms have cell specialization. Specialization allows some cells to do only digestion while others do respiration or circulation. Therefore, the organism is more efficient. Being multicellular also means that an organism may grow larger than a unicellular organism. Size is an advantage because, in general, the larger the organism is, the fewer predators it faces. Finally, being unicellular means that when your one cell dies, you are dead. In a multicellular organism, the death of one cell does not mean the death of the organism.

Teacher's Note: In fact, only multicellular organisms can have an efficient vascular system, which is the key to efficient delivery of materials to cells and removal of wastes from cells. Most students will probably not know this, but some advanced or interested students may grasp this idea.

Elephant-Sized Amoebas?

Teacher's Notes

Time Required
Two 45-minute class periods

Lab Ratings

Teacher Prep 2
Student Set-Up 2
Concept Level 3
Clean Up 1

Safety Caution
Remind students to review all safety cautions and icons before beginning this lab activity.

Preparation Notes
Some students may find it difficult to work with a nonspecific unit of measurement. If so, the cube models easily convert to centimeters. You may want to add some small items, such as peas, beans, popcorn, or peppercorns, to the sand to represent organelles floating in the cytoplasm. Some students may need to review what a ratio is and how ratios are used.

Model-Making Lab

OBJECTIVES

Explore why a single-celled organism cannot grow to the size of an elephant.

Create a model of a cell to illustrate the concept of surface area–to-volume ratio.

MATERIALS

- calculator (optional)
- cubic cell patterns
- heavy paper or poster board
- sand, fine
- scale or balance
- scissors
- tape, transparent

SAFETY

Elephant-Sized Amoebas?

An amoeba is a single-celled organism. Like most cells, amoebas are microscopic. Why can't amoebas grow as large as elephants? If an amoeba grew to the size of a quarter, the amoeba would starve to death. To understand how this can be true, build a model of a cell and see for yourself.

Procedure

1. Use heavy paper or poster board to make four cube-shaped cell models from the patterns supplied by your teacher. Cut out each cell model, fold the sides to make a cube, and tape the tabs on the sides. The smallest cell model has sides that are each one unit long. The next larger cell has sides of two units. The next cell has sides of three units, and the largest cell has sides of four units. These paper models represent the cell membrane, the part of a cell's exterior through which food and wastes pass.

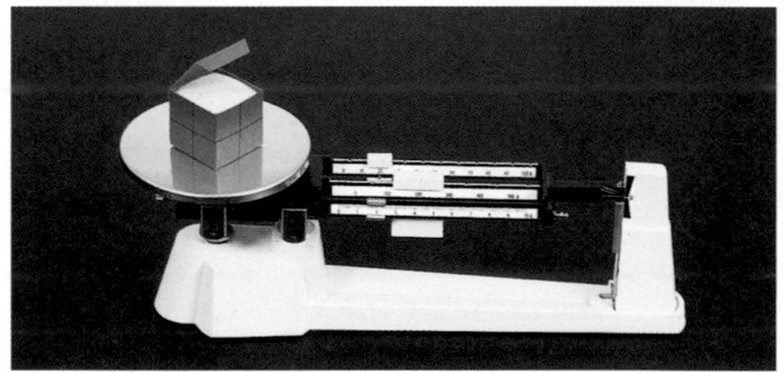

CHAPTER RESOURCES

Chapter Resource File

- Datasheet for Chapter Lab
- Lab Notes and Answers

Technology

Classroom Videos
- Lab Video

LabBook
- Cells Alive!

CHAPTER RESOURCES

Workbooks

Whiz-Bang Demonstrations
- Grand Strand GENERAL

Labs You Can Eat
- The Incredible Edible Cell GENERAL

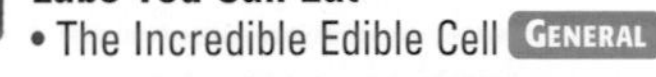

Long-Term Projects & Research Ideas
- Ewe Again, Dolly? ADVANCED

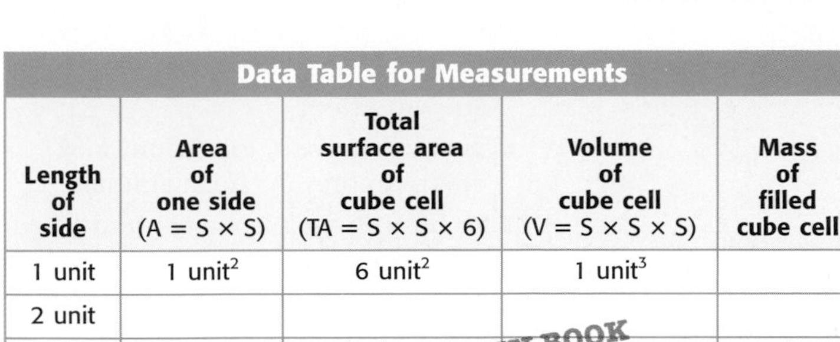

Data Table for Measurements

Length of side	Area of one side (A = S × S)	Total surface area of cube cell (TA = S × S × 6)	Volume of cube cell (V = S × S × S)	Mass of filled cube cell
1 unit	1 unit^2	6 unit^2	1 unit^3	
2 unit				
3 unit				
4 unit				

Key to Formula Symbols

S = the length of one side

A = area

6 = number of sides

V = volume

TA = total area

2. Copy the data table shown above. Use each formula to calculate the data about your cell models. Record your calculations in the table. Calculations for the smallest cell have been done for you.
3. Carefully fill each model with fine sand until the sand is level with the top edge of the model. Find the mass of the filled models by using a scale or a balance. What does the sand in your model represent?
4. Record the mass of each filled cell model in your Data Table for Measurements. (Always remember to use the appropriate mass unit.)

Analyze the Results

1. **Constructing Tables** Make a data table like the one shown at right.
2. **Organizing Data** Use the data from your Data Table for Measurements to find the ratios for each of your cell models. For each of the cell models, fill in the Data Table for Ratios .

Draw Conclusions

3. **Interpreting Information** As a cell grows larger, does the ratio of total surface area to volume increase, decrease, or stay the same?
4. **Interpreting Information** As a cell grows larger, does the total surface area–to-mass ratio increase, decrease, or stay the same?
5. **Drawing Conclusions** Which is better able to supply food to all the cytoplasm of the cell: the cell membrane of a small cell or the cell membrane of a large cell? Explain your answer.
6. **Evaluating Data** In the experiment, which is better able to feed all of the cytoplasm of the cell: the cell membrane of a cell that has high mass or the cell membrane of a cell that has low mass? You may explain your answer in a verbal presentation to the class, or you may choose to write a report and illustrate it with drawings of your models.

Data Table for Ratios

Length of side	Ratio of total surface area to volume	Ratio of total surface area to mass
1 unit		
2 unit		
3 unit		
4 unit		

Data Table for Measurements

Length of side S	Area of one side (square units)	Total surface area of cube cell (square units)	Volume of cube cell (cubic units)	Mass of cube cell (sample answer, in grams)
1	1	6	1	4.5
2	4	24	8	30
3	9	54	27	105
4	16	96	64	230

Data Table for Ratios

Length of side S	Total surface area–to-volume ratio	Total surface area–to-mass ratio (sample answer)
1	6:1	6:4.5 = 1.33:1
2	24:8 = 3:1	24:30 = 0.80:1
3	54:27 = 2:1	54:105 = 0.51:1
4	96:64 = 1.5:1	96:230 = 0.42:1

Cell Model Template

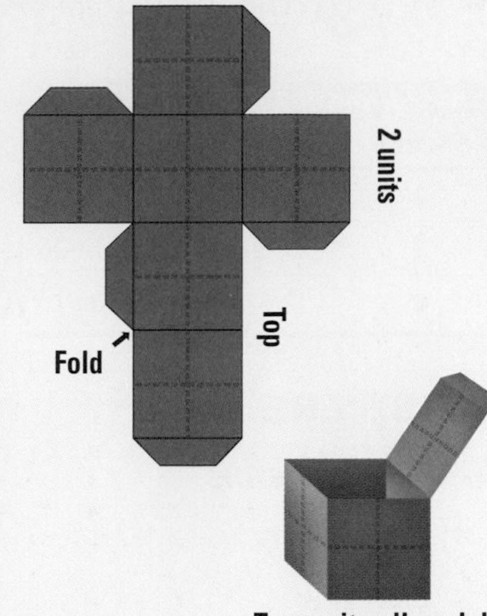

Two-unit cell model

Using the template above, prepare four patterns for students to use to make their cubes. Make one cube 1 unit wide, one cube 2 units wide, one cube 3 units wide, and one cube 4 units wide. The unit can be the size of your choosing.

Procedure

3. The sand represents cytoplasm.
4. Masses may vary.

Analyze the Results

2. See the tables below.

Draw Conclusions

3. decreases
4. decreases
5. the cell membrane of a small cell. A small cell has a higher surface-area-to-volume ratio than a large cell has, so more nutrients per cubic unit of volume can enter a small cell.
6. the cell membrane of a cell with low mass

Terry Rakes
Elmwood Junior High School
Rogers, Arkansas

Chapter Review

Assignment Guide	
Section	**Questions**
1	1, 4, 10–13, 23
2	3, 6, 9, 16–19, 22, 24–26
3	2, 5, 7–8, 14–15, 20–21

ANSWERS

Using Key Terms

1. cell
2. function
3. organelles
4. eukaryote
5. tissue
6. cell wall

Understanding Key Ideas

7. c
8. d
9. a
10. b
11. b
12. c

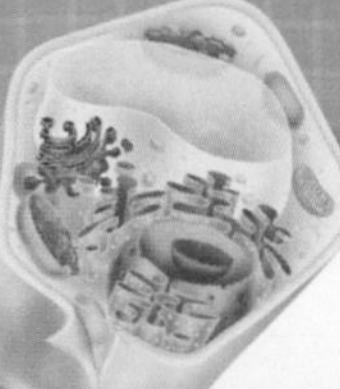

Chapter Review

USING KEY TERMS

Complete each of the following sentences by choosing the correct term from the word bank.

cell	organ
cell membrane	prokaryote
organelles	eukaryote
cell wall	tissue
structure	function

1. A(n) ___ is the most basic unit of all living things.
2. The job that an organ does is the ___ of that organ.
3. Ribosomes and mitochondria are types of ___.
4. A(n) ___ is an organism whose cells have a nucleus.
5. A group of cells working together to perform a specific function is a(n) ___.
6. Only plant cells have a(n) ___.

UNDERSTANDING KEY IDEAS

Multiple Choice

7. Which of the following best describes an organ?
   a. a group of cells that work together to perform a specific job
   b. a group of tissues that belong to different systems
   c. a group of tissues that work together to perform a specific job
   d. a body structure, such as muscles or lungs
8. The benefits of being multicellular include
   a. small size, long life, and cell specialization.
   b. generalized cells, longer life, and ability to prey on small animals.
   c. larger size, more enemies, and specialized cells.
   d. longer life, larger size, and specialized cells.
9. In eukaryotic cells, which organelle contains the DNA?
   a. nucleus **c.** smooth ER
   b. Golgi complex **d.** vacuole
10. Which of the following statements is part of the cell theory?
   a. All cells suddenly appear by themselves.
   b. All cells come from other cells.
   c. All organisms are multicellular.
   d. All cells have identical parts.
11. The surface area–to-volume ratio of a cell limits
   a. the number of organelles that the cell has.
   b. the size of the cell.
   c. where the cell lives.
   d. the types of nutrients that a cell needs.
12. Two types of organisms whose cells do not have a nucleus are
   a. prokaryotes and eukaryotes.
   b. plants and animals.
   c. eubacteria and archaebacteria.
   d. single-celled and multicellular organisms.

13. Cells must be small in order to have a large enough surface area–to-volume ratio to get sufficient nutrients to survive and to get rid of wastes.

14. Cells are the smallest unit of all living things. Cells combine to make tissues. Different tissues combine to make organs, which have specialized jobs in the body. Organs work together in organ systems, which perform body functions.

Short Answer

13. Explain why most cells are small.

14. Describe the four levels of organization in living things.

15. What is the difference between the structure of an organ and the function of the organ?

16. Name two functions of a cell membrane.

17. What are the structure and function of the cytoskeleton in a cell?

CRITICAL THINKING

18. **Concept Mapping** Use the following terms to create a concept map: *cells, organisms, Golgi complex, organ systems, organs, nucleus, organelle,* and *tissues.*

19. **Making Comparisons** Compare and contrast the functions of the endoplasmic reticulum and the Golgi complex.

20. **Identifying Relationships** Explain how the structure and function of an organism's parts are related. Give an example.

21. **Evaluating Hypotheses** One of your classmates states a hypothesis that all organisms must have organ systems. Is your classmate's hypothesis valid? Explain your answer.

22. **Predicting Consequences** What would happen if all of the ribosomes in your cells disappeared?

23. **Expressing Opinions** Scientists think that millions of years ago the surface of the Earth was very hot and that the atmosphere contained a lot of methane. In your opinion, which type of organism, a eubacterium or an archaebacterium, is the older form of life? Explain your reasoning.

INTERPRETING GRAPHICS

Use the diagram below to answer the questions that follow.

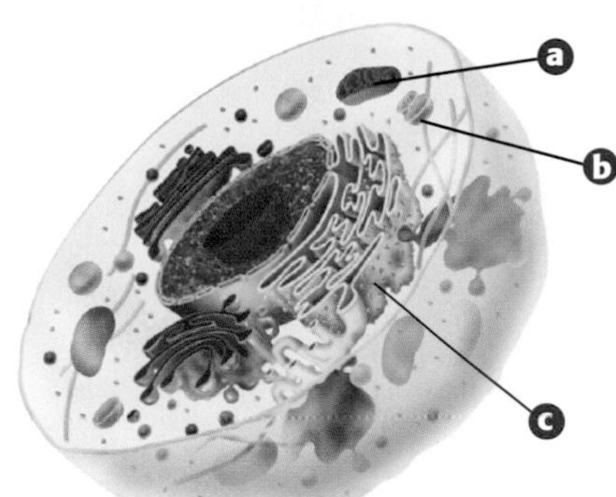

24. What is the name of the structure identified by the letter *a*?

25. Which letter identifies the structure that digests food particles and foreign invaders?

26. Which letter identifies the structure that makes proteins, lipids, and other materials and that contains tubes and passageways that enable substances to move to different places in the cell?

15. Structure is the shape of a part. Function is the job a part does.

16. The cell membrane separates the cell's contents from the outside environment and controls the flow of nutrients, wastes, and other materials into and out of the cell.

17. The cytoskeleton is a web of tubular and stringy proteins. The cytoskeleton helps give the cell shape and helps the cell move.

CHAPTER RESOURCES

Chapter Resource File

- Chapter Review GENERAL
- Chapter Test A GENERAL
- Chapter Test B ADVANCED
- Chapter Test C SPECIAL NEEDS
- Vocabulary Activity GENERAL

Workbooks

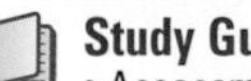

Study Guide
- Assessment resources are also available in Spanish.

Critical Thinking

18. 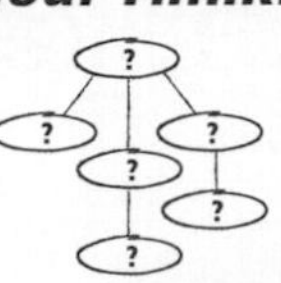 An answer to this exercise can be found at the end of this book.

19. Sample answer: The endoplasmic reticulum (ER) is a series of folded membranes within a cell where many proteins, lipids, and other materials are made in the cell. The smooth ER also helps break down toxic materials. The ER is the internal delivery system in a cell. The Golgi complex modifies, packages, and distributes proteins to other parts of the cell. It takes materials from the ER and encloses them in a small bubble of membrane. Then, it delivers them to where they are needed in other parts of the cell as well as outside the cell.

20. Sample answer: The structure of a part is its shape and the material it is made of. The function of a part is what that shape and material enable that part to do in the body; for example, alveoli are tiny sacs in the lungs that hold gases. Alveoli are made of a membrane that enables oxygen and carbon dioxide to pass into and out of the blood.

21. Sample answer: not valid; Some organisms are unicellular and have no tissues, organ, or organ systems.

22. Ribosomes make proteins, which all cells and all organisms need to survive. If your ribosomes disappeared, you would die.

23. Sample answer: Archaebacteria are older because there are many types of methane-making archaebacteria and because many types of archaebacteria live in very hot places.

Interpreting Graphics

24. endoplasmic reticulum

25. c

26. a

Standardized Test Preparation

Teacher's Note

To provide practice under more realistic testing conditions, give students 20 minutes to answer all of the questions in this Standardized Test Preparation.

MISCONCEPTION ALERT

Answers to the standardized test preparation can help you identify student misconceptions and misunderstandings.

READING

Passage 1

1. D
2. G
3. B

Question 1: Students may select incorrect answer B if they misread the part of the passage about snottites eventually becoming rock. Snottites themselves are a mixture of bacteria, sticky fluids, and minerals.

Question 2: Students may select incorrect answer I if, again, they misread the part of the passage about snottites eventually hardening into rock structures. Snottites do not create other structures in caves. The best answer is that snottite bacteria do not need sunlight because snottites live deep underground and are acidophiles that do not depend on sunlight for food.

Standardized Test Preparation

READING

Read each of the passages below. Then, answer the questions that follow each passage.

Passage 1 Exploring caves can be dangerous but can also lead to interesting discoveries. For example, deep in the darkness of Cueva de Villa Luz, a cave in Mexico, are slippery formations called *snottites*. They were named snottites because they look just like a two-year-old's runny nose. If you use an electron microscope to look at them, you see that snottites are bacteria; thick, sticky fluids; and small amounts of minerals produced by the bacteria. As tiny as they are, these bacteria can build up snottite structures that may eventually turn into rock. Formations in other caves look like hardened snottites. The bacteria in snottites are acidophiles. Acidophiles live in environments that are highly acidic. Snottite bacteria produce sulfuric acid and live in an environment that is similar to the inside of a car battery.

1. Which statement best describes snottites?
- **A** Snottites are bacteria that live in car batteries.
- **B** Snottites are rock formations found in caves.
- **C** Snottites were named for a cave in Mexico.
- **D** Snottites are made of bacteria, sticky fluids, and minerals.

2. Based on this passage, which conclusion about snottites is most likely to be correct?
- **F** Snottites are found in caves everywhere.
- **G** Snottite bacteria do not need sunlight.
- **H** You could grow snottites in a greenhouse.
- **I** Snottites create other bacteria in caves.

3. What is the main idea of this passage?
- **A** Acidophiles are unusual organisms.
- **B** Snottites are strange formations.
- **C** Exploring caves is dangerous.
- **D** Snottites are large, slippery bacteria.

Passage 2 The world's smallest mammal may be a bat about the size of a jelly bean. The scientific name for this tiny animal, which was unknown until 1974, is *Craseonycteris thonglongyai*. It is so small that it is sometimes called the *bumblebee bat*. Another name for this animal is the *hog-nosed bat*. Hog-nosed bats were given their name because one of their distinctive features is a piglike muzzle. Hog-nosed bats differ from other bats in another way: they do not have a tail. But, like other bats, hog-nosed bats do eat insects that they catch in mid-air. Scientists think that the bats eat small insects that live on the leaves at the tops of trees. Hog-nosed bats live deep in limestone caves and have been found in only one country, Thailand.

1. According to the passage, which statement about hog-nosed bats is most accurate?
- **A** They are the world's smallest animal.
- **B** They are about the size of a bumblebee.
- **C** They eat leaves at the tops of trees.
- **D** They live in hives near caves in Thailand.

2. Which of the following statements describes distinctive features of hog-nosed bats?
- **F** The bats are very small and eat leaves.
- **G** The bats live in caves and have a tail.
- **H** The bats live in Thailand and are mammals.
- **I** The bats have a piglike muzzle and no tail.

3. From the information in this passage, which conclusion is most likely to be correct?
- **A** Hog-nosed bats are similar to other bats.
- **B** Hog-nosed bats are probably rare.
- **C** Hog-nosed bats can sting like a bumblebee.
- **D** Hog-nosed bats probably eat fruit.

Passage 2

1. B
2. I
3. B

Question 1: Students may select incorrect answer A if they misread "world's smallest mammal" as being "world's smallest animal."

Question 3: Students may select incorrect answer A if they overlook information in the passage that describes how hog-nosed bats are both similar to and different from other bats.

INTERPRETING GRAPHICS

The diagrams below show two kinds of cells. Use these cell diagrams to answer the questions that follow.

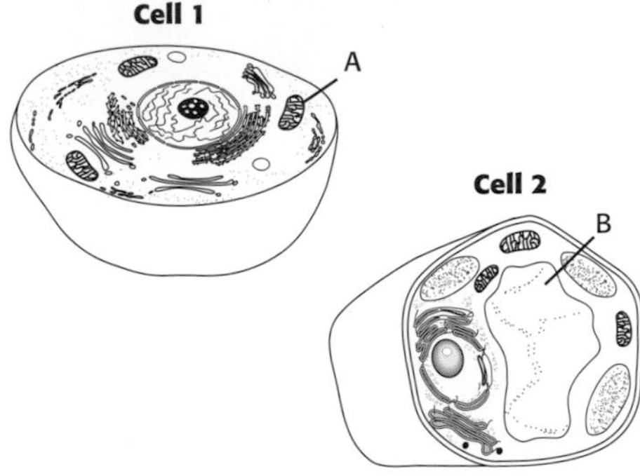

1. What is the name of the organelle labeled *A* in Cell 1?
   - **A** endoplasmic reticulum
   - **B** mitochondrion
   - **C** vacuole
   - **D** nucleus

2. What type of cell is Cell 1?
   - **F** a bacterial cell
   - **G** a plant cell
   - **H** an animal cell
   - **I** a prokaryotic cell

3. What is the name and function of the organelle labeled *B* in Cell 2?
   - **A** The organelle is a vacuole, and it stores water and other materials.
   - **B** The organelle is the nucleus, and it contains the DNA.
   - **C** The organelle is the cell wall, and it gives shape to the cell.
   - **D** The organelle is a ribosome, where proteins are put together.

4. What type of cell is Cell 2? How do you know?
   - **F** prokaryotic; because it does not have a nucleus
   - **G** eukaryotic; because it does not have a nucleus
   - **H** prokaryotic; because it has a nucleus
   - **I** eukaryotic; because it has a nucleus

MATH

Read each question below, and choose the best answer.

1. What is the surface area–to-volume ratio of the rectangular solid shown in the diagram below?

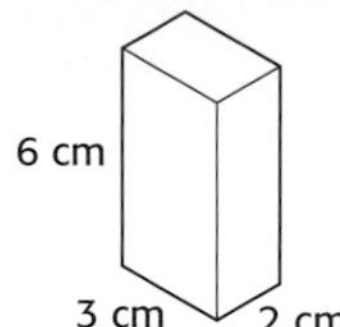

   - **A** 0.5:1
   - **B** 2:1
   - **C** 36:1
   - **D** 72:1

2. Look at the diagram of the cell below. Three molecules of food per cubic unit of volume per minute are required for the cell to survive. One molecule of food can enter through each square unit of surface area per minute. What will happen to this cell?

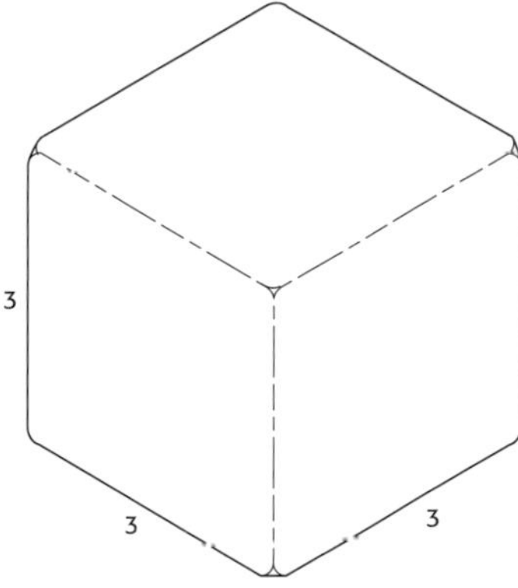

   - **F** The cell is too small, and it will starve.
   - **G** The cell is too large, and it will starve.
   - **H** The cell is at a size that will allow it to survive.
   - **I** There is not enough information to determine the answer.

INTERPRETING GRAPHICS

1. B
2. H
3. A
4. I

Question 2: The cell is not a bacterium or a prokaryotic cell because it has a nucleus, and it is not a plant cell because it has no cell wall.

Question 4: The cell has a nucleus, and only eukaryotic cells have a nucleus. Answer I is the only answer that has that combination of facts.

MATH

1. D
2. G

Question 2: When students calculate the cell's surface area–to-volume ratio, they will find that it is 2.33:1. Therefore, only 2.33 food molecules can enter per minute. Because the cell needs 3 molecules of food per minute, the cell is too large, and it will starve. It may help students to understand how the surface area–to-volume ratio affects survival by showing students a variety of three-dimensional models.

CHAPTER RESOURCES

Chapter Resource File

 • **Standardized Test Preparation** GENERAL

State Resources

 For specific resources for your state, visit **go.hrw.com** and type in the keyword **HSMSTR.**

Weird Science

Background

Within the last 20 years, biologists' ideas of which environments would be suitable for life have increased dramatically. Discoveries of organisms that live under extreme conditions of temperature and pressure have led to the use of the word *extremophile* for these life-forms.

Scientific Discoveries

Background

It is important to note that stem cells are very different from other cells. All stem cells—embryonic or adult—have three unique characteristics. First, stem cells are unspecialized. So, a stem cell in its original form cannot perform the function of a muscle cell or a blood cell. But these unspecialized cells can give rise to specialized cells and then can perform these functions. Second, stem cells can divide and renew for long periods of time. Scientists have shown that as long as a stem cell remains unspecialized, it can continue to divide for an extended period of time. Third, stem cells can evolve into specialized cells through the process of differentiation.

Science in Action

Scientific Discoveries

Discovery of the Stem Cell

What do Parkinson's disease, diabetes, aplastic anemia, and Alzheimer's disease have in common? All of these diseases are diseases for which stem cells may provide treatment or a cure. Stem cells are unspecialized cells from which all other kinds of cells can grow. And research on stem cells has been going on almost since microscopes were invented. But scientists have been able to culture, or grow, stem cells in laboratories for only about the last 20 years. Research during these 20 years has shown scientists that stem cells can be useful in treating—and possibly curing—a variety of diseases.

Language Arts ACTIVITY

WRITING SKILL Imagine that you are a doctor who treats diseases such as Parkinson's disease. Design and create a pamphlet or brochure that you could use to explain what stem cells are. Include in your pamphlet a description of how stem cells might be used to treat one of your patients who has Parkinson's disease. Be sure to include information about Parkinson's disease.

Weird Science

Extremophiles

Are there organisms on Earth that can give scientists clues about possible life elsewhere? Yes, there are! These organisms are called *extremophiles,* and they live where the environment is extreme. For example, some extremophiles live in the hot volcanic thermal vents deep in the ocean. Other extremophiles live in the extreme cold of Antarctica. But these organisms do not live only in extreme environments. Research shows that extremophiles may be abundant in plankton in the ocean. And not all extremophiles are archaebacteria; some extremophiles are eubacteria.

Social Studies ACTIVITY

Choose one of the four types of extremophiles. Do some research about the organism you have chosen and make a poster showing what you learned about it, including where it can be found, under what conditions it lives, how it survives, and how it is used.

Answer to Social Studies Activity

Students' posters should reflect the research they have done. For example, a student who chooses methanogens may show that these extremophiles live in a wide variety of places and in a large number of geographical locations. The poster may also include an explanation of how these organisms get nutrients and how the metabolism of methanogens is different from human metabolism. Students should also show any commercial, industrial, or medical uses of whichever organism they have chosen.

Answer to Language Arts Activity

Students' pamphlets or brochures should present a basic explanation of what stem cells are, where they come from, why they are useful, and how they may be used specifically to treat Parkinson's disease. So, the student will also have to include a little information about Parkinson's disease.

People in Science

Caroline Schooley

Microscopist Imagine that your assignment is the following: Go outside. Look at 1 ft^2 of the ground for 30 min. Make notes about what you observe. Be prepared to describe what you see. If you look at the ground with just your naked eyes, you may quickly run out of things to see. But what would happen if you used a microscope to look? How much more would you be able to see? And how much more would you have to talk about? Caroline Schooley could tell you.

Caroline Schooley joined a science club in middle school. That's when her interest in looking at things through a microscope began. Since then, Schooley has spent many years studying life through a microscope. She is a microscopist. A *microscopist* is someone who uses a microscope to look at small things. Microscopists use their tools to explore the world of small things that cannot be seen by the naked eye. And with today's powerful electron microscopes, microscopists can study things we could never see before, things as small as atoms.

Math Activity

An average bacterium is about 0.000002 m long. A pencil point is about 0.001 m wide. Approximately how many bacteria would fit on a pencil point?

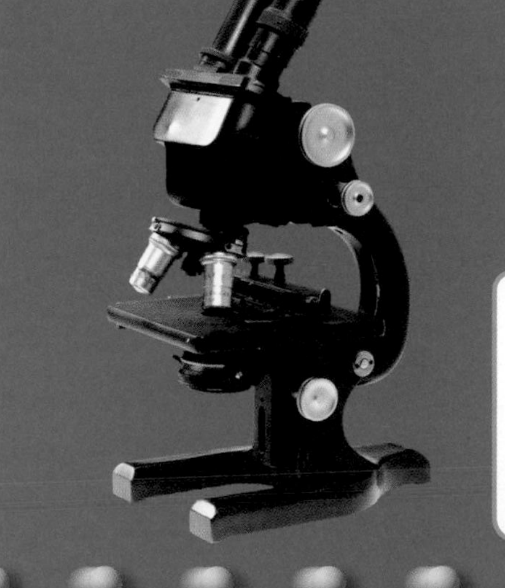

To learn more about these Science in Action topics, visit **go.hrw.com** and type in the keyword **HL5CELF.**

Current Science

Check out Current Science® articles related to this chapter by visiting go.hrw.com. Just type in the keyword HL5CS03.

Answer to Math Activity

0.001 m (size of pencil point) ÷ 0.000002 m (size of bacteria) = 500

So, approximately 500 bacteria could fit on a pencil point.

People in Science

Background

Caroline Schooley wants students to think about microscopes and microscopy. The field is changing. One of the newest uses of microscopy is in nanotechnology. *Nanotechnology* is the science of manipulating materials on an atomic or molecular level to build microscopic devices. To do so, scientists will develop tiny machines (called *assemblers*) that can manipulate atoms and molecules as directed. Tiny nanomachines (called *replicators*) will be then programmed to build more assemblers. Nanotechnology can be thought of as molecular manufacturing.

The Cell in Action

Chapter Planning Guide

Compression guide:
To shorten instruction because of time limitations, omit the Chapter Lab.

OBJECTIVES	LABS, DEMONSTRATIONS, AND ACTIVITIES	TECHNOLOGY RESOURCES
PACING • 135 min pp. 32–37 **Chapter Opener**	**SE Start-up Activity,** p. 33 ◆ GENERAL	**OSP Parent Letter** ■ GENERAL **CD Student Edition on CD-ROM** **CD Guided Reading Audio CD** ■ **TR Chapter Starter Transparency*** **VID Brain Food Video Quiz**
Section 1 Exchange with the Environment • Explain the process of diffusion. • Describe how osmosis occurs. • Compare passive transport with active transport. • Explain how large particles get into and out of cells.	**TE Demonstration** Membrane Model, p. 34 ◆ GENERAL **SE Quick Lab** Bead Diffusion, p. 35 ◆ GENERAL **CRF Datasheet for Quick Lab*** **TE Demonstration** Crossing Membranes, p. 35 ◆ GENERAL **SE Inquiry Labs** The Perfect Taters Mystery, p. 46 ◆ GENERAL **CRF Datasheet for Chapter Lab*** **LB Inquiry Labs** Fish Farms in Space* ◆ GENERAL **LB Whiz-Bang Demonstrations** It's in the Bag!* ◆ BASIC	**CRF Lesson Plans*** **TR Bellringer Transparency*** **TR** Passive Transport and Active Transport* **TR** Endocytosis and Exocytosis* **CRF SciLinks Activity*** GENERAL **VID Lab Videos for Life Science** **CD Interactive Explorations CD-ROM** The Nose Knows GENERAL
PACING • 45 min pp. 38–41 **Section 2 Cell Energy** • Describe photosynthesis and cellular respiration. • Compare cellular respiration with fermentation.	**TE Demonstration** Leaves and Light, p. 38 GENERAL **SE Connection to Chemistry** Earth's Early Atmosphere, p. 39 GENERAL **TE Group Activity** Recycling Carbon, p. 39 GENERAL **TE Group Activity** Photosynthesis and Cellular Respiration, p. 40 GENERAL **SE Skills Practice Lab** Stayin' Alive!, p. 189 GENERAL **CRF Datasheet for LabBook***	**CRF Lesson Plans*** **TR Bellringer Transparency*** **TR** The Connection Between Photosynthesis and Respiration* **TR** ***LINK TO PHYSICAL SCIENCE*** Solar Heating Systems*
PACING • 45 min pp. 42–45 **Section 3 The Cell Cycle** • Explain how cells produce more cells. • Describe the process of mitosis. • Explain how cell division differs in animals and plants.	**TE Activity** Making Models, p. 42 GENERAL **SE Connection to Language Arts** Picking Apart Vocabulary, p. 43 GENERAL **TE Connection Activity** Math, p. 43 ADVANCED **TE Activity** Four Phases of Mitosis, p. 45 ADVANCED **LB Labs You Can Eat** The Mystery of the Runny Gelatin* ◆ GENERAL **LB Whiz-Bang Demonstrations** Stop Picking on My Enzyme* ◆ BASIC **LB Long-Term Projects & Research Ideas** Taming the Wild Yeast* ◆ ADVANCED **SE Science in Action** Math, Social Studies, and Language Arts Activities, pp. 52–53 GENERAL	**CRF Lesson Plans*** **TR Bellringer Transparency*** **TR** The Cell Cycle* **TE Internet Activity,** p. 44 GENERAL

PACING • 90 min

CHAPTER REVIEW, ASSESSMENT, AND STANDARDIZED TEST PREPARATION

CRF Vocabulary Activity* GENERAL
SE Chapter Review, pp. 48–49 GENERAL
CRF Chapter Review* ■ GENERAL
CRF Chapter Tests A* ■ GENERAL, **B*** ADVANCED, **C*** SPECIAL NEEDS
SE Standardized Test Preparation, pp. 50–51 GENERAL
CRF Standardized Test Preparation* GENERAL
CRF Performance-Based Assessment* GENERAL
OSP Test Generator GENERAL
CRF Test Item Listing* GENERAL

Online and Technology Resources

Visit **go.hrw.com** for a variety of free resources related to this textbook. Enter the keyword **HL5ACT.**

Students can access interactive problem-solving help and active visual concept development with the *Holt Science and Technology* Online Edition available at **www.hrw.com.**

Guided Reading Audio CD

A direct reading of each chapter using instructional visuals as guideposts. For auditory learners, reluctant readers, and Spanish-speaking students. Available in English and Spanish.

KEY			
SE Student Edition	**CRF** Chapter Resource File	**SS** Science Skills Worksheets	* Also on One-Stop Planner
TE Teacher Edition	**OSP** One-Stop Planner	**MS** Math Skills for Science Worksheets	◆ Requires advance prep
	LB Lab Bank	**CD** CD or CD-ROM	■ Also available in Spanish
	TR Transparencies	**VID** Classroom Video/DVD	

SKILLS DEVELOPMENT RESOURCES	SECTION REVIEW AND ASSESSMENT	STANDARDS CORRELATIONS
SE Pre-Reading Activity, p. 32 GENERAL **OSP Science Puzzlers, Twisters & Teasers** GENERAL		National Science Education Standards UCP 3; SAI 1; SPSP 5; LS 1b
CRF Directed Reading A* ■ BASIC, **B*** SPECIAL NEEDS **CRF Vocabulary and Section Summary*** ■ GENERAL **SE Reading Strategy** Reading Organizer, p. 34 GENERAL **TE Inclusion Strategies,** p. 35 **CRF Reinforcement Worksheet** Into and Out of the Cell* BASIC **SS Science Skills** Doing a Lab Write-Up* BASIC **SS Science Skills** Taking Notes* BASIC **MS Math Skills for Science** Multiplying Whole Numbers* BASIC **MS Math Skills for Science** Dividing Whole Numbers with Long Division* BASIC	**SE Reading Checks,** pp. 35, 37 GENERAL **TE Reteaching,** p. 36 BASIC **TE Quiz,** p. 36 GENERAL **TE Alternative Assessment,** p. 36 ADVANCED **TE Homework,** p. 36 GENERAL **SE Section Review,*** p. 37 ■ GENERAL **CRF Section Quiz*** ■ GENERAL	UCP 1, 2, 3, 4; LS 1c; *Chapter Lab:* UCP 2; SAI 1
CRF Directed Reading A* ■ BASIC, **B*** SPECIAL NEEDS **CRF Vocabulary and Section Summary*** ■ GENERAL **SE Reading Strategy** Discussion, p. 38 GENERAL **CRF Reinforcement Worksheet** Activities of the Cell* BASIC **CRF Critical Thinking** A Celluloid Thriller* ADVANCED **SS Science Skills** Using Logic* BASIC **SS Science Skills** Identifying Bias* GENERAL	**SE Reading Checks,** pp. 39, 41 GENERAL **TE Reteaching,** p. 40 BASIC **TE Quiz,** p. 40 GENERAL **TE Alternative Assessment,** p. 40 GENERAL **TE Homework,** p. 40 ADVANCED **SE Section Review,*** p. 41 ■ GENERAL **CRF Section Quiz*** ■ GENERAL	UCP 1, 3, 4, 5; SAI 2; ST 2; SPSP 4; LS 1c, 4c; *LabBook:* SAI 1; SPSP 1; LS 1c
CRF Directed Reading A* ■ BASIC, **B*** SPECIAL NEEDS **CRF Vocabulary and Section Summary*** ■ GENERAL **SE Reading Strategy** Paired Summarizing, p. 42 GENERAL **TE Connection to Math** Cell Multiplication, p. 42 BASIC **TE Reading Strategy** Prediction Guide, p. 43 GENERAL **TE Inclusion Strategies,** p. 44 **MS Math Skills for Science** Multiplying Whole Numbers* GENERAL **MS Math Skills for Science** Grasping Graphing* GENERAL **SS Science Skills** Organizing Your Research* GENERAL **SS Science Skills** Researching on the Web* BASIC **CRF Reinforcement Worksheet** This Is Radio KCEL* BASIC	**SE Reading Checks,** pp. 43, 44 GENERAL **TE Reteaching,** p. 44 BASIC **TE Quiz,** p. 44 GENERAL **TE Alternative Assessment,** p. 44 GENERAL **SE Section Review,*** p. 45 ■ GENERAL **CRF Section Quiz*** ■ GENERAL	SAI 1; LS 1c, 2d

This convenient CD-ROM includes:

- **Lab Materials QuickList Software**
- **Holt Calendar Planner**
- **Customizable Lesson Plans**
- **Printable Worksheets**
- **ExamView® Test Generator**

CNN Student News

cnnstudentnews.com

Find the latest news, lesson plans, and activities related to important scientific events.

www.scilinks.org

Maintained by the **National Science Teachers Association.** See Chapter Enrichment pages for a complete list of topics.

Current Science®

Check out ***Current Science*** articles and activities by visiting the HRW Web site at **go.hrw.com.** Just type in the keyword **HL5CS04T**.

Classroom Videos

- **Lab Videos** demonstrate the chapter lab.
- **Brain Food Video Quizzes** help students review the chapter material.

Chapter Resources

Visual Resources

CHAPTER STARTER TRANSPARENCY

The Cell in Action — CHAPTER STARTER

What If . . . ?

How long would you like to live? What if you could live to be 120 years old? or 150 and beyond? Since ancient times, people have searched in vain for a magical fountain or potion that could give them eternal youth. No one has yet found the secret of immortality, but scientists have recently made a startling discovery that may help extend people's lives.

In January of 1998, researchers at the University of Texas reported that they had found an enzyme in the body that acts like a "cellular fountain of youth." In the laboratory, the enzyme enables human cells to stay young and multiply long past the time when cells would normally stop dividing and die. Researchers hope that the enzyme can someday be used to understand and treat certain cancers and other incurable diseases. Although the so-called immortalizing enzyme won't help people live forever, it may help them live longer, healthier lives.

Every living thing is made of cells. In this chapter you will learn how cells grow and how they make more cells. You will also learn how cells transport materials and obtain the energy they need to survive.

BELLRINGER TRANSPARENCIES

The Cell in Action — BELLRINGER TRANSPARENCY

Section: Exchange with the Environment
Which of the following best describes a living cell:
a) building block
b) a living organism
c) a complex factory
d) all of the above

Write a paragraph in your **science journal** defending your choice.

Section: Cell Energy
Make a list of all the different types of cells that you can think of and the jobs they do. Then make a list of all the reasons that a cell needs energy.

Write your answers in your **science journal.**

TEACHING TRANSPARENCIES

The Cell in Action — Passive and Active Transport — TEACHING TRANSPARENCY

Passive transport
Cell membrane
ATP Energy
Active transport

The Cell in Action — TEACHING TRANSPARENCY

The Connection Between Photosynthesis and Respiration

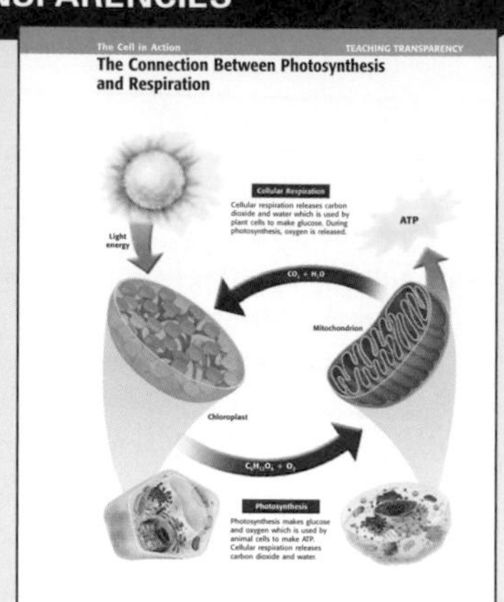

TEACHING TRANSPARENCIES

The Cell in Action — Endocytosis — TEACHING TRANSPARENCY

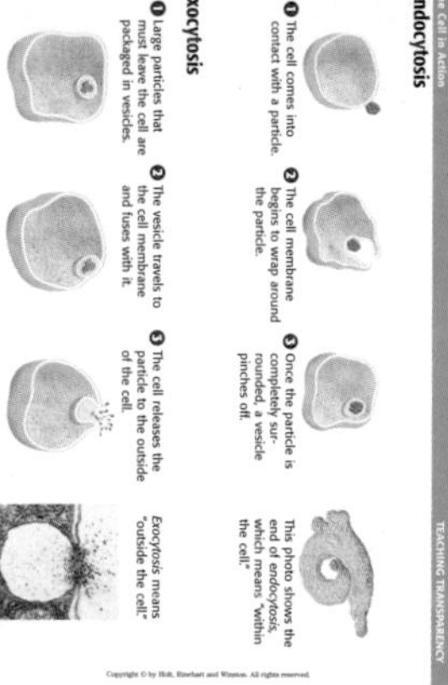

The Cell in Action — TEACHING TRANSPARENCY

The Cell Cycle

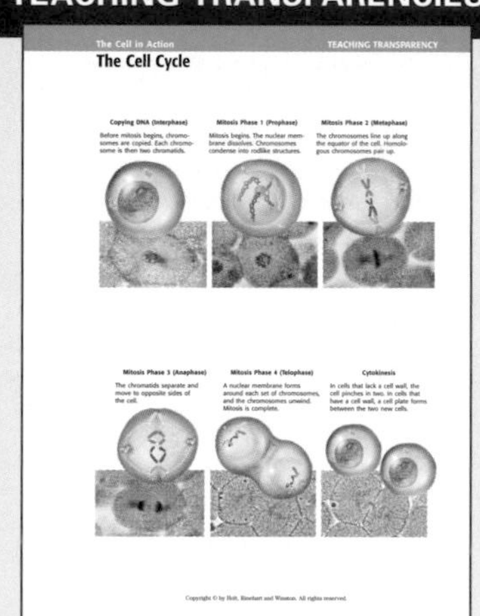

Heat and Heat Technology — Solar Heating Systems — TEACHING TRANSPARENCY

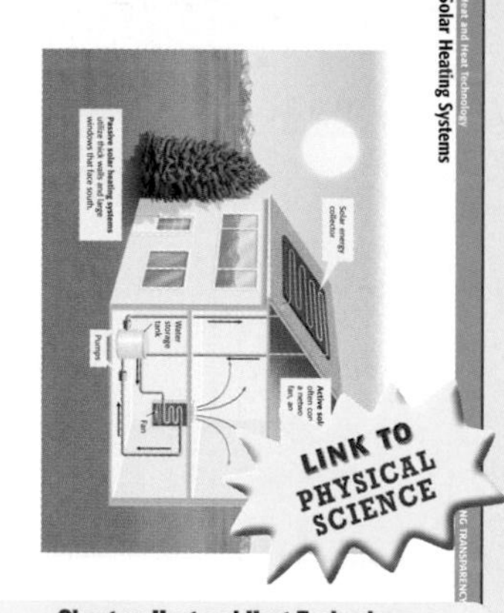

Chapter: Heat and Heat Technology

CONCEPT MAPPING TRANSPARENCY

The Cell in Action — CONCEPT MAPPING TRANSPARENCY

Use the following terms to complete the concept map below:
ATP, photosynthesis, oxygen, water, consumers, lactic acid, producers, respiration, energy

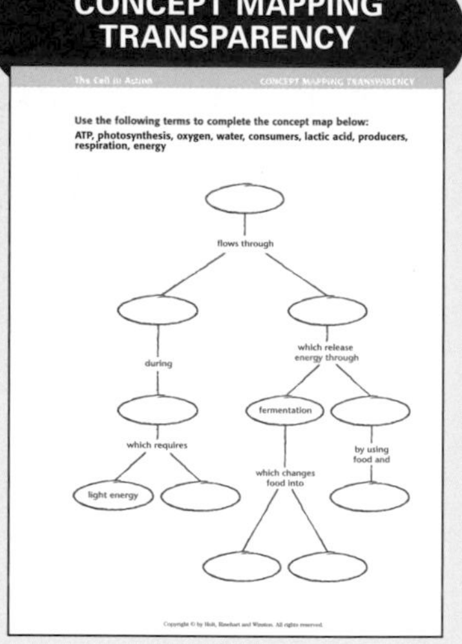

Planning Resources

LESSON PLANS

Lesson Plan SAMPLE

Section: Waves

Pacing
Regular Schedule: with lab(s):2 days without lab(s)2 days
Block Schedule: with lab(s):1 1/2 days without lab(s)1 day

Objectives
1. Relate the seven properties of life to a living organism.
2. Describe seven themes that can help you to organize what you learn about biology.
3. Identify the tiny structures that make up all living organisms.
4. Differentiate between reproduction and heredity and between metabolism and homeostasis.

National Science Education Standards Covered

LSInter6:Cells have particular structures that underlie their functions.

LSMat1:Most cell functions involve chemical reactions.

LSBeh1:Cells store and use information to guide their functions.

UCP1:Cell functions are regulated.

SI1: Cells can differentiate and form complete multicellular organisms.

PS1:Species evolve over time.

ESS1: The great diversity of organisms is the result of more than 3.5 billion years of evolution.

ESS2: Natural selection and its evolutionary consequences provide a scientific explanation for the fossil record of ancient life forms as well as for the striking molecular similarities observed among the diverse species of living organisms.

ST1: The millions of different species of plants, animals, and microorganisms that live on Earth today are related by descent from common ancestors.

ST2: The energy for life primarily comes from the sun.

SPSP1: The complexity and organization of organisms accommodates the need for obtaining, transforming, transporting, releasing, and eliminating the matter and energy used to sustain the organism.

SPSP6: As matter and energy flows through different levels of organization of living systems—cells, organs, communities—and between living systems and the physical environment, chemical elements are recombined in different ways.

HNS1: Organisms have behavioral responses to internal changes and to external stimuli.

PARENT LETTER

SAMPLE

Dear Parent,

Your son's or daughter's science class will soon begin exploring the chapter entitled "The World of Physical Science." In this chapter, students will learn about how the scientific method applies to the world of physical science and the role of physical science in the world. By the end of the chapter, students should demonstrate a clear understanding of the chapter's main ideas and be able to discuss the following topics:

1. physical science as the study of energy and matter (Section 1)
2. the role of physical science in the world around them (Section 1)
3. careers that rely on physical science (Section 1)
4. the steps used in the scientific method (Section 2)
5. examples of technology (Section 2)
6. how the scientific method is used to answer questions and solve problems (Section 2)
7. how our knowledge of science changes over time (Section 2)
8. how models represent real objects or systems (Section 3)
9. examples of different ways models are used in science (Section 3)
10. the importance of the International System of Units (Section 4)
11. the appropriate units to use for particular measurements (Section 4)
12. how area and density are derived quantities (Section 4)

Questions to Ask Along the Way

You can help your son or daughter learn about these topics by asking interesting questions such as the following:

- What are some surprising careers that use physical science?
- What is a characteristic of a good hypothesis?
- When is it a good idea to use a model?
- Why do Americans measure things in terms of inches and yards and meters ?

ALSO IN SPANISH

TEST ITEM LISTING

TEST ITEM LISTING
The World of Science SAMPLE

MULTIPLE CHOICE

1. A limitation of models is that
   a. they are large enough to see.
   b. they do not act exactly like the things that they model.
   c. they are smaller than the things that they model.
   d. they model unfamiliar things.
   Answer: B Difficulty: I Section: 3 Objective: 2
2. The length 10 m is equal to
   a. 100 cm. c. 10,000 mm.
   b. 1,000 cm. d. Both (b) and (c)
   Answer: B Difficulty: I Section: 3 Objective: 2
3. To be valid, a hypothesis must be
   a. testable. c. made into a law.
   b. supported by evidence. d. Both (a) and (b)
   Answer: B Difficulty: I Section: 3 Objective: 2 1
4. The statement "Sheila has a stain on her shirt" is an example of a(n)
   a. law. c. observation.
   b. hypothesis. d. prediction.
   Answer: B Difficulty: I Section: 3 Objective: 2
5. A hypothesis is often developed out of
   a. observations. c. laws.
   b. experiments. d. Both (a) and (b)
   Answer: B Difficulty: I Section: 3 Objective: 2
6. How many milliliters are in 3.5 kL?
   a. 3,500 mL c. 3,500, 000 mL
   b. 0.0035 mL d. 35,000 mL
   Answer: B Difficulty: I Section: 3 Objective: 2
7. A map of Seattle is an example of a
   a. law. c. model.
   b. theory. d. unit.
   Answer: B Difficulty: I Section: 3 Objective: 2
8. A lab has the safety icons shown below. These icons mean that you should wear
   a. only safety goggles. c. safety goggles and a lab apron.
   b. only a lab apron. d. safety goggles, a lab apron, and gloves.
   Answer: B Difficulty: I Section: 3 Objective: 2
9. The law of conservation of mass says the tot al mass before a chemical change is
   a. more than the total mass after the change.
   b. less than the total mass after the change.
   c. the same as the total mass after the change.
   d. not the same as the total mass after the change.
   Answer: B Difficulty: I Section: 3 Objective: 2
10. In which of the following areas might you find a geochemist at work?
    a. studying the chemistry of rocks c. studying fishes
    b. studying forestry d. studying the atmosphere
    Answer: B Difficulty: I Section: 3 Objective: 2

One-Stop Planner® CD-ROM

This CD-ROM includes all of the resources shown here and the following time-saving tools:

- *Lab Materials QuickList Software*
- *Customizable lesson plans*
- *Holt Calendar Planner*
- *The powerful ExamView® Test Generator*

For a preview of available worksheets covering math and science skills, see pages T12–T19. All of these resources are also on the One-Stop Planner®.

Meeting Individual Needs

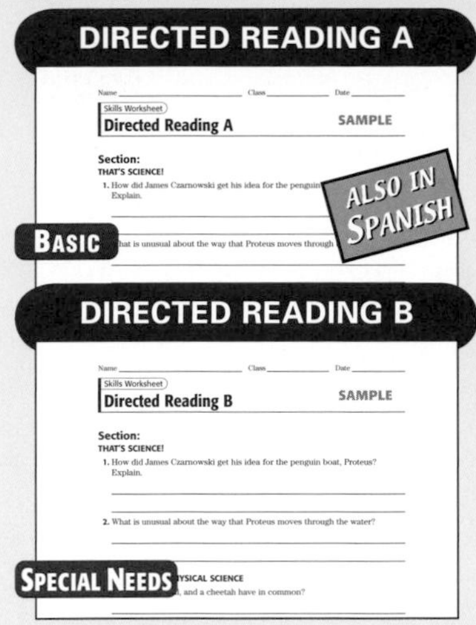

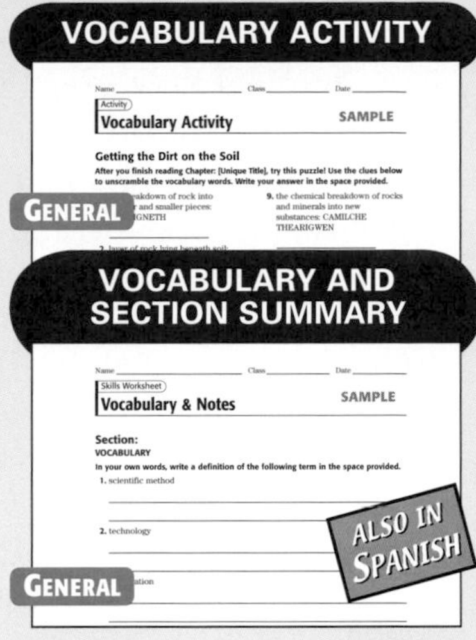

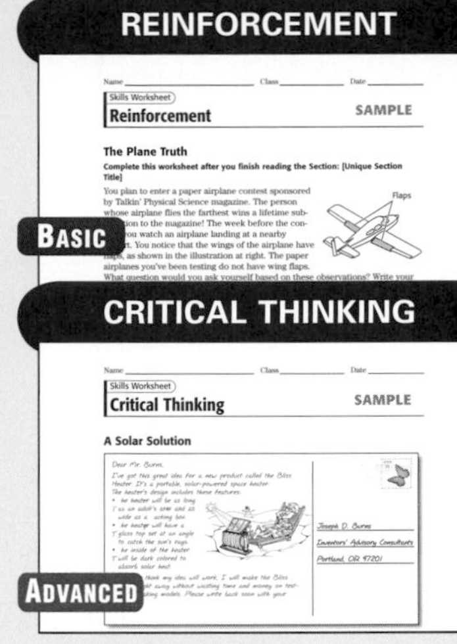

Labs and Activities

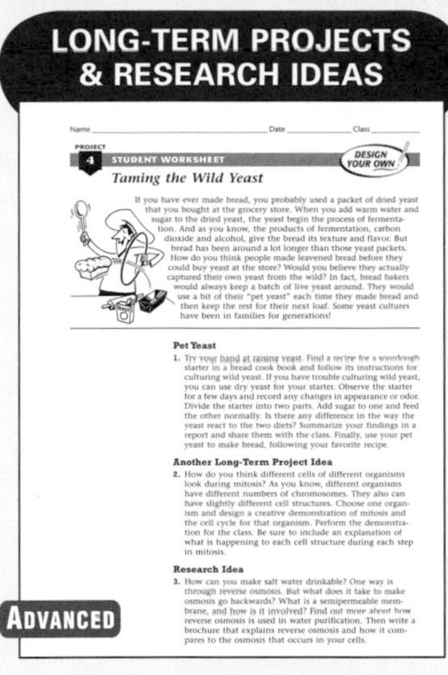

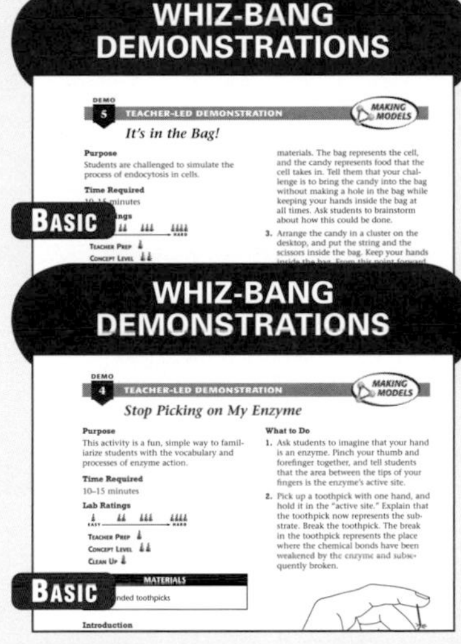

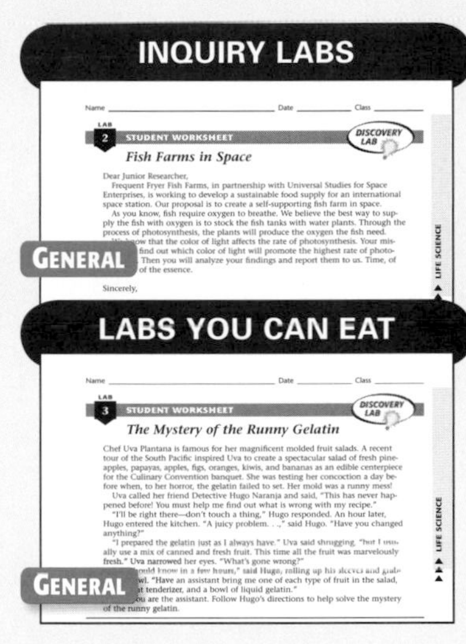

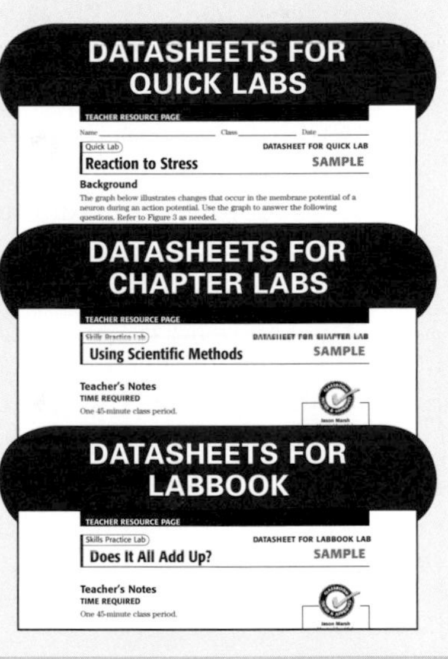

Review and Assessments

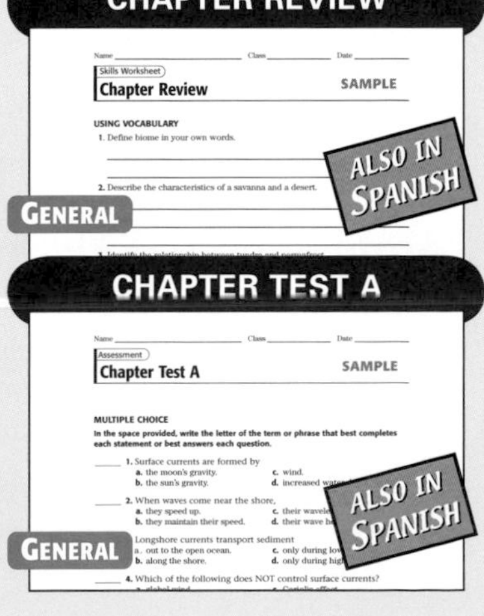

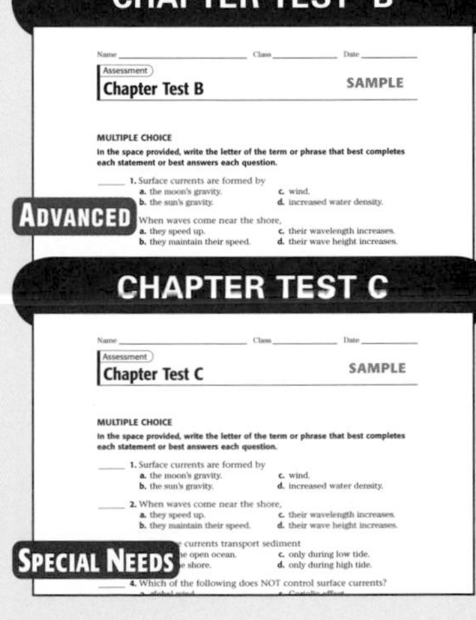

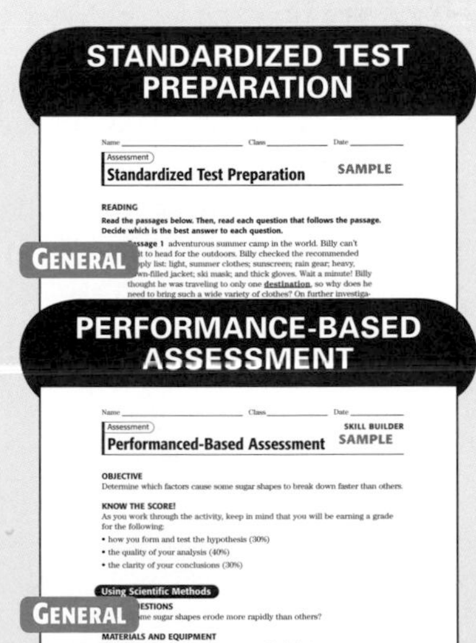

2 Chapter Enrichment

This Chapter Enrichment provides relevant and interesting information to expand and enhance your presentation of the chapter material.

Section 1

Exchange with the Environment

Endocytosis

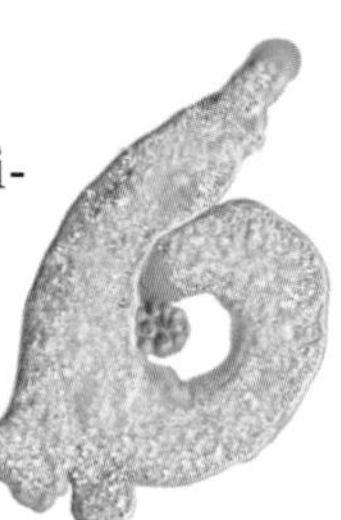

- There are three different mechanisms of endocytosis: phagocytosis, receptor-mediated endocytosis, and pinocytosis. These processes allow a substance to enter a cell without passing through the cell membrane. The substance involved determines which method is used.
- Large particles such as bacteria enter the cell by phagocytosis. The host cell changes shape, and the membrane sends out projections called *pseudopods,* meaning "false feet," which surround the particle, bringing it inside the cell.
- In receptor-mediated endocytosis, receptors on the membrane that are specific for a given substance bind to the substance before the endocytotic process begins. This method is used during cholesterol metabolism.
- In pinocytosis, the cell membrane surrounds the substance and forms a vesicle to bring the material into the cell. Pinocytosis usually involves material that is dissolved in water.

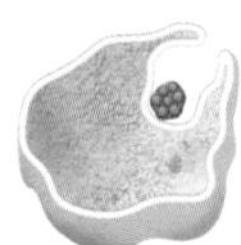
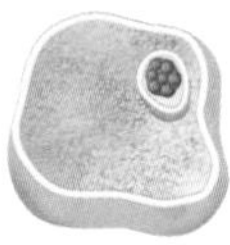

Reverse Osmosis

- Reverse osmosis is a process that forces water across semipermeable membranes under high pressure. The high pressure reverses the natural tendency of the solutes on the more concentrated side of the membrane to pass through to the less-concentrated side. In this way, water passing through the membrane is purified.

Is That a Fact!

- ◆ The largest single-celled organism that ever lived was a protozoan that measured 20 cm in diameter. It is now extinct.

Section 2

Cell Energy

Early Plant Scientists

- Jan Baptista Van Helmont (1580–1644) was a Belgian chemist, physiologist, and physician who coined the word *gas.* Van Helmont was the first scientist to comprehend the existence of gases separate from the atmospheric air. Although he didn't know that it was carbon dioxide, van Helmont stated that the *spiritus sylvestre,* or "wild spirit," emitted by burning charcoal was the same as that given off by fermenting grape juice. He applied chemistry to the study of physiological processes, and for this he is known as the "father of biochemistry."

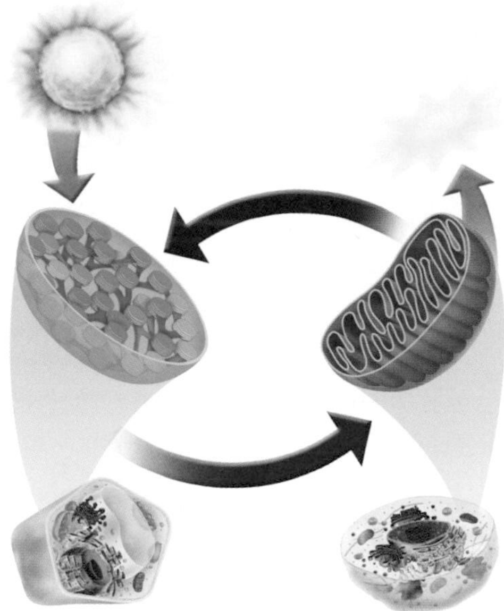

- Joseph Priestley (1733–1804) was an English clergyman and physical scientist who was one of the discoverers of oxygen. He also observed that light was vital for plant growth and that green leaves released oxygen.
- Jan Ingenhousz (1730–1799), a Dutch-born British physician and scientist, discovered photosynthesis.

Carotenoids and Photosynthesis

- Carotenoids are responsible for the orange colors in plants. Carotenoids are usually masked by chlorophyll. They are sensitive to wavelengths of light to which chlorophyll cannot respond. Carotenoids can absorb the light waves and transfer the energy to chlorophyll, which then incorporates that energy into the photosynthetic pathway.

For background information about teaching strategies and issues, refer to the ***Professional Reference for Teachers.***

Section 3

The Cell Cycle

Cytogenetics

- Cytogeneticists study the role of human chromosomes in health and disease. Chromosome studies can reveal abnormalities such as whether a person is carrying the genetic material for a genetically linked disease.

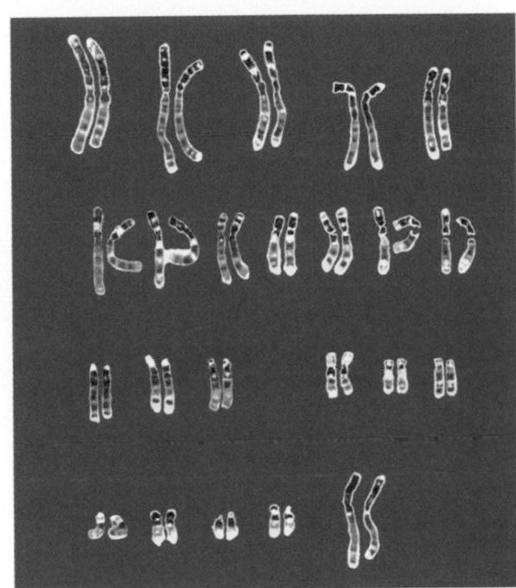

Cell Division

- The frequency of cell division varies a great deal. Fruit-fly embryo cells divide about every eight minutes. Human liver cells may not divide for up to one year. Scientists are still trying to determine what orchestrates growth and regulates cell division. This information would help scientists understand diseases of unregulated cell division, such as cancer.
- DNA and chromosomes are related but are not the same thing. A chromosome is made up of DNA that has been wound up and organized with proteins that hold it all together. For much of the cell cycle, DNA is loose and not very visible.

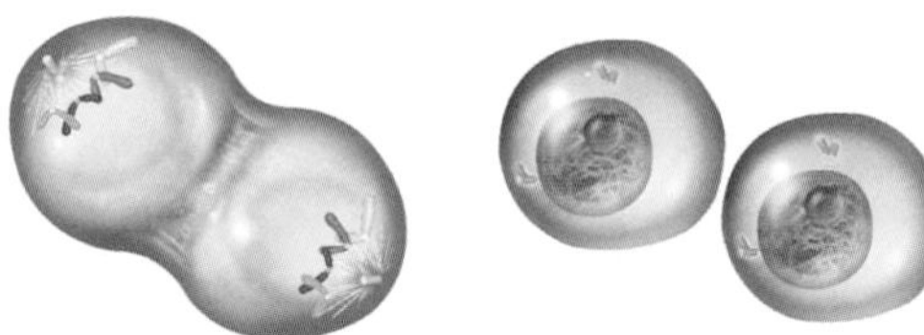

Is That a Fact!

◆ In an adult human body, cell division happens at least 10 million times every second.

Cell Adhesion

- Blood cells exist individually in the body, but most other cells are connected to each other. Usually this involves special adhesion proteins, such as adherins, cadherins, catenins, and integrins. These proteins connect adjoining cells by physically locking the cells together, fastening one cell to the next. Sometimes these junctions are outside the cell, and sometimes they are inside. Adhesion proteins can span the cell membranes and connect the inside of one cell to the inside of its neighbor cell.

Is That a Fact!

◆ In a healthy body, cells reproduce at exactly the same rate at which cells die. However, some agents make cells reproduce uncontrollably, causing a disease known as cancer. One of these carcinogenic agents is ultraviolet radiation, which is emitted by the sun and ultraviolet lamps. People who spend excessive amounts of time in the sun run the risk of developing skin cancer.

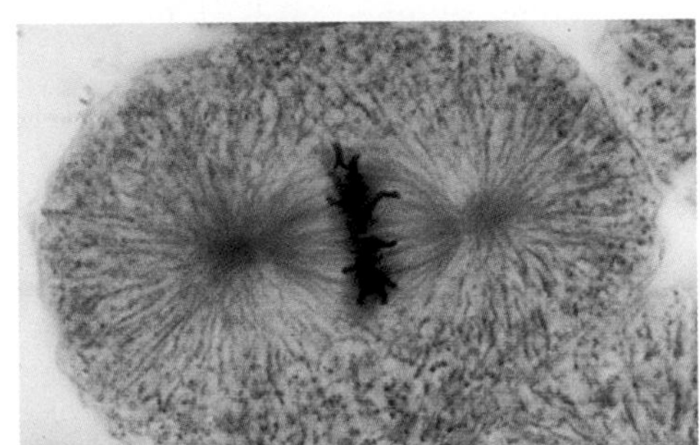

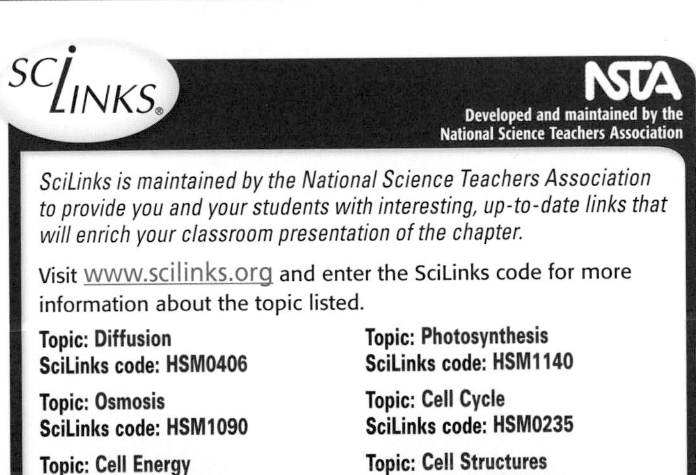

SciLinks is maintained by the National Science Teachers Association to provide you and your students with interesting, up-to-date links that will enrich your classroom presentation of the chapter.

Visit www.scilinks.org and enter the SciLinks code for more information about the topic listed.

Topic: Diffusion
SciLinks code: HSM0406

Topic: Osmosis
SciLinks code: HSM1090

Topic: Cell Energy
SciLinks code: HSM0237

Topic: Photosynthesis
SciLinks code: HSM1140

Topic: Cell Cycle
SciLinks code: HSM0235

Topic: Cell Structures
SciLinks code: HSM0240

Overview

In this chapter, students will learn about how cells interact with their environment, how cells get nutrients and get rid of wastes, and where cells get the energy from to carry out all the activities of life. Students will also learn about how cells produce more cells.

Assessing Prior Knowledge

Students should be familiar with the following topic:

- cells as the basic units of life

Identifying Misconceptions

Students may not think of cells as self-contained units of life. It is important for students to realize that cells, just like multicellular organisms, live in an environment and must perform all the activities—such as taking in nutrients, producing energy, and getting rid of wastes—necessary to stay alive and reproduce. Students may also be confused about the difference between cells and molecules. Emphasize the relationship between cells and molecules, and that proteins, carbohydrates, and other substances are made of molecules. These molecules must be smaller than the cells they enter and leave. Many students believe that proteins and other molecules are bigger than cells.

2

The Cell in Action

About the PHOTO

This adult katydid is emerging from its last immature, or nymph, stage. As the katydid changed from a nymph to an adult, every structure of its body changed. To grow and change, an organism must produce new cells. When a cell divides, it makes a copy of its genetic material.

PRE-READING ACTIVITY

FOLDNOTES **Tri-Fold** Before you read the chapter, create the FoldNote entitled "Tri-Fold" described in the **Study Skills** section of the Appendix. Write what you know about the actions of cells in the column labeled "Know." Then, write what you want to know in the column labeled "Want." As you read the chapter, write what you learn about the actions of cells in the column labeled "Learn."

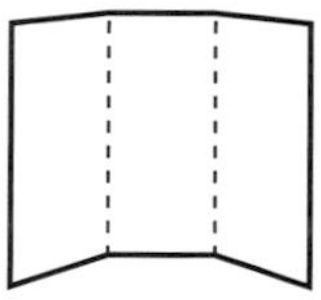

Standards Correlations

National Science Education Standards

The following codes indicate the National Science Education Standards that correlate to this chapter. The full text of the standards is at the front of the book.

Chapter Opener
UCP 3; SAI 1; SPSP 5; LS 1b

Section 1 Exchange with the Environment
UCP 1, 2, 3, 4; LS 1c

Section 2 Cell Energy
UCP 1, 3, 4, 5; SAI 2; ST 2; SPSP 4; LS 1c, 4c; *LabBook:* SAI 1; SPSP1; LS 1c

Section 3 The Cell Cycle
SAI 1; LS 1c, 2d

Chapter Lab
UCP 2; SAI 1

Chapter Review
UCP 1, 2, 3, 4, 5; SAI 1, 2; ST 2; SPSP 4, 5; LS 1b, 1c, 2d, 4c

Science in Action
UCP 1, 2, 3, 5; SAI 1, 2; ST 1, 2; SPSP 5; LS 1d, 1e, 1f

START-UP ACTIVITY

Cells in Action

Yeast are single-celled fungi that are an important ingredient in bread. Yeast cells break down sugar molecules to release energy. In the process, carbon dioxide gas is produced, which causes bread dough to rise.

Procedure

1. Add **4 mL of a sugar solution** to **10 mL of a yeast-and-water mixture**. Use a **stirring rod** to thoroughly mix the two liquids.
2. Pour the stirred mixture into a small test tube.
3. Place a slightly **larger test tube** over the **small test tube.** The top of the small test tube should touch the bottom of the larger test tube.
4. Hold the test tubes together, and quickly turn both test tubes over. Place the test tubes in a test-tube rack.
5. Use a **ruler** to measure the height of the fluid in the large test tube. Wait 20 min, and then measure the height of the liquid again.

Analysis

1. What is the difference between the first height measurement and the second height measurement?
2. What do you think caused the change in the fluid's height?

START-UP ACTIVITY

MATERIALS

For Each Student

- cup, small plastic
- ruler
- stirring rod
- sugar solution
- test tube, large plastic
- test-tube rack
- test tube, small plastic
- yeast-and-water mixture

Safety Caution: Remind students to review all safety cautions and icons before beginning this lab activity. Students should wear safety goggles at all times and wash their hands when they are finished. Students should not taste the solutions.

Teacher's Notes: The yeast suspension is prepared by mixing one package of dry yeast in 250 mL of water. The sugar solution is prepared by dissolving 30 mL (2 tbsp) of sugar in 100 mL of water.

Answers

1. Answers may vary. Students should subtract the first measurement from the second measurement.
2. When the yeast cells released the energy in sugar, the CO_2 that the cells produced increased the volume of air in the smaller tube and pushed more yeast-and-sugar mixture into the larger tube, increasing the height of the liquid in the larger tube.

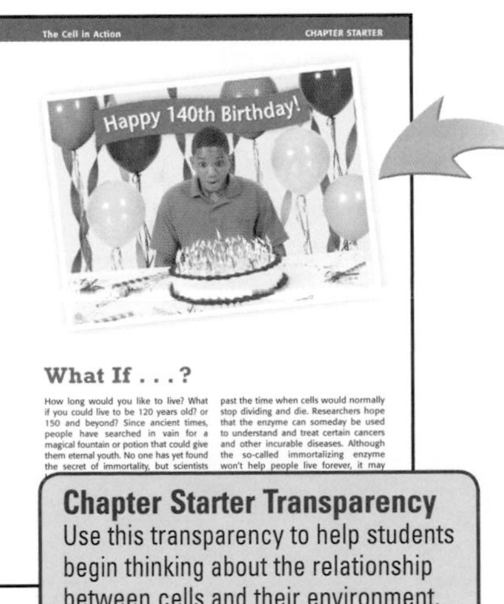

Chapter Starter Transparency
Use this transparency to help students begin thinking about the relationship between cells and their environment.

CHAPTER RESOURCES

Technology

Transparencies
- Chapter Starter Transparency

Student Edition on CD-ROM

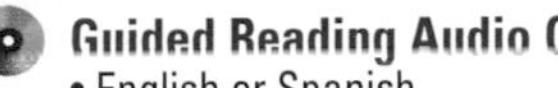

Guided Reading Audio CD
- English or Spanish

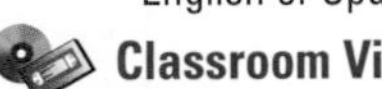

Classroom Videos
- Brain Food Video Quiz

Workbooks

Science Puzzlers, Twisters & Teasers
- The Cell in Action GENERAL

Focus

Overview

This section explains the processes of diffusion and osmosis. Students will compare the passive and active transport of particles into and out of cells.

Bellringer

Write the following on the board:

Which of the following best describes a living cell: a building block, a living organism, a complex factory, or all of the above? Explain your choice.

Motivate

Demonstration — GENERAL

Membrane Model Blow soap bubbles for the class. Explain that soap bubbles have properties, such as flexibility, that are similar to biological membranes. Components of soap film and of cell membranes move around freely. Soap bubbles and membranes are self-sealing. If two bubbles or membranes collide, they fuse. If one is cut in half, two smaller but whole bubbles or membranes form. English Language Learners
LS Visual

SECTION 1

Exchange with the Environment

READING WARM-UP

Objectives

- Explain the process of diffusion.
- Describe how osmosis occurs.
- Compare passive transport with active transport.
- Explain how large particles get into and out of cells.

Terms to Learn

diffusion
osmosis
passive transport
active transport
endocytosis
exocytosis

READING STRATEGY

Reading Organizer As you read this section, make a table comparing active transport and passive transport.

What would happen to a factory if its power were shut off or its supply of raw materials never arrived? What would happen if the factory couldn't get rid of its garbage?

Like a factory, an organism must be able to obtain energy and raw materials and get rid of wastes. An organism's cells perform all of these functions. These functions keep cells healthy so that they can divide. Cell division allows organisms to grow and repair injuries.

The exchange of materials between a cell and its environment takes place at the cell's membrane. To understand how materials move into and out of the cell, you need to know about diffusion.

What Is Diffusion?

What happens if you pour dye on top of a layer of gelatin? At first, it is easy to see where the dye ends and the gelatin begins. But over time, the line between the two layers will blur, as shown in **Figure 1.** Why? Everything, including the gelatin and the dye, is made up of tiny moving particles. Particles travel from where they are crowded to where they are less crowded. This movement from areas of high concentration (crowded) to areas of low concentration (less crowded) is called **diffusion** (di FYOO zhuhn). Dye particles diffuse from where they are crowded (near the top of the glass) to where they are less crowded (in the gelatin). Diffusion also happens within and between living cells. Cells do not need to use energy for diffusion.

diffusion the movement of particles from regions of higher density to regions of lower density

Figure 1 *The particles of the dye and the gelatin slowly mix by diffusion.*

CHAPTER RESOURCES

Chapter Resource File

- Lesson Plan
- Directed Reading A BASIC
- Directed Reading B SPECIAL NEEDS

Technology

Transparencies
- Bellringer

Workbooks

Science Skills
- Doing a Lab Write-up GENERAL
- Taking Notes GENERAL

Math Skills for Science
- Multiplying Whole Numbers BASIC
- Dividing Whole Numbers with Long Division BASIC

CONNECTION to Math — GENERAL

Gas Diffusion Have students solve the following problem in class or as part of their homework:

Gases diffuse about 10,000 times faster in air than in water. If a gas diffuses to fill a room completely in 6 min, how long would it take the gas to fill a similar volume of still water? (60,000 min) How many hours would that be? (1,000 h) How many days? (41.67 days) **LS Logical**

Figure 2 Osmosis

1 The side that holds only pure water has the higher concentration of water particles.

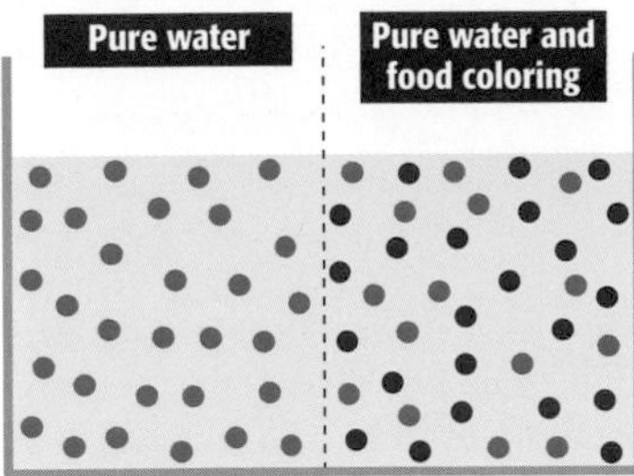

2 During osmosis, water particles move to where they are less concentrated.

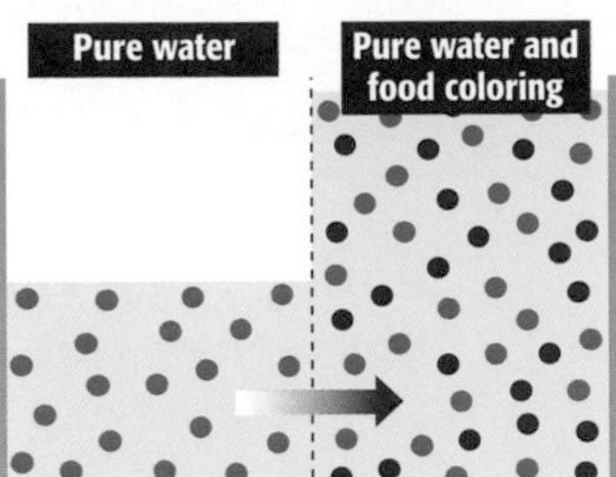

Diffusion of Water

The cells of organisms are surrounded by and filled with fluids that are made mostly of water. The diffusion of water through cell membranes is so important to life processes that it has been given a special name—**osmosis** (ahs MOH sis).

osmosis the diffusion of water through a semipermeable membrane

Water is made up of particles, called *molecules*. Pure water has the highest concentration of water molecules. When you mix something, such as food coloring, sugar, or salt, with water, you lower the concentration of water molecules. **Figure 2** shows how water molecules move through a membrane that is semipermeable (SEM i PUHR mee uh buhl). *Semipermeable* means that only certain substances can pass through. The picture on the left in **Figure 2** shows liquids that have different concentrations of water. Over time, the water molecules move from the liquid with the high concentration of water molecules to the liquid with the lower concentration of water molecules.

The Cell and Osmosis

Osmosis is important to cell functions. For example, red blood cells are surrounded by plasma. Plasma is made up of water, salts, sugars, and other particles. The concentration of these particles is kept in balance by osmosis. If red blood cells were in pure water, water molecules would flood into the cells and cause them to burst. When red blood cells are put into a salty solution, the concentration of water molecules inside the cell is higher than the concentration of water outside. This difference makes water move out of the cells, and the cells shrivel up. Osmosis also occurs in plant cells. When a wilted plant is watered, osmosis makes the plant firm again.

Reading Check **Why would red blood cells burst if you placed them in pure water?** (*See the Appendix for answers to Reading Checks.*)

Bead Diffusion

1. Put three groups of **colored beads** on the bottom of a **plastic bowl.** Each group should be made up of five beads of the same color.
2. Stretch some **clear plastic wrap** tightly over the top of the bowl. Gently shake the bowl for 10 seconds while watching the beads.
3. How is the scattering of the beads like the diffusion of particles? How is it different from the diffusion of particles?

Teach

Demonstration — GENERAL

MATERIALS

- bag, plastic, sealable (or dialysis membrane)
- beakers, 500 mL (2)
- cornstarch
- eyedropper
- graduated cylinder
- tincture of iodine
- twist tie

Crossing Membranes Fill one beaker with 250 mL of water. Add 20 drops of iodine. Fill a second beaker with 250 mL of water, and stir in 15 mL (1 tbsp) of cornstarch. Pour one-half of the starch-and-water mixture into the plastic bag or dialysis membrane. Secure the top of the bag with the twist tie. Rinse off any starch-and-water mixture on the outside of the bag. Place the bag in the iodine-and-water mixture. Check immediately for any changes. Look again after 30 min. Iodine is used to test for the presence of starch; it will turn the starch-and-water mixture black. Iodine particles are small enough to move through the tiny holes in the plastic bag; starch molecules are too large. English Language Learners
LS Visual

MATERIALS

For Each Group

- colored beads, 3 groups
- plastic bowl
- plastic wrap, clear

Answers

3. Beads moved from areas of more-concentrated colors to areas of less-concentrated colors. Eventually, different-colored beads were mixed somewhat evenly. Mixing beads required the use of students' energy and occurred more quickly than diffusion normally occurs.

INCLUSION Strategies

- *Visually Impaired*
- *Attention Deficit Disorder*

Organize the class into teams of four or five. Ask each team to use clay to show the steps in endocytosis and exocytosis. Point out that the first step in exocytosis and the last step in endocytosis are basically the same because the two procedures are essentially the opposite of each other. English Language Learners
LS Kinesthetic

Answer to Reading Check

Red cells would burst in pure water because water particles move from outside, where particles were dense, to inside the cell, where particles were less dense. This movement of water would cause red cells to fill up and burst.

Close

Reteaching — BASIC

Writing **Cell Transport Instructions** Have students write an instruction manual that tells a cell how to transport both a large molecule and a small molecule through the cell membrane. LS **Logical**

Quiz — GENERAL

1. What part of the cell do materials pass through to get into and out of the cell? (the cell membrane)
2. What is osmosis? (the diffusion of water through the semipermeable cell membrane)

Alternative Assessment — ADVANCED

Writing **Science Biography** Have students write a brief biography of Albert Claude (1898–1983), who used the electron microscope to study cells. (Claude shared the 1974 Nobel Prize for physiology with his student George Palade and with Christian de Duve.) LS **Verbal**

Figure 3 *In passive transport, particles travel through proteins to areas of lower concentration. In active transport, cells use energy to move particles, usually to areas of higher concentration.*

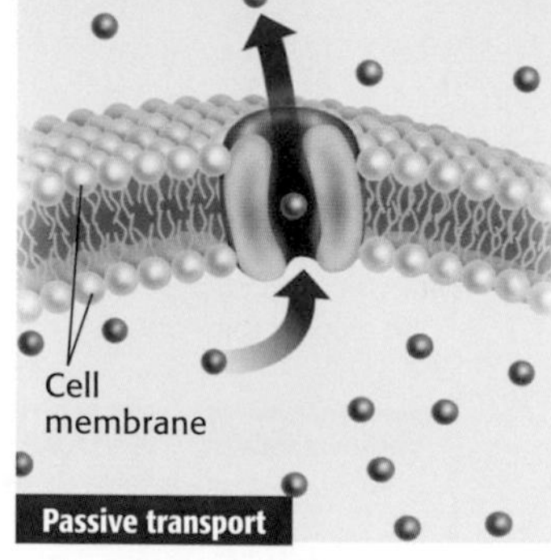

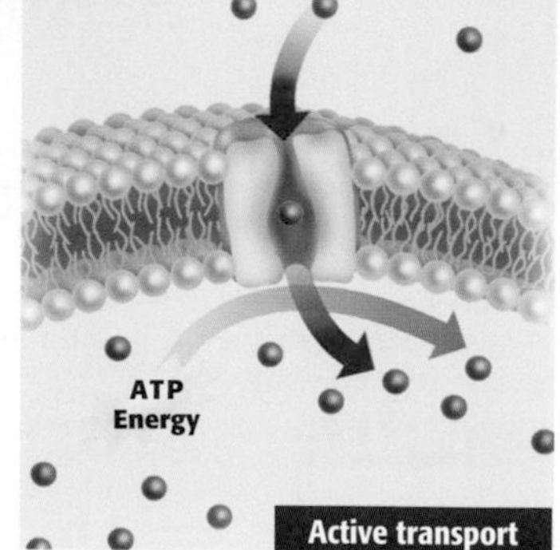

passive transport the movement of substances across a cell membrane without the use of energy by the cell

active transport the movement of substances across the cell membrane that requires the cell to use energy

endocytosis the process by which a cell membrane surrounds a particle and encloses the particle in a vesicle to bring the particle into the cell

Moving Small Particles

Small particles, such as water and sugars, cross the cell membrane through passageways called *channels*. These channels are made up of proteins in the cell membrane. Particles travel through these channels by either passive or active transport. The movement of particles across a cell membrane without the use of energy by the cell is called **passive transport**, and is shown in **Figure 3.** During passive transport, particles move from an area of high concentration to an area of low concentration. Diffusion and osmosis are examples of passive transport.

A process of transporting particles that requires the cell to use energy is called **active transport.** Active transport usually involves the movement of particles from an area of low concentration to an area of high concentration.

Moving Large Particles

Small particles cross the cell membrane by diffusion, passive transport, and active transport. Large particles move into and out of the cell by processes called *endocytosis* and *exocytosis*.

Endocytosis

The active-transport process by which a cell surrounds a large particle and encloses the particle in a vesicle to bring the particle into the cell is called **endocytosis** (EN doh sie TOH sis). *Vesicles* are sacs formed from pieces of cell membrane. **Figure 4** shows a cell taking up a large particle through endocytosis.

Figure 4 **Endocytosis**

1 The cell comes into contact with a particle.

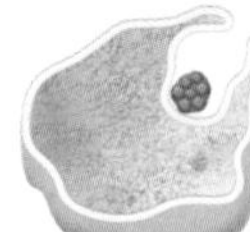

2 The cell membrane begins to wrap around the particle.

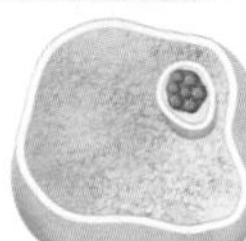

3 Once the particle is completely surrounded, a vesicle pinches off.

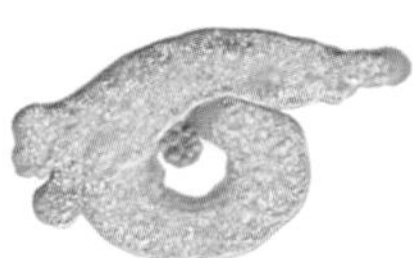

This photo shows the end of *endocytosis*, which means "within the cell."

CHAPTER RESOURCES

Technology

Transparencies
- Passive Transport and Active Transport
- Endocytosis/Exocytosis

Homework — GENERAL

Writing **Transport** Ask students to describe how each of the following materials would get through a cell membrane and into a cell. Which of the materials require active transport?

a. pure water
b. sugar entering a cell that already contains a high concentration of particles
c. sugar entering a cell that has a low concentration of particles
d. a large protein (b, d)

LS **Logical**

Figure 5 **Exocytosis**

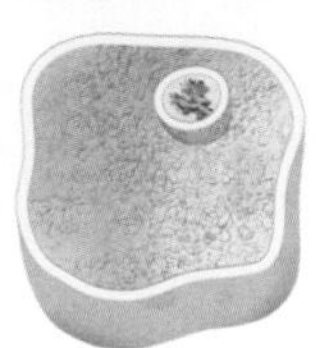

1. Large particles that must leave the cell are packaged in vesicles.

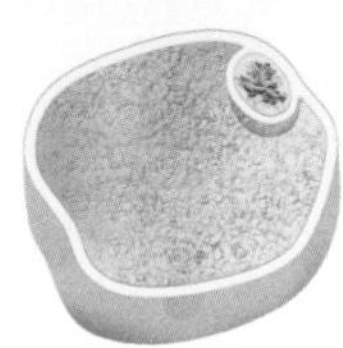

2. The vesicle travels to the cell membrane and fuses with it.

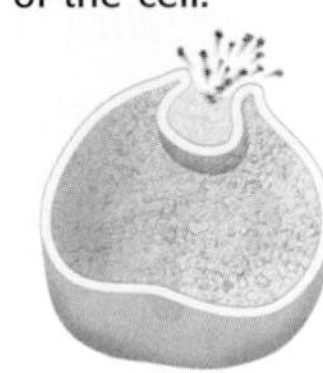

3. The cell releases the particle to the outside of the cell.

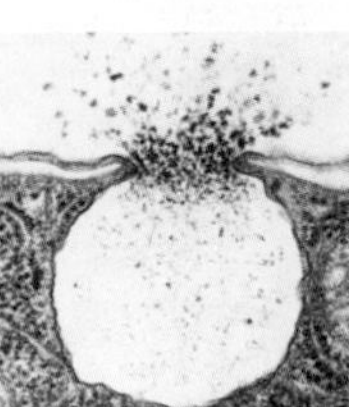

Exocytosis means "outside the cell."

Exocytosis

When a large particle leaves the cell, the cell uses an active-transport process called **exocytosis** (EK soh sie TOH sis). During exocytosis, a vesicle forms around a large particle within the cell. The vesicle carries the particle to the cell membrane. The vesicle then fuses with the cell membrane and releases the particle to the outside of the cell. Exocytosis is shown in **Figure 5.**

exocytosis the process in which a cell releases a particle by enclosing the particle in a vesicle that then moves to the cell surface and fuses with the cell membrane

Reading Check What is exocytosis?

SECTION Review

Summary

- Diffusion is the movement of particles from an area of high concentration to an area of low concentration.
- Osmosis is the diffusion of water through a semipermeable membrane.
- Cells move small particles by diffusion, which is an example of passive transport, and by active transport.
- Large particles enter the cell by endocytosis, and exit the cell by exocytosis.

Using Key Terms

For each pair of terms, explain how the meanings of the terms differ.

1. *diffusion* and *osmosis*
2. *active transport* and *passive transport*
3. *endocytosis* and *exocytosis*

Understanding Key Ideas

4. The movement of particles from a less crowded area to a more crowded area requires
   a. sunlight.
   b. energy.
   c. a membrane.
   d. osmosis.
5. What structures allow particles to move through cell membranes?

Math Skills

6. The area of particle 1 is 2.5 mm^2. The area of particle 2 is 0.5 mm^2. The area of particle 1 is how many times as big as the area of particle 2?

Critical Thinking

7. **Predicting Consequences** What would happen to a cell if its channel proteins were damaged and unable to transport particles? What would happen to the organism if many of its cells were damaged in this way? Explain your answer.
8. **Analyzing Ideas** Why does active transport require energy?

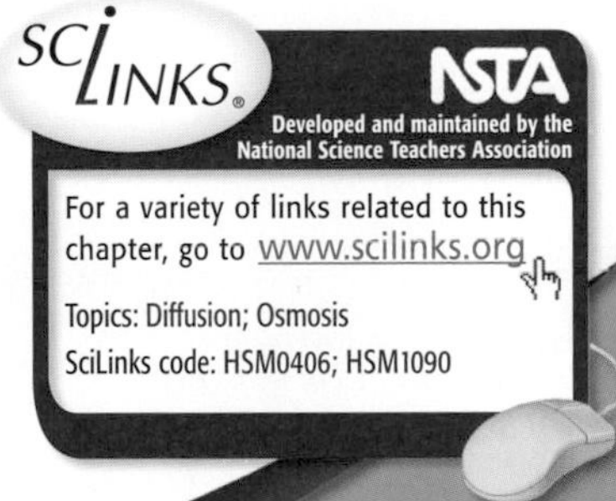

SCILINKS® NSTA Developed and maintained by the National Science Teachers Association

For a variety of links related to this chapter, go to www.scilinks.org

Topics: Diffusion; Osmosis
SciLinks code: HSM0406; HSM1090

Answers to Section Review

1. Sample answer: Diffusion is when any kind of particles move from a crowded area to a less crowded area. Osmosis is diffusion of water through a semipermeable membrane.
2. Sample answer Active transport requires energy, while passive transport does not require energy.
3. Sample answer: Endocytosis is the process which brings things into a cell, and exocytosis takes things out of a cell.
4. b
5. Small particles move through channels in the cell membrane.
6. 5
7. Sample answer: If a cell was unable to transport particles, it would not be able to function properly and would most likely die. If many of an organism's cells were damaged in this way, the organism would become sick and most likely die.
8. Active transport requires energy because the cell must work against the flow of particles.

Answer to Reading Check

Exocytosis is the process by which a cell moves large particles to the outside of the cell.

CHAPTER RESOURCES

Chapter Resource File

- Section Quiz GENERAL
- Section Review GENERAL
- Vocabulary and Section Summary GENERAL
- Reinforcement Worksheet BASIC
- SciLinks Activity GENERAL
- Datasheet for Quick Lab

Technology

Interactive Explorations CD-ROM
- The Nose Knows GENERAL

SECTION 2

Focus

Overview

This section introduces energy and the cell. Students learn about solar energy and the process of photosynthesis. Finally, students learn about cellular respiration and fermentation.

Ask students to make a list of all the reasons why a cell might need energy. Remind students that there are many types of cells doing many different jobs.

Motivate

Demonstration — GENERAL

Leaves and Light Ask students what they think would happen if a plant could not get sunlight. A few days before teaching this section, cut out a square from black construction paper. Fold the square over a leaf of any common plant, such as a geranium. Affix the square with a paper clip. Be sure the leaf does not receive any sunlight. Leave the leaf covered for about one week. Remove the black square. The leaf will be paler than the other leaves. In the absence of sunlight, chlorophyll is depleted and not replenished. The leaf's green color will have faded. English Language Learners

LS Visual

SECTION 2

Cell Energy

READING WARM-UP

Objectives

- Describe photosynthesis and cellular respiration.
- Compare cellular respiration with fermentation.

Terms to Learn

photosynthesis
cellular respiration
fermentation

READING STRATEGY

Discussion Read this section silently. Write down questions that you have about this section. Discuss your questions in a small group.

Why do you get hungry? Feeling hungry is your body's way of telling you that your cells need energy.

All cells need energy to live, grow, and reproduce. Plant cells get their energy from the sun. Many animal cells get the energy they need from food.

From Sun to Cell

Nearly all of the energy that fuels life comes from the sun. Plants capture energy from the sun and change it into food through a process called **photosynthesis.** The food that plants make supplies them with energy. This food also becomes a source of energy for the organisms that eat the plants.

Photosynthesis

Plant cells have molecules that absorb light energy. These molecules are called *pigments*. Chlorophyll (KLAWR uh FIL), the main pigment used in photosynthesis, gives plants their green color. Chlorophyll is found in chloroplasts.

Plants use the energy captured by chlorophyll to change carbon dioxide and water into food. The food is in the form of the simple sugar glucose. Glucose is a carbohydrate. When plants make glucose, they convert the sun's energy into a form of energy that can be stored. The energy in glucose is used by the plant's cells. Photosynthesis also produces oxygen. Photosynthesis is summarized in **Figure 1.**

photosynthesis the process by which plants, algae, and some bacteria use sunlight, carbon dioxide, and water to make food

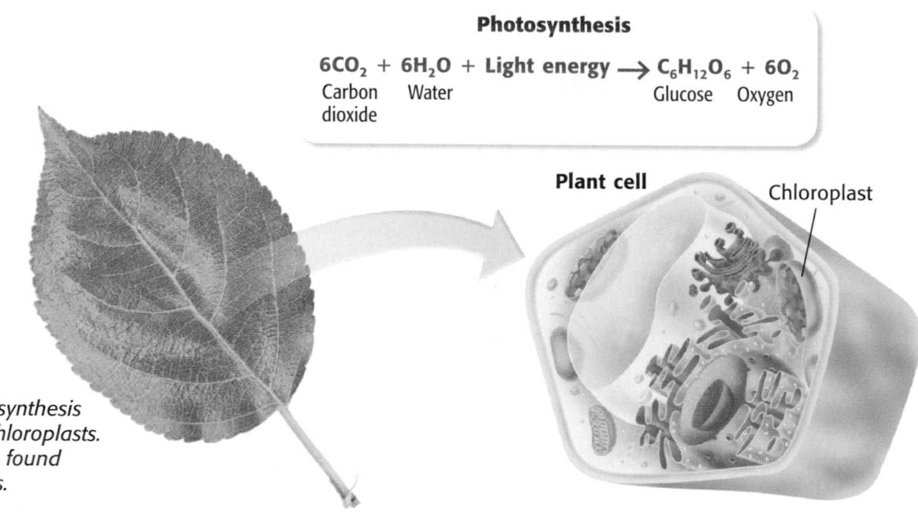

Figure 1 *Photosynthesis takes place in chloroplasts. Chloroplasts are found inside plant cells.*

CHAPTER RESOURCES

Chapter Resource File

- Lesson Plan
- Directed Reading A BASIC
- Directed Reading B SPECIAL NEEDS

Technology

Transparencies

- Bellringer
- ***LINK TO PHYSICAL SCIENCE*** Solar Heating Systems

Workbooks

Science Skills

- Using Logic GENERAL
- Identifying Bias GENERAL

MISCONCEPTION ALERT

Not the Only Steps The processes of photosynthesis and respiration are complex chemical reactions that involve several steps shown by many chemical reactions. The much-simpler equations shown for the processes of respiration and photosynthesis in this chapter are the *net* equations for those reactions.

Getting Energy from Food

Animal cells have different ways of getting energy from food. One way, called **cellular respiration,** uses oxygen to break down food. Many cells can get energy without using oxygen through a process called **fermentation.** Cellular respiration will release more energy from a given food than fermentation will.

Cellular Respiration

The word *respiration* means "breathing," but cellular respiration is different from breathing. Breathing supplies the oxygen needed for cellular respiration. Breathing also removes carbon dioxide, which is a waste product of cellular respiration. But cellular respiration is a chemical process that occurs in cells.

Most complex organisms, such as the cow in **Figure 2,** obtain energy through cellular respiration. During cellular respiration, food (such as glucose) is broken down into CO_2 and H_2O, and energy is released. Most of the energy released maintains body temperature. Some of the energy is used to form adenosine triphosphate (ATP). ATP supplies energy that fuels cell activities.

Most of the process of cellular respiration takes place in the cell membrane of prokaryotic cells. But in the cells of eukaryotes, cellular respiration takes place mostly in the mitochondria. The process of cellular respiration is summarized in **Figure 2.** Does the equation in the figure remind you of the equation for photosynthesis? **Figure 3** on the next page shows how photosynthesis and respiration are related.

Reading Check **What is the difference between cellular respiration and breathing?** *(See the Appendix for answers to Reading Checks.)*

CONNECTION TO Chemistry

Earth's Early Atmosphere

Scientists think that Earth's early atmosphere lacked oxygen. Because of this lack of oxygen, early organisms used fermentation to get energy from food. When organisms began to photosynthesize, the oxygen they produced entered the atmosphere. How do you think this oxygen changed how other organisms got energy?

cellular respiration the process by which cells use oxygen to produce energy from food

fermentation the breakdown of food without the use of oxygen

Cellular Respiration

$$C_6H_{12}O_6 + 6O_2 \rightarrow 6CO_2 + 6H_2O + \text{energy (ATP)}$$

Glucose · Oxygen · Carbon dioxide · Water

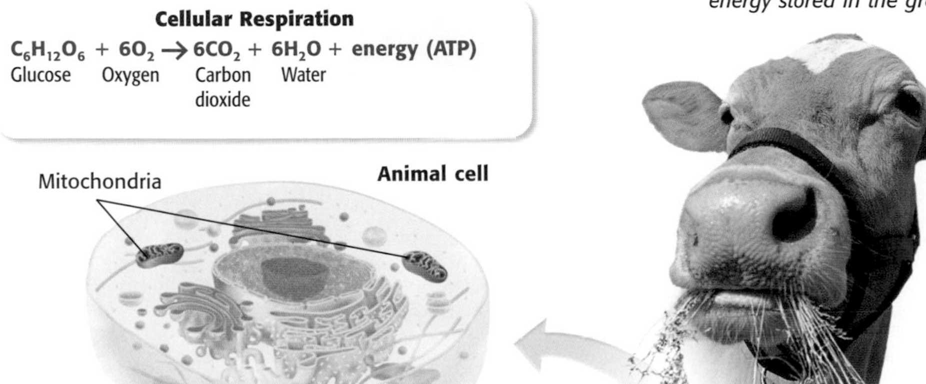

Figure 2 *The mitochondria in the cells of this cow will use cellular respiration to release the energy stored in the grass.*

Teach

Group ACTIVITY—GENERAL

Recycling Carbon Organize the class into groups of three or four. Have each group write the story of a carbon atom as it is used throughout time. Stories should begin with a molecule of carbon dioxide. Ask: "What plant uses it for photosynthesis? What animals swallow it and use it to fuel respiration?" Have students share their stories if time allows. **LS Verbal/Interpersonal**

Answer to Connection to Chemistry

The presence of oxygen gave other organisms the opportunity to get energy in other ways besides fermentation. Most organisms would eventually use cellular respiration to get energy from food.

CONNECTION to *Physical Science*—ADVANCED

Solar Heating Conventional solar heating is a much simpler process than photosynthesis. The sun's energy heats either the house itself, or it heats water, which then circulates through the house. If students have ever felt the warm water from a hose that has been left in the sun, they have felt stored solar energy. Use the teaching transparency titled "Solar Heating Systems" to illustrate how solar energy can be used to heat a home. **LS Visual**

Q: How do cells communicate with each other?

A: by cell phone, of course

Answer to Reading Check

Cellular respiration is a chemical process by which cells produce energy from food. Breathing supplies the body with the raw materials needed for cellular respiration.

Close

Reteaching — Basic

Concept Mapping Have students draw a concept map of energy transfer using the following images:

sunshine; tree, for firewood; sugar cane; yeast consuming sugar, making bread rise; person chopping firewood, for baking oven; person eating bread

Students should note on their maps which organisms use photosynthesis, which use respiration, and which use fermentation. **LS Visual**

Quiz — General

Ask students whether the following statements are true or false.

1. Plants and animals capture their energy from the sun. (false)
2. Cellular respiration describes how a cell breathes. (false)
3. Fermentation produces ATP and lactic acid. (true)

Alternative Assessment — General

Lungs of the Earth Tell students that plants are sometimes called the "lungs of the Earth." Ask students to think about this and to prepare an illustrated presentation for the class. Students may want to research the role that rain forests play as Earth's "lungs" and explain the contributions rain forests make to the health of the planet. **LS Verbal/Visual**

Figure 3 The Connection Between Photosynthesis and Respiration

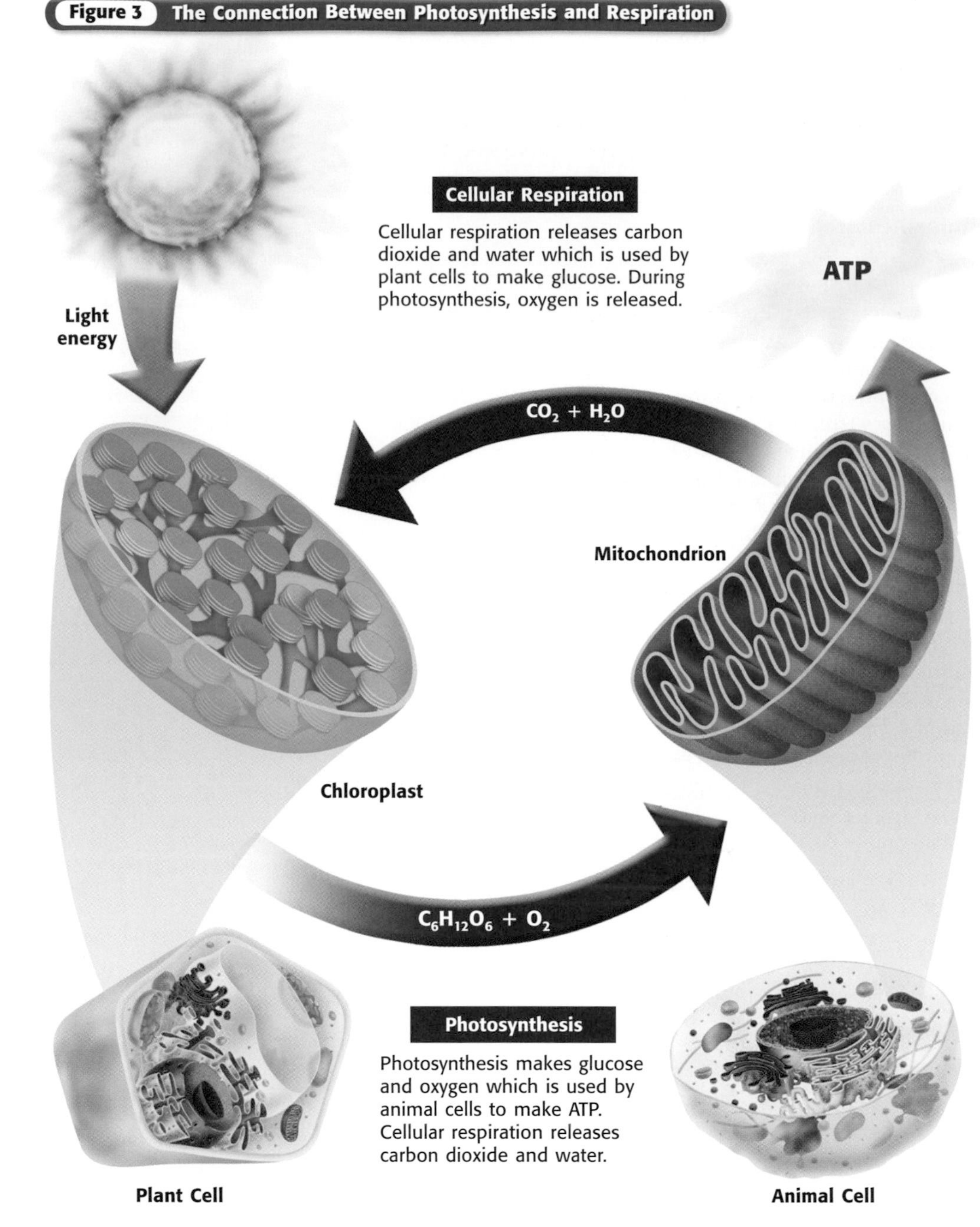

Group Activity — General

Photosynthesis and Cellular Respiration Have students work in pairs and refer to the diagram on this page. Have each pair compare and contrast photosynthesis and respiration. Ask students to answer the following questions: "What happens to the ATP? Where does the ATP go? How is ATP used by the cell? How is the cell's use of CO_2 and H_2O similar to people's recycling of paper and glass bottles?" **LS Interpersonal/Logical**

Homework — Advanced

Comparing Cell Processes Newer kinds of solar cells simulate photosynthesis more closely than older solar cells do. Just as plant cells use energy from the sun to change water and carbon dioxide into energy-rich sugars, these new solar cells use the sun's energy to convert water into energy-rich hydrogen gas, which can be used as fuel. The byproduct of this process is oxygen. Have students research these newer solar cells and make a poster showing how they work. **LS Visual**

Connection Between Photosynthesis and Respiration

As shown in **Figure 3,** photosynthesis transforms energy from the sun into glucose. During photosynthesis, cells take in CO_2 and release O_2. During cellular respiration, cells use O_2 to break down glucose and release energy and CO_2. Each process makes the materials that are needed for the other process to occur elsewhere.

Fermentation

Have you ever felt a burning sensation in your leg muscles while you were running? When muscle cells can't get the oxygen needed for cellular respiration, they use the process of fermentation to get energy. One kind of fermentation happens in your muscles and produces lactic acid. The buildup of lactic acid contributes to muscle fatigue and causes a burning sensation. This kind of fermentation also happens in the muscle cells of other animals and in some fungi and bacteria. Another type of fermentation occurs in some types of bacteria and in yeast as described in **Figure 4.**

Figure 4 *Yeast forms carbon dioxide during fermentation. The bubbles of CO_2 gas cause the dough to rise and leave small holes in bread after it is baked.*

Reading Check What are two kinds of fermentation?

SECTION Review

Summary

- Most of the energy that fuels life processes comes from the sun.
- The sun's energy is converted into food by the process of photosynthesis.
- Cellular respiration breaks down glucose into water, carbon dioxide, and energy.
- Fermentation is a way that cells get energy from their food without using oxygen.

Using Key Terms

1. In your own words, write a definition for the term *fermentation.*

Understanding Key Ideas

2. O_2 is released during
   a. cellular respiration.
   b. photosynthesis.
   c. breathing.
   d. fermentation.
3. How are photosynthesis and cellular respiration related?
4. How are respiration and fermentation similar? How are they different?

Math Skills

5. Cells of plant A make 120 molecules of glucose an hour. Cells of plant B make half as much glucose as plant A does. How much glucose does plant B make every minute?

Critical Thinking

6. **Analyzing Relationships** Why are plants important to the survival of all other organisms?
7. **Applying Concepts** You have been given the job of restoring life to a barren island. What types of organisms would you put on the island? If you want to have animals on the island, what other organisms must you bring? Explain your answer.

For a variety of links related to this chapter, go to www.scilinks.org

Topic: Cell Energy; Photosynthesis
SciLinks code: HSM0237; HSM1140

Answers to Section Review

1. Sample answer: Fermentation is how some organisms get energy from food without using oxygen.
2. b
3. Photosynthesis uses the waste materials of cellular respiration, and cellular respiration uses the waste materials of photosynthesis.
4. Cellular respiration and fermentation both release the energy stored in food. Fermentation does not use oxygen, and cellular respiration does use oxygen.
5. The cells of plant B make an average of 1 glucose molecule per minute.
6. Sample answer: Plants turn energy from the sun into chemical energy. Animals that eat the plants use the stored energy. Plants also produce O_2.
7. Sample answer: Animals such as birds, insects, and mammals would be good on the island. Plants must be on the island in order to provide a source of food for the animals.

Answer to Reading Check

the kind that happens in the muscle cells of animals and the kind that occurs in bacteria

CHAPTER RESOURCES

Chapter Resource File

- **Section Quiz** GENERAL
- **Section Review** GENERAL
- **Vocabulary and Section Summary** GENERAL
- **Reinforcement Worksheet** BASIC
- **Critical Thinking** ADVANCED

Technology

Transparencies
- The Connection Between Photosynthesis and Respiration

SECTION 3

Focus

Overview

This section introduces the life cycle of a cell. Students will learn how cells reproduce and how mitosis is important. Finally, students will learn how cell division differs in plants and animals.

Bellringer

On the board, write the following:

> Biology is the only science in which multiplication means the same thing as division.

Have students write an explanation of this sentence. (When cells divide, they are multiplying. Some students may point out that multiplying a number by a fraction is the same as division.)

Motivate

ACTIVITY — GENERAL

Making Models Have pairs of students use string for the cell membrane and pieces of pipe cleaners for chromosomes to demonstrate the basic steps of mitosis, as described in this section. LS **Visual/Interpersonal**

SECTION 3

The Cell Cycle

In the time that it takes you to read this sentence, your body will have made millions of new cells! Making new cells allows you to grow and replace cells that have died.

READING WARM-UP

Objectives

- Explain how cells produce more cells.
- Describe the process of mitosis.
- Explain how cell division differs in animals and plants.

Terms to Learn

cell cycle
chromosome
homologous chromosomes
mitosis
cytokinesis

READING STRATEGY

Paired Summarizing Read this section silently. In pairs, take turns summarizing the material. Stop to discuss ideas that seem confusing.

The environment in your stomach is so acidic that the cells lining your stomach must be replaced every few days. Other cells are replaced less often, but your body is constantly making new cells.

The Life of a Cell

As you grow, you pass through different stages in life. Your cells also pass through different stages in their life cycle. The life cycle of a cell is called the **cell cycle.**

The cell cycle begins when the cell is formed and ends when the cell divides and forms new cells. Before a cell divides, it must make a copy of its deoxyribonucleic acid (DNA). DNA is the hereditary material that controls all cell activities, including the making of new cells. The DNA of a cell is organized into structures called **chromosomes.** Copying chromosomes ensures that each new cell will be an exact copy of its parent cell. How does a cell make more cells? It depends on whether the cell is prokaryotic (with no nucleus) or eukaryotic (with a nucleus).

cell cycle the life cycle of a cell

chromosome in a eukaryotic cell, one of the structures in the nucleus that are made up of DNA and protein; in a prokaryotic cell, the main ring of DNA

Making More Prokaryotic Cells

Prokaryotic cells are less complex than eukaryotic cells are. Bacteria, which are prokaryotes, have ribosomes and a single, circular DNA molecule but don't have membrane-enclosed organelles. Cell division in bacteria is called *binary fission,* which means "splitting into two parts." Binary fission results in two cells that each contain one copy of the circle of DNA. A few of the bacteria in **Figure 1** are undergoing binary fission.

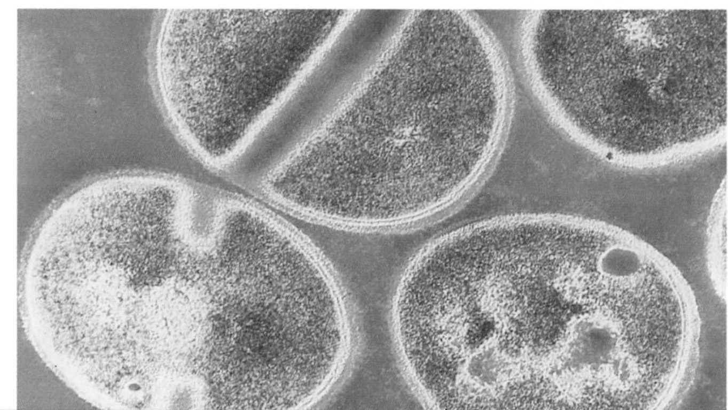

Figure 1 *Bacteria reproduce by binary fission.*

CHAPTER RESOURCES

Chapter Resource File

- **Lesson Plan**
- **Directed Reading A** BASIC
- **Directed Reading B** SPECIAL NEEDS

Technology

Transparencies
- Bellringer
- The Cell Cycle

Workbooks

Math Skills for Science
- Multiplying Whole Numbers BASIC
- Grasping Graphing GENERAL

CONNECTION to Math — BASIC

Cell Multiplication It takes Cell A 1 h to complete its cell cycle and produce two cells. The cell cycle of Cell B takes 2 h. How many more cells would be formed from Cell A than from Cell B in 6 h?

(After 6 h, Cell A would have formed 64 cells, and Cell B would have formed 8 cells. Cell A would have formed 56 cells more than Cell B.)
LS **Logical/Verbal**

Eukaryotic Cells and Their DNA

Eukaryotic cells are more complex than prokaryotic cells are. The chromosomes of eukaryotic cells contain more DNA than those of prokaryotic cells do. Different kinds of eukaryotes have different numbers of chromosomes. More-complex eukaryotes do not necessarily have more chromosomes than simpler eukaryotes do. For example, fruit flies have 8 chromosomes, potatoes have 48, and humans have 46. **Figure 2** shows the 46 chromosomes of a human body cell lined up in pairs. These pairs are made up of similar chromosomes known as **homologous chromosomes** (hoh MAHL uh guhs KROH muh SOHMZ).

Reading Check Do more-complex organisms always have more chromosomes than simpler organisms do? *(See the Appendix for answers to Reading Checks.)*

Figure 2 *Human body cells have 46 chromosomes, or 23 pairs of chromosomes.*

Making More Eukaryotic Cells

The eukaryotic cell cycle includes three stages. In the first stage, called *interphase,* the cell grows and copies its organelles and chromosomes. After each chromosome is duplicated, the two copies are called *chromatids*. Chromatids are held together at a region called the *centromere*. The joined chromatids twist and coil and condense into an X shape, as shown in **Figure 3.** After this step, the cell enters the second stage of the cell cycle.

In the second stage, the chromatids separate. The complicated process of chromosome separation is called **mitosis.** Mitosis ensures that each new cell receives a copy of each chromosome. Mitosis is divided into four phases, as shown on the following pages.

In the third stage, the cell splits into two cells. These cells are identical to each other and to the original cell.

homologous chromosomes chromosomes that have the same sequence of genes and the same structure

mitosis in eukaryotic cells, a process of cell division that forms two new nuclei, each of which has the same number of chromosomes

Figure 3 *This duplicated chromosome consists of two chromatids. The chromatids are joined at the centromere.*

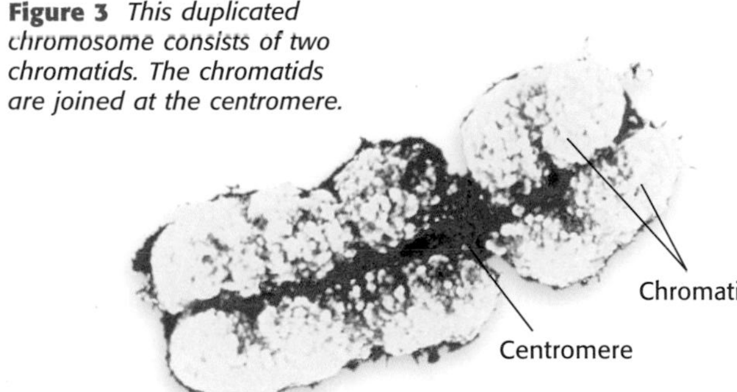

CONNECTION TO Language Arts

Picking Apart Vocabulary
Brainstorm what words are similar to the parts of the term *homologous chromosome.* What can you guess about the meaning of the term's root words? Look up the roots of the words, and explain how they help describe the concept. **ACTIVITY**

Teach

READING STRATEGY — GENERAL

Prediction Guide Have students guess whether the following statement is true or false before they read this page:

The number of chromosomes in eukaryotic cells differs because some organisms are more complex than others. (false)

Have students evaluate their answer after they read the page. **LS Visual**

Answer to Reading Check

No, the number of chromosomes is not always related to the complexity of organisms.

CONNECTION ACTIVITY
Math — ADVANCED

Graphing Cell Division Have students make a bar graph to represent the following information about different cell types:

- Cell Type A divides every 3 h
- Cell Type B divides every 6 h
- Cell Type C divides every 8 h

If you start with one cell of each type, how many A, B, and C cells will be produced in a 24-hour period? On a graph, compare the number of cells produced by each type. (Cell A, 256; Cell B, 16; Cell C, 8) **LS Visual/Logical**

Answer to Connection to Language Arts

The Greek root of the word *homologous* means "corresponding in structure or character." The roots for the word *chromosome* mean "color" and "body." *Homologous chromosome* mean "chromosomes that have the same structure."

WEIRD SCIENCE

A fern called *Ophioglossum reticulatum* has 1,260 chromosomes per cell, more than any other organism.

Is That a Fact!

Before sophisticated microscopes were available, scientists could not see cells pinching and dividing. Many scientists believed that cells came into existence spontaneously—as though crystallizing out of bodily fluids.

Close

Reteaching — Basic

Writing **Biography of a Cell** Have students write and illustrate the biography of a cell. It can be humorous or serious, but it should include accurate descriptions of how materials are transported into and out of the cell and how cells reproduce. **LS Visual/Verbal**

Quiz — General

1. What is cell division? (It is the process by which cells reproduce themselves.)
2. How do prokaryotic cells make more cells? (binary fission)
3. How do eukaryotic cells make more cells? (mitosis and cytokinesis)

Alternative Assessment — General

Writing **Mitosis and Cancer** Have students research the role of mitosis in cancer and write a report or create a poster or other visual presentation on what they learn. Students' reports should include information about various cancer treatments, such as radiation, chemotherapy, and surgery. **LS Verbal/Visual**

Answer to Reading Check

During mitosis in plant cells, a cell plate is formed. During mitosis in animal cells, a cell plate does not form.

Figure 4 The Cell Cycle

Copying DNA (Interphase)

Before mitosis begins, chromosomes are copied. Each chromosome is then two chromatids.

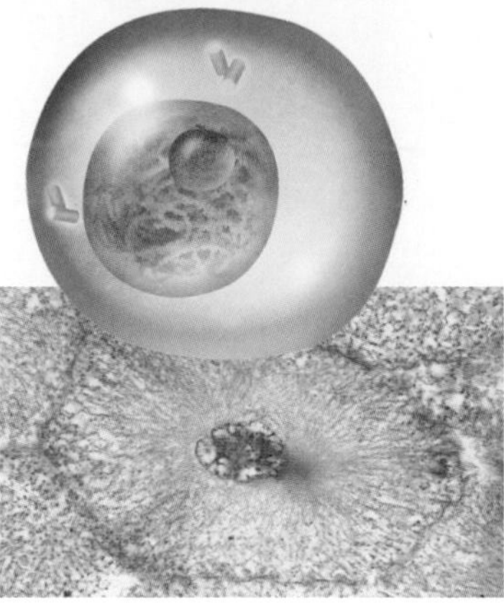

Mitosis Phase 1 (Prophase)

Mitosis begins. The nuclear membrane dissolves. Chromosomes condense into rodlike structures.

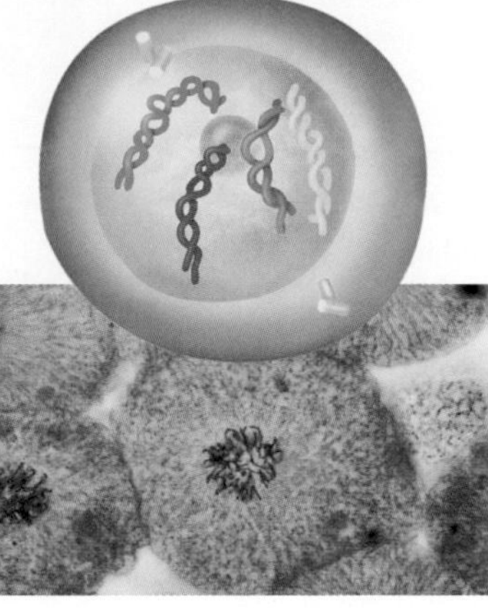

Mitosis Phase 2 (Metaphase)

The chromosomes line up along the equator of the cell. Homologous chromosomes pair up.

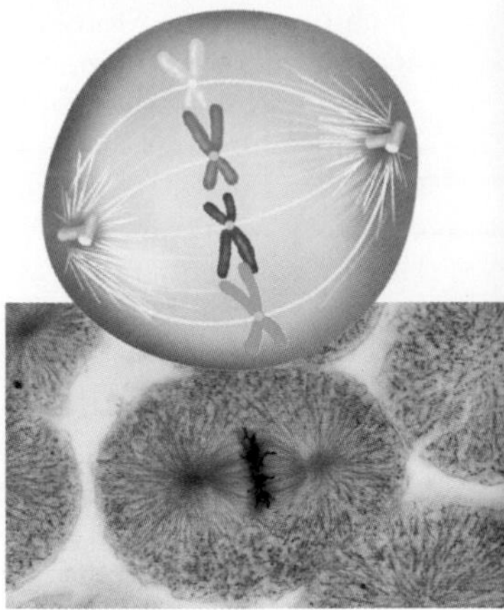

Mitosis and the Cell Cycle

Figure 4 shows the cell cycle and the phases of mitosis in an animal cell. Mitosis has four phases that are shown and described above. This diagram shows only four chromosomes to make it easy to see what's happening inside the cell.

cytokinesis the division of the cytoplasm of a cell

Cytokinesis

In animal cells and other eukaryotes that do not have cell walls, division of the cytoplasm begins at the cell membrane. The cell membrane begins to pinch inward to form a groove, which eventually pinches all the way through the cell, and two daughter cells form. The division of cytoplasm is called **cytokinesis** and is shown at the last step of **Figure 4.**

Eukaryotic cells that have a cell wall, such as the cells of plants, algae, and fungi, reproduce differently. In these cells, a *cell plate* forms in the middle of the cell. The cell plate becomes the new cell membranes that separate the new cells. After the cell splits into two, a new cell wall forms where the cell plate was. The cell plate and a late stage of cytokinesis in a plant cell are shown in **Figure 5.**

Reading Check What is the difference between mitosis in an animal cell and mitosis in a plant cell?

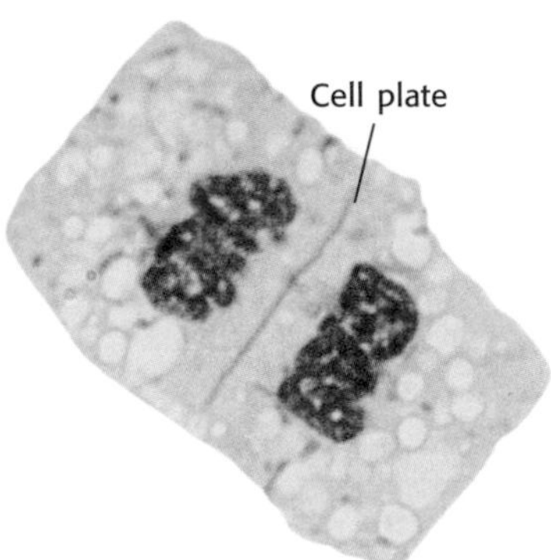

Figure 5 *When a plant cell divides, a cell plate forms and the cell splits into two cells.*

Inclusion Strategies

- *Developmentally Delayed* • *Hearing Impaired*
- *Learning Disabled*

Make a three-column table with these column headings: *Characteristics, Prokaryotic Cells,* and *Eukaryotic Cells.* Under Characteristics, use these five row headings: *Small or large?, Complex or simple?, More or less DNA?, Has organelles?,* and *Number of stages in cell division.* Have students copy the table and fill it in. **LS Logical/Visual**

Internet Activity

Sequence Board — General

For an internet activity related to this chapter, have students go to **go.hrw.com** and type in the keyword **HL5ACTW.**

Mitosis Phase 3 (Anaphase)

The chromatids separate and move to opposite sides of the cell.

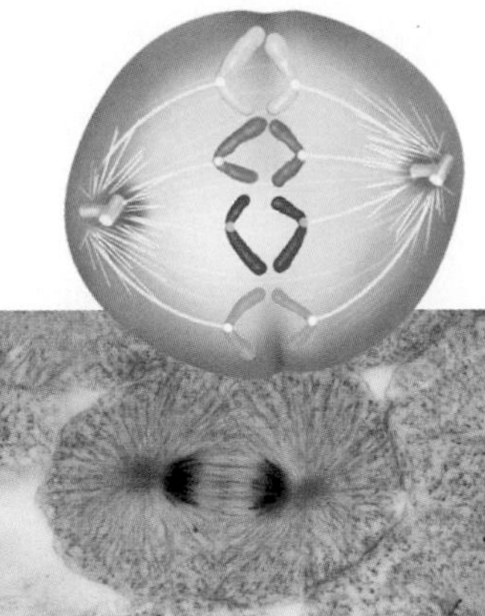

Mitosis Phase 4 (Telophase)

A nuclear membrane forms around each set of chromosomes, and the chromosomes unwind. Mitosis is complete.

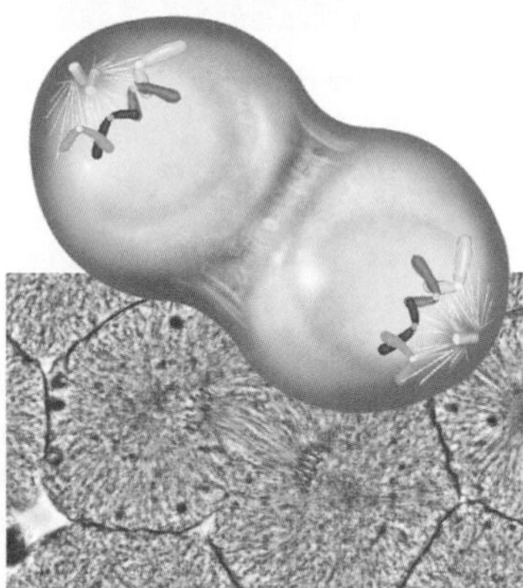

Cytokinesis

In cells that lack a cell wall, the cell pinches in two. In cells that have a cell wall, a cell plate forms between the two new cells.

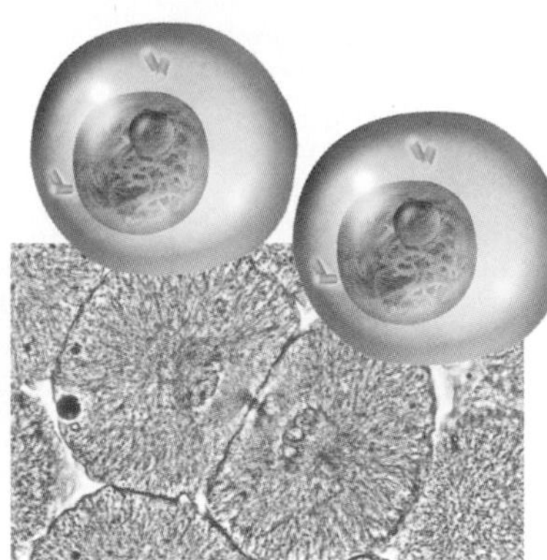

SECTION Review

Summary

- A cell produces more cells by first copying its DNA.
- Eukaryotic cells produce more cells through the four phases of mitosis.
- Mitosis produces two cells that have the same number of chromosomes as the parent cell.
- At the end of mitosis, a cell divides the cytoplasm by cytokinesis.
- In plant cells, a cell plate forms between the two new cells during cytokinesis.

Using Key Terms

1. In your own words, write a definition for each of the following terms: *cell cycle* and *cytokinesis*.

Understanding Key Ideas

2. A eukaryotic cell
   a. has no nucleus.
   b. undergoes binary fission.
   c. has a nucleus.
   d. has a cell wall.
3. Why is it important for chromosomes to be copied before cell division?
4. How are binary fission and mitosis similar? How do they differ?

Math Skills

5. Cell A takes 6 h to complete mitosis. Cell B takes 8 h to complete mitosis. After 24 h, how many more copies of cell A would there be than cell B?

Critical Thinking

6. **Predicting Consequences** What would happen if cytokinesis occurred without mitosis?
7. **Applying Concepts** How does mitosis ensure that a new cell is just like its parent cell?
8. **Making Comparisons** Compare the processes that animal cells and plant cells use to make new cells. How are the processes different?

For a variety of links related to this chapter, go to www.scilinks.org

Topic: Cell Cycle
SciLinks code: HSM0235

Answers to Section Review

1. Sample answer: The cell cycle describes all of the stages a cell goes through in its life. Cytokinesis is the last stage of cell reproduction when a cell's cytoplasm is split between the two new cells.
2. c
3. Chromosomes need to be copied so that the two new cells have the same genetic material as the parent cell.
4. Binary fission and mitosis are similar in that they are both forms of cell reproduction. Their duplication processes, however, differ from one another. Binary fission is a simpler form of cell reproduction. In binary fission, only a ring of genetic material is copied. In mitosis, the chromosomes must be copied before the cell reproduces.
5. 8
6. If cytokinesis occurred without mitosis, each cell would only have half of the parent cell's genetic material or less.
7. Mitosis ensures that each new cell receives a copy of each chromosome, and hence, an exact copy of the parent cell's genetic material.
8. The processes of animal and plant cells are different because plant cells have cell walls. Cytokinesis is different in plant cells, but all other stages of mitosis are essentially the same as they are in animal cells.

ACTIVITY — ADVANCED

Four Phases of Mitosis Have students research in high school or college texts to find the names of the four phases of mitosis. (prophase, metaphase, anaphase, and telophase) Have students find the name of the time between cell divisions. (interphase) Finally, have students create a circular diagram and use this information to illustrate the cell cycle. LS **Visual/Logical**

CHAPTER RESOURCES

Chapter Resource File

- Section Quiz GENERAL
- Section Review GENERAL
- Vocabulary and Section Summary GENERAL
- Reinforcement Worksheet BASIC

Workbooks

Science Skills
- Researching on the Web BASIC
- Organizing Your Research GENERAL

Inquiry Lab

The Perfect Taters Mystery

Teacher's Notes

Time Required
Two 45-minute class periods

Lab Ratings

EASY → HARD

Teacher Prep 2 of 4
Student Set-Up 2 of 4
Concept Level 3 of 4
Clean Up 1 of 4

MATERIALS

The materials listed on the student pages are enough for one class of students. You will need one or two potatoes per class. Do not allow students to cut or peel potatoes. You will need to do this ahead of time. Allow students to choose the number of containers they will need for the experiment. They may wish to test several salt concentrations.

Safety Caution
Remind students to review all safety cautions and icons before beginning this lab activity.

Avoid including green or discolored parts of the potato in the pieces students work with. These could cause illness.

Using Scientific Methods

Inquiry Lab

OBJECTIVES

Examine osmosis in potato cells.

Design a procedure that will give the best results.

MATERIALS

- cups, clear plastic, small
- potato pieces, freshly cut
- potato samples (A, B, and C)
- salt
- water, distilled

SAFETY

The Perfect Taters Mystery

You are the chief food detective at Perfect Taters Food Company. The boss, Mr. Fries, wants you to find a way to keep his potatoes fresh and crisp before they are cooked. His workers have tried several methods, but these methods have not worked. Workers in Group A put the potatoes in very salty water, and something unexpected happened to the potatoes. Workers in Group B put the potatoes in water that did not contain any salt, and something else happened! Workers in Group C didn't put the potatoes in any water, and that didn't work either. Now, you must design an experiment to find out what can be done to make the potatoes stay crisp and fresh.

- Before you plan your experiment, review what you know. You know that potatoes are made of cells. Plant cells contain a large amount of water. Cells have membranes that hold water and other materials inside and keep some things out. Water and other materials must travel across cell membranes to get into and out of the cell.
- Mr. Fries has told you that you can obtain as many samples as you need from the workers in Groups A, B, and C. Your teacher will have these samples ready for you to observe.
- Make a data table like the one below. List your observations in the data table. Make as many observations as you can about the potatoes tested by workers in Groups A, B, and C.

Observations	
Group A	
Group B	
Group C	

Ask a Question

1. Now that you have made your observations, state Mr. Fries's problem in the form of a question that can be answered by your experiment.

Lab Notes
Osmosis is often a confusing and misunderstood concept in life science. Quite often, students can repeat the definition of the process but are unable to apply the concept to explain the movement of water in different osmotic environments. In this lab, students will have an opportunity to observe osmosis in a model and obtain measurable results. This lab can be done as a class demonstration if materials and space are limited. The purpose of this lab is to reinforce comprehension of osmosis and to practice the scientific method.

CHAPTER RESOURCES

Chapter Resource File

- Datasheet for Chapter Lab
- Lab Notes and Answers

Technology

Classroom Videos
- Lab Video

- Stayin' Alive!

Form a Hypothesis

2. Form a hypothesis based on your observations and your questions. The hypothesis should be a statement about what causes the potatoes not to be crisp and fresh. Based on your hypothesis, make a prediction about the outcome of your experiment. State your prediction in an if-then format.

Test the Hypothesis

3. Once you have made a prediction, design your investigation. Check your experimental design with your teacher before you begin. Mr. Fries will give you potato pieces, water, salt, and no more than six containers.

4. Keep very accurate records. Write your plan and procedure. Make data tables. To be sure your data is accurate, measure all materials carefully and make drawings of the potato pieces before and after the experiment.

Analyze the Results

1. **Explaining Events** Explain what happened to the potato cells in Groups A, B, and C in your experiment. Include a discussion of the cell membrane and the process of osmosis.

Draw Conclusions

2. **Analyzing Results** Write a letter to Mr. Fries that explains your experimental method, results, and conclusion. Then, make a recommendation about how he should handle the potatoes so that they will stay fresh and crisp.

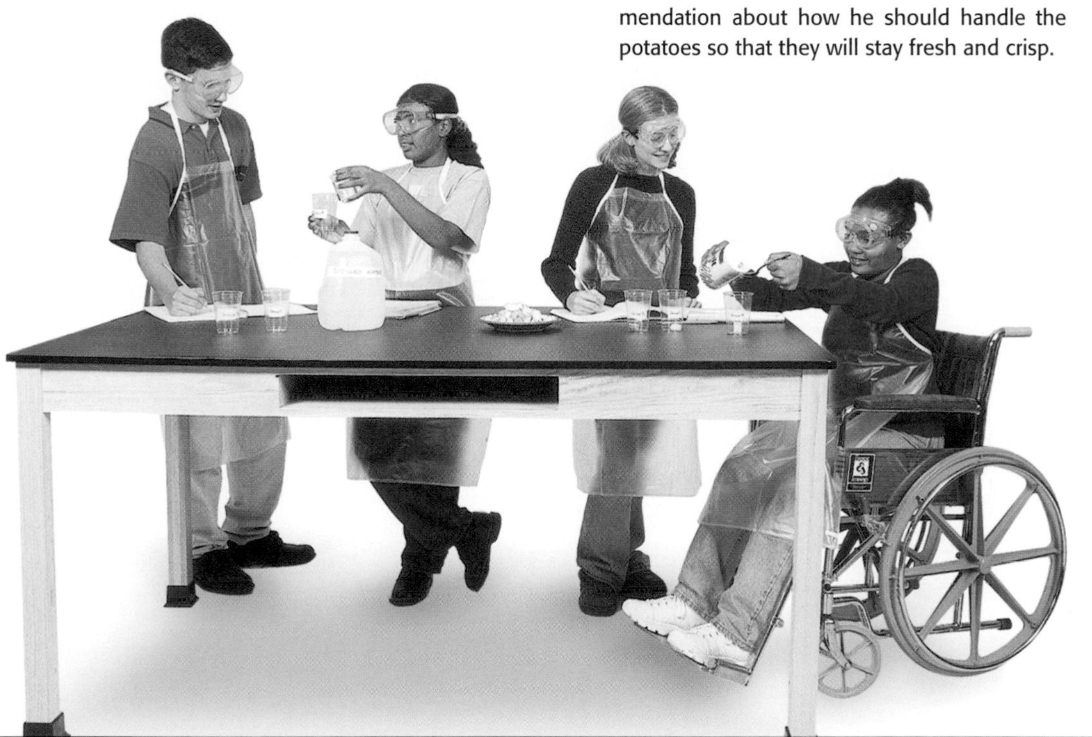

CHAPTER RESOURCES

Workbooks

- **Whiz-Bang Demonstrations**
  - It's in the Bag! BASIC
  - Stop Picking on My Enzyme BASIC
- **Labs You Can Eat**
  - The Mystery of the Runny Gelatin GENERAL
- **Inquiry Labs**
  - Fish Farms in Space GENERAL
- **Long-Term Projects & Research Ideas**
  - Taming the Wild Yeast ADVANCED

Susan Gorman
North Ridge Middle School
North Richland Hills, Texas

Analyze the Results

1. The potato cells in Group A were placed in very salty water. The potatoes shriveled up because water moved out of the cell and into the salty water (from an area of high concentration of water to an area of low concentration of water). This may be confusing to some students, who may think that because the concentration of salt is high outside the potato, the salt should move to the area of lower concentration. Explain that although water can move through a cell membrane by osmosis, salt must be moved across a cell membrane by a process that requires energy.

   The potato cells in Group B were placed in water with no salt. The potatoes swelled because the concentration of water was lower inside the cell. (The concentration of salt and other molecules was higher inside the potato cell.)

   The potato cells in Group C turned brown and dried up because the water concentration outside the cell was low. In fact, there wasn't any water at all. The water evaporated as soon as it left the cell membrane. The potato cells turned brown because of chemical reactions with the air.

Draw Conclusions

2. Letters to Mr. Fries will vary according to each student's results. However, all students should explain that through trial and error they found one salt concentration that was closest to the concentration of salt and other molecules inside the potato. This is the concentration that should be used to maintain an osmotic balance in the potato. Furthermore, some students will realize that the potatoes must be kept in water to prevent them from turning brown.

Chapter Review

Assignment Guide	
Section	**Questions**
1	1, 2, 6, 8, 12, 16
2	3, 4, 7, 10, 13, 17
3	5, 9, 11, 14, 15, 18–22

ANSWERS

Using Key Terms

1. Sample answer: Osmosis is the diffusion of water through a semipermeable membrane.
2. Sample answer: Exocytosis is the process cells use to remove large particles; endocytosis is the process cells use to move large particles into a cell
3. photosynthesis
4. cellular respiration
5. Cytokinesis is the division of just the cytoplasm. Mitosis is the process in eukaryotic cells in which the nuclear material splits to form two new nuclei.
6. Active transport requires the cell to use energy to move substances. Passive transport does not require the cell to use any energy.
7. Cellular respiration releases stored energy by using oxygen. Fermentation releases stored energy without using oxygen.

Understanding Key Ideas

8. c
9. a
10. d
11. c

Chapter Review

USING KEY TERMS

1. Use the following terms in the same sentence: *diffusion* and *osmosis*.
2. In your own words, write a definition for each of the following terms: *exocytosis* and *endocytosis*.

Complete each of the following sentences by choosing the correct term from the word bank.

cellular respiration
photosynthesis
fermentation

3. Plants use ___ to make glucose.
4. During ___, oxygen is used to break down food molecules releasing large amounts of energy.

For each pair of terms, explain how the meanings of the terms differ.

5. *cytokinesis* and *mitosis*
6. *active transport* and *passive transport*
7. *cellular respiration* and *fermentation*

UNDERSTANDING KEY IDEAS

Multiple Choice

8. The process in which particles move through a membrane from a region of low concentration to a region of high concentration is
   - **a.** diffusion.
   - **b.** passive transport.
   - **c.** active transport.
   - **d.** fermentation.
9. What is the result of mitosis?
   - **a.** two identical cells
   - **b.** two nuclei
   - **c.** chloroplasts
   - **d.** two different cells
10. Before the energy in food can be used by a cell, the energy must first be transferred to molecules of
    - **a.** proteins.
    - **b.** carbohydrates.
    - **c.** DNA.
    - **d.** ATP.
11. Which of the following cells would form a cell plate during the cell cycle?
    - **a.** a human cell
    - **b.** a prokaryotic cell
    - **c.** a plant cell
    - **d.** All of the above

Short Answer

12. Are exocytosis and endocytosis examples of active or passive transport? Explain your answer.
13. Name the cell structures that are needed for photosynthesis and the cell structures that are needed for cellular respiration.
14. Describe the three stages of the cell cycle of a eukaryotic cell.

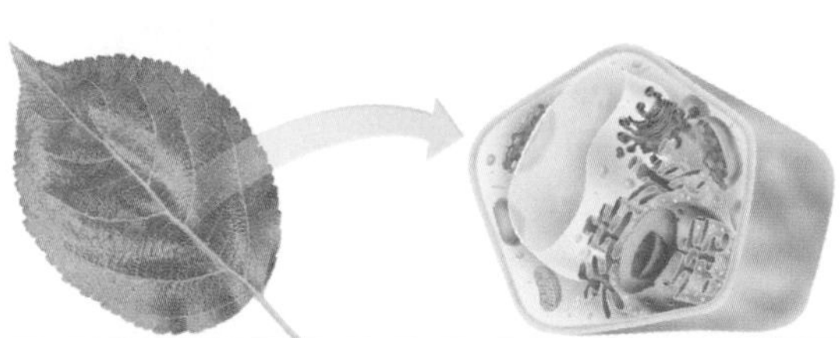

12. Endocytosis and exocytosis are examples of active transport. In both processes the cell must change shape, wrap around a particle, and make other movements that require the cell to use energy.
13. Chloroplasts are needed for photosynthesis. Cellular respiration requires mitochondria.
14. The first stage is cell growth and copying of DNA (duplication). The second stage is mitosis, which involves separating the duplicated chromosomes. The third stage is cytokinesis (cell division), which results in two separate, identical cells.

CRITICAL THINKING

15 Concept Mapping Use the following terms to create a concept map: *chromosome duplication, cytokinesis, prokaryote, mitosis, cell cycle, binary fission,* and *eukaryote.*

16 Making Inferences Which one of the plants pictured below was given water mixed with salt, and which one was given pure water? Explain how you know, and be sure to use the word *osmosis* in your answer.

17 Identifying Relationships Why would your muscle cells need to be supplied with more food when there is a lack of oxygen than when there is plenty of oxygen present?

18 Applying Concepts A parent cell has 10 chromosomes.

a. Will the cell go through binary fission or mitosis and cytokinesis to produce new cells?

b. How many chromosomes will each new cell have after the parent cell divides?

INTERPRETING GRAPHICS

The picture below shows a cell. Use the picture below to answer the questions that follow.

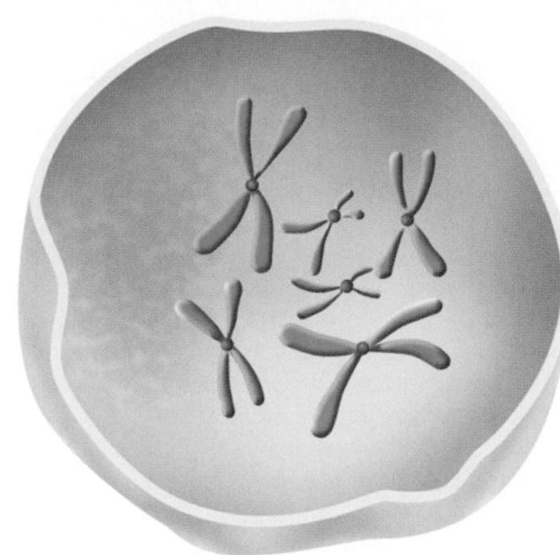

19 Is the cell prokaryotic or eukaryotic?

20 Which stage of the cell cycle is this cell in?

21 How many chromatids are present? How many pairs of homologous chromosomes are present?

22 How many chromosomes will be present in each of the new cells after the cell divides?

CHAPTER RESOURCES

Chapter Resource File

- Chapter Review GENERAL
- Chapter Test A GENERAL
- Chapter Test B ADVANCED
- Chapter Test C SPECIAL NEEDS
- Vocabulary Activity GENERAL

Workbooks

Study Guide

- Assessment resources are also available in Spanish.

Critical Thinking

15. 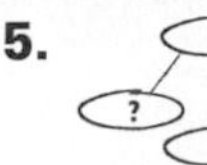

An answer to this exercise can be found at the end of this book.

16. The plant on the left was given pure water. Its stems and leaves are standing up straight. The wilted plant on the right was given salt water. Osmosis occurred, and water in the plant moved into the soil, where the concentration of water was lower.

17. When there is plenty of oxygen, the cells can get energy from cellular respiration. When there is a lack of oxygen, the cells must use fermentation, which doesn't produce as much energy. For fermentation to produce more energy, more food would be required.

18. a. The cell is a eukaryotic cell and will go through mitosis and cytokinesis. Prokaryotic cells have only one chromosome.

b. Each new cell will receive a copy of each chromosome, so each new cell will have 10 chromosomes.

Interpreting Graphics

19. The cell is eukaryotic because it shows sister chromatids linked at centromeres.

20. The cell is in mitosis because the chromosomes have already duplicated.

21. There are 12 chromatids. There are three pairs of homologous chromosomes.

22. There will be six chromosomes in each new cell.

Standardized Test Preparation

Teacher's Note

To provide practice under more realistic testing conditions, give students 20 minutes to answer all of the questions in this Standardized Test Preparation.

MISCONCEPTION ALERT

Answers to the standardized test preparation can help you identify student misconceptions and misunderstandings.

Passage 1

1. C
2. G
3. A

TEST DOCTOR

Question 1: Students may choose wrong answers A and B if they mistakenly read the passage to say that burning a log is the same as the release of energy in a cell during cellular respiration.

Question 3: Students may choose wrong answer B if they already know that heat is released during cellular respiration. While this may be true, the information is not contained anywhere in this passage. The correct answer, based on the passage, is A.

Standardized Test Preparation

READING

Read each of the passages below. Then, answer the questions that follow each passage.

Passage 1 Perhaps you have heard that jogging or some other kind of exercise "burns" a lot of Calories. The word *burn* is often used to describe what happens when your cells release stored energy from food. The burning of food in living cells is not the same as the burning of logs in a campfire. When logs burn, the energy stored in wood is released as thermal energy and light in a single reaction. But this kind of reaction is not the kind that happens in cells. Instead, the energy that cells get from food molecules is released at each step of a series of chemical reactions.

1. According to the passage, how do cells release energy from food?
   - **A** in a single reaction
   - **B** as thermal energy and light
   - **C** in a series of reactions
   - **D** by burning

2. Which of the following statements is a fact in the passage?
   - **F** Wood burns better than food does.
   - **G** Both food and wood have stored energy.
   - **H** Food has more stored energy than wood does.
   - **I** When it is burned, wood releases only thermal energy.

3. According to the passage, why might people be confused between what happens in a living cell and what happens in a campfire?
   - **A** The word *burn* may describe both processes.
   - **B** Thermal energy is released during both processes.
   - **C** Wood can be burned and broken down by living cells.
   - **D** Jogging and other exercises use energy.

Passage 2 The word *respiration* means "breathing," but cellular respiration is different from breathing. Breathing supplies your cells with the oxygen that they need for cellular respiration. Breathing also rids your body of carbon dioxide, which is a waste product of cellular respiration. Cellular respiration is the chemical process that releases energy from food. Most organisms obtain energy from food through cellular respiration. During cellular respiration, oxygen is used to break down food (glucose) into CO_2 and H_2O, and energy is released. In humans, most of the energy released is used to maintain body temperature.

1. According to the passage, what is glucose?
   - **A** a type of chemical process
   - **B** a type of waste product
   - **C** a type of organism
   - **D** a type of food

2. According to the passage, how does cellular respiration differ from breathing?
   - **F** Breathing releases carbon dioxide, but cellular respiration releases oxygen.
   - **G** Cellular respiration is a chemical process that uses oxygen to release energy from food, but breathing supplies cells with oxygen.
   - **H** Cellular respiration requires oxygen, but breathing does not.
   - **I** Breathing rids your body of waste products, but cellular respiration stores wastes.

3. According to the passage, how do humans use most of the energy released?
   - **A** to break down food
   - **B** to obtain oxygen
   - **C** to maintain body temperature
   - **D** to get rid of carbon dioxide

Passage 2

1. D
2. G
3. C

TEST DOCTOR

Question 2: Students may choose wrong answer F if they mistakenly read the passage to say that cellular respiration "releases" oxygen instead of "requires" oxygen. Cellular respiration releases carbon dioxide and water.

Question 3: The passage talks about the release of energy from food. Students may choose wrong answer A if they mistakenly read the passage to say that most of the energy is used to break down food. Most of the food energy is used to maintain body temperature.

INTERPRETING GRAPHICS

The graph below shows the cell cycle. Use this graph to answer the questions that follow.

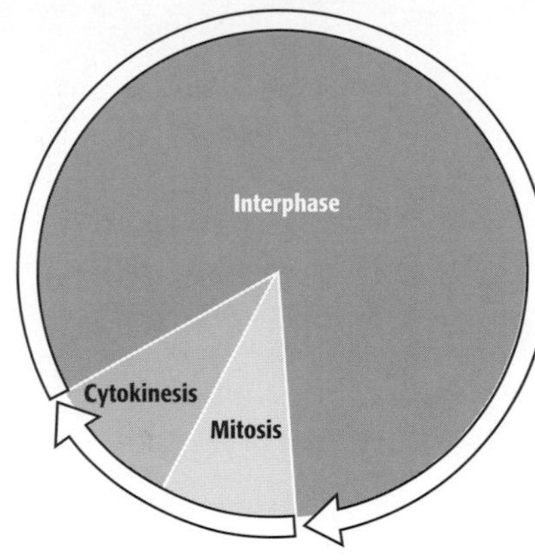

1. Which part of the cell cycle lasts longest?
   A interphase
   B mitosis
   C cytokinesis
   D There is not enough information to determine the answer.

2. Which of the following lists the parts of the cell cycle in the proper order?
   F mitosis, cytokinesis, mitosis
   G interphase, cytokinesis, mitosis
   H interphase, mitosis, interphase
   I mitosis, cytokinesis, interphase

3. Which part of the cell cycle is the briefest?
   A interphase
   B cell division
   C cytokinesis
   D There is not enough information to determine the answer.

4. Why is the cell cycle represented by a circle?
   F The cell cycle is a continuous process that begins again after it finishes.
   G The cell cycle happens only in cells that are round.
   H The cell cycle is a linear process.
   I The cell is in interphase for more than half of the cell cycle.

MATH

Read each question below, and choose the best answer.

1. A normal cell spends 90% of its time in interphase. How is 90% expressed as a fraction?
   A 3/4
   B 4/5
   C 85/100
   D 9/10

2. If a cell lived for 3 weeks and 4 days, how many days did it live?
   F 7
   G 11
   H 21
   I 25

3. How is $2 \times 3 \times 3 \times 3 \times 3$ expressed in exponential notation?
   A 3×2^4
   B 2×3^3
   C 3^4
   D 2×3^4

4. Cell A has 3 times as many chromosomes as cell B has. After cell B's chromosomes double during mitosis, cell B has 6 chromosomes. How many chromosomes does cell A have?
   F 3
   G 6
   H 9
   I 18

5. If $x + 2 = 3$, what does $x + 1$ equal?
   A 4
   B 3
   C 2
   D 1

6. If $3x + 2 = 26$, what does $x + 1$ equal?
   F 7
   G 8
   H 9
   I 10

Standardized Test Preparation

INTERPRETING GRAPHICS

1. A
2. I
3. D
4. F

TEST DOCTOR

Question 2: Students may select incorrect answers F and H if they follow the arrow but skip a step on the graph. They may select incorrect answer G if they ignore the direction of the arrow. Only answer I has the steps of the cell cycle in proper sequence.

MATH

1. D
2. I
3. D
4. H
5. C
6. H

Question 4: Students may have trouble converting this word problem into a numerical statement because they may confuse what is happening to cell B (its chromosomes are doubling in number to 6, which means that it starts with 3) with what is happening to cell A (nothing). Students may select incorrect answer I because 3 times 6 is 18. Students who are struggling may want to create a small data table that shows what they "know" (the information given in the word problem) and what they are trying to find out (what the question asks). Word problems are a challenge for many students, and often a table or chart will help them keep the information straight.

CHAPTER RESOURCES

Chapter Resource File

• Standardized Test Preparation GENERAL

State Resources

For specific resources for your state, visit **go.hrw.com** and type in the keyword **HSMSTR.**

Scientific Discovery

Background

The release of energy from food is called *cellular respiration.* Cellular respiration takes place in two stages. The end result of the process is that energy is stored in the cell in the form of ATP (adenosine triphosphate) molecules.

In the microbial battery, scientists harvest some of this energy and transfer it into electricity that can be readily used.

One of the benefits of the microbial battery is its ability to make use of waste products. Ask students to consider the effect this might have on the energy demands of nations that have limited access to fossil fuels.

Science Fiction

Teaching Strategy—Basic

This is a relatively long story, containing quite a few medical terms. Students may find it easier to read if the class discusses some of the unfamiliar terms before they start reading the story.

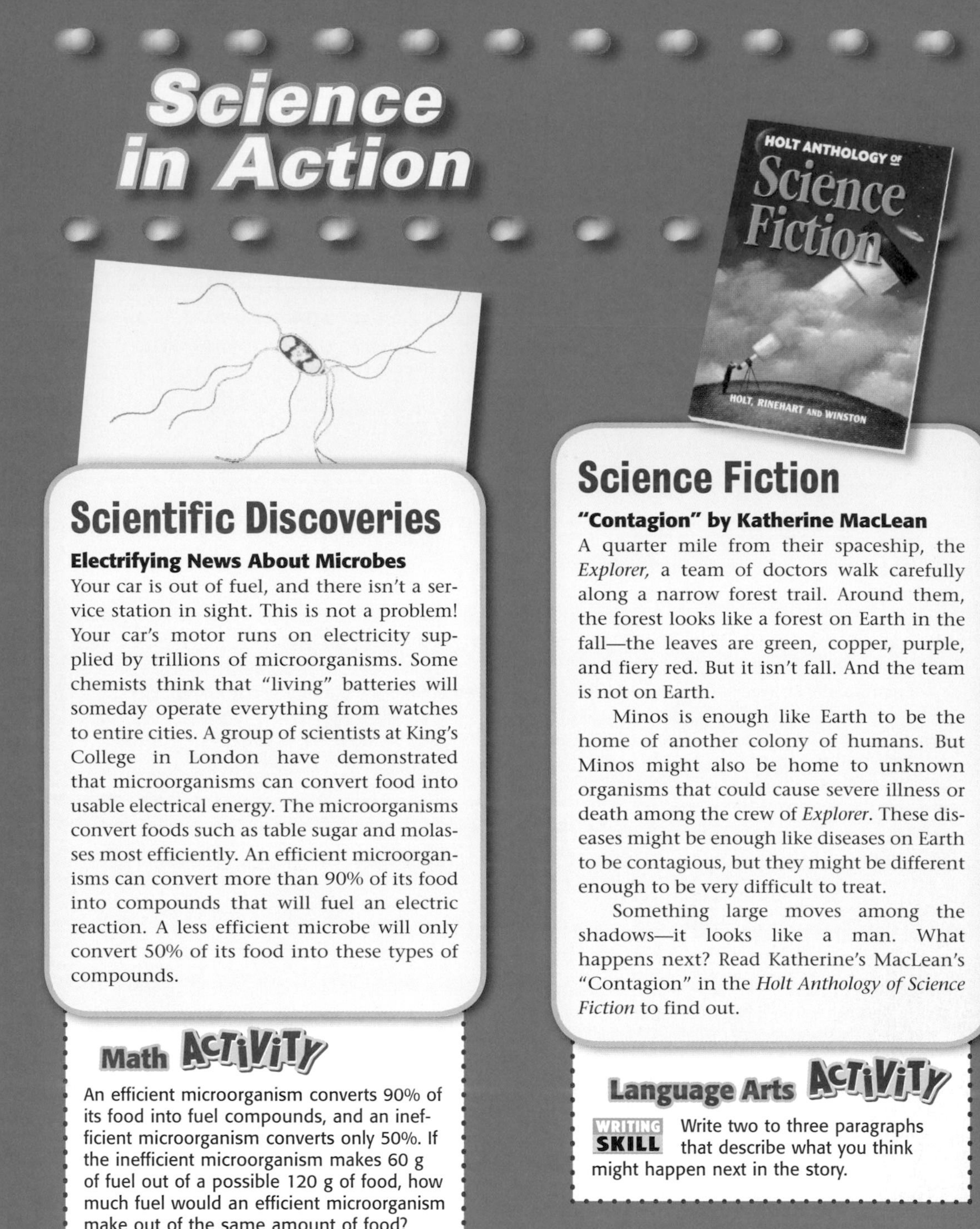

Science in Action

Scientific Discoveries

Electrifying News About Microbes

Your car is out of fuel, and there isn't a service station in sight. This is not a problem! Your car's motor runs on electricity supplied by trillions of microorganisms. Some chemists think that "living" batteries will someday operate everything from watches to entire cities. A group of scientists at King's College in London have demonstrated that microorganisms can convert food into usable electrical energy. The microorganisms convert foods such as table sugar and molasses most efficiently. An efficient microorganisms can convert more than 90% of its food into compounds that will fuel an electric reaction. A less efficient microbe will only convert 50% of its food into these types of compounds.

Math Activity

An efficient microorganism converts 90% of its food into fuel compounds, and an inefficient microorganism converts only 50%. If the inefficient microorganism makes 60 g of fuel out of a possible 120 g of food, how much fuel would an efficient microorganism make out of the same amount of food?

Science Fiction

"Contagion" by Katherine MacLean

A quarter mile from their spaceship, the *Explorer,* a team of doctors walk carefully along a narrow forest trail. Around them, the forest looks like a forest on Earth in the fall—the leaves are green, copper, purple, and fiery red. But it isn't fall. And the team is not on Earth.

Minos is enough like Earth to be the home of another colony of humans. But Minos might also be home to unknown organisms that could cause severe illness or death among the crew of *Explorer*. These diseases might be enough like diseases on Earth to be contagious, but they might be different enough to be very difficult to treat.

Something large moves among the shadows—it looks like a man. What happens next? Read Katherine's MacLean's "Contagion" in the *Holt Anthology of Science Fiction* to find out.

Language Arts Activity

WRITING SKILL Write two to three paragraphs that describe what you think might happen next in the story.

Answer to Math Activity

An efficient microbe converts 90% of its food to fuel compounds; 90% of 120 g is 108 g of fuel compounds.

Answer to Language Arts Activity

Students' predictions will vary. Whatever a student predicts, the prediction should be reasonably related to the information that the student has from reading this introductory paragraph.

Careers

Jerry Yakel

Neuroscientist Jerry Yakel credits a sea slug for making him a neuroscientist. In a college class studying neurons, or nerve cells, Yakel got to see firsthand how ions move across the cell membrane of *Aplysia californica,* also known as a sea hare. He says, "I was totally hooked. I knew that I wanted to be a neurophysiologist then and there. I haven't wavered since."

Today, Yakel is a senior investigator for the National Institutes of Environmental Health Sciences, which is part of the U.S. government's National Institutes of Health. "We try to understand how the normal brain works," says Yakel of his team. "Then, when we look at a diseased brain, we train to understand where the deficits are. Eventually, someone will have an idea about a drug that will tweak the system in this or that way."

Yakel studies the ways in which nicotine affects the human brain. "It is one of the most prevalent and potent neurotoxins in the environment," says Yakel. "I'm amazed that it isn't higher on the list of worries for the general public."

Social Studies ACTIVITY

WRITING SKILL Research a famous or historical figure in science. Write a short report that outlines how he or she became interested in science.

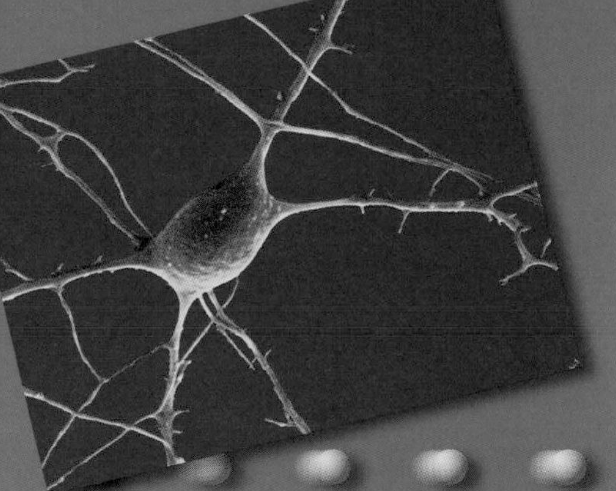

To learn more about these Science in Action topics, visit **go.hrw.com** and type in the keyword **HL5ACTF**.

Current Science

Check out Current Science® articles related to this chapter by visiting go.hrw.com. Just type in the keyword HL5CS04.

Careers

Background

Jerry Yakel grew up in Ventura County, California. After graduating from high school, he attended a nearby community college "ostensibly to continue running track, figuring out life." Eventually, he relocated to Oregon State University, where he obtained a B.S. in 1982. He was accepted into UCLA in 1983 and received a Ph.D. in 1988.

Working for the NIH was not something Yakel originally expected to do. "Most of us trained in universities think we will work there," he says. He does enjoy some aspects of being outside the typical university setting. "In the NIH, we are supposed to take more risks in our research." He also enjoys the focus he is able to bring to his work. "I miss having students to teach, but then again I get to spend more time doing research," Yakel says. His choice of environment hasn't affected his passion. "Honestly, the type of research I [would] do actually is the same."

Answer to Social Studies Activity

Students may write about any historical figure in science. Some students may go back as far as Archimedes; others may choose Hypatia (the first woman to be a true astronomer), Benjamin Franklin, Marie Curie, Albert Einstein, Rosalind Franklin, or one of hundreds of other people. The important issues for the student are why the person is important to science and how the person became interested in science.

Heredity
Chapter Planning Guide

Compression guide:
To shorten instruction because of time limitations, omit the Chapter Lab.

OBJECTIVES	LABS, DEMONSTRATIONS, AND ACTIVITIES	TECHNOLOGY RESOURCES
PACING • 90 min pp. 54–61 **Chapter Opener**	**SE Start-up Activity,** p. 55 ◆ GENERAL	**OSP Parent Letter** ■ GENERAL **CD Student Edition on CD-ROM** **CD Guided Reading Audio CD** ■ **TR Chapter Starter Transparency*** **VID Brain Food Video Quiz**
Section 1 Mendel and His Peas • Explain the relationship between traits and heredity. • Describe the experiments of Gregor Mendel. • Explain the difference between dominant and recessive traits.	**TE Activity** Trait Trends, p. 56 GENERAL **SE School-to-Home Activity** Describing Traits, p. 57 GENERAL **TE Demonstration** Flower Dissection, p. 58 ◆ BASIC **TE Activity** Mendelian Crosses, p. 58 ADVANCED **SE Science in Action** Math, Science, and Social Studies Activities, pp. 82–83 GENERAL	**CRF Lesson Plans*** **TR Bellringer Transparency*** **CRF SciLinks Activity*** GENERAL
PACING • 90 min pp. 62–67 **Section 2 Traits and Inheritance** • Explain how genes and alleles are related to genotype and phenotype. • Use the information in a Punnett square. • Explain how probability can be used to predict possible genotypes in offspring. • Describe three exceptions to Mendel's observations.	**TE Demonstration,** p. 62 ◆ BASIC **SE Quick Lab** Making a Punnett Square, p. 63 GENERAL **CRF Datasheet for Quick Lab*** **SE Quick Lab** Taking Your Chances, p. 64 ◆ GENERAL **CRF Datasheet for Quick Lab*** **TE Connection Activity** Math, p. 64 ADVANCED **SE Connection to Chemistry** Round and Wrinkled, p. 65 GENERAL **SE Model-Making Lab** Bug Builders, Inc., p. 76 ◆ GENERAL **CRF Datasheet for Chapter Lab***	**CRF Lesson Plans*** **TR Bellringer Transparency*** **TR** Punnett Squares **TR** ***LINK TO PHYSICAL SCIENCE*** The Periodic Table of the Elements* **VID Lab Videos for Life Science**
PACING • 45 min pp. 68–75 **Section 3 Meiosis** • Explain the difference between mitosis and meiosis. • Describe how chromosomes determine sex. • Explain why sex-linked disorders occur in one sex more often than in the other. • Interpret a pedigree.	**TE Activity** Crosses, p. 68 GENERAL **TE Connection Activity** Math, p. 68 ADVANCED **TE Activity** Describing Meiosis, p. 71 BASIC **TE Connection Activity** Math, p. 71 GENERAL **TE Group Activity** Comparing Mitosis and Meiosis, p. 72 GENERAL **TE Connection Activity** Language Arts, p. 73 ADVANCED **SE Inquiry Lab** Tracing Traits, p. 187 GENERAL **CRF Datasheet for LabBook*** **LB Long-Term Projects & Research Ideas** Portrait of a Dog* ADVANCED	**CRF Lesson Plans*** **TR Bellringer Transparency*** **TR** The Steps of Meiosis: A* **TR** The Steps of Meiosis: B* **TR** Meiosis and Dominance* **TE Internet Activity,** p. 75 GENERAL

PACING • 90 min

CHAPTER REVIEW, ASSESSMENT, AND STANDARDIZED TEST PREPARATION

CRF Vocabulary Activity* GENERAL
SE Chapter Review, pp. 78–79 GENERAL
CRF Chapter Review* ■ GENERAL
CRF Chapter Tests A* ■ GENERAL, **B*** ADVANCED, **C*** SPECIAL NEEDS
SE Standardized Test Preparation, pp. 80–81 GENERAL
CRF Standardized Test Preparation* GENERAL
CRF Performance-Based Assessment* GENERAL
OSP Test Generator GENERAL
CRF Test Item Listing* GENERAL

Online and Technology Resources

Visit **go.hrw.com** for a variety of free resources related to this textbook. Enter the keyword **HL5HER.**

Students can access interactive problem-solving help and active visual concept development with the *Holt Science and Technology* Online Edition available at **www.hrw.com.**

Guided Reading Audio CD

A direct reading of each chapter using instructional visuals as guideposts. For auditory learners, reluctant readers, and Spanish-speaking students. Available in English and Spanish.

KEY

SE Student Edition	CRF Chapter Resource File	SS Science Skills Worksheets	* Also on One-Stop Planner
TE Teacher Edition	OSP One-Stop Planner	MS Math Skills for Science Worksheets	◆ Requires advance prep
	LB Lab Bank	CD CD or CD-ROM	■ Also available in Spanish
	TR Transparencies	VID Classroom Video/DVD	

SKILLS DEVELOPMENT RESOURCES	SECTION REVIEW AND ASSESSMENT	STANDARDS CORRELATIONS
SE Pre-Reading Activity, p. 54 GENERAL **OSP Science Puzzlers, Twisters & Teasers*** GENERAL		National Science Education Standards UCP 2, 3; LS 1d, 2c
CRF Directed Reading A* ■ BASIC, **B*** SPECIAL NEEDS **CRF Vocabulary and Section Summary*** ■ GENERAL **SE Reading Strategy** Brainstorming, p. 56 GENERAL **SE Math Practice** Understanding Ratios, p. 60 GENERAL **TE Reading Strategy** Paired Reading, p. 57 BASIC **TE Inclusion Strategies,** p. 59 ◆ **MS Math Skills for Science** What Is a Ratio?* GENERAL **SS Science Skills** Finding Useful Sources* GENERAL **CRF Critical Thinking** A Bittersweet Solution* ADVANCED	**SE Reading Checks,** pp. 56, 59, 60 GENERAL **TE Reteaching,** p. 60 BASIC **TE Quiz,** p. 118 GENERAL **TE Alternative Assessment,** p. 60 ADVANCED **SE Section Review,*** p. 61 ■ GENERAL **TE Homework,** p. 61 GENERAL **CRF Section Quiz*** ■ GENERAL	UCP 1, 2; SAI 1, 2; ST 2; SPSP 5; HNS 1, H2, 3; LS 2b, 2e; *Chapter Lab:* SAI 1; HNS 2; LS 2c, 2e; *LabBook:* UCP 2; SAI 1; HNS 2; LS 2b, 2c, 2e
CRF Directed Reading A* ■ BASIC, **B*** SPECIAL NEEDS **CRF Vocabulary and Section Summary*** ■ GENERAL **SE Reading Strategy** Paired Summarizing, p. 62 GENERAL **SE Math Focus** Probability, p. 65 GENERAL **MS Math Skills for Science** Punnett Square Popcorn* GENERAL **CRF Reinforcement Worksheet** Dimples and DNA* BASIC	**SE Reading Checks,** pp. 62, 64, 66 GENERAL **TE Homework,** p. 65 GENERAL **TE Reteaching,** p. 66 BASIC **TE Quiz,** p. 66 GENERAL **TE Alternative Assessment,** p. 67 GENERAL **SE Section Review,*** p. 67 ■ GENERAL **CRF Section Quiz*** ■ GENERAL	UCP 2, 3; LS 2a, 2b, 2c, 2d, 2e
CRF Directed Reading A* ■ BASIC, **B*** SPECIAL NEEDS **CRF Vocabulary and Section Summary*** ■ GENERAL **SE Reading Strategy** Reading Organizer, p. 68 GENERAL **SE Connection to Language Arts** Greek Roots, p. 69 GENERAL **TE Reading Strategy** Prediction Guide, p. 70 GENERAL **TE Inclusion Strategies,** p. 72 **CRF Reinforcement Worksheet** Vocabulary Garden* BASIC	**SE Reading Checks,** pp. 69, 70 GENERAL **TE Reteaching,** p. 74 BASIC **TE Quiz,** p. 74 GENERAL **TE Alternative Assessment,** p. 74 GENERAL **TE Homework,** p. 74 ADVANCED **SE Section Review,*** p. 75 ■ GENERAL **CRF Section Quiz*** ■ GENERAL	UCP 4, 5; SAI 1; SPSP 5; HNS 2, 3; LS 1c, 1d, 2a, 2b, 2c, 2d

This convenient CD-ROM includes:
- **Lab Materials QuickList Software**
- **Holt Calendar Planner**
- **Customizable Lesson Plans**
- **Printable Worksheets**
- **ExamView® Test Generator**

CNN Student News

cnnstudentnews.com

Find the latest news, lesson plans, and activities related to important scientific events.

www.scilinks.org

Maintained by the **National Science Teachers Association.** See Chapter Enrichment pages for a complete list of topics.

Current Science®

Check out ***Current Science*** articles and activities by visiting the HRW Web site at **go.hrw.com.** Just type in the keyword **HL5CS05T.**

Classroom Videos
- **Lab Videos** demonstrate the chapter lab.
- **Brain Food Video Quizzes** help students review the chapter material.

3 Chapter Resources

Visual Resources

CHAPTER STARTER TRANSPARENCY

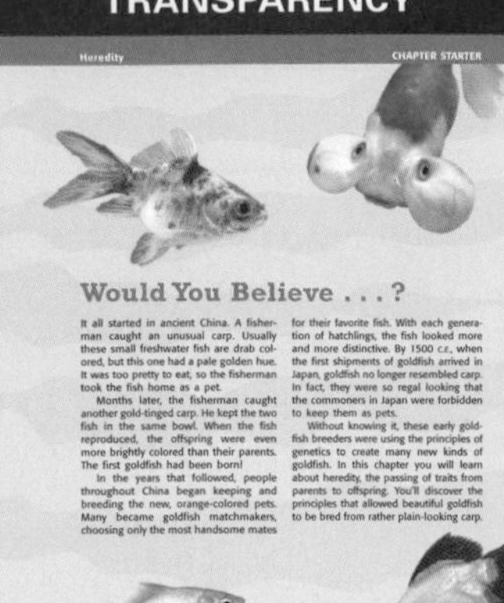

BELLRINGER TRANSPARENCIES

Heredity — BELLRINGER TRANSPARENCY

Section: Mendel and His Peas

You have probably noticed that different people have different characteristics, such as eye color, hair color, or whether or not their ear lobes attach directly to their head or hang down loosely. These characteristics are called traits. Where do you think people get these different traits? How do you think they are passed from one generation to the next?

Write your answers in your **science journal.**

Section: Traits and Inheritance

If you flip a coin, what are the chances that it will land on heads? tails? Suppose that you flip the coin, get heads, and then flip again. What are the chances that you will get heads again? What are the chances you will get heads two times in a row? five times?

Record your answers in your **science journal.**

TEACHING TRANSPARENCIES

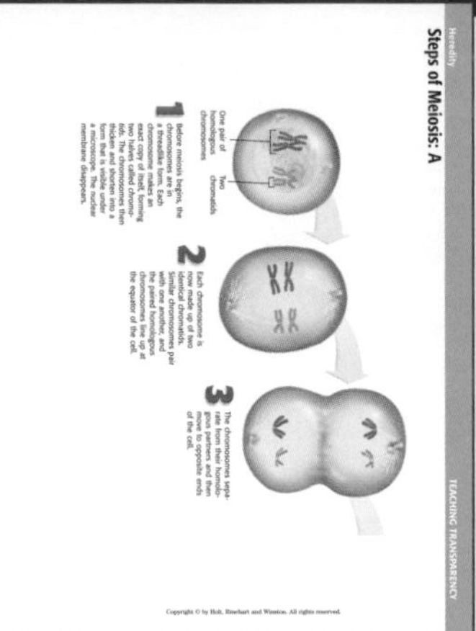

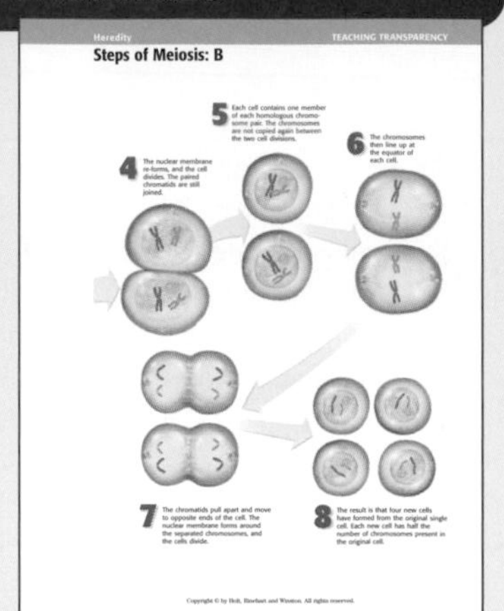

TEACHING TRANSPARENCIES

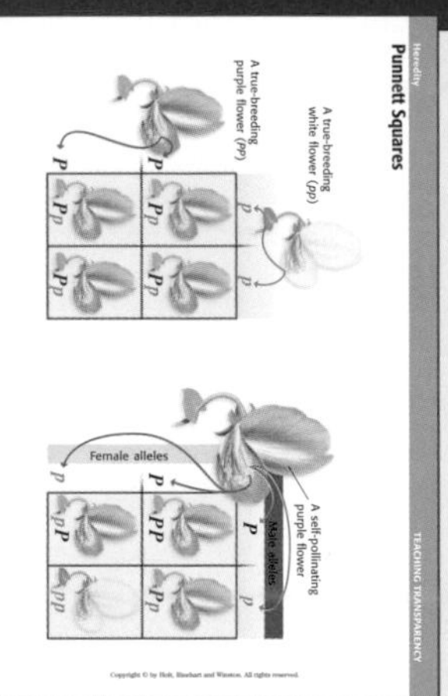

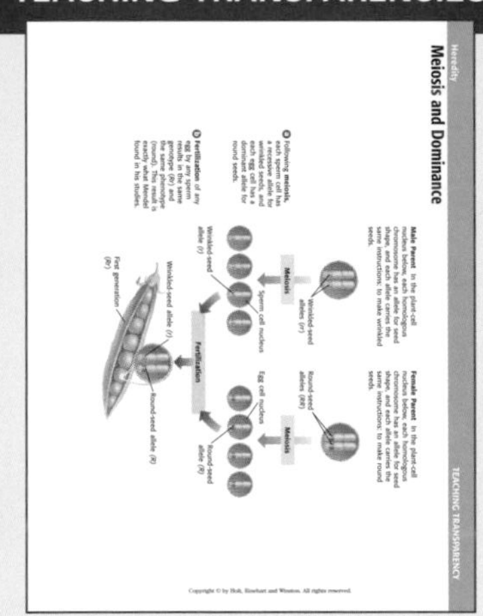

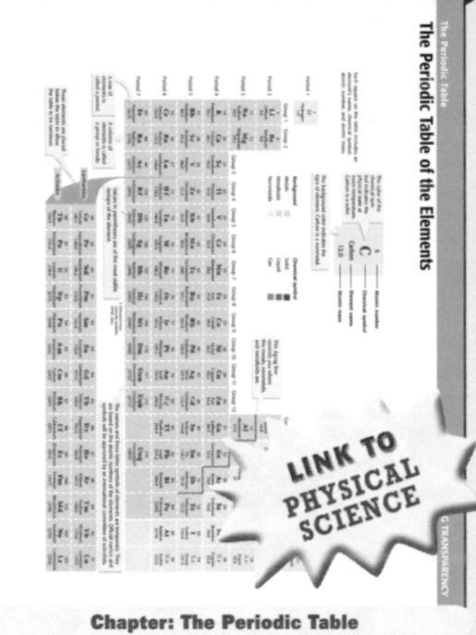

CONCEPT MAPPING TRANSPARENCY

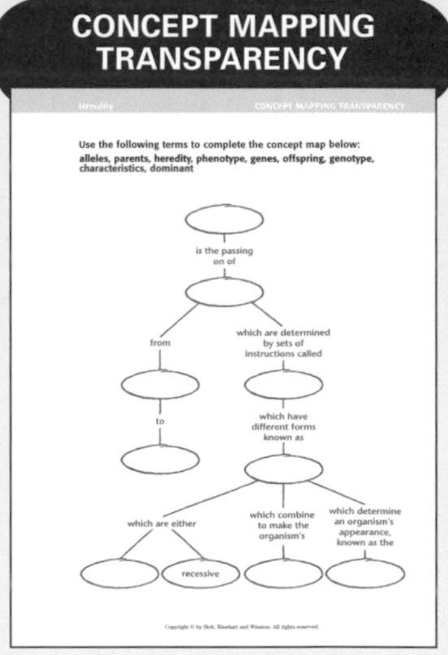

Planning Resources

LESSON PLANS

Lesson Plan — SAMPLE

Section: Waves

Pacing

Regular Schedule: with lab(s):2 days without lab(s)2 days
Block Schedule: with lab(s):1 1/2 days without lab(s)1 day

Objectives

1. Relate the seven properties of life to a living organism.
2. Describe seven themes that can help you to organize what you learn about biology.
3. Identify the tiny structures that make up all living organisms.
4. Differentiate between reproduction and heredity and between metabolism and homeostasis.

National Science Education Standards Covered

LSInter6: Cells have particular structures that underlie their functions.

LSMat1: Most cell functions involve chemical reactions.

LSBeh1: Cells store and use information to guide their functions.

UCP1: Cell functions are regulated.

SI1: Cells can differentiate and form complete multicellular organisms.

PS1: Species evolve over time.

ESS1: The great diversity of organisms is the result of more than 3.5 billion years of evolution.

ESS2: Natural selection and its evolutionary consequences provide a scientific explanation for the fossil record of ancient life forms as well as for the striking molecular similarities observed among the diverse species of living organisms.

ST1: The millions of different species of plants, animals, and microorganisms that live on Earth today are related by descent from common ancestors.

ST2: The energy for life primarily comes from the sun.

SPSP1: The complexity and organization of organisms accommodates the need for obtaining, transforming, transporting, releasing, and eliminating the matter and energy used to sustain the organism.

SPSP6: As matter and energy flows through different levels of organization of living systems—cells, organs, communities—and between living systems and the physical environment, chemical elements are recombined in different ways.

HNS1: Organisms have behavioral responses to internal changes and to external stimuli.

PARENT LETTER

SAMPLE

Dear Parent,

Your son's or daughter's science class will soon begin exploring the chapter entitled "The World of Physical Science." In this chapter, students will learn about how the scientific method applies to the world of physical science and the role of physical science in the world. By the end of the chapter, students should demonstrate a clear understanding of the chapter's main ideas and be able to discuss the following topics:

1. physical science as the study of energy and matter (Section 1)
2. the role of physical science in the world around them (Section 1)
3. careers that rely on physical science (Section 1)
4. the steps used in the scientific method (Section 2)
5. examples of technology (Section 2)
6. how the scientific method is used to answer questions and solve problems (Section 2)
7. how our knowledge of science changes over time (Section 2)
8. how models represent real objects or systems (Section 3)
9. examples of different ways models are used in science (Section 3)
10. the importance of the International System of Units (Section 4)
11. the appropriate units to use for particular measurements (Section 4)
12. how area and density are derived quantities (Section 4)

Questions to Ask Along the Way

You can help your son or daughter learn about these topics by asking interesting questions such as the following:

- What are some surprising careers that use physical science?
- What is a characteristic of a good hypothesis?
- When is it a good idea to use a model?
- Why do Americans measure things in terms of inches and yards and meters ?

ALSO IN SPANISH

TEST ITEM LISTING

TEST ITEM LISTING
The World of Science — SAMPLE

MULTIPLE CHOICE

1. A limitation of models is that
   a. they are large enough to see.
   b. they do not act exactly like the things that they model.
   c. they are smaller than the things that they model.
   d. they model unfamiliar things.
   Answer: B Difficulty: I Section: 3 Objective: 2
2. The length 10 m is equal to
   a. 100 cm. c. 10,000 mm.
   b. 1,000 cm. d. Both (b) and (c)
   Answer: B Difficulty: I Section: 3 Objective: 2
3. To be valid, a hypothesis must be
   a. testable. c. made into a law.
   b. supported by evidence. d. Both (a) and (b)
   Answer: B Difficulty: I Section: 3 Objective: 2 1
4. The statement "Sheila has a stain on her shirt" is an example of a(n)
   a. law. c. observation.
   b. hypothesis. d. prediction.
   Answer: B Difficulty: I Section: 3 Objective: 2
5. A hypothesis is often developed out of
   a. observations. c. laws.
   b. experiments. d. Both (a) and (b)
   Answer: B Difficulty: I Section: 3 Objective: 2
6. How many milliliters are in 3.5 kL?
   a. 3,500 mL c. 3,500, 000 mL
   b. 0.0035 mL d. 35,000 mL
   Answer: B Difficulty: I Section: 3 Objective: 2
7. A map of Seattle is an example of a
   a. law. c. model.
   b. theory. d. unit.
   Answer: B Difficulty: I Section: 3 Objective: 2
8. A lab has the safety icons shown below. These icons mean that you should wear
   a. only safety goggles. c. safety goggles and a lab apron.
   b. only a lab apron. d. safety goggles, a lab apron, and gloves.
   Answer: B Difficulty: I Section: 3 Objective: 2
9. The law of conservation of mass says the tot al mass before a chemical change is
   a. more than the total mass after the change.
   b. less than the total mass after the change.
   c. the same as the total mass after the change.
   d. not the same as the total mass after the change.
   Answer: B Difficulty: I Section: 3 Objective: 2
10. In which of the following areas might you find a geochemist at work?
   a. studying the chemistry of rocks c. studying fishes
   b. studying forestry d. studying the atmosphere
   Answer: B Difficulty: I Section: 3 Objective: 2

One-Stop Planner® CD-ROM

This CD-ROM includes all of the resources shown here and the following time-saving tools:

- *Lab Materials QuickList Software*
- *Customizable lesson plans*
- *Holt Calendar Planner*
- *The powerful ExamView® Test Generator*

For a preview of available worksheets covering math and science skills, see pages T12–T19. All of these resources are also on the One-Stop Planner®.

Meeting Individual Needs

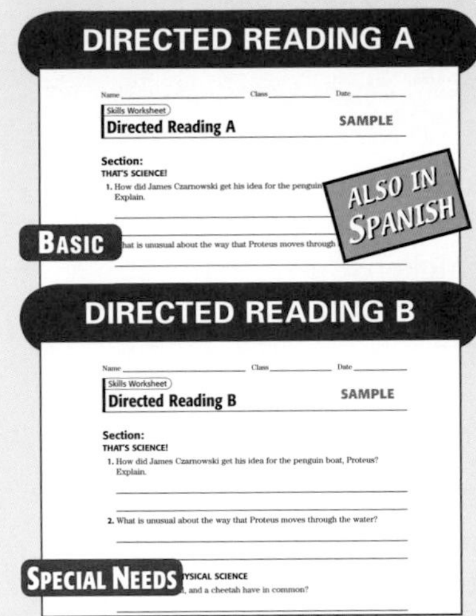

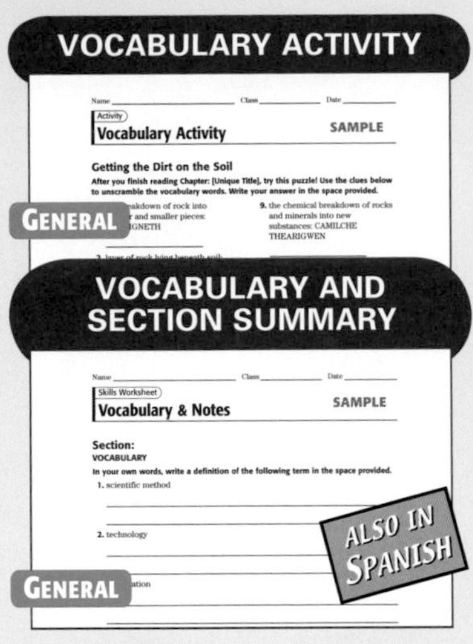

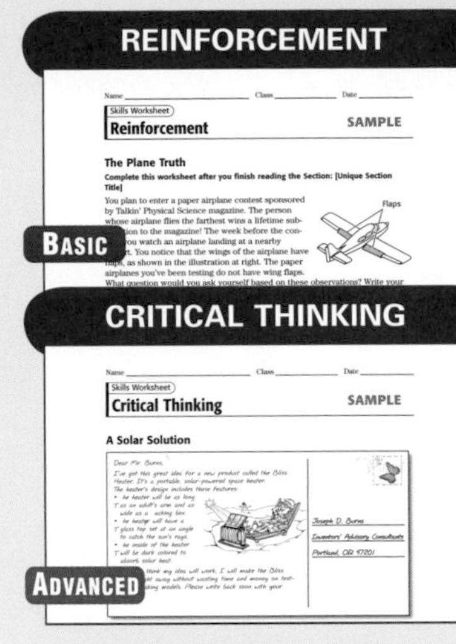

Labs and Activities

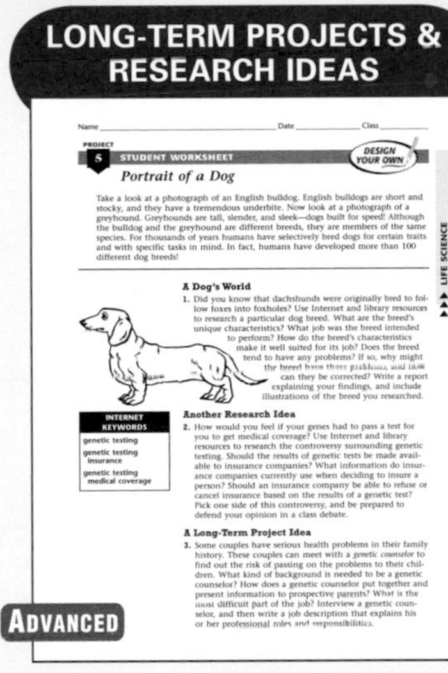

Review and Assessments

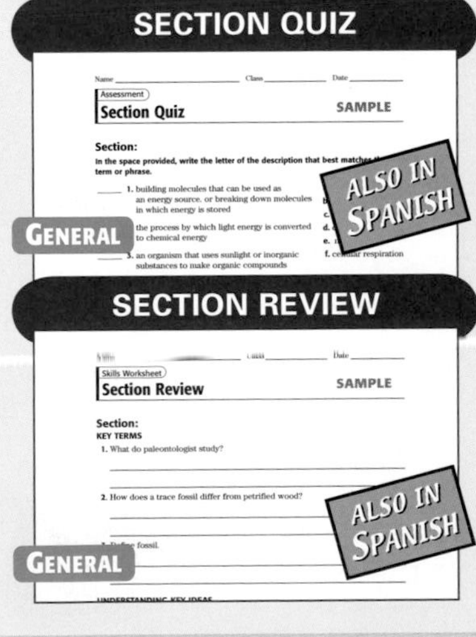

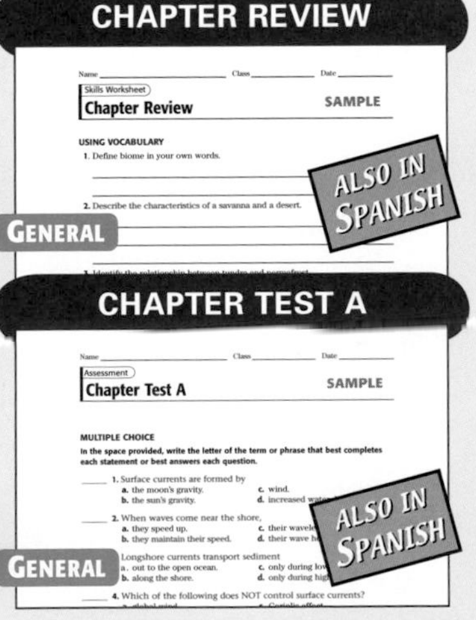

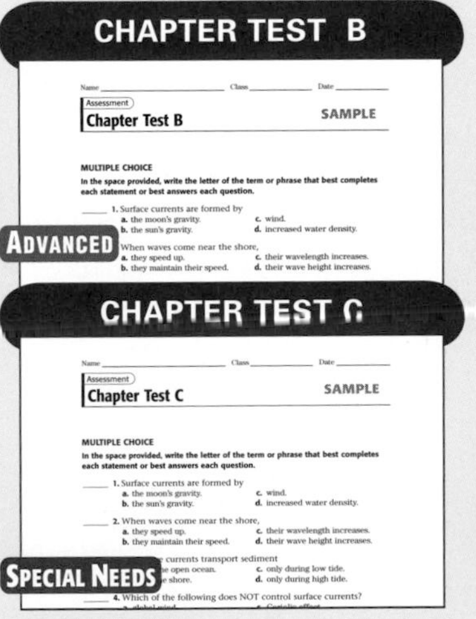

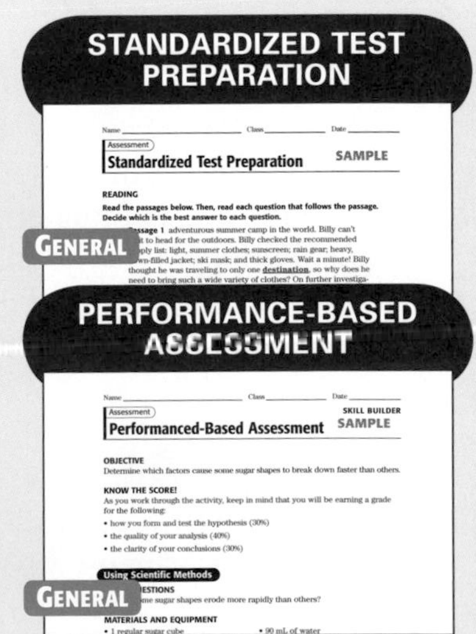

3 Chapter Enrichment

This Chapter Enrichment provides relevant and interesting information to expand and enhance your presentation of the chapter material.

Section 1

Mendel and His Peas

Gregor Mendel

- In 1843, in the city of Brünn, Austria (which is now Brno, a city in the Czech Republic), Gregor Mendel (1822–1884) entered a monastery. In 1865, Mendel published the results of his garden-pea experiments. Although Mendel's ideas are widespread today, few scientists learned of his work during his lifetime because there were few ways to distribute information. Mendel presented his findings in two lectures, and only 40 copies of his work were printed in his lifetime.
- When Mendel was elected abbot of the monastery in 1868, his duties prevented him from visiting other scientists or attending conferences where he could have discussed his results. Not until 1900, when Mendel's work was rediscovered by scientists in Holland, Germany, and Austria-Hungary, were his theories spread through the scientific community.

- Mendel's work was used to support Darwin's theory of evolution by natural selection and is considered to be the foundation of modern genetics. Mendel also made contributions to beekeeping, horticulture, and meteorology. In 1877, Mendel became interested in weather and began issuing weather reports to local farmers.

Is That a Fact!

◆ From 1856 to 1863, while studying inheritance, Mendel grew almost 30,000 pea plants!

Section 2

Traits and Inheritance

Punnett and His Squares

- Punnett squares are named after their inventor, R. C. Punnett. Punnett explored inheritance by crossing different breeds of chickens in the early 1900s, soon after Mendel's work was rediscovered.

Pollination

- Pollen can be transferred between plants by wind, insects, and a variety of animals. Some common pollinators are bees, butterflies, moths, flies, bats, and birds. Animals are attracted to the color of the flower, the patterns found on the petals, or the flower's fragrance. Pollen is an excellent food for some animals.

Is That a Fact!

◆ Male bees have only half the number of chromosomes that female bees have.

Section 3

Meiosis

Chromosomes

- Chromosomes are composed of genes, the sequences of DNA that provide the instructions for making all the proteins in an organism. During cell division, the duplicated chromosomes separate so that one copy of each chromosome is present in the two new cells.

Walther Flemming

- Walther Flemming (1843–1905), a German physician and anatomist, was the first to use a microscope and special dyes to study cell division. Flemming used the term *mitosis* to describe the process he observed.

Mitosis

- In mitosis, a cell divides to form two identical cells. The steps of the process are similar in almost all living organisms. In addition to enabling growth, mitosis allows organisms to replace cells that have died or malfunctioned. Mitosis can take anywhere from a few minutes to a few hours, and it may be affected by characteristics of the environment, such as light and temperature.

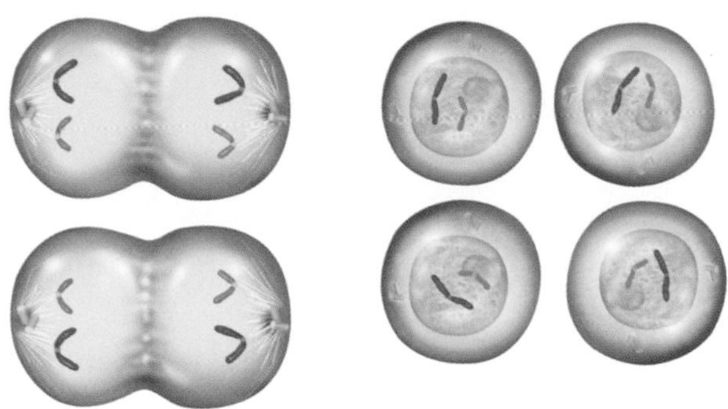

Meiosis

- Meiosis is not the same in all organisms. In humans, meiosis is very different in males and females. In males, meiosis results in four similar sperm cells. In females, however, only one functional egg is produced. The other resulting cells, which are known as *polar bodies,* are formed during the division of the original cell but do not mature.

Genetic Disorders

- A genetic disorder results from an inherited disruption in an organism's DNA. These inherited disruptions can take several forms, including a change in the number of chromosomes and the deletion or duplication of entire chromosomes or parts of chromosomes. Often, the change responsible for a disorder is the alteration of a single specific gene. However, some genetic disorders result from several of these genetic alterations occurring simultaneously. Diseases resulting from these alterations cause a wide variety of physical malfunctions and developmental problems.

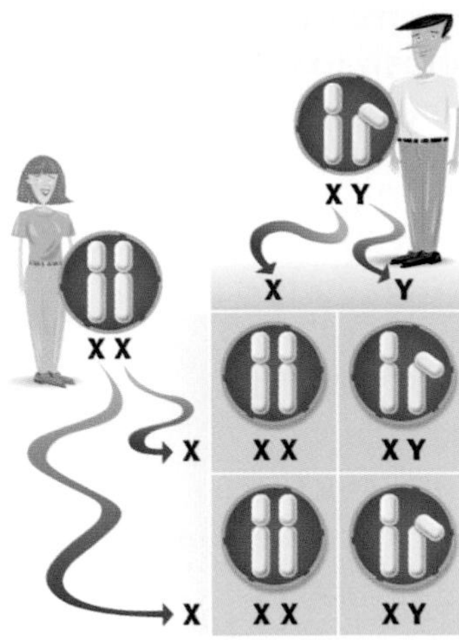

- Cystic fibrosis (CF) is a disease for which one in 31 Americans carries a recessive trait. If two of these people have children together, there is a 25% chance that any child born to them will have the disease. CF affects the intestinal, bronchial, and sweat glands. In people with CF, these glands secrete thick, sticky fluids that are difficult for the body to process, impeding breathing and digestion. Due to improvements in diagnosis and treatment, median life expectancy for those with CF has improved from under 10 years in 1960 to an estimated 40 years for those born in 1990.
- Rubinstein-Taybi syndrome (RTS) is a complex genetic disorder whose characteristics include broad thumbs and toes, mental retardation, and distinctive facial features. This wide range of characteristics is believed to be linked to any one of a number of mutations in a gene responsible for providing the body with a protein called *CBP.* CBP is thought to be vital to the body's delicate metabolism. Because CBP greatly influences body processes, people with a problem producing CBP have a wide range of difficulties. Children with RTS can benefit from proper nutrition and early intervention with therapies and special education.

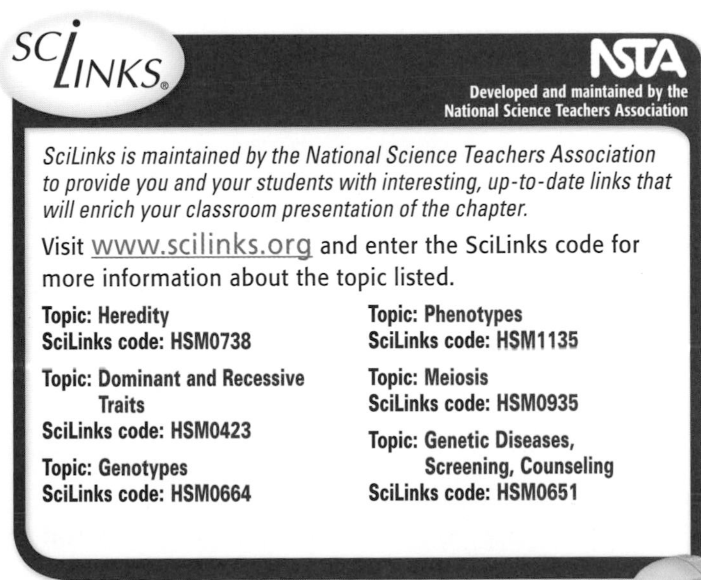

SciLinks is maintained by the National Science Teachers Association to provide you and your students with interesting, up-to-date links that will enrich your classroom presentation of the chapter.

Visit www.scilinks.org and enter the SciLinks code for more information about the topic listed.

Topic: Heredity
SciLinks code: HSM0738

Topic: Dominant and Recessive Traits
SciLinks code: HSM0423

Topic: Genotypes
SciLinks code: HSM0664

Topic: Phenotypes
SciLinks code: HSM1135

Topic: Meiosis
SciLinks code: HSM0935

Topic: Genetic Diseases, Screening, Counseling
SciLinks code: HSM0651

Overview

Tell students that this chapter will introduce heredity—the ways that traits are passed from parents to offspring. The chapter describes the ways scientists study heredity and the role of sexual reproduction.

Assessing Prior Knowledge

Students should be familiar with the following topics:

- scientific methods
- cells
- mitosis

Identifying Misconceptions

Students often hold onto misconceptions about inheritance, even after instruction. For example, they may believe that traits are inherited from only one parent or that environmentally caused characteristics may be passed on to offspring. Students tend to understand phenotype (physical traits) more easily than genotype. Finally, the process of meiosis, as it relates to the structure and location of chromosomes, is very complex. Most students require time and repeated exposure in order to comprehend all the parts and steps of meiosis. Assure students that the concepts of heredity are a foundation that will be built upon throughout their studies of life science.

3

Heredity

About the PHOTO

The guinea pig in the middle has dark fur, and the other two have light orange fur. The guinea pig on the right has longer hair than the other two. Why do these guinea pigs look different from one another? The length and color of their fur was determined before they were born. These are just two of the many traits determined by genetic information. Genetic information is passed on from parents to their offspring.

PRE-READING ACTIVITY

FOLDNOTES **Key-Term Fold** Before you read the chapter, create the FoldNote entitled "Key-Term Fold" described in the **Study Skills** section of the Appendix. Write a key term from the chapter on each tab of the key-term fold. Under each tab, write the definition of the key term.

Standards Correlations

National Science Education Standards

The following codes indicate the National Science Education Standards that correlate to this chapter. The full text of the standards is at the front of the book.

Chapter Opener
UCP 2, 3; LS 1d, 2c

Section 1 Mendel and His Peas
UCP 1, 2; SAI 1, 2; ST 2; SPSP 5; HNS 1, 2, 3; LS 2b, 2e

Section 2 Traits and Inheritance
UCP 2, 3; LS 2a, 2b, 2c, 2d, 2e

Section 3 Meiosis
UCP 4, 5; SAI 1; SPSP 5; HNS 2, 3; LS 1c, 1d, 2a, 2b, 2c, 2d

Chapter Lab
SAI 1; HNS 2; LS 2c, 2e

Chapter Review
LS 1c, 2a, 2b, 2c, 2d, 2e

Science in Action
ST 2; SPSP 5

START-UP ACTiViTy

Clothing Combos

How do the same parents have children with many different traits?

Procedure

1. Gather **three boxes**. Put **five hats** in the first box, **five gloves** in the second, and **five scarves** in the third.
2. Without looking in the boxes, select one item from each box. Repeat this process, five students at a time, until the entire class has picked "an outfit." Record what outfit each student chooses.

Analysis

1. Were any two outfits exactly alike? Did you see all possible combinations? Explain your answer.
2. Choose a partner. Using your outfits, how many different combinations could you make by giving a third person one hat, one glove, and one scarf? How is this process like parents passing traits to their children?
3. After completing this activity, why do you think parents often have children who look very different from each other?

START-UP ACTiViTy

MATERIALS

For Each Group

- boxes large, (3)
- gloves different types, (5)
- hats different types, (5)
- scarves different types, (5)

Safety Caution: Infestations of head lice are a common problem in schools. Sharing hats should be avoided during such a period. Jackets or sweatshirts could be substituted for hats in this exercise.

Answers

1. Answers may vary. There should be many different combinations. It is not likely that students will see all of the possible combinations.
2. Sample answer: eight new combinations (taken from the outfits of the two "parents") would be possible for the third person ("offspring"). This process is like inheritance because you are choosing combinations of hats, scarves, and gloves randomly. Traits are also passed from parent to offspring randomly. By combining the traits (outfits) of two "parents" (partners), there are many possible combinations of traits in the "offspring" (third person).
3. Sample answer: The number of possible genetic combinations is huge because we have so many genes.

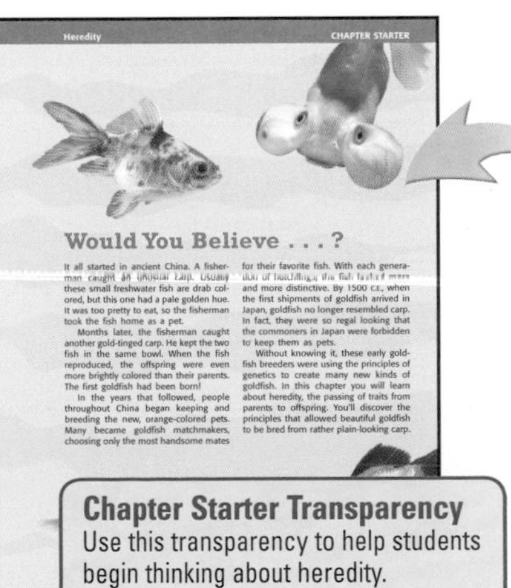

Heredity CHAPTER STARTER

Would You Believe . . . ?

It all started in ancient China. A fisherman caught an unusual carp. Usually these small freshwater fish are drab colored, but this one had a pale golden hue. It was too pretty to eat, so the fisherman took the fish home as a pet.

Months later, the fisherman caught another gold-tinged carp. He kept the two fish in the same bowl. When the fish reproduced, the offspring were even more brightly colored than their parents. The first goldfish had been born!

In the years that followed, people throughout China began keeping and breeding the new, orange-colored pets. Many became goldfish matchmakers, choosing only the most handsome mates for their favorite fish. With each generation of [illegible] and more distinctive. By 1500 C.E., when the first shipments of goldfish arrived in Japan, goldfish no longer resembled carp. In fact, they were so regal looking that the commoners in Japan were forbidden to keep them as pets.

Without knowing it, these early goldfish breeders were using the principles of genetics to create many new kinds of goldfish. In this chapter you will learn about heredity, the passing of traits from parents to offspring. You'll discover the principles that allowed beautiful goldfish to be bred from rather plain-looking carp.

Chapter Starter Transparency
Use this transparency to help students begin thinking about heredity.

CHAPTER RESOURCES

Technology

- **Transparencies** 
  - Chapter Starter Transparency
- **Student Edition on CD-ROM**
- **Guided Reading Audio CD**
  - English or Spanish
- **Classroom Videos**
  - Brain Food Video Quiz

Workbooks

- **Science Puzzlers, Twisters & Teasers**
  - Heredity 

SECTION 1

Focus

Overview

This section introduces the genetic experiments of Gregor Mendel. Students explore how crosses between different parent plants produce different offspring. Students are also introduced to genetic probability.

Bellringer

Present the following prompt to your students: "You have probably noticed that different people have different traits, such as eye color, hair color, and ear lobes that do or do not attach directly to their head. Where do people get these different traits?" (Many traits are inherited from parents and passed from parents to offspring through genes.)

Motivate

ACTIVITY — GENERAL

Trait Trends Create a large table to record the number of students with the following traits: widow's peak, ability to roll tongue, and attached earlobes. Have pairs of students enter data for each other by adding tick marks on the table. Ask students if they can see any trends in the class data. If possible, compile data from several classes. LS **Kinesthetic/Interpersonal**

SECTION 1

READING WARM-UP

Objectives

- Explain the relationship between traits and heredity.
- Describe the experiments of Gregor Mendel.
- Explain the difference between dominant and recessive traits.

Terms to Learn

heredity
dominant trait
recessive trait

READING STRATEGY

Brainstorming The key idea of this section is heredity. Brainstorm words and phrases related to heredity.

heredity the passing of genetic traits from parent to offspring

Mendel and His Peas

Why don't you look like a rhinoceros? The answer to this question seems simple: Neither of your parents is a rhinoceros. But there is more to this answer than meets the eye.

As it turns out, **heredity,** or the passing of traits from parents to offspring, is more complicated than you might think. For example, you might have curly hair, while both of your parents have straight hair. You might have blue eyes even though both of your parents have brown eyes. How does this happen? People have investigated this question for a long time. About 150 years ago, Gregor Mendel performed important experiments. His discoveries helped scientists begin to find some answers to these questions.

Reading Check **What is heredity?** (*See the Appendix for answers to Reading Checks.*)

Who Was Gregor Mendel?

Gregor Mendel, shown in **Figure 1,** was born in 1822 in Heinzendorf, Austria. Mendel grew up on a farm and learned a lot about flowers and fruit trees.

When he was 21 years old, Mendel entered a monastery. The monks taught science and performed many scientific experiments. From there, Mendel was sent to Vienna where he could receive training in teaching. However, Mendel had trouble taking tests. Although he did well in school, he was unable to pass the final exam. He returned to the monastery and put most of his energy into research. Mendel discovered the principles of heredity in the monastery garden.

Figure 1 *Gregor Mendel discovered the principles of heredity while studying pea plants.*

Unraveling the Mystery

From working with plants, Mendel knew that the patterns of inheritance were not always clear. For example, sometimes a trait that appeared in one generation (parents) was not present in the next generation (offspring). In the generation after that, though, the trait showed up again. Mendel noticed these kinds of patterns in several other living things, too. Mendel wanted to learn more about what caused these patterns.

To keep his investigation simple, Mendel decided to study only one kind of organism. Because he had studied garden pea plants before, they seemed like a good choice.

CHAPTER RESOURCES

Chapter Resource File

- **Lesson Plan**
- **Directed Reading A** BASIC
- **Directed Reading B** SPECIAL NEEDS

Technology

Transparencies

- Bellringer

Answer to Reading Check

the passing of traits from parents to offspring

Self-Pollinating Peas

In fact, garden peas were a good choice for several reasons. Pea plants grow quickly, and there are many different kinds available. They are also able to self-pollinate. A *self-pollinating plant* has both male and female reproductive structures. So, pollen from one flower can fertilize the ovule of the same flower or the ovule of another flower on the same plant. The flower on the right side of **Figure 2** is self-pollinating.

Why is it important that pea plants can self-pollinate? Because eggs (in an ovule) and sperm (in pollen) from the same plant combine to make a new plant, Mendel was able to grow true-breeding plants. When a *true-breeding plant* self-pollinates, all of its offspring will have the same trait as the parent. For example, a true-breeding plant with purple flowers will always have offspring with purple flowers.

Pea plants can also cross-pollinate. In *cross-pollination,* pollen from one plant fertilizes the ovule of a flower on a different plant. There are several ways that this can happen. Pollen may be carried by insects to a flower on a different plant. Pollen can also be carried by the wind from one flower to another. The left side of **Figure 2** shows these kinds of cross-pollination.

SCHOOL to HOME

Describing Traits

How would you describe yourself? Would you say that you are tall or short, have curly hair or straight hair? Make a list of some of your physical traits. Make a second list of traits that you were not born with, such as "caring" or "good at soccer." Talk to your family about your lists. Do they agree with your descriptions?

ACTIVITY

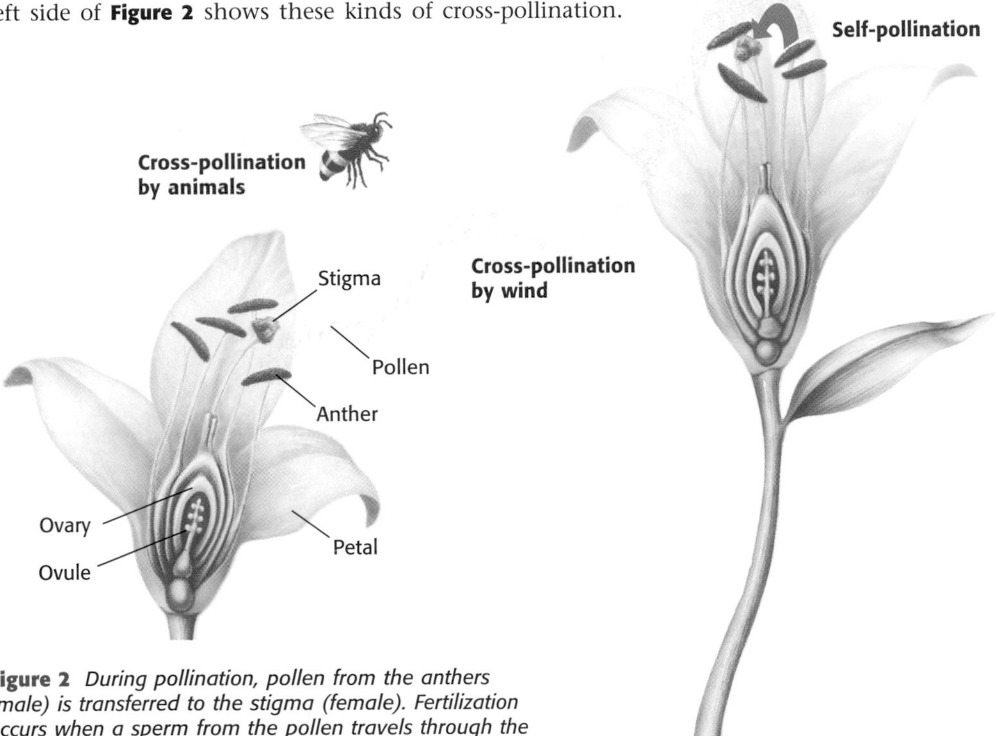

Figure 2 *During pollination, pollen from the anthers (male) is transferred to the stigma (female). Fertilization occurs when a sperm from the pollen travels through the stigma and enters the egg in an ovule.*

Teach

READING STRATEGY — BASIC

Paired Reading Have students read the section silently. As they read, students should make notes or write questions about any section that is confusing or hard to understand. Then, have students discuss the section with a partner, and allow students to help each other understand the material from this section. **LS Verbal/Interpersonal**

Using the Figure — BASIC

Flower Fertilization Discuss the physical processes involved in the fertilization of the flowers illustrated in **Figure 2.** These flowers can be fertilized by another flower or can fertilize themselves. Compare this figure with **Figure 4** on the next page, and point out that removing the anthers from the flower makes it impossible for the plant to self-pollinate. **LS Visual/Verbal**

CONNECTION to *Real World* — GENERAL

Rapidly Growing Organisms Mendel favored the garden pea because it grows quickly, allowing him to produce many generations within a short time span. Modern scientists favor yeast, bacteria, fruit flies, and mice for studies of heredity and genetics. Each of these organisms has a rapid rate of reproduction. However, rapidly-growing organisms can pose problems. For example, medical scientists face ongoing threats from strains of bacteria that develop resistance to common antibiotics. In some cases, medications that were once widely prescribed are no longer effective. **LS Logical/Intrapersonal**

Is That a Fact!

Although Mendel was brilliant, he had difficulty learning from scientific texts. In the monastery gardens, Mendel explored the scientific ideas he had trouble with in school. While trying to grow better peas, he discovered genetics, an entirely new field of science!

CONNECTION to *Physical Science* — GENERAL

Seeing Flower Color Flower color depends on which wavelength of light is reflected by the petals. For example, red absorbs all of the wavelengths except those for red. Display a color spectrum to illustrate that white light is composed of a "rainbow" gradient of colors. Show students different colors, and ask them which colors of light are being absorbed and which are being reflected to our eye. **LS Visual**

Teach, *continued*

Discussion — GENERAL

Scientific Methods Have students identify the use of scientific methods in Mendel's work.

- **Ask a question:** How are traits inherited?
- **Form a hypothesis:** Inheritance has a pattern.
- **Test the hypothesis:** Cross true-breeding plants and offspring.
- **Analyze the results:** Identify patterns in inherited traits.
- **Draw conclusions:** Traits are inherited in predictable patterns.
- **Communicate the results:** Publish the results for peer review.

Ask students, "Why weren't Mendel's ideas accepted for so many years?" (because of problems with the last step—other scientists could not easily read or understand his findings)

LS Logical/Verbal

Demonstration — BASIC

Flower Dissection Obtain a flower that has anthers and a stigma, such as a pea flower, a tulip, or a lily. Be careful because pollen can stain clothing and cause allergic reactions. Dissect the flower, and show students the anthers and the stigma. Ask students if this flower could self-pollinate. (yes, because it has both anthers and a stigma) Demonstrate how Mendel removed the anthers of his flowers and then used a small brush to transfer pollen from plant to plant.

LS Kinesthetic English Language Learners

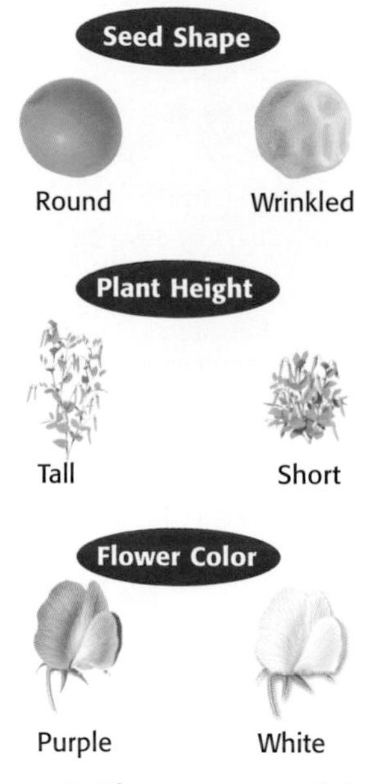

Figure 3 *These are some of the plant characteristics that Mendel studied.*

Characteristics

Mendel studied only one characteristic at a time. A *characteristic* is a feature that has different forms in a population. For example, hair color is a characteristic in humans. The different forms, such as brown or red hair, are called *traits*. Mendel used plants that had different traits for each of the characteristics he studied. For instance, for the characteristic of flower color, he chose plants that had purple flowers and plants that had white flowers. Three of the characteristics Mendel studied are shown in **Figure 3.**

Mix and Match

Mendel was careful to use plants that were true breeding for each of the traits he was studying. By doing so, he would know what to expect if his plants were to self-pollinate. He decided to find out what would happen if he bred, or crossed, two plants that had different traits of a single characteristic. To be sure the plants cross-pollinated, he removed the anthers of one plant so that the plant could not self-pollinate. Then, he used pollen from another plant to fertilize the plant, as shown in **Figure 4.** This step allowed Mendel to select which plants would be crossed to produce offspring.

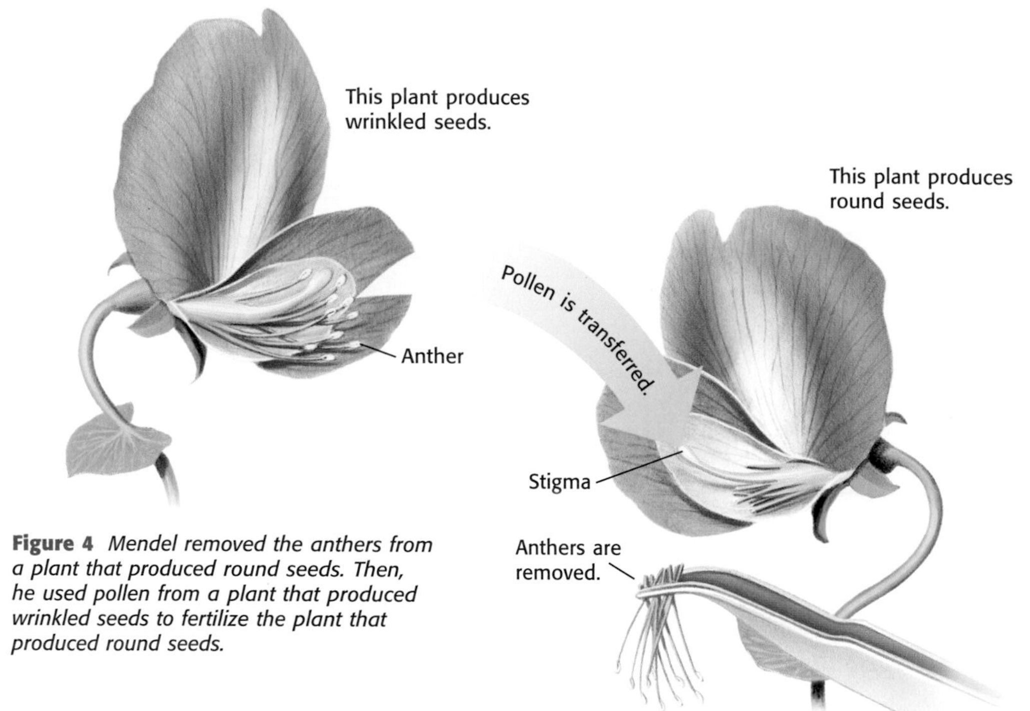

Figure 4 *Mendel removed the anthers from a plant that produced round seeds. Then, he used pollen from a plant that produced wrinkled seeds to fertilize the plant that produced round seeds.*

ACTIVITY — ADVANCED

Mendelian Crosses Give each student a purple bead (*P*) and a white bead (*p*), and ask students to perform a Mendelian cross. Tell students to begin the first generation with the allele combination *Pp*. Have students randomly "pollinate" with 10 other members of the class. To pollinate, one student should hide one bead in each hand. The partner should pick a hand. That hand holds the allele from one parent. Partners should switch roles and repeat this step to determine the allele from the second parent. Students should record the genotype for each pollination. Have students tally the results and determine the ratio of white-flowering plants to purple-flowering plants that results from the matches.

LS Kinesthetic/Interpersonal Co-op Learning

Mendel's First Experiments

In his first experiments, Mendel crossed pea plants to study seven different characteristics. In each cross, Mendel used plants that were true breeding for different traits for each characteristic. For example, he crossed plants that had purple flowers with plants that had white flowers. This cross is shown in the first part of **Figure 5.** The offspring from such a cross are called *first-generation plants*. All of the first-generation plants in this cross had purple flowers. Are you surprised by the results? What happened to the trait for white flowers?

Mendel got similar results for each cross. One trait was always present in the first generation, and the other trait seemed to disappear. Mendel chose to call the trait that appeared the **dominant trait.** Because the other trait seemed to fade into the background, Mendel called it the **recessive trait.** (To *recede* means "to go away or back off.") To find out what might have happened to the recessive trait, Mendel decided to do another set of experiments.

Mendel's Second Experiments

Mendel allowed the first-generation plants to self-pollinate. **Figure 5** also shows what happened when a first-generation plant with purple flowers was allowed to self-pollinate. As you can see, the recessive trait for white flowers reappeared in the second generation.

Mendel did this same experiment on each of the seven characteristics. In each case, some of the second-generation plants had the recessive trait.

Reading Check Describe Mendel's second set of experiments.

dominant trait the trait observed in the first generation when parents that have different traits are bred

recessive trait a trait that reappears in the second generation after disappearing in the first generation when parents with different traits are bred

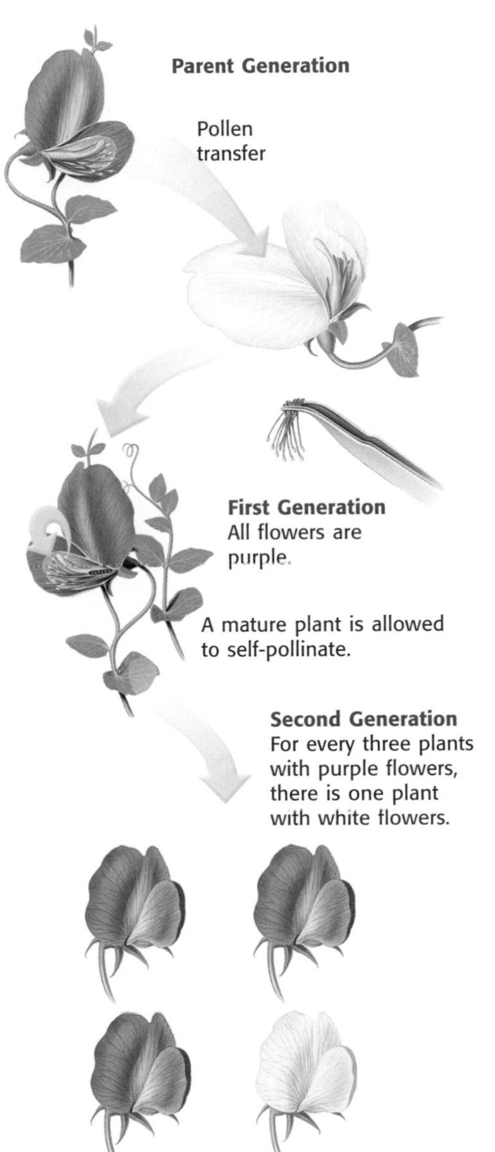

Figure 5 *Mendel used the pollen from a plant with purple flowers to fertilize a plant with white flowers. Then, he allowed the offspring to self-pollinate.*

MISCONCEPTION ALERT

Recessive Traits Students may believe that recessive traits are rare. Point out that either dominant *or* recessive traits may be more common in a population. For example, blond hair and blue eyes are recessive traits, yet they are very common in parts of Scandinavia. Conversely, dominant traits are not always the most common. For example, the trait of having six fingers on one hand is dominant!

Answer to Reading Check

During his second set of experiments, Mendel allowed the first-generation plants, which resulted from his first set of experiments, to self-pollinate.

INCLUSION Strategies

- ***Learning Disabled***
- ***Attention Deficit Disorder***
- ***Developmentally Delayed***

Use this activity to physically model the abstract concepts of *recessive* and *dominant*. To prepare, gather sheets of two kinds of transparent film: clear and purple. Cut the sheets into rectangles that are small enough to handle. Make one rectangle of each type for each student in the class.

1. Tell students that they are going to serve as models of Mendel's experiment in **Figure 5.** Give half of the students two purple rectangles each. Announce that these students have purple flowers. Give the other half of the students two clear rectangles each. Announce that these students have white flowers. Announce that the class now represents Mendel's parent generation.
2. Have each student trade one rectangle with another student. Announce that these new combinations represent Mendel's first generation. Tell students to hold the rectangles together in front of a light. Tell students that the purple gene is dominant and that those who see purple through the rectangles have purple flowers. This generation all has purple flowers.
3. Finally, have students trade one rectangle randomly with another student. Announce that the class now represents Mendel's second generation. Have students count the number of "flowers" of each type, and compare these results to Mendel's.

English Language Learners
LS Kinesthetic

Close

Answers for Table 1 Ratios

Seed color	3.00:1
Seed shape	2.96:1
Pod color	2.82:1
Pod shape	2.95:1
Flower position	3.14:1
Plant height	2.84:1

Reteaching — BASIC

Mendel's Experiments Have students re-enact Mendel's experiments using cups (to represent a plant), colored buttons or chips (to represent various alleles or genotypes), and colored strips of paper (to represent visible traits or phenotypes). Have students perform crosses by taking alleles from "parent" cups and creating "offspring" cups, deciding which traits would then become visible.

LS Kinesthetic/Logical English Language Learners

Quiz — GENERAL

1. What did Mendel call the trait that appeared in all of his first-generation plants? (the dominant trait)
2. What is the probability of getting heads in a coin toss? (1/2)

Alternative Assessment — ADVANCED

Story of a Scientist Have students create a comic book or short video drama about Mendel's life and work. Tell students to highlight his use of the scientific method and his habits as a scientist. **LS Interpersonal**

Understanding Ratios

A ratio is a way to compare two numbers. Look at **Table 1.** The ratio of plants with purple flowers to plants with white flowers can be written as 705 to 224 or 705:224. This ratio can be reduced, or simplified, by dividing the first number by the second as follows:

$$\frac{705}{224} = \frac{3.15}{1}$$

which is the same thing as a ratio of 3.15:1.

For every 3 plants with purple flowers, there will be roughly 1 plant with white flowers. Try this problem:

In a box of chocolates, there are 18 nougat-filled chocolates and 6 caramel-filled chocolates. What is the ratio of nougat-filled chocolates to caramel-filled chocolates?

Ratios in Mendel's Experiments

Mendel then decided to count the number of plants with each trait that turned up in the second generation. He hoped that this might help him explain his results. Take a look at Mendel's results, shown in **Table 1.**

As you can see, the recessive trait did not show up as often as the dominant trait. Mendel decided to figure out the ratio of dominant traits to recessive traits. A *ratio* is a relationship between two different numbers that is often expressed as a fraction. Calculate the dominant-to-recessive ratio for each characteristic. (If you need help, look at the Math Practice at left.) Do you notice anything interesting about the ratios? Round to the nearest whole number. Are the ratios all the same, or are they different?

Reading Check What is a ratio?

Table 1 Mendel's Results

Characteristic	Dominant traits	Recessive traits	Ratio
Flower color	705 purple	224 white	3.15:1
Seed color	6,002 yellow	2,001 green	?
Seed shape	5,474 round	1,850 wrinkled	?
Pod color	428 green	152 yellow	?
Pod shape	882 smooth	299 bumpy	?
Flower position	651 along stem	207 at tip	?
Plant height	787 tall	277 short	?

Answer to Math Practice

The ratio of nougat-filled chocolates to caramel-filled chocolates is 18:6, or 18/6, which can be reduced to 3/1. This fraction can be rewritten as 3:1 or 3 to 1.

Answers to questions on student page

All the ratios are about the same. They can be rounded to 3:1.

Answer to Reading Check

A ratio is a relationship between two different numbers that is often expressed as a fraction.

Gregor Mendel—Gone but Not Forgotten

Mendel realized that his results could be explained only if each plant had two sets of instructions for each characteristic. Each parent would then donate one set of instructions. In 1865, Mendel published his findings. But good ideas are sometimes overlooked or misunderstood at first. It wasn't until after his death, more than 30 years later, that Mendel's work was widely recognized. Once Mendel's ideas were rediscovered and understood, the door was opened to modern genetics. Genetic research, as shown in **Figure 6,** is one of the fastest changing fields in science today.

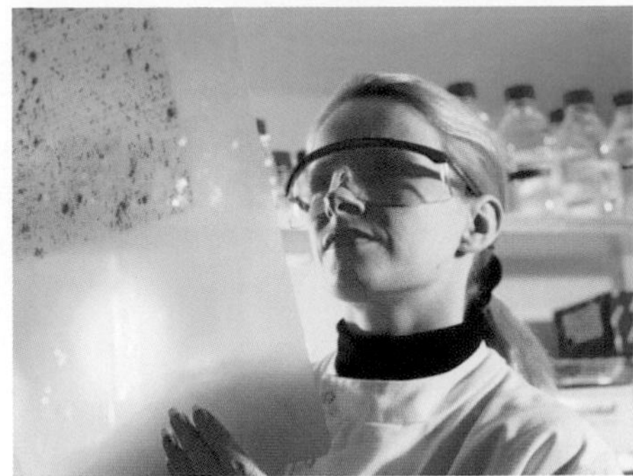

Figure 6 *This researcher is continuing the work started by Gregor Mendel more than 100 years ago.*

SECTION Review

Summary

- Heredity is the passing of traits from parents to offspring.
- Gregor Mendel made carefully planned experiments using pea plants that could self-pollinate.
- When parents with different traits are bred, dominant traits are always present in the first generation. Recessive traits are not visible in the first generation but reappear in the second generation.
- Mendel found a 3:1 ratio of dominant-to-recessive traits in the second generation.

Using Key Terms

1. Use each of the following terms in a separate sentence: *heredity, dominant trait,* and *recessive trait.*

Understanding Key Ideas

2. A plant that has both male and female reproductive structures is able to
   a. self-replicate.
   b. self-pollinate.
   c. breed true.
   d. None of the above
3. Explain the difference between self-pollination and cross-pollination.
4. What is the difference between a trait and a characteristic? Give one example of each.
5. Describe Mendel's first set of experiments.
6. Describe Mendel's second set of experiments.

Math Skills

7. In a bag of chocolate candies, there are 21 brown candies and 6 green candies. What is the ratio of brown to green? What is the ratio of green to brown?

Critical Thinking

8. **Predicting Consequences** Gregor Mendel used only true-breeding plants. If he had used plants that were not true breeding, do you think he would have discovered dominant and recessive traits? Explain.
9. **Applying Concepts** In cats, there are two types of ears: normal and curly. A curly-eared cat mated with a normal-eared cat, and all of the kittens had curly ears. Are curly ears a dominant or recessive trait? Explain.
10. **Identifying Relationships** List three other fields of study that use ratios.

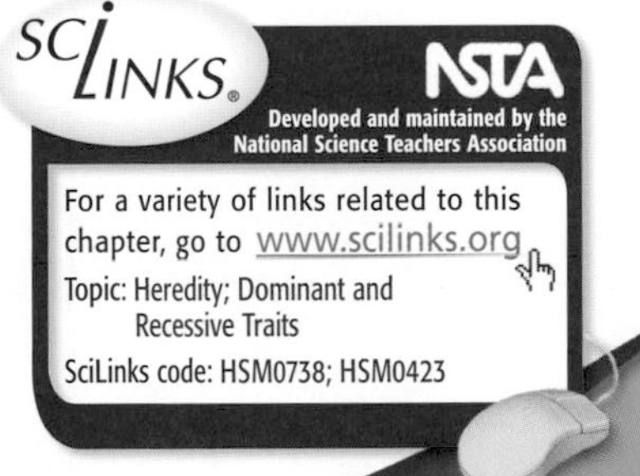

For a variety of links related to this chapter, go to www.scilinks.org
Topic: Heredity; Dominant and Recessive Traits
SciLinks code: HSM0738; HSM0423

Homework — GENERAL

Poster Project Have students create posters to illustrate Mendel's first and second experiments. Have each student demonstrate one of the seven traits that Mendel studied. Encourage students to use materials such as flowers, yellow and green seeds, or wrinkled and round peas. Each project should clearly identify the parents, the first generation, and the second generation. **LS Visual/Logical**

CHAPTER RESOURCES

Chapter Resource File

- Section Quiz GENERAL
- Section Review GENERAL
- Vocabulary and Section Summary GENERAL
- Critical Thinking Worksheet ADVANCED
- SciLinks Activity GENERAL

Workbooks

Science Skills
- Finding Useful Sources GENERAL

Math Skills for Science
- What is a Ratio? GENERAL

Answers to Section Review

1. Sample answer: Heredity is the passing of traits from parents to their offspring. A dominant trait is a trait that is present in the first generation when parents with different traits produce offspring. A recessive trait is a trait that is not present in the first generation but often reappears in the second generation.
2. b
3. Self-pollination occurs when pollen from a particular plant is deposited on a stigma from the same plant. Cross-pollination occurs when the pollen and stigma are from two different plants.
4. Sample answer: A characteristic is something that has different forms in a population, and a trait is each one of the possible forms. For example, eye color is a characteristic in humans, and brown eyes, green eyes, and blue eyes are all possible traits.
5. Sample answer: During Mendel's first experiments, he crossed two plants that were true breeding for different traits. In each case, the offspring had the dominant trait.
6. During Mendel's second experiments, he allowed the plants that were the offspring from his first experiments to self-pollinate. In these cases, some of the second-generation plants had the recessive trait.
7. 7:2 brown to green; 2:7 green to brown
8. Sample answer: If Mendel had used plants that were not true breeding, the dominant trait would not have been as clear for each characteristic, and he would not have gotten such a clear 3:1 ratio. The concept of dominant and recessive may have stayed hidden for a longer period of time.
9. Curly ears are dominant because it is the trait that is represented in the first generation.
10. Sample answer: sociology, physics, and chemistry

SECTION 2

Focus

Overview

In this section, students distinguish between genotype and phenotype and use mathematical models to predict the results of genetic crosses. They also learn some exceptions to Mendel's rules of inheritance.

Bellringer

Have students respond to the following prompts: "If you flip a coin, what are the chances that it will land on heads?" (1/2 or 50%) "tails?" (same) "Suppose you flip the coin once, get heads, and then flip it again. What are the chances that you will get heads again?" (still 1/2 or 50%) "Explain." (Each flip of the coin is independent of the last. The chances are the same on each flip.)

Motivate

Demonstration—BASIC

Ratios To review fractions and ratios, display three pennies and one nickel, and then ask students the following questions: "How many coins are there in all?" (4) "What fraction of the coins are pennies?" (3/4) "What fraction of the coins are nickels?" (1/4) "What is the ratio of pennies to nickels?" (3 to 1)
LS Visual/Verbal

SECTION 2

Traits and Inheritance

Mendel calculated the ratio of dominant traits to recessive traits. He found a ratio of 3:1. What did this tell him about how traits are passed from parents to offspring?

READING WARM-UP

Objectives

- Explain how genes and alleles are related to genotype and phenotype.
- Use the information in a Punnett square.
- Explain how probability can be used to predict possible genotypes in offspring.
- Describe three exceptions to Mendel's observations.

Terms to Learn

gene, allele, phenotype, genotype, probability

READING STRATEGY

Paired Summarizing Read this section silently. In pairs, take turns summarizing the material. Stop to discuss ideas that seem confusing.

A Great Idea

Mendel knew from his experiments with pea plants that there must be two sets of instructions for each characteristic. The first-generation plants carried the instructions for both the dominant trait and the recessive trait. Scientists now call these instructions for an inherited trait **genes.** Each parent gives one set of genes to the offspring. The offspring then has two forms of the same gene for every characteristic—one from each parent. The different forms (often dominant and recessive) of a gene are known as **alleles** (uh LEELZ). Dominant alleles are shown with a capital letter. Recessive alleles are shown with a lowercase letter.

Reading Check **What is the difference between a gene and an allele?** (*See the Appendix for answers to Reading Checks.*)

Phenotype

Genes affect the traits of offspring. An organism's appearance is known as its **phenotype** (FEE noh TIEP). In pea plants, possible phenotypes for the characteristic of flower color would be purple flowers or white flowers. For seed color, yellow and green seeds are the different phenotypes.

Phenotypes of humans are much more complicated than those of peas. Look at **Figure 1** below. The man has an inherited condition called *albinism* (AL buh NIZ uhm). Albinism prevents hair, skin, and eyes from having normal coloring.

gene one set of instructions for an inherited trait

allele one of the alternative forms of a gene that governs a characteristic, such as hair color

phenotype an organism's appearance or other detectable characteristic

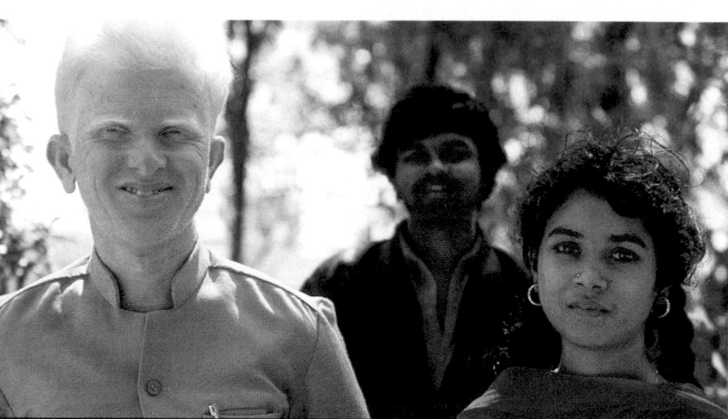

Figure 1 *Albinism is an inherited disorder that affects a person's phenotype in many ways.*

CHAPTER RESOURCES

Chapter Resource File

- Lesson Plan
- Directed Reading A BASIC
- Directed Reading B SPECIAL NEEDS

Technology

Transparencies
- Bellringer
- Punnett Squares
- LINK TO PHYSICAL SCIENCE The Periodic Table of the Elements

Answer to Reading Check

A gene contains the instructions for an inherited trait. the different versions of a gene are called *alleles.*

Genotype

Both inherited alleles together form an organism's **genotype.** Because the allele for purple flowers (*P*) is dominant, only one *P* allele is needed for the plant to have purple flowers. A plant with two dominant or two recessive alleles is said to be *homozygous* (HOH moh ZIE guhs). A plant that has the genotype *Pp* is said to be *heterozygous* (HET uhr OH ZIE guhs).

Punnett Squares

A Punnett square is used to organize all the possible combinations of offspring from particular parents. The alleles for a true-breeding, purple-flowered plant are written as *PP*. The alleles for a true-breeding, white-flowered plant are written as *pp*. The Punnett square for this cross is shown in **Figure 2.** All of the offspring have the same genotype: *Pp*. The dominant allele, *P*, in each genotype ensures that all of the offspring will be purple-flowered plants. The recessive allele, *p*, may be passed on to the next generation. This Punnett square shows the results of Mendel's first experiments.

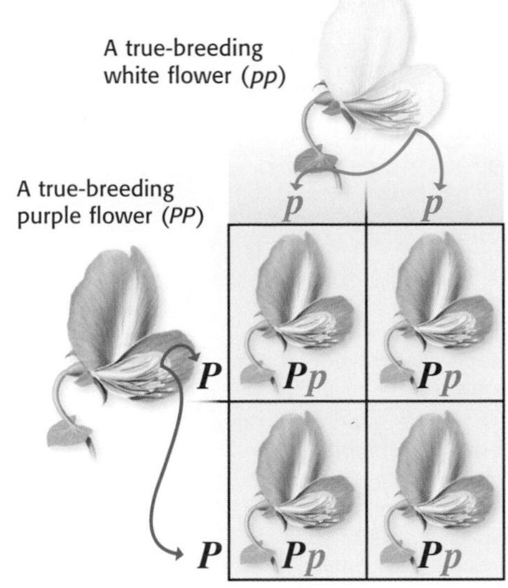

Figure 2 *All of the offspring for this cross have the same genotype—*Pp.

genotype the entire genetic makeup of an organism; also the combination of genes for one or more specific traits

Making a Punnett Square

1. Draw a square, and divide it into four sections.
2. Write the letters that represent alleles from one parent along the top of the box.
3. Write the letters that represent alleles from the other parent along the side of the box.
4. The cross shown at right is between two plants that produce round seeds. The genotype for each is *Rr*. Round seeds are dominant, and wrinkled seeds are recessive. Follow the arrows to see how the inside of the box was filled. The resulting alleles inside the box show all the possible genotypes for the offspring from this cross. What would the phenotypes for these offspring be?

Teach

CONNECTION to Physical Science—ADVANCED

Mathematical Models The Punnett square and the periodic Table are both mathematical models that were developed by scientists who observed numerical patterns. These models are used to organize scientific understanding of patterns and to make predictions. Show students the teaching transparency entitled "The Periodic Table of the Elements." Discuss the ways that a Punnett square is similar. **LS Visual/Logical**

MISCONCEPTION ALERT

Invisible Phenotypes Students may overgeneralize the idea that a phenotype can be a visible trait. This idea may help students to differentiate phenotype from genotype, but remind students that phenotype is any trait that is inherited (in other words, a result of the genotype). However, not all such traits may be visible. Most traits are fundamentally expressed as chemicals produced by cells.

Answer to Quick Lab

Three of the offspring would have round seeds, and one would have wrinkled seeds.

CHAPTER RESOURCES

Workbooks

Math Skills for Science
- Punnett Square Popcorn GENERAL

SCIENCE HUMOR

Q: What do you get when you cross a bridge with a bicycle?

A: to the other side

Teach, continued

Answers to Quick Lab

4. Students should get *bb* on average 1/4 or 25% of the time.
5. 1/4 or 25%
6. 1/4 (If brown fur results from genotype *Bb*, then brown fur is dominant, and white fur will result from the genotype *bb*.)

CONNECTION ACTIVITY
Math — ADVANCED

Probability of Independent Events The probability of two or more independent events is the product of the individual probabilities. For example, the probability of getting heads in a coin toss is 1/2, but the probability of getting heads twice in a row is 1/2 × 1/2, or 1/4. Have students consider the following parent genotypes for pea plants: *PpRr* and *Pprr*. Work out and discuss the probability of each possible combined phenotype. (For example, the probability of a plant with white flowers and round seeds is 1/4 × 1/2 = 1/8.) LS Logical

Answer to Reading Check

Probability is the mathematical chance that something will happen.

Taking Your Chances

You have two guinea pigs. Each has brown fur and the genotype *Bb*. You want to predict what their offspring might look like. Try this to find out.

1. Stick a **piece of masking tape** on each side of **two quarters.**
2. Label one side with a capital *B* and the other side with a lowercase *b*.
3. Toss both coins 10 times, making note of your results each time.
4. How many times did you get the *bb* combination?
5. What is the probability that the next toss will result in *bb*?
6. What are the chances that the guinea pigs' offspring will have white fur (with the genotype *bb*)?

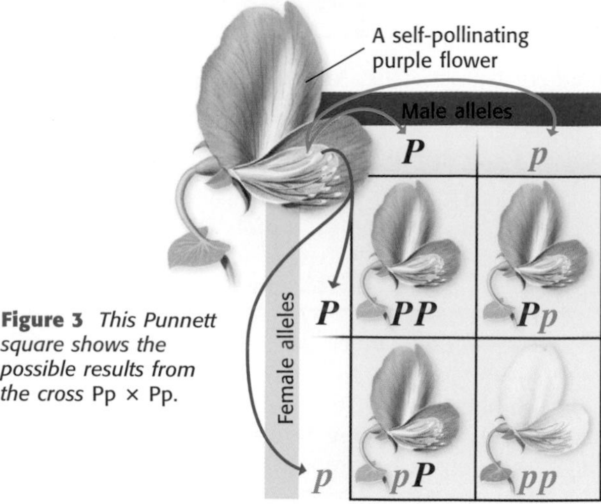

Figure 3 *This Punnett square shows the possible results from the cross* Pp × Pp.

More Evidence for Inheritance

In Mendel's second experiments, he allowed the first generation plants to self-pollinate. **Figure 3** shows a self-pollination cross of a plant with the genotype *Pp*. What are the possible genotypes of the offspring?

Notice that one square shows the genotype *Pp,* while another shows *pP*. These are exactly the same genotype. The other possible genotypes of the offspring are *PP* and *pp*. The combinations *PP, Pp,* and *pP* have the same phenotype—purple flowers. This is because each contains at least one dominant allele (*P*).

Only one combination, *pp,* produces plants that have white flowers. The ratio of dominant to recessive is 3:1, just as Mendel calculated from his data.

What Are the Chances?

Each parent has two alleles for each gene. When these alleles are different, as in *Pp,* offspring are equally likely to receive either allele. Think of a coin toss. There is a 50% chance you'll get heads and a 50% chance you'll get tails. The chance of receiving one allele or another is as random as a coin toss.

Probability

The mathematical chance that something will happen is known as **probability.** Probability is most often written as a fraction or percentage. If you toss a coin, the probability of tossing tails is 1/2—you will get tails half the time.

probability the likelihood that a possible future event will occur in any given instance of the event

Reading Check What is probability?

MISCONCEPTION ALERT

The Role of Chance Students' lack of understanding of mathematical probability may block their understanding of the random and independent sorting of genes that occurs during meiosis. Be careful that students do not overextend mathematical probabilities to predict the outcome of single events. It is correct to predict that an average of many outcomes will be similar to, but not exactly match, a probability ratio.

Q: What do you get when you cross a crocodile with an abalone?

A: a crocabaloney

Probability If you roll a pair of dice, what is the probability that you will roll 2 threes?

Step 1: Count the number of faces on a single die. Put this number in the denominator: 6.

Step 2: Count how many ways you can roll a three with one die. Put this number in the numerator: 1/6.

Step 3: To find the probability that you will throw 2 threes, multiply the probability of throwing the first three by the probability of throwing the second three: $1/6 \times 1/6 = 1/36$.

Now It's Your Turn

If you roll a single die, what is the probability that you will roll an even number?

Calculating Probabilities

To find the probability that you will toss two heads in a row, multiply the probability of tossing the first head (1/2) by the probability of tossing the second head (1/2). The probability of tossing two heads in a row is 1/4.

Genotype Probability

To have white flowers, a pea plant must receive a *p* allele from each parent. Each offspring of a $Pp \times Pp$ cross has a 50% chance of receiving either allele from either parent. So, the probability of inheriting two *p* alleles is $1/2 \times 1/2$, which equals 1/4, or 25%. Traits in pea plants are easy to predict because there are only two choices for each trait, such as purple or white flowers and round or wrinkled seeds. Look at **Figure 4.** Do you see only two distinct choices for fur color?

Figure 4 *These kittens inherited one allele from their mother for each trait.*

CONNECTION TO Chemistry

Round and Wrinkled Round seeds may look better, but wrinkled seeds taste sweeter. The dominant allele for seed shape, *R*, causes sugar to be changed into starch (which is a storage molecule for sugar). This change makes the seed round. Seeds with the genotype *rr* do not make or store this starch. Because the sugar has not been changed into starch, the seed tastes sweeter. If you had a pea plant with round seeds (*Rr*), what would you cross it with to get some offspring with wrinkled seeds? Draw a Punnett square showing your cross.

ACTIVITY

Answer to Math Focus

3/6 or 1/2

Answer to Connection to Chemistry

You would cross it with a plant with wrinkled seeds (*rr*). Students should draw a Punnett square showing this cross.

	R	*r*
r	*Rr*	*rr*
r	*Rr*	*rr*

MISCONCEPTION ALERT

Exception to Mendel's Rules Caution students not to assume that all inherited traits follow the examples studied by Mendel. For instance, a cross between a red-haired horse and a white-haired horse can produce a horse with both red and white hair. Such a horse is said to have a roan coat. This is an example of *codominance*—the mixed expression of two traits within the same organism. As in the case of incomplete dominance, both alleles are visible in the offspring, and therefore neither allele is purely dominant.

Homework GENERAL

Punnett Squares Have students create Punnett squares for each of the different crosses in Mendel's experiments. Students should include the genotype and phenotype of each parent and each set of possible offspring. **LS Visual/Logical**

WEIRD SCIENCE

Many ordinary fruits and vegetables carry recessive genes for bizarre traits. For instance, a recessive gene in tomatoes causes the skin to be covered with fuzzy hair!

Close

Reteaching BASIC

Exceptions Have students describe three exceptions to Mendel's heredity principles in their **science journal.** LS Verbal

Quiz GENERAL

In rabbits, the allele for black fur, *B*, is dominant over the allele for white fur, *b*. Suppose two black parents produce one white and three black bunnies.

1. What are the genotypes of the parents? (The parents must both have the recessive allele, so they are both genotype *Bb*.)
2. What are the possible genotypes of all four siblings? (White has genotype *bb*, and black may have *BB* or *Bb*.)

Alternative Assessment GENERAL

PORTFOLIO **Tracing Traits** Ask students to imagine two pure-bred animal parents that have different genetic traits. Have them assign three characteristics, such as tall or short and red nosed or blue nosed, to each parent. Have students label each characteristic as either dominant or recessive. Then, have students use Punnett squares to determine the possible genotypes and phenotypes for each trait in the parents' offspring and in a possible second generation.

LS Logical/Interpersonal English Language Learners

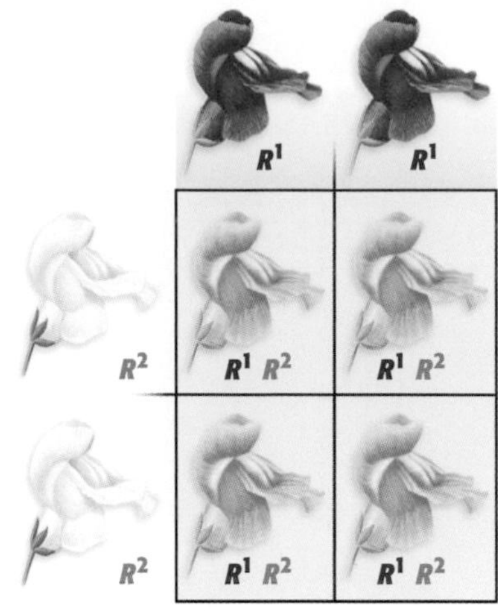

Figure 5 *Cross-breeding two true-breeding snapdragons provides a good example of incomplete dominance.*

More About Traits

As you may have already discovered, things are often more complicated than they first appear to be. Gregor Mendel uncovered the basic principles of how genes are passed from one generation to the next. But as scientists learned more about heredity, they began to find exceptions to Mendel's principles. A few of these exceptions are explained below.

Incomplete Dominance

Since Mendel's discoveries, researchers have found that sometimes one trait is not completely dominant over another. These traits do not blend together, but each allele has its own degree of influence. This is known as *incomplete dominance.*

One example of incomplete dominance is found in the snapdragon flower. **Figure 5** shows a cross between a true-breeding red snapdragon (R^1R^1) and a true-breeding white snapdragon (R^2R^2). As you can see, all of the possible phenotypes for their offspring are pink because both alleles of the gene have some degree of influence.

Reading Check What is incomplete dominance?

One Gene, Many Traits

Sometimes one gene influences more than one trait. An example of this phenomenon is shown by the white tiger in **Figure 6.** The white fur is caused by a single gene, but this gene influences more than just fur color. Do you see anything else unusual about the tiger? If you look closely, you'll see that the tiger has blue eyes. Here, the gene that controls fur color also influences eye color.

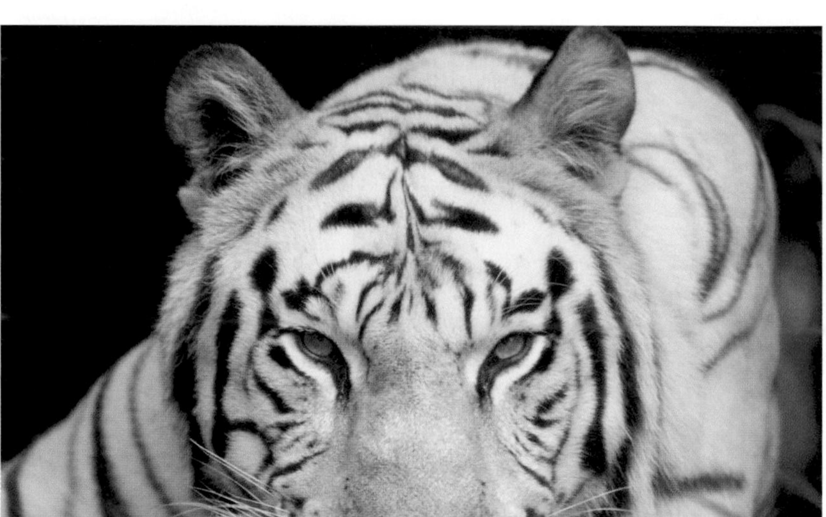

Figure 6 *The gene that gave this tiger white fur also influenced its eye color.*

Round Peas Mendel found that round seeds were dominant over wrinkled seeds. However, at the microscopic level, this is a case of incomplete dominance. The *R* and *r* alleles actually seem to affect the amount of starch produced in the pea. *RR* seeds have many starch grains that give them a full, round shape, but *rr* seeds have few starch grains and a wrinkled shape. *Rr* seeds have an intermediate number of starch grains—but enough for the pea to be full and round.

Many Genes, One Trait

Some traits, such as the color of your skin, hair, and eyes, are the result of several genes acting together. Therefore, it's difficult to tell if some traits are the result of a dominant or a recessive gene. Different combinations of alleles result in different eye-color shades, as shown in **Figure 7.**

The Importance of Environment

Genes aren't the only influences on traits. A guinea pig could have the genes for long fur, but its fur could be cut. In the same way, your environment influences how you grow. Your genes may make it possible that you will grow to be tall, but you need a healthy diet to reach your full potential height.

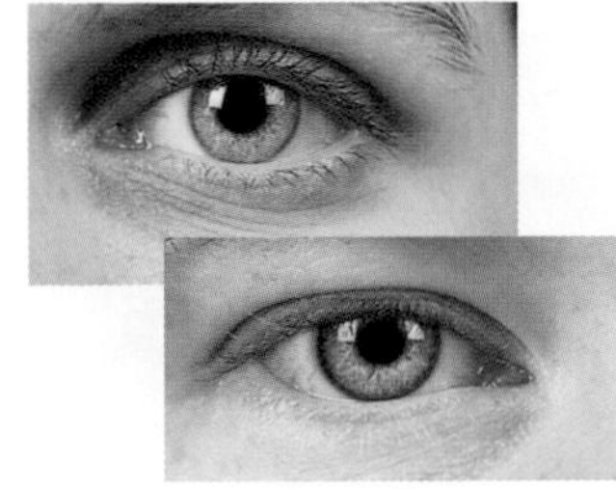

Figure 7 *At least two genes determine human eye color. That's why many shades of a single color are possible.*

SECTION Review

Summary

- Instructions for an inherited trait are called *genes*. For each gene, there are two alleles, one inherited from each parent. Both alleles make up an organism's genotype. Phenotype is an organism's appearance.
- Punnett squares show all possible offspring genotypes.
- Probability can be used to describe possible outcomes in offspring and the likelihood of each outcome.
- Incomplete dominance occurs when one allele is not completely dominant over the other allele.
- Some genes influence more than one trait.

Using Key Terms

1. Use the following terms in the same sentence: *gene* and *allele*.
2. In your own words, write a definition for each of the following terms: *genotype* and *phenotype*.

Understanding Key Ideas

3. Use a Punnett square to determine the possible genotypes of the offspring of a *BB* × *Bb* cross.
   a. all *BB*
   b. *BB*, *Bb*
   c. *BB*, *Bb*, *bb*
   d. all *bb*
4. How are genes and alleles related to genotype and phenotype?
5. Describe three exceptions to Mendel's observations.

Math Skills

6. What is the probability of rolling a five on one die three times in a row?

Critical Thinking

7. **Applying Concepts** The allele for a cleft chin, *C*, is dominant among humans. What are the results of a cross between parents with genotypes *Cc* and *cc*?

Interpreting Graphics

The Punnett square below shows the alleles for fur color in rabbits. Black fur, *B*, is dominant over white fur, *b*.

8. Given the combinations shown, what are the genotypes of the parents?
9. If black fur had incomplete dominance over white fur, what color would the offspring be?

Answer to Reading Check

In incomplete dominance, one trait is not completely dominant over another.

Answers to Section Review

1. Sample answer: Each gene can have several alleles.
2. Sample answer: Genotype is the set of alleles an organism has inherited from its parents. Phenotype is the way the genes are expressed physically.
3. b
4. The genotype of an organism contains the two alleles for each characteristic. One allele of each pair was inherited from each of the organism's parents. The phenotype of the organism is the way the genotype affects the organism physically. For example, if an organism inherits one dominant allele for brown fur and one recessive allele for white fur, its phenotype will be brown fur.
5. incomplete dominance, one gene influencing more than one trait, and one trait being influenced by many genes
6. 1/6 × 1/6 × 1/6 = 1/216
7. Approximately half of the offspring will have the phenotype of cleft chins (genotypes Cc), and half will not (genotypes cc).
8. BB, bb
9. Sample answer: a shade of gray

MISCONCEPTION ALERT

Nature Versus Nurture Many students believe that characteristics acquired through the environment may be inherited, or believe that learned skills and behavioral similarities (perhaps learned from parents) are necessarily inherited. Although environment may influence the expression of genes, an organism may only pass on those genes that it was born with (unless there is a mutation in the genes of the sex cells).

CHAPTER RESOURCES

Chapter Resource File

- Section Quiz GENERAL
- Section Review GENERAL
- Vocabulary and Section Summary GENERAL
- Reinforcement Worksheet BASIC
- Datasheet for Quick Lab

SECTION 3

Focus

Overview

In this section, students are introduced to meiosis and relate it to Mendel's findings. Students also learn about sex chromosomes and hereditary disorders.

Bellringer

Ask students to write a sentence for each of the following terms: *heredity, genotype, phenotype.* (Sample answer: Heredity is the passing of traits from parents to offspring. The combination of an organism's alleles is its genotype. All of an organism's physical traits are its phenotype.)

Motivate

ACTIVITY — GENERAL

Crosses Have students model a cross between an organism with one pair of chromosomes and a member of the opposite sex of its species. Show the chromosomes in the cross as "$F_1F_2 \times M_1M_2$." Explain that F_1 and F_2 represent the father's chromosomes, and M_1 and M_2 represent the mother's chromosomes. Ask students, "If each parent contributes only one chromosome from his or her own pair to the offspring, what are the possible combinations in the offspring?" (F_1M_1, F_1M_2, F_2M_1, and F_2M_2) LS Logical/Visual

SECTION 3

Meiosis

Where are genes located, and how do they pass information? Understanding reproduction is the first step to finding the answers.

READING WARM-UP

Objectives

- Explain the difference between mitosis and meiosis.
- Describe how chromosomes determine sex.
- Explain why sex-linked disorders occur in one sex more often than in the other.
- Interpret a pedigree.

Terms to Learn

homologous chromosomes
meiosis
sex chromosome
pedigree

READING STRATEGY

Reading Organizer As you read this section, make a flowchart of the steps of meiosis.

There are two kinds of reproduction: asexual and sexual. Asexual reproduction results in offspring with genotypes that are exact copies of their parent's genotype. Sexual reproduction produces offspring that share traits with their parents but are not exactly like either parent.

Asexual Reproduction

In *asexual reproduction,* only one parent cell is needed. The structures inside the cell are copied, and then the parent cell divides, making two exact copies. This type of cell reproduction is known as *mitosis*. Most of the cells in your body and most single-celled organisms reproduce in this way.

Sexual Reproduction

In sexual reproduction, two parent cells join together to form offspring that are different from both parents. The parent cells are called *sex cells.* Sex cells are different from ordinary body cells. Human body cells have 46, or 23 pairs of, chromosomes. One set of human chromosomes is shown in **Figure 1.** Chromosomes that carry the same sets of genes are called **homologous** (hoh MAHL uh guhs) **chromosomes.** Imagine a pair of shoes. Each shoe is like a homologous chromosome. The pair represents a homologous pair of chromosomes. But human sex cells are different. They have 23 chromosomes—half the usual number. Each sex cell has only one of the chromosomes from each homologous pair. Sex cells have only one "shoe."

homologous chromosomes chromosomes that have the same sequence of genes and the same structure

meiosis a process in cell division during which the number of chromosomes decreases to half the original number by two divisions of the nucleus, which results in the production of sex cells

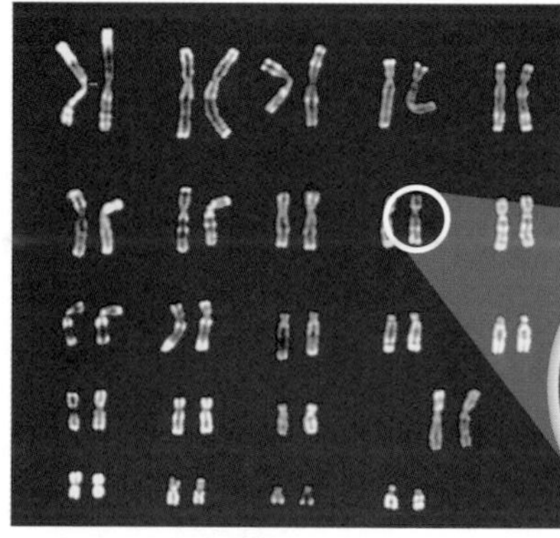

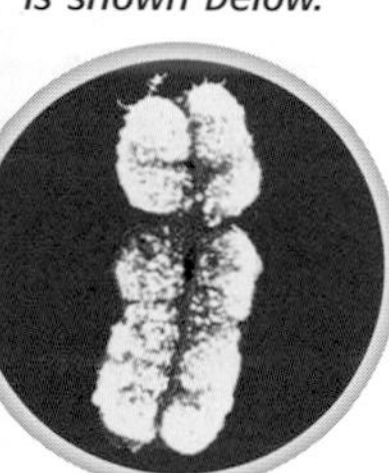

Figure 1 *Human body cells have 23 pairs of chromosomes. One member of a pair of homologous chromosomes is shown below.*

CHAPTER RESOURCES

Chapter Resource File

- Lesson Plan
- Directed Reading A BASIC
- Directed Reading B SPECIAL NEEDS

Technology

Transparencies
- Bellringer

CONNECTION ACTIVITY — Math — ADVANCED

Crosses In algebraic multiplication, some students use the mnemonic device FOIL (**f**irst, **o**uter, **i**nner, **l**ast). This device can be used to calculate genotype crosses. For example, the cross $X_1X_2 \times Y_1Y_2$ yields:

First: X_1X_2

Outer: X_1Y_2

Inner: X_2Y_1

Last: Y_1Y_2

LS Logical/Auditory

Meiosis

Sex cells are made during meiosis (mie OH sis). **Meiosis** is a copying process that produces cells with half the usual number of chromosomes. Each sex cell receives one-half of each homologous pair. For example, a human egg cell has 23 chromosomes, and a sperm cell has 23 chromosomes. The new cell that forms when an egg cell and a sperm cell join has 46 chromosomes.

Reading Check **How many chromosomes does a human egg cell have?** *(See the Appendix for answers to Reading Checks.)*

Genes and Chromosomes

What does all of this have to do with the location of genes? Not long after Mendel's work was rediscovered, a graduate student named Walter Sutton made an important observation. Sutton was studying sperm cells in grasshoppers. Sutton knew of Mendel's studies, which showed that the egg and sperm must each contribute the same amount of information to the offspring. That was the only way the 3:1 ratio found in the second generation could be explained. Sutton also knew from his own studies that although eggs and sperm were different, they did have something in common: Their chromosomes were located inside a nucleus. Using his observations of meiosis, his understanding of Mendel's work, and some creative thinking, Sutton proposed something very important:

Genes are located on chromosomes!

Understanding meiosis was critical to finding the location of genes. Before you learn about meiosis, review mitosis, shown in **Figure 2.** Meiosis is outlined in **Figure 3** on the next two pages.

CONNECTION TO Language Arts

Greek Roots The word *mitosis* is related to a Greek word that means "threads." Threadlike spindles are visible during mitosis. The word *meiosis* comes from a Greek word that means "to make smaller." How do you think meiosis got its name?

Figure 2 Mitosis Revisited

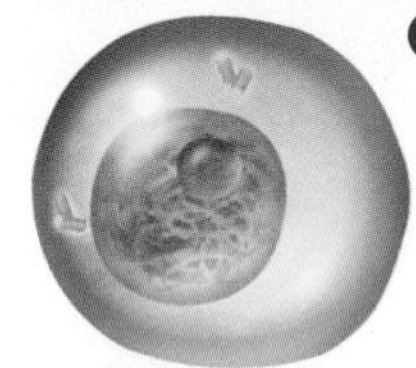

1 Each chromosome is copied.

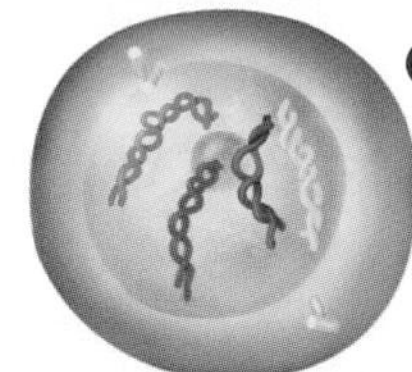

2 The chromosomes thicken and shorten. Each chromosome consists of two identical copies, called *chromatids.*

3 The nuclear membrane dissolves. The chromatids line up along the equator (center) of the cell.

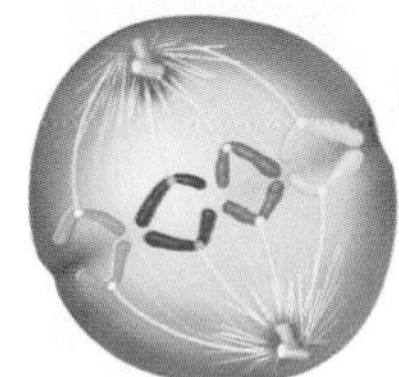

4 The chromatids pull apart.

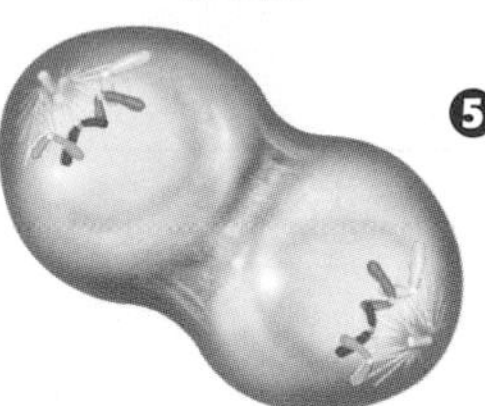

5 The nuclear membrane forms around the separated chromatids. The chromosomes unwind, and the cell divides.

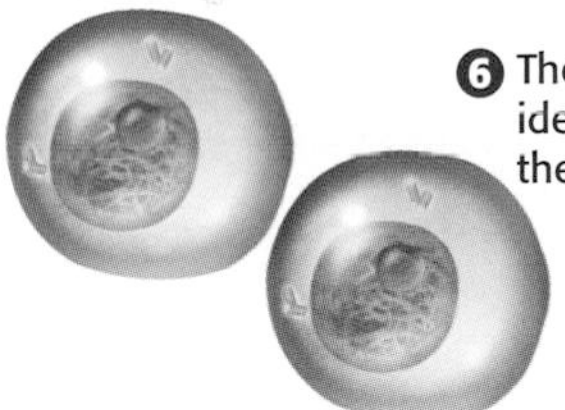

6 The result is two identical copies of the original cell.

Teach

Using the Figure—BASIC

Mitosis and Meiosis Have students examine **Figures 2** and **3** to compare what happens in each type of cell division. On the board, draw two identical cells, each of which contains four chromosomes. Label one cell "Mitosis," and label the other "Meiosis." Have students describe what happens in each stage of mitosis, and have them illustrate the stages on the board and in their **science journal.** (Using colored chalk might help distinguish between the dividing chromosomes.) Repeat the process for meiosis. Point out that mitosis results in two identical cells, each of which contains four chromosomes, and meiosis results in four cells, each of which contains two chromosomes. **LS Visual**

Answer to Reading Check

23 chromosomes

CONNECTION to Real World—ADVANCED

Aging and Cell Division

Research suggests a connection between aging, cell division, and mitosis. The ends of the chromosomes are protected by special sequences of DNA that do not seem to code for proteins but rather serve a function similar to that of the plastic tips on the ends of shoelaces. These structures, called *telomeres,* act as protective caps on the ends of the long strand of DNA that makes up each chromosome. However, with each cell division, the telomeres lose a little bit of material. At some point, the telomeres become so short that the cell can no longer divide. Eventually, the cell dies, which brings the organism one step closer to its inevitable end.

Answer to Connection to Language Arts

Sample answer: Meiosis makes each of the daughter cells smaller than the parent cell. Also, there are fewer chromosomes on the daughter cells than in the parent cell.

Science Bloopers

Wrong Number In 1918, a prominent scientist miscounted the number of chromosomes in a human cell. He counted 48. For almost 40 years, scientists thought this number was correct. In fact, not until 1956 were chromosomes counted correctly and found to number only 46.

READING STRATEGY — GENERAL

Prediction Guide Before students read the passage about meiosis, ask them whether the following statements are true or false. Students will discover the answers as they explore the rest of the section.

- Mitosis is the only type of cell division. (false)
- Only cells that produce sex cells undergo meiosis. (true)
- Sex cells contain half the number of chromosomes that other body cells do. (true)

LS Verbal/Auditory

Answer to Reading Check

During meiosis, one parent cell makes four new cells.

Discussion — GENERAL

Predicting Problems Ask students what they think would happen if something went wrong during cell division and the sperm or egg cell ended up with either too few or too many chromosomes? (The fertilized egg, with too few or too many chromosomes, may die, or the growing embryo may have birth defects. Down syndrome occurs in humans when the offspring receives an extra twenty-first chromosome.)

LS Verbal/Logical

The Steps of Meiosis

During mitosis, chromosomes are copied once, and then the nucleus divides once. During meiosis, chromosomes are copied once, and then the nucleus divides twice. The resulting sperm and eggs have half the number of chromosomes of a normal body cell. **Figure 3** shows all eight steps of meiosis. Read about each step as you look at the figure. Different types of living things have different numbers of chromosomes. In this illustration, only four chromosomes are shown.

Reading Check **How many cells are made from one parent cell during meiosis?**

Figure 3 Steps of Meiosis

Read about each step as you look at the diagram. Different types of living things have different numbers of chromosomes. In this diagram, only four chromosomes are shown.

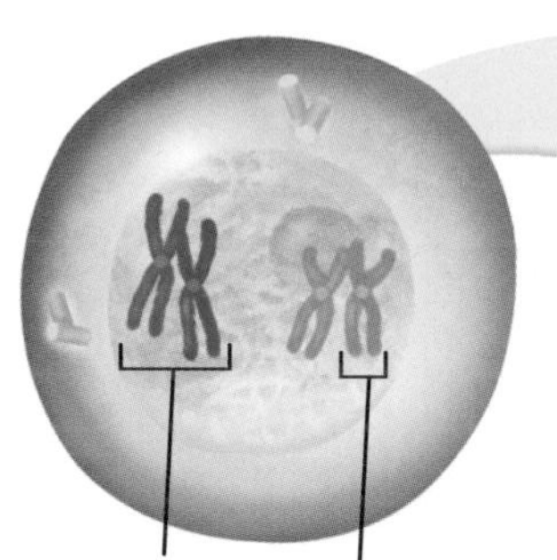

One pair of homologous chromosomes

Two chromatids

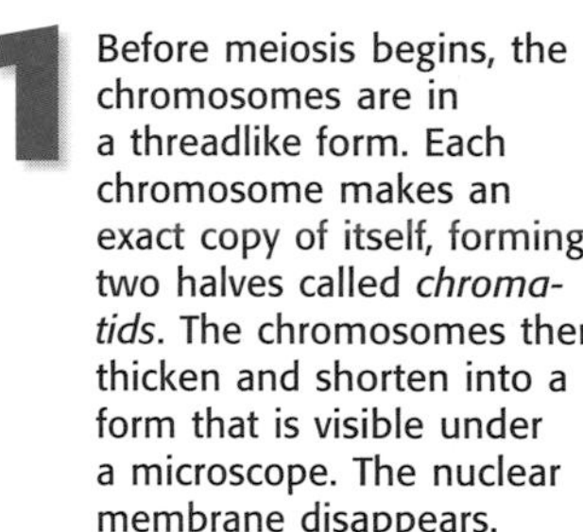

1 Before meiosis begins, the chromosomes are in a threadlike form. Each chromosome makes an exact copy of itself, forming two halves called *chromatids*. The chromosomes then thicken and shorten into a form that is visible under a microscope. The nuclear membrane disappears.

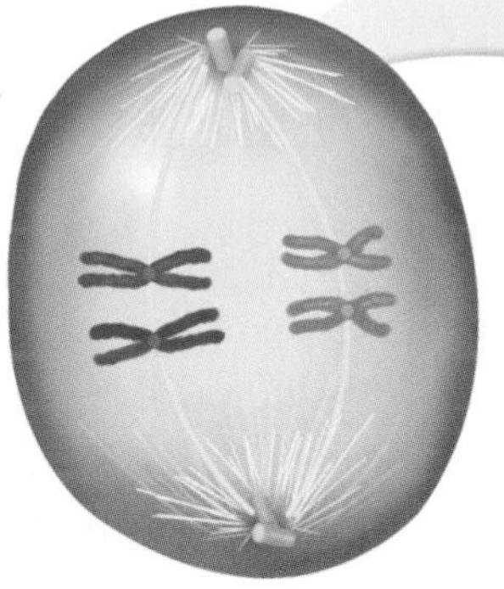

2 Each chromosome is now made up of two identical chromatids. Similar chromosomes pair with one another, and the paired homologous chromosomes line up at the equator of the cell.

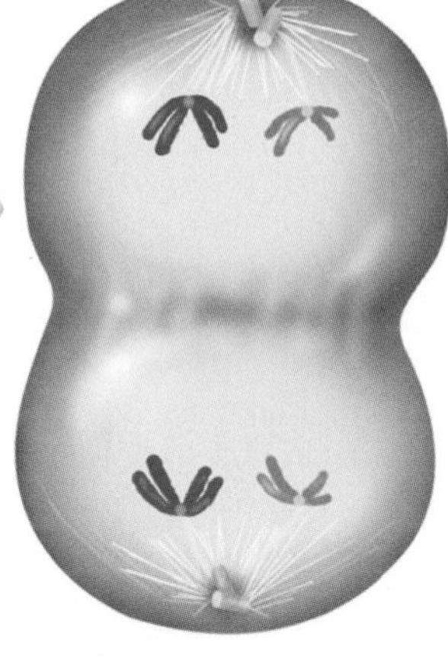

3 The chromosomes separate from their homologous partners and then move to opposite ends of the cell.

CHAPTER RESOURCES

Technology

Transparencies
- The Steps of Meiosis: A
- The Steps of Meiosis: B

Is That a Fact!

There are many organisms that have more chromosomes than humans do.

4 The nuclear membrane re-forms, and the cell divides. The paired chromatids are still joined.

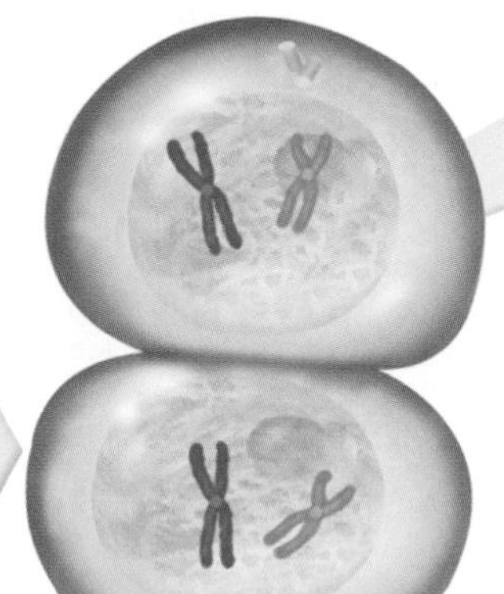

5 Each cell contains one member of each homologous chromosome pair. The chromosomes are not copied again between the two cell divisions.

6 The chromosomes then line up at the equator of each cell.

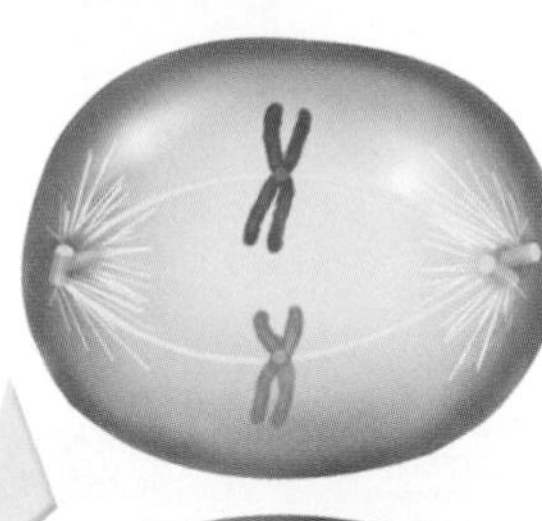

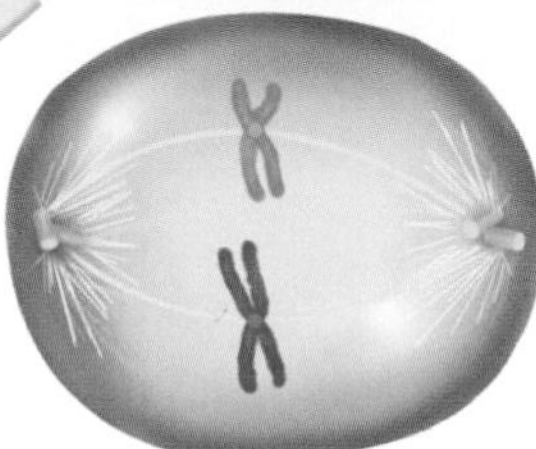

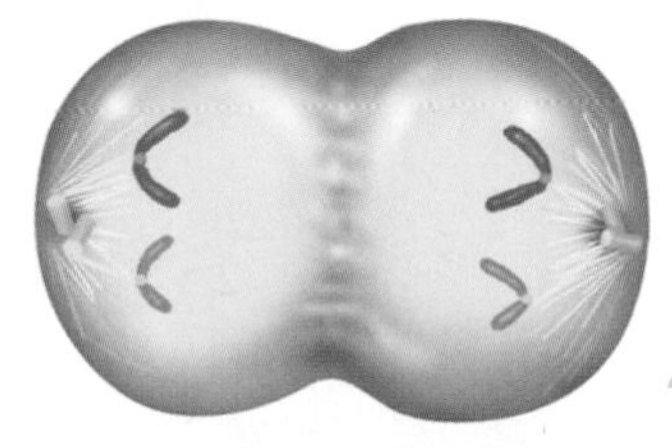

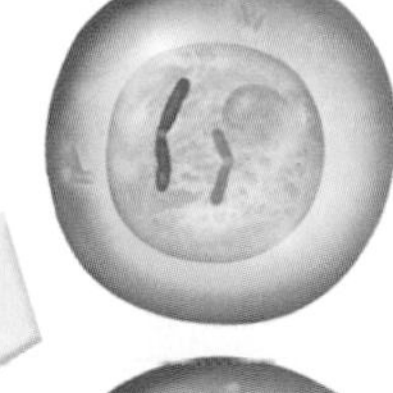

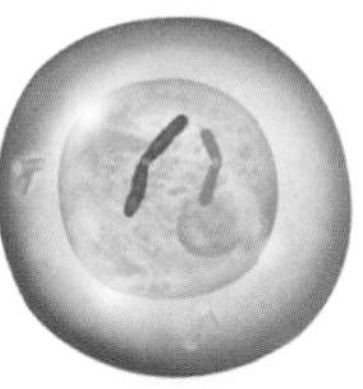

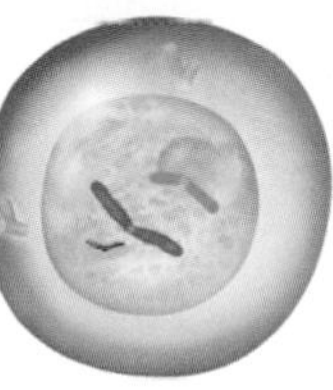

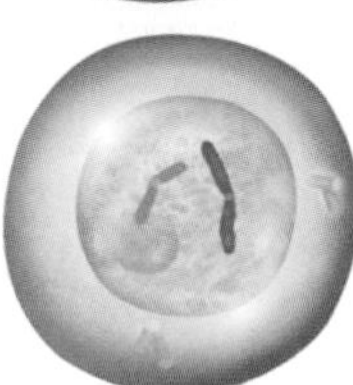

7 The chromatids pull apart and move to opposite ends of the cell. The nuclear membrane forms around the separated chromosomes, and the cells divide.

8 The result is that four new cells have formed from the original single cell. Each new cell has half the number of chromosomes present in the original cell.

SCIENCE HUMOR

Q: If human sex cells are created by meiosis, how are cat sex cells produced?

A: by meow-sis

MISCONCEPTION ALERT

Chromatids and Chromosome Pairing Students often have difficulty keeping track of the differences between the way that chromatids and chromosome pairs move during mitosis as compared to meiosis. Caution students to note these differences as they compare mitosis and meiosis, and to analyze the ways that these differences are critical to each process.

ACTIVITY — BASIC

Describing Meiosis Have students write their own captions for the steps of meiosis illustrated here. They should use language and descriptions that will help them understand and remember the material.

LS Verbal/Visual

CONNECTION ACTIVITY Math — GENERAL

Chromosome Number Meiosis and sexual reproduction have benefits for organisms because these processes maintain a variety of traits within a population. Meiosis and sexual recombination reshuffle the genetic material in each generation. Furthermore, the division of chromosomes during meiosis ensures that when the egg and sperm combine, the new organism has the same number of chromosomes as its parents. To explore these concepts, ask students the following questions:

- If the normal number of chromosomes for a certain organism is 30, how many chromosomes would be found in the egg or sperm cells? (15)
- What would happen if eggs and sperm were produced by mitosis instead of by meiosis? (The organism would produce sex cells with a full set of 30 chromosomes.)
- If the organism described above were to have offspring that also produced sex cells by mitosis, how many chromosomes would be found in the descendants after four generations? (first generation: 60; second generation: 120; third generation: 240; fourth generation: 400)

LS Verbal/Logical

INCLUSION Strategies

- *Learning Disabled*
- *Attention Deficit Disorder*

Have students make a flip book that animates the phases of meiosis. First, have students draw the events of meiosis in at least 15 sketches on sturdy cards. Explain that each drawing should vary only slightly from the one before it. When the book is flipped through quickly, the images should appear to be in motion, and students will be able to watch meiosis in action. This activity could be repeated to demonstrate mitosis. English Language Learners
LS Visual

Group ACTIVITY—GENERAL

Comparing Mitosis and Meiosis Organize the class into small groups. Instruct each group to create a table listing the similarities and differences between mitosis and meiosis. Challenge groups to make the longest list possible in a limited time period. After their time is up, have groups report items from their lists. Discuss and correct items as you compile a single, large table for display in the classroom. English Language Learners
LS Visual/Verbal

Meiosis and Mendel

As Walter Sutton figured out, the steps in meiosis explained Mendel's results. **Figure 4** shows what happens to a pair of homologous chromosomes during meiosis and fertilization. The cross shown is between a plant that is true breeding for round seeds and a plant that is true breeding for wrinkled seeds.

Each fertilized egg in the first generation had one dominant allele and one recessive allele for seed shape. Only one genotype was possible because all sperm formed by the male parent during meiosis had the wrinkled-seed allele, and all of the female parent's eggs had the round-seed allele. Meiosis also helped explain other inherited characteristics.

Figure 4 **Meiosis and Dominance**

Male Parent In the plant-cell nucleus below, each homologous chromosome has an allele for seed shape, and each allele carries the same instructions: to make wrinkled seeds.

Female Parent In the plant-cell nucleus below, each homologous chromosome has an allele for seed shape, and each allele carries the same instructions: to make round seeds.

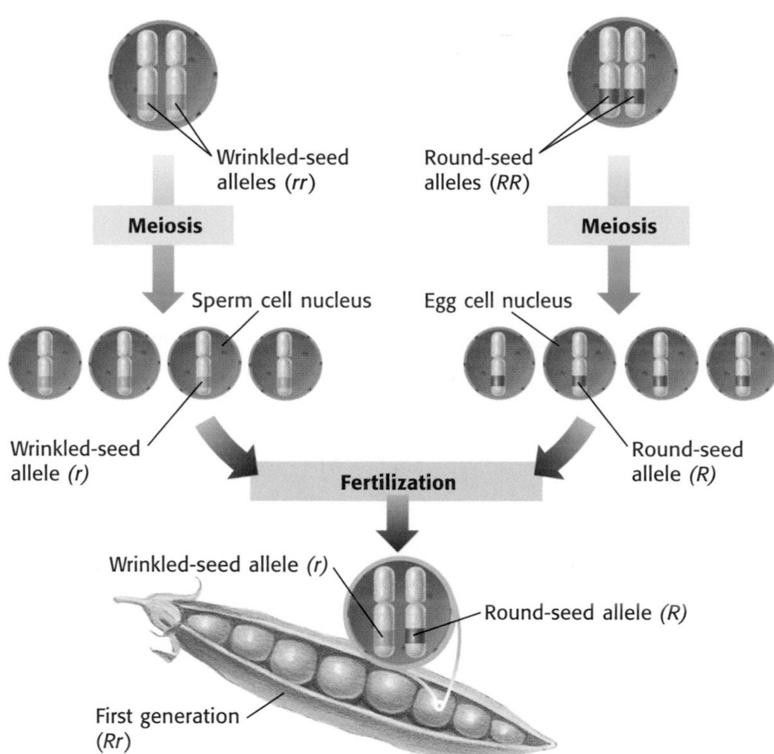

a Following **meiosis,** each sperm cell has a recessive allele for wrinkled seeds, and each egg cell has a dominant allele for round seeds.

b **Fertilization** of any egg by any sperm results in the same genotype (*Rr*) and the same phenotype (round). This result is exactly what Mendel found in his studies.

CHAPTER RESOURCES

Technology

Transparencies
- Meiosis and Dominance

Is That a Fact!

Martin-Bell syndrome is a genetic disorder also known as *Fragile X syndrome*. It is one of the most common forms of inherited mental retardation. This disorder is a genetic condition associated with mental retardation and autism. The disorder is identified by flaws apparent in the long arm of the X chromosome.

Sex Chromosomes

Information contained on chromosomes determines many of our traits. **Sex chromosomes** carry genes that determine sex. In humans, females have two X chromosomes. But human males have one X chromosome and one Y chromosome.

During meiosis, one of each of the chromosome pairs ends up in a sex cell. Females have two X chromosomes in each body cell. When meiosis produces the egg cells, each egg gets one X chromosome. Males have both an X chromosome and a Y chromosome in each body cell. Meiosis produces sperm with either an X or a Y chromosome. An egg fertilized by a sperm with an X chromosome will produce a female. If the sperm contains a Y chromosome, the offspring will be male, as shown in **Figure 5.**

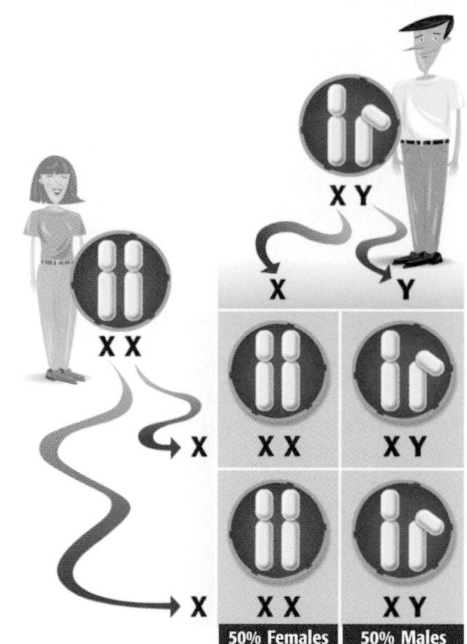

Figure 5 *Egg and sperm combine to form either the XX or XY combination.*

Sex-Linked Disorders

The Y chromosome does not carry all of the genes of an X chromosome. Females have two X chromosomes, so they carry two copies of each gene found on the X chromosome. This makes a backup gene available if one becomes damaged. Males have only one copy of each gene on their one X chromosome. The genes for certain disorders, such as colorblindness, are carried on the X chromosome. These disorders are called *sex-linked disorders*. Because the gene for such disorders is recessive, men are more likely to have sex-linked disorders.

People who are colorblind can have trouble distinguishing between shades of red and green. To help the colorblind, some cities have added shapes to their street lights, as shown in **Figure 6.** Hemophilia (HEE moh FIL ee uh) is another sex-linked disorder. Hemophilia prevents blood from clotting, and people with hemophilia bleed for a long time after small cuts. Hemophilia can be fatal.

sex chromosome one of the pair of chromosomes that determine the sex of an individual

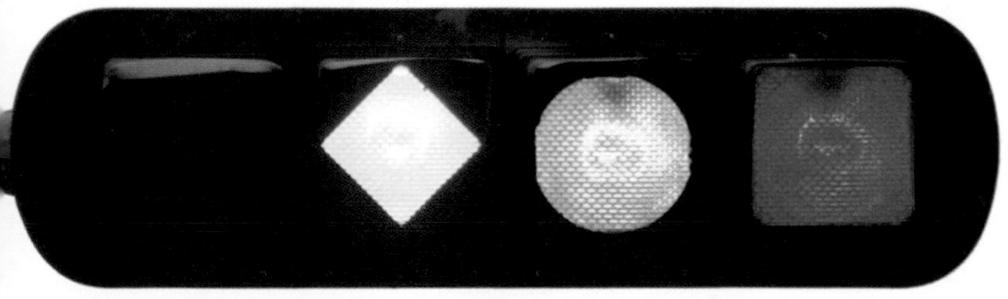

Figure 6 *This stoplight in Canada is designed to help the colorblind see signals easily. This photograph was taken over a few minutes to show all three shapes.*

CONNECTION ACTIVITY
Language Arts—GENERAL

Chromosome Chronicle Have students write an entry in their **science journal** to chronicle events in the "life" of a chromosome containing an allele for a specific trait. Have students describe the chromosome's role in the parent organism, the first-generation offspring, and the second-generation offspring. Descriptions should define whether the trait is dominant or recessive, and they should include an analysis of the factors that determine the genotype and phenotype of the parent and the offspring. LS **Verbal**

CONNECTION to
Real World—ADVANCED

Other Types of Genetic Disorders Other types of genetic disorders include chromosomal disorders and somatic gene disorders. Chromosomal disorders are abnormalities in the number or structure of the chromosomes. These disorders are often severe and include Down syndrome, Turner's syndrome, and Klinefelter's syndrome. Somatic gene disorders are conditions in which gene abnormalities develop only in certain cells. Many forms of cancer are caused by a somatic gene disorder.

BRAIN FOOD

In human males, meiosis and sperm production take about nine weeks and occur continuously after puberty begins. In females, meiosis and egg production begin before birth and then stop until puberty. From puberty until menopause, one egg each month resumes meiosis and finishes developing. So, production of a mature egg may take up to 50 years!

MISCONCEPTION ALERT

Cells That Undergo Meiosis A common misconception is that all or many types of cells can undergo meiosis. Make sure students understand that meiosis occurs only during the formation of sex cells (egg or sperm cells).

Close

Reteaching — BASIC

Modeling Mates Have students use Punnett squares to model several possible combinations of parents with sex-linked traits that are variously dominant and recessive.
LS Logical/Kinesthetic English Language Learners

Quiz — GENERAL

Are the following statements true or false?

1. Every one of the chromosomes is different between men and women. (false)
2. Men and women each have different numbers of chromosomes in their sex cells. (false)
3. If you looked inside a cell during mitosis and you could see the chromosomes lining up, you could tell whether the cell belongs to a man or a woman. (true)

Alternative Assessment — GENERAL

Writing **Meiosis versus Mitosis** Tell students that there will be a mock debate to decide whether mitosis or meiosis is "better." First, have the class discuss and agree upon a definition of "better." Then, have students choose a "side" and prepare a written argument that is supported by scientific facts. You may wish to allow volunteers to act out such a debate. **LS Verbal**

Figure 7 **Pedigree for a Recessive Disease**

Males   Females

Vertical lines connect children to their parents.

A solid square or circle indicates that the person has a certain trait.

A half-filled square or circle indicates that the person is a carrier of the trait.

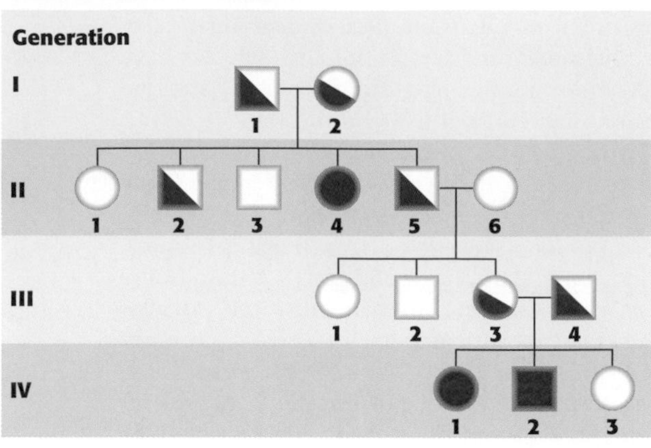

Genetic Counseling

Hemophilia and other genetic disorders can be traced through a family tree. If people are worried that they might pass a disease to their children, they may consult a genetic counselor. These counselors often make use of a diagram known as a **pedigree,** which is a tool for tracing a trait through generations of a family. By making a pedigree, a counselor can often predict whether a person is a carrier of a hereditary disease. The pedigree shown in **Figure 7** traces a disease called *cystic fibrosis* (SIS tik FIE broh sis). Cystic fibrosis causes serious lung problems. People with this disease have inherited two recessive alleles. Both parents need to be carriers of the gene for the disease to show up in their children.

pedigree a diagram that shows the occurrence of a genetic trait in several generations of a family

Pedigrees can be drawn up to trace any trait through a family tree. You could even draw a pedigree that would show how you inherited your hair color. Many different pedigrees could be drawn for a typical family.

Selective Breeding

For thousands of years, humans have seen the benefits of the careful breeding of animals. In *selective breeding,* organisms with desirable characteristics are mated. You have probably enjoyed the benefits of selective breeding, although you may not have realized it. For example, you have probably eaten an egg from a chicken that was bred to produce a larger number of eggs. Your pet dog might even be a result of selective breeding. Roses, like the one shown in **Figure 8,** have been selectively bred to produce large flowers. Wild roses are much smaller and have fewer petals than roses that you could buy at a nursery.

Figure 8 *Roses have been selectively bred to create large, bright flowers.*

WEIRD SCIENCE

Gene therapy is an experimental field of medical research in which defective genes are replaced with healthy genes. One way to insert healthy genes involves using a delivery system called a *gene gun* to inject microscopic gold bullets coated with genetic material.

Homework — ADVANCED

Pet Pedigrees Have students obtain a copy of the pedigree of a thoroughbred animal from a professional breeder of dogs, cats, horses, or other animals. Ask students to write a paragraph explaining what information the pedigree provides about the animal and its ancestors.
LS Verbal/Interpersonal

SECTION Review

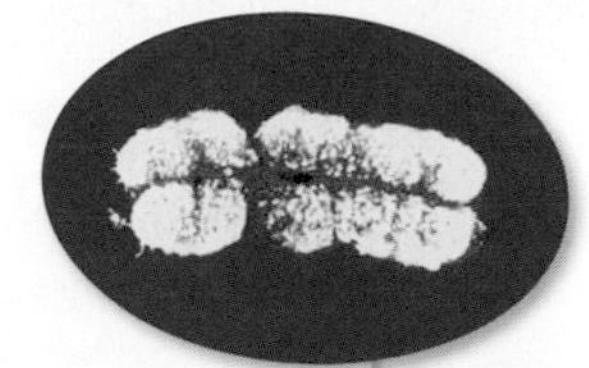

Summary

- In mitosis, chromosomes are copied once, and then the nucleus divides once. In meiosis, chromosomes are copied once, and then the nucleus divides twice.
- The process of meiosis produces sex cells, which have half the number of chromosomes. These two halves combine during reproduction.
- In humans, females have two X chromosomes. So, each egg contains one X chromosome. Males have both an X and a Y chromosome. So, each sperm cell contains either an X or a Y chromosome.
- Sex-linked disorders occur in males more often than in females. Colorblindness and hemophilia are examples of sex-linked disorders.
- A pedigree is a diagram used to trace a trait through many generations of a family.

Using Key Terms

1. Use each of the following terms in the same sentence: *meiosis* and *sex chromosomes*.

In each of the following sentences, replace the incorrect term with the correct term from the word bank.

pedigree	homologous chromosomes
meiosis	mitosis

2. During fertilization, chromosomes are copied, and then the nucleus divides twice.

3. A Punnett square is used to show how inherited traits move through a family.

4. During meiosis, sex cells line up in the middle of the cell.

Understanding Key Ideas

5. Genes are found on

a. chromosomes.

b. proteins.

c. alleles.

d. sex cells.

6. If there are 14 chromosomes in pea plant cells, how many chromosomes are present in a sex cell of a pea plant?

7. Draw the eight steps of meiosis. Label one chromosome, and show its position in each step.

Interpreting Graphics

Use this pedigree to answer the question below.

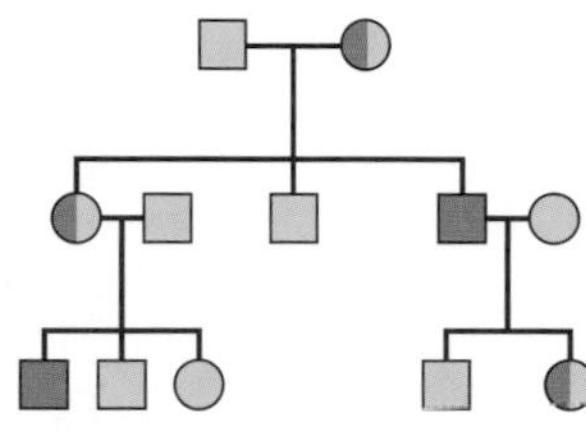

8. Is this disorder sex linked? Explain your reasoning.

Critical Thinking

9. Identifying Relationships Put the following in order of smallest to largest: chromosome, gene, and cell.

10. Applying Concepts A pea plant has purple flowers. What alleles for flower color could the sex cells carry?

For a variety of links related to this chapter, go to www.scilinks.org

Topic: Meiosis; Genetic Diseases, Screening, Counseling

SciLinks code: HSM0935; HSM0651

Answers to Section Review

1. Sample answer: At the end of meiosis, each sex cell will contain only one sex chromosome (either X or Y).

2. During meiosis, chromosomes are copied, and then the nucleus divides twice.

3. A pedigree is used to show how inherited traits move through a family.

4. During meiosis, homologous chromosomes line up in the middle of the cell.

5. a

6. 7

7. Answers may vary. Students' drawings should be similar to the diagram of meiosis in the student text.

8. Sample answer: yes; The disorder seems to be sex linked because the females are carriers of the disease but only males have the disease itself.

9. gene, chromosome, cell

10. Sample answer: Because the purple gene (*P*) is dominant over the white gene (*p*), the genotype of the purple-flowered pea plant could be either *PP* or *Pp*. Thus, the possible alleles carried by the sex cells would be *P* or *p*.

INTERNET ACTIVITY

Essay GENERAL

For an internet activity related to this chapter, have students goto **go.hrw.com** and type in the keyword **HL5DNAW.**

CHAPTER RESOURCES

Chapter Resource File

- **Section Quiz** GENERAL
- **Section Review** GENERAL
- **Vocabulary and Section Summary** GENERAL
- **Reinforcement Worksheet** BASIC

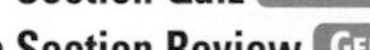

Model-Making Lab

Bug Builders, Inc.

Teacher's Notes

Time Required
Two 45-minute class periods

Lab Ratings

EASY → HARD

Teacher Prep 3
Student Set-Up 2
Concept Level 3
Clean Up 1

MATERIALS
The materials listed on the student page are enough for a group of 3–4 students. For step 3, prepare 14 small paper sacks—representing paired parent alleles for each of seven characteristics—as follows:

1. Cut 1 in. squares of paper to represent alleles. Use seven colors of paper—a different color for each characteristic. Cut enough squares so that each student will receive two alleles for each characteristic.
2. Mark half the squares for each characteristic with capital letters to indicate dominant alleles. Mark the other half with lowercase letters to indicate recessive alleles.
3. Label each pair of sacks with one of the seven characteristics. Place an equal number of alleles in each sack, but randomly mix both dominant and recessive alleles.
4. For each characteristic, label one sack "Mom" and the other sack "Dad." Have students draw one allele from each sack.

Using Scientific Methods

Model-Making Lab

Bug Builders, Inc.

OBJECTIVES

Build models to further your understanding of inheritance.

Examine the traits of a population of offspring.

MATERIALS

- allele sacks (14) (supplied by your teacher)
- gumdrops, green and black (feet)
- map pins (eyes)
- marshmallows, large (head and body segments)
- pipe cleaners (tails)
- pushpins, green and blue (noses)
- scissors
- toothpicks, red and green (antennae)

SAFETY

Imagine that you are a designer for a toy company that makes toy alien bugs. The president of Bug Builders, Inc., wants new versions of the wildly popular Space Bugs, but he wants to use the bug parts that are already in the warehouse. It's your job to come up with a new bug design. You have studied how traits are passed from one generation to another. You will use this knowledge to come up with new combinations of traits and assemble the bug parts in new ways. Model A and Model B, shown below, will act as the "parent" bugs.

Ask a Question

1. If there are two forms of each of the seven traits, then how many possible combinations are there?

Form a Hypothesis

2. Write a hypothesis that is a possible answer to the question above. Explain your reasoning.

Test the Hypothesis

3. Your teacher will display 14 allele sacks. The sacks will contain slips of paper with capital or lowercase letters on them. Take one piece of paper from each sack. (Remember: Capital letters represent dominant alleles, and lowercase letters represent recessive alleles.) One allele is from "Mom," and one allele is from "Dad." After you have recorded the alleles you have drawn, place the slips of paper back into the sack.

Model A ("Mom")
- red antennae
- 3 body segments
- curly tail
- 2 pairs of legs
- green nose
- black feet
- 3 eyes

Model B ("Dad")
- green antennae
- 2 body segments
- straight tail
- 3 pairs of legs
- blue nose
- green feet
- 2 eyes

Safety Caution
Remind students to review all safety cautions and icons before beginning this lab activity. Students should use caution with toothpicks and should not eat any of the materials used.

CHAPTER RESOURCES

Chapter Resource File

- Datasheet for Chapter Lab
- Lab Notes and Answers

Technology

Classroom Videos
- Lab Video

- Tracing Traits

Bug Family Traits				
Trait	**Model A "Mom" allele**	**Model B "Dad" allele**	**New model "Baby" genotype**	**New model "Baby" phenotype**
Antennae color				
Number of body segments				
Tail shape				
Number of leg pairs				
Nose color				
Foot color				
Number of eyes				

4. Create a table like the one above. Fill in the first two columns with the alleles that you selected from the sacks. Next, fill in the third column with the genotype of the new model ("Baby").

5. Use the information below to fill in the last column of the table.

Genotypes and Phenotypes	
RR or *Rr*—red antennae	*rr*—green antennae
SS or *Ss*—3 body segments	*ss*—2 body segments
CC or *Cc*—curly tail	*cc*—straight tail
LL or *Ll*—3 pairs of legs	*ll*—2 pairs of legs
BB or *Bb*—blue nose	*bb*—green nose
GG or *Gg*—green feet	*gg*—black feet
EE or *Ee*—2 eyes	*ee*—3 eyes

6. Now that you have filled out your table, you are ready to pick the parts you need to assemble your bug. (Toothpicks can be used to hold the head and body segments together and as legs to attach the feet to the body.)

Analyze the Results

1. **Organizing Data** Take a poll of the traits of the offspring. What are the ratios for each trait?

2. **Examining Data** Do any of the new models look exactly like the parents? Explain.

Draw Conclusions

3. **Interpreting Information** What are the possible genotypes of the parent bugs?

4. **Making Predictions** How many different genotypes are possible in the offspring?

Applying Your Data

Find a mate for your "Baby" bug. What are the possible genotypes and phenotypes of the offspring from this match?

Ask a Question

1. There are 128 possible combinations. (Calculation: There are two forms of each of seven characteristics, so, $2 \times 2 \times 2 \times 2 \times 2 \times 2 \times 2 = 2^7 = 128$)

Analyze the Results

1. Student ratios should be similar to the ratios determined when the alleles were selected by the teacher.
2. If any students have offspring bugs that look like one of the parents, have students compare the genotype of the offspring with the genotype of the parents. The offspring and parents look alike but still have different genotypes for some traits.

Draw Conclusions

3. Student answers should reflect the data on parent alleles that were recorded in step 6.
4. Students' answers should include Punnett squares based on the parental traits. Except for the results obtained by parental genotypes that are all homozygous recessive, students will see other possibilities for genotypes and phenotypes from the same parents.

Applying Your Data

Students should create Punnett squares to show the possible genotypes and describe phenotypes that follow the rules of dominance for each characteristic.

CHAPTER RESOURCES

Workbooks

Long-Term Projects & Research Ideas
- Portrait of a Dog ADVANCED

Kathy LaRoe
East Valley Middle School
East Helena, Montana

Chapter Review

Assignment Guide	
Section	**Questions**
1	7, 13, 18
2	2, 4, 5, 8, 9, 11, 19–23
3	1, 3, 6, 10, 12, 14–17

ANSWERS

Using Key Terms

1. sex cells
2. phenotype, genotype
3. Meiosis
4. alleles

Understanding Key Ideas

5. d
6. c
7. b
8. b
9. c
10. c
11. b

Chapter Review

USING KEY TERMS

Complete each of the following sentences by choosing the correct term from the word bank.

sex cells	genotype
sex chromosomes	alleles
phenotype	meiosis

1. Sperm and eggs are known as _____.
2. The _____ is the expression of a trait and is determined by the combination of alleles called the _____.
3. _____ produces cells with half the normal number of chromosomes.
4. Different versions of the same genes are called _____.

UNDERSTANDING KEY IDEAS

Multiple Choice

5. Genes carry information that determines
   - **a.** alleles.
   - **b.** ribosomes.
   - **c.** chromosomes.
   - **d.** traits.
6. The process that produces sex cells is
   - **a.** mitosis.
   - **b.** photosynthesis.
   - **c.** meiosis.
   - **d.** probability.
7. The passing of traits from parents to offspring is called
   - **a.** probability.
   - **b.** heredity.
   - **c.** recessive.
   - **d.** meiosis.
8. If you cross a white flower with the genotype *pp* with a purple flower with the genotype *PP,* the possible genotypes in the offspring are
   - **a.** *PP* and *pp*.
   - **b.** all *Pp*.
   - **c.** all *PP*.
   - **d.** all *pp*.
9. For the cross in item 8, what would the phenotypes be?
   - **a.** all white
   - **b.** 3 purple and 1 white
   - **c.** all purple
   - **d.** half white, half purple
10. In meiosis,
    - **a.** chromosomes are copied twice.
    - **b.** the nucleus divides once.
    - **c.** four cells are produced from a single cell.
    - **d.** two cells are produced from a single cell.
11. When one trait is not completely dominant over another, it is called
    - **a.** recessive.
    - **b.** incomplete dominance.
    - **c.** environmental factors.
    - **d.** uncertain dominance.

Short Answer

12. Which sex chromosomes do females have? Which do males have?

13. In one or two sentences, define the term *recessive trait* in your own words.

14. How are sex cells different from other body cells?

15. What is a sex-linked disorder? Give one example of a sex-linked disorder that is found in humans.

CRITICAL THINKING

16. **Concept Mapping** Use the following terms to create a concept map: *meiosis, eggs, cell division, X chromosome, mitosis, Y chromosome, sperm,* and *sex cells.*

17. **Identifying Relationships** If you were a carrier of one allele for a certain recessive disorder, how could genetic counseling help you prepare for the future?

18. **Applying Concepts** If a child has blue eyes and both her parents have brown eyes, what does that tell you about the allele for blue eyes? Explain.

19. **Applying Concepts** What is the genotype of a pea plant that is true-breeding for purple flowers?

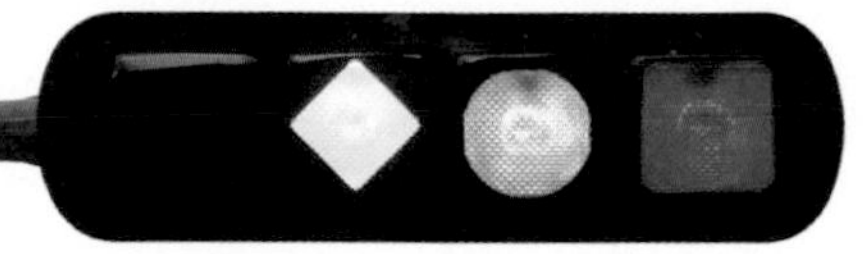

INTERPRETING GRAPHICS

Use the Punnett square below to answer the questions that follow.

	?	?
T	***TT***	***TT***
t	***Tt***	***Tt***

20. What is the unknown genotype?

21. If *T* represents the allele for tall pea plants and *t* represents the allele for short pea plants, what is the phenotype of each parent and of the offspring?

22. If each of the offspring were allowed to self-fertilize, what are the possible genotypes in the next generation?

23. What is the probability of each genotype in item 22?

Critical Thinking

16.  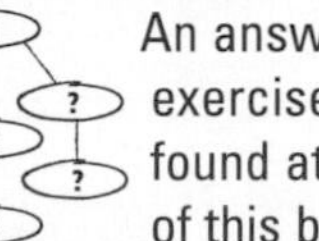

An answer to this exercise can be found at the end of this book.

17. Sample answer: A genetic counselor could test my spouse to see if my spouse is also a carrier of the recessive allele. The counselor could then predict what the chances are that we could have a child with the recessive disorder.

18. The allele for blue eyes is recessive.

19. *PP*

Interpreting Graphics

20. *TT*

21. All the parents and offspring are tall pea plants.

22. Students should make two new Punnett squares. Self-fertilization of *TT* (*TT* × *TT*) will yield offspring that are all *TT*. Self fertilization of *Tt* (*Tt* × *Tt*) will yield offspring that are *TT*, *Tt*, and *tt*.

23. *TT* has a 100% probability with a *TT* parent and a 25% probability with a *Tt* parent. *Tt* has a 50% probability with a *Tt* parent and a 0% probability with a *TT* parent. The genotype *tt* has a 25% probability with a *Tt* parent and a 0% probability with a *TT* parent.

12. Females have two X chromosomes. Males have one X and one Y chromosome.

13. Sample answer: A recessive trait is a genetic trait that is expressed only if there are two recessive alleles for the gene. A recessive trait is not expressed if an allele for a dominant trait is present.

14. Sex cells have half the number of chromosomes as other body cells.

15. Sample answer: A sex-linked disorder is a disorder that is caused by a gene on one of the sex chromosomes and so is expressed in one sex more than the other. Color blindness is a sex-linked disorder found in humans.

CHAPTER RESOURCES

Chapter Resource File

- Chapter Review GENERAL
- Chapter Test A GENERAL
- Chapter Test B ADVANCED
- Chapter Test C SPECIAL NEEDS
- Vocabulary Activity GENERAL

Workbooks

Study Guide

- Assessment resources are also available in Spanish.

Standardized Test Preparation

Teacher's Note

To provide practice under more realistic testing conditions, give students 20 minutes to answer all of the questions in this Standardized Test Preparation.

MISCONCEPTION ALERT

Answers to the standardized test preparation can help you identify student misconceptions and misunderstandings.

READING

Passage 1

1. C
2. F
3. C

Question 2: This question primarily requires the reader to re-read the sentence in which the word is used, which clearly serves to define the word. Then, the reader must look among the possible answers for the one that most closely matches the meaning given in the sentence.

Question 3: This question requires a simple deduction from the final two sentences of the passage. The uses of "if," "then," and "therefore" are clear indicators of logical reasoning. Remind students to look for these kinds of indicators for these types of test questions.

Standardized Test Preparation

READING

Read the passages below. Then, answer the questions that follow each passage.

Passage 1 The different versions of a gene are called *alleles*. When two different alleles occur together, one is often expressed while the other has no obvious effect on the organism's appearance. The expressed form of the trait is dominant. The trait that was not expressed when the dominant form of the trait was present is called *recessive*. Imagine a plant that has both purple and white alleles for flower color. If the plant blooms purple, then purple is the dominant form of the trait. Therefore, white is the recessive form.

1. According to the passage, which of the following statements is true?

A All alleles are expressed all of the time.
B All traits for flower color are dominant.
C When two alleles are present, the expressed form of the trait is dominant.
D A recessive form of a trait is always expressed.

2. According to the passage, a trait that is not expressed when the dominant form is present is called

F recessive.
G an allele.
H heredity.
I a gene.

3. According to the passage, which allele for flower color is dominant?

A white
B pink
C purple
D yellow

Passage 2 Sickle cell anemia is a recessive genetic disorder. People inherit this disorder only when they inherit the disease-causing recessive allele from both parents. The disease causes the body to make red blood cells that bend into a sickle (or crescent moon) shape. The sickle-shaped red blood cells break apart easily. Therefore, the blood of a person with sickle cell anemia carries less oxygen. Sickle-shaped blood cells also tend to get stuck in blood vessels. When a blood vessel is blocked, the blood supply to organs can be cut off. But the sickle-shaped blood cells can also protect a person from malaria. Malaria is a disease caused by an organism that invades red blood cells.

1. According to the passage, sickle cell anemia is a

A recessive genetic disorder.
B dominant genetic disorder.
C disease caused by an organism that invades red blood cells.
D disease also called *malaria*.

2. According to the passage, sickle cell anemia can help protect a person from

F blocked blood vessels.
G genetic disorders.
H malaria.
I low oxygen levels.

3. Which of the following is a fact in the passage?

A When blood vessels are blocked, vital organs lose their blood supply.
B When blood vessels are blocked, it causes the red blood cells to bend into sickle shapes.
C The blood of a person with sickle cell anemia carries more oxygen.
D Healthy red blood cells never get stuck in blood vessels.

Passage 2

1. A
2. H
3. A

Question 2: The answer to this question comes from the second-to-last sentence in the passage. Weak readers often miss details from the middle parts of passages, and standardized tests sometimes probe for this kind of mistake with such questions. One strategy for this type of question is to form a question such as "From what problem can sickle cell anemia protect a person?" and then re-read or skim the passage with this question in mind.

INTERPRETING GRAPHICS

The Punnett square below shows a cross between two flowering plants. Use this Punnett square to answer the questions that follow.

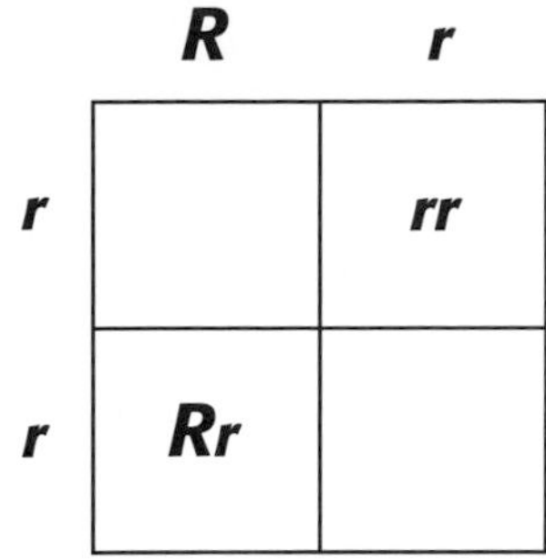

1. What is the genotype of the offspring represented in the upper left-hand box of the Punnett square?
   A *RR*
   B *Rr*
   C *rr*
   D *rrr*

2. What is the genotype of the offspring represented in the lower right-hand box of the Punnett square?
   F *RR*
   G *Rr*
   H *rr*
   I *rrr*

3. What is the ratio of *Rr* (purple-flowered plants) to *rr* (white-flowered plants) in the offspring?
   A 1:3
   B 2:2
   C 3:1
   D 4:0

MATH

Read each question below, and choose the best answer.

1. What is another way to write $4 \times 4 \times 4$?
   A 4^2
   B 4^3
   C 3^3
   D 3^4

2. Jane was making a design on top of her desk with pennies. She put 4 pennies in the first row, 7 pennies in the second row, and 13 pennies in the third row. If Jane continues this pattern, how many pennies will she put in the sixth row?
   F 25
   G 49
   H 97
   I 193

3. In which of the following lists are the numbers in order from smallest to greatest?
   A 0.012, 0.120, 0.123, 1.012
   B 1.012, 0.123, 0.120, 0.012
   C 0.123, 0.120, 0.012, 1.012
   D 0.123, 1.012, 0.120, 0.012

4. In which of the following lists are the numbers in order from smallest to greatest?
   F −12.0, −15.5, 2.2, 4.0
   G −15.5, −12.0, 2.2, 4.0
   H −12.0, −15.5, 4.0, 2.2
   I 2.2, 4.0, −12.0, −15.5

5. Which of the following is equal to −11?
   A $7 + 4$
   B $-4 + 7$
   C $-7 + 4$
   D $-7 + -4$

6. Catherine earned $75 for working 8.5 h. How much did she earn per hour?
   F $10.12
   G $9.75
   H $8.82
   I $8.01

INTERPRETING GRAPHICS

1. B
2. H
3. B

TEST DOCTOR

Questions 1 and 2: These questions require understanding of the term *genotype* and the ability to complete a Punnett square. Students who miss these questions may need to review these concepts.

Question 3: This question asks for the ratio of the genotype *Rr* to the genotype *rr*. If completed, the Punnett square would show 2 *Rr* and 2 *rr* genotypes. Thus, the ratio would be 2:2 (answer B). Students who miss this question may need to review the concept of ratios.

MATH

1. B
2. H
3. A
4. G
5. D
6. H

TEST DOCTOR

Question 6: This question is essentially a simple long-division problem, but students may get confused or discouraged by long division when the calculation extends for many decimal places. For this problem, students can save time if they recognize that they need only to find the answer in dollars and cents. Thus, they need to calculate to the thousandths place ($8.823) and then round their answer to the nearest cent.

CHAPTER RESOURCES

Chapter Resource File

• **Standardized Test Preparation** GENERAL

State Resources

For specific resources for your state, visit **go.hrw.com** and type in the keyword **HSMSTR.**

Science in Action

Science, Technology, and Society

Background

Genetic research has spawned a flurry of debate over ethical, social, and legal issues surrounding the use of genetic information. These issues include the privacy and ownership of personal genetic information and the possibility that people will selectively breed or control the birth of their children based on genetic knowledge.

Weird Science

Teaching Strategy-GENERAL

Offer the following analogies to help students grasp the concepts discussed in this article.

- Blueprints: Show students sample construction blueprints. Explain that genes are like these plans for a building and that mutations are like mistakes in copying, reading, or building from the blueprints.
- Recipes: Show students a book of cake recipes. Genes are like recipes, and an organism is like a cake made according to a recipe. A mutation is like using a different ingredient or a different amount of an ingredient. The mutation may or may not "ruin" the "cake."

Science in Action

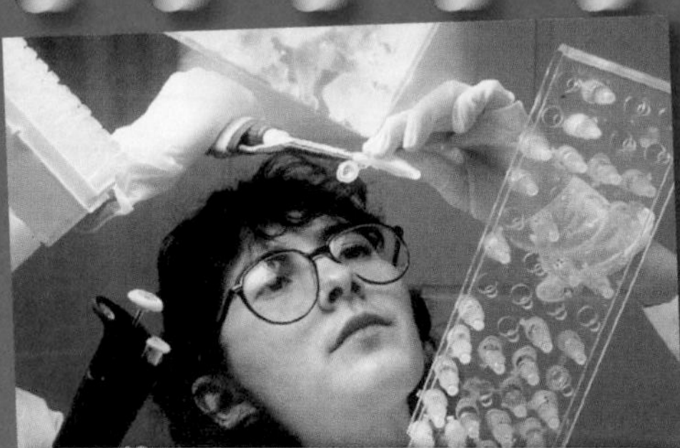

Science, Technology, and Society

Mapping the Human Genome

In 2003, scientists finished one of the most ambitious research projects ever. Researchers with the Human Genome Project (HGP) mapped the human body's complete set of genetic instructions, which is called the *genome*. You might be wondering whose genome the scientists are decoding. Actually, it doesn't matter—only 1% of each person's genetic material is unique. The researchers' goals are to identify how tiny differences in that 1% make each of us who we are and to begin to understand how some differences can cause disease. Scientists are already using the map to think of new ways to treat genetic diseases, such as asthma, diabetes, and kidney disease.

Social Studies ACTIVITY

WRITING SKILL Research DNA fingerprinting. Write a short report describing how DNA fingerprinting has affected the way criminals are caught.

This is a normal fruit fly under a scanning electron microscope.

This fruit fly has legs growing where its antennae should be.

Weird Science

Lab Rats with Wings

Drosophila melanogaster (droh SAHF i luh muh LAN uh GAS tuhr) is the scientific name for the fruit fly. This tiny insect has played a big role in helping scientists understand many illnesses. Because fruit flies reproduce every 2 weeks, scientists can alter a fruit fly gene and see the results of the experiment very quickly. Another important reason for using these "lab rats with wings" is that their genetic code is simple and well understood. Fruit flies have 12,000 genes, but humans have more than 30,000. Scientists use fruit flies to find out about diseases like cancer, Alzheimer's, and muscular dystrophy.

Language Arts ACTIVITY

WRITING SKILL The mythical creature called the *Chimera* (kie MIR uh) was said to be part lion, part goat, and part serpent. According to legend, the Chimera terrorized people for years until it was killed by a brave hero. The word *chimera* now refers to any organism that has parts from many organisms. Write a short story about the Chimera that describes what it looks like and how it came to be.

Answer to Social Studies Activity

Sample answer: DNA fingerprinting has made it much easier to match genetic material (evidence) at a crime scene to the genetic information of one particular individual. DNA can be found in hair, saliva, blood, and small skin cells. The DNA is sequenced and then compared to the DNA fingerprint of particular individuals. When the DNA fingerprints match, police can be sure that the person was at the scene of the crime.

Answer to Language Arts Activity

The Chimera (or Chimaera) was said to be a savage beast that spat fire from its mouth. In classical Greco-Roman stories, it wreaked havoc on the ancient lands until it was killed by the hero Bellerophon, who rode his winged horse Pegasus. This basic story is among the most ancient myths and appears in many texts from Homer's *Iliad* to traditional fairy tales.

Careers

Stacey Wong

Genetic Counselor If your family had a history of a particular disease, what would you do? Would you eat healthier foods, get more exercise, or visit your doctor regularly? All of those are good ideas, but Stacey Wong went a step farther. Her family's history of cancer helped her decide to become a genetic counselor. "Genetic counselors are usually part of a team of health professionals," she says, which can include physicians, nurses, dieticians, social workers, laboratory personnel, and others. "If a diagnosis is made by the geneticist," says Wong, "then I provide genetic counseling." When a patient visits a genetic counselor, the counselor asks many questions and builds a family medical history. Although counseling involves discussing what it means to have a genetic condition, Wong says "the most important part is to get to know the patient or family we are working with, listen to their concerns, gain an understanding of their values, help them to make decisions, and be their advocate."

Math ACTIVITY

The probability of inheriting genetic disease *A* is 1/10,000. The probability of inheriting genetic disease *B* is also 1/10,000. What is the probability that one person would inherit both genetic diseases *A* and *B*?

go.hrw.com

To learn more about these Science in Action topics, visit **go.hrw.com** and type in the keyword **HL5HERF.**

Current Science

Check out Current Science® articles related to this chapter by visiting go.hrw.com. Just type in the keyword HL5CS05.

Careers

Background

Stacey Wong was born in Oakland, California, and grew up in the nearby suburb of Alameda. She received a B.S. in cell and molecular biology from UCLA and an M.S. in genetic counseling from California State University Northridge. More information about genetic-counseling careers can be obtained from the National Society of Genetic Counselors.

Answer to Math Activity

1/10,000 × 1/10,000 = 1/100,000,000

Genes and DNA
Chapter Planning Guide

Compression guide:
To shorten instruction because of time limitations, omit the Chapter Lab.

OBJECTIVES	LABS, DEMONSTRATIONS, AND ACTIVITIES	TECHNOLOGY RESOURCES
PACING • 90 min pp. 84–89 **Chapter Opener**	**SE Start-up Activity,** p. 85 ◆ GENERAL	**OSP Parent Letter** ■ GENERAL **CD Student Edition on CD-ROM** **CD Guided Reading Audio CD** ■ **TR Chapter Starter Transparency*** **VID Brain Food Video Quiz**
Section 1 What Does DNA Look Like? • List three important events that led to understanding the structure of DNA. • Describe the basic structure of a DNA molecule. • Explain how DNA molecules can be copied.	**TE Activity** Modeling Code, p. 86 GENERAL **TE Group Activity** A Place in History, p. 87 GENERAL **SE Quick Lab** Making a Model of DNA, p. 88 ◆ GENERAL **CRF Datasheet for Quick Lab*** **SE Science in Action** Math, Social Studies, and Language Arts Activities, pp. 104–105 GENERAL **SE Model-Making Lab** Base-Pair Basics, p. 98 ◆ GENERAL **CRF Datasheet for Chapter Lab*** **LB Whiz-Bang Demonstrations** Grand Strand* GENERAL	**CRF Lesson Plans*** **TR Bellringer Transparency*** **TR** DNA Structure* **CRF SciLinks Activity*** GENERAL **VID Lab Videos for Life Science**
PACING • 45 min pp. 90–97 **Section 2 How DNA Works** • Explain the relationship between DNA, genes, and proteins. • Outline the basic steps in making a protein. • Describe three types of mutations, and provide an example of a gene mutation. • Describe two examples of uses of genetic knowledge.	**TE Demonstration** A Tight Fit, p. 90 ◆ GENERAL **TE Connection Activity** Chemistry, p. 92 ADVANCED **TE Group Activity** Skit, p. 92 GENERAL **TE Connection Activity** Math, p. 93 ◆ GENERAL **TE Activity** Complementary Code, p. 94 GENERAL **SE School-to-Home Activity** An Error in the Message, p. 95 GENERAL **TE Connection Activity** Social Studies, p. 96 ADVANCED **LB Long-Term Projects & Research Ideas** The Antifreeze Protein* ADVANCED **LB Long-Term Projects & Research Ideas** Ewe Again, Dolly?* ADVANCED	**CRF Lesson Plans*** **TR Bellringer Transparency*** **TR** Unraveling DNA* **TR** The Making of a Protein: A* **TR** The Making of a Protein: B* **TR** *LINK TO EARTH SCIENCE* The Formation of Smog* **TR** How Sickle Cell Anemia Results from a Mutation* **SE Internet Activity,** p. 92 GENERAL **CD Interactive Explorations CD-ROM** DNA Pawprints GENERAL

PACING • 90 min

CHAPTER REVIEW, ASSESSMENT, AND STANDARDIZED TEST PREPARATION

CRF Vocabulary Activity* GENERAL
SE Chapter Review, pp. 100–101 GENERAL
CRF Chapter Review* ■ GENERAL
CRF Chapter Tests A* ■ GENERAL, **B*** ADVANCED, **C*** SPECIAL NEEDS
SE Standardized Test Preparation, pp. 102–103 GENERAL
CRF Standardized Test Preparation* GENERAL
CRF Performance-Based Assessment* GENERAL
OSP Test Generator GENERAL
CRF Test Item Listing* GENERAL

Online and Technology Resources

Visit **go.hrw.com** for a variety of free resources related to this textbook. Enter the keyword **HL5DNA.**

Students can access interactive problem-solving help and active visual concept development with the *Holt Science and Technology* Online Edition available at **www.hrw.com.**

Guided Reading Audio CD

A direct reading of each chapter using instructional visuals as guideposts. For auditory learners, reluctant readers, and Spanish-speaking students. Available in English and Spanish.

KEY			
SE Student Edition	CRF Chapter Resource File	SS Science Skills Worksheets	* Also on One-Stop Planner
TE Teacher Edition	OSP One-Stop Planner	MS Math Skills for Science Worksheets	◆ Requires advance prep
	LB Lab Bank	CD CD or CD-ROM	■ Also available in Spanish
	TR Transparencies	VID Classroom Video/DVD	

SKILLS DEVELOPMENT RESOURCES	SECTION REVIEW AND ASSESSMENT	STANDARDS CORRELATIONS
SE Pre-Reading Activity, p. 84 GENERAL **OSP Science Puzzlers, Twisters & Teasers** GENERAL		National Science Education Standards UCP 5; SAI 1, 2; ST 2; LS 2e
CRF Directed Reading A* ■ BASIC, **B*** SPECIAL NEEDS **CRF Vocabulary and Section Summary*** ■ GENERAL **SE Reading Strategy** Prediction Guide, p. 86 GENERAL **SE Connection to Chemistry** Linus Pauling, p. 87 GENERAL **TE Reading Strategy** Mnemonics, p. 87 BASIC **TE Inclusion Strategies,** p. 87 **MS Math Skills for Science** A Shortcut for Multiplying Large Numbers* GENERAL **SS Science Skills** Science Drawing* GENERAL	**SE Reading Checks,** pp. 87, 89 GENERAL **TE Reteaching,** p. 88 BASIC **TE Quiz,** p. 88 GENERAL **TE Alternative Assessment,** p. 88 GENERAL **SE Section Review,*** p. 89 ■ GENERAL **CRF Section Quiz*** ■ GENERAL	UCP 2, 5; SAI 1, 2; ST 1, 2; SPSP 5; HNS 1, 2, 3; LS 1a, 2d, 5a; *Chapter Lab:* UCP 2, 5; SAI 1, 2; HNS 1; LS 1a
CRF Directed Reading A* ■ BASIC, **B*** SPECIAL NEEDS **CRF Vocabulary and Section Summary*** ■ GENERAL **SE Reading Strategy** Reading Organizer, p. 90 GENERAL **SE Math Practice** Code Combinations, p. 93 GENERAL **TE Inclusion Strategies,** p. 95 **SE Connection to Social Studies** Genetic Property, p. 96 GENERAL **CRF Reinforcement Worksheet** DNA Mutations* BASIC **CRF Critical Thinking** The Perfect Parrot* ADVANCED	**SE Reading Checks,** pp. 90, 93, 94, 95, 96 GENERAL **TE Homework,** p. 92 GENERAL **TE Homework,** p. 95 GENERAL **TE Reteaching,** p. 96 BASIC **SE Section Review,*** p. 97 ■ GENERAL **TE Quiz,** p. 96 GENERAL **TE Alternative Assessment,** p. 96 GENERAL **CRF Section Quiz*** ■ GENERAL	UCP 1, 4, 5; SAI 1, 2; ST 2; SPSP 4, 5; LS 1c, 1e, 1f, 2b, 2c, 2d, 2e, 5b

One-Stop Planner® CD-ROM

This convenient CD-ROM includes:
- **Lab Materials QuickList Software**
- **Holt Calendar Planner**
- **Customizable Lesson Plans**
- **Printable Worksheets**
- **ExamView® Test Generator**

CNN Student News

cnnstudentnews.com

Find the latest news, lesson plans, and activities related to important scientific events.

SciLinks NSTA

www.scilinks.org

Maintained by the **National Science Teachers Association.** See Chapter Enrichment pages for a complete list of topics.

Current Science®

Check out ***Current Science*** articles and activities by visiting the HRW Web site at **go.hrw.com.** Just type in the keyword **HL5CS06T.**

Classroom Videos
- **Lab Videos** demonstrate the chapter lab.
- **Brain Food Video Quizzes** help students review the chapter material.

4 Chapter Resources

Visual Resources

CHAPTER STARTER TRANSPARENCY

Genes and DNA — CHAPTER STARTER

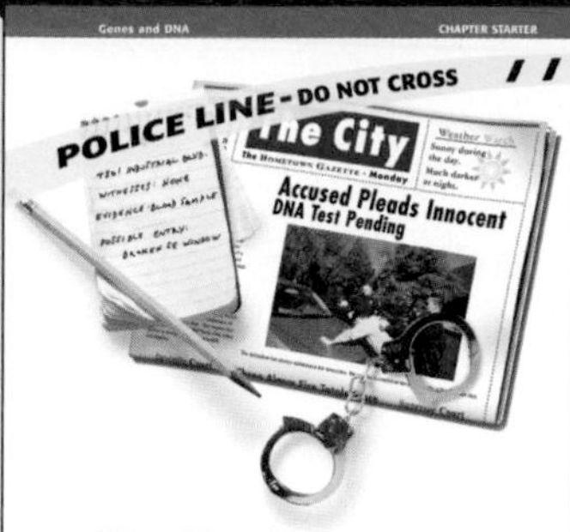

What If . . . ?

Imagine that you are wrongly accused of a crime. A witness at the scene picks you out of a lineup. Your blood type matches blood found at the scene. You were really at home when the crime occurred, but you have no way to prove it. However, there is a witness that can help you—your DNA. DNA is a substance found in nearly every one of your cells.

A technique known as DNA fingerprinting can produce an image of the patterns made by your DNA. Unless you have an identical twin, your DNA fingerprint is like no one else's and can be used to identify you. But what is DNA? Why is yours unique? What does DNA have to do with who you are? In this chapter you will learn the answers to these questions.

BELLRINGER TRANSPARENCIES

Genes and DNA — BELLRINGER TRANSPARENCY

Section: What Does DNA Look Like?

Can you explain the difference between traits and characteristics? Which is more closely associated with DNA and genes? Where do you think DNA and genes are usually found?

Write your answers in your **science journal.**

Section: How DNA Works

Unscramble the following words:
tpsoneir
ineesg

Now think of three words you associate with each of the above words and use them all in a paragraph that highlights what you know about DNA.

Write your paragraph in your **science journal.**

TEACHING TRANSPARENCIES

Genes and DNA — DNA Structure — TEACHING TRANSPARENCY

Genes and DNA — TEACHING TRANSPARENCY

The Making of a Protein: A

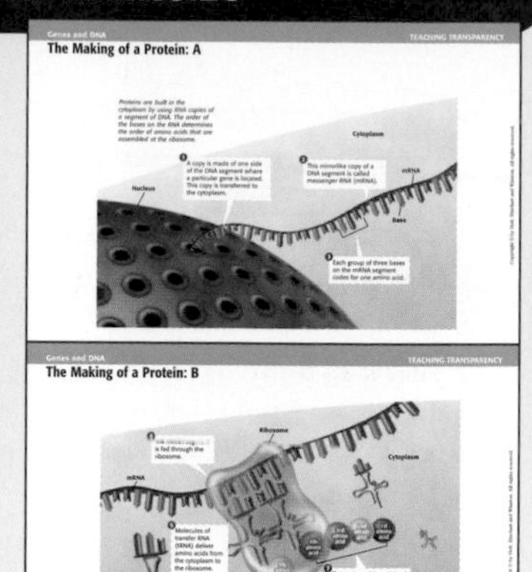

Genes and DNA — TEACHING TRANSPARENCY

The Making of a Protein: B

TEACHING TRANSPARENCIES

Genes and DNA — TEACHING TRANSPARENCY

How Sickle Cell Anemia Results from a Mutation

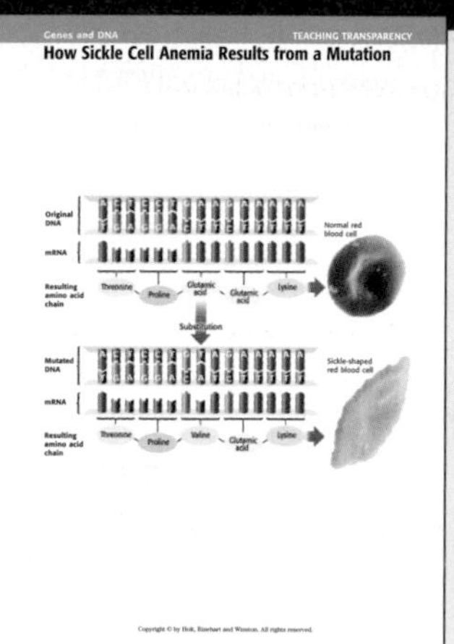

Genes and DNA — Unraveling DNA — TEACHING TRANSPARENCY

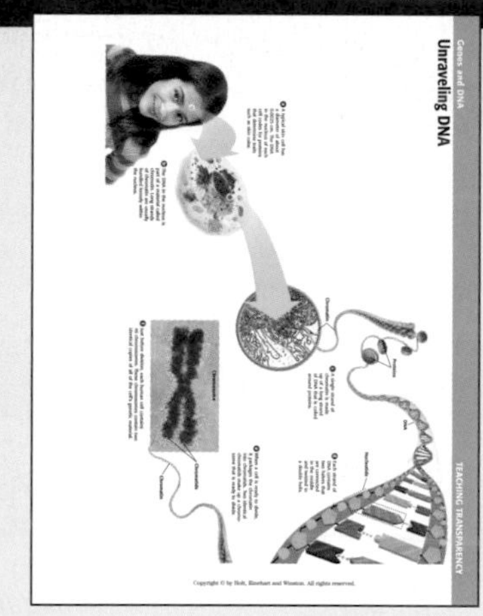

The Atmosphere — TEACHING TRANSPARENCY

The Formation of Smog

Chapter: The Atmosphere

CONCEPT MAPPING TRANSPARENCY

Genes and DNA — CONCEPT MAPPING TRANSPARENCY

Use the following terms to complete the concept map below:
mutation, amino acids, nucleotide, DNA, adenine, genes, guanine, proteins, chromosomes, cytosine

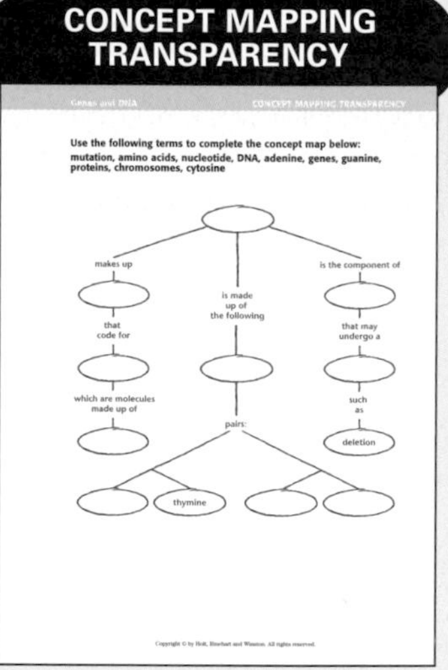

Planning Resources

LESSON PLANS

Lesson Plan — SAMPLE

Section: Waves

Pacing

Regular Schedule:	**with lab(s):**2 days	**without lab(s)**2 days
Block Schedule:	**with lab(s):**1 1/2 days	**without lab(s)**1 day

Objectives

1. Relate the seven properties of life to a living organism.
2. Describe seven themes that can help you to organize what you learn about biology.
3. Identify the tiny structures that make up all living organisms.
4. Differentiate between reproduction and heredity and between metabolism and homeostasis.

National Science Education Standards Covered

LSInter6: Cells have particular structures that underlie their functions.

LSMat1: Most cell functions involve chemical reactions.

LSBeh1: Cells store and use information to guide their functions.

UCP1: Cell functions are regulated.

SI1: Cells can differentiate and form complete multicellular organisms.

PS1: Species evolve over time.

ESS1: The great diversity of organisms is the result of more than 3.5 billion years of evolution.

ESS2: Natural selection and its evolutionary consequences provide a scientific explanation for the fossil record of ancient life forms as well as for the striking molecular similarities observed among the diverse species of living organisms.

ST1: The millions of different species of plants, animals, and microorganisms that live on Earth today are related by descent from common ancestors.

ST2: The energy for life primarily comes from the sun.

SPSP1: The complexity and organization of organisms accommodates the need for obtaining, transforming, transporting, releasing, and eliminating the matter and energy used to sustain the organism.

SPSP6: As matter and energy flows through different levels of organization of living systems—cells, organs, communities—and between living systems and the physical environment, chemical elements are recombined in different ways.

HNS1: Organisms have behavioral responses to internal changes and to external stimuli.

PARENT LETTER

SAMPLE

Dear Parent,

Your son's or daughter's science class will soon begin exploring the chapter entitled "The World of Physical Science." In this chapter, students will learn about how the scientific method applies to the world of physical science and the role of physical science in the world. By the end of the chapter, students should demonstrate a clear understanding of the chapter's main ideas and be able to discuss the following topics:

1. physical science as the study of energy and matter (Section 1)
2. the role of physical science in the world around them (Section 1)
3. careers that rely on physical science (Section 1)
4. the steps used in the scientific method (Section 2)
5. examples of technology (Section 2)
6. how the scientific method is used to answer questions and solve problems (Section 2)
7. how our knowledge of science changes over time (Section 2)
8. how models represent real objects or systems (Section 3)
9. examples of different ways models are used in science (Section 3)
10. the importance of the International System of Units (Section 4)
11. the appropriate units to use for particular measurements (Section 4)
12. how area and density are derived quantities (Section 4)

Questions to Ask Along the Way

You can help your son or daughter learn about these topics by asking interesting questions such as the following:

- What are some surprising careers that use physical science?
- What is a characteristic of a good hypothesis?
- When is it a good idea to use a model?
- Why do Americans measure things in terms of inches and yards and meters ?

ALSO IN SPANISH

TEST ITEM LISTING

TEST ITEM LISTING
The World of Science — SAMPLE

MULTIPLE CHOICE

1. A limitation of models is that
   a. they are large enough to see.
   b. they do not act exactly like the things that they model.
   c. they are smaller than the things that they model.
   d. they model unfamiliar things.
   Answer: B Difficulty: I Section: 3 Objective: 2
2. The length 10 m is equal to
   a. 100 cm. c. 10,000 mm.
   b. 1,000 cm. d. Both (b) and (c)
   Answer: B Difficulty: I Section: 3 Objective: 2
3. To be valid, a hypothesis must be
   a. testable. c. made into a law.
   b. supported by evidence. d. Both (a) and (b)
   Answer: B Difficulty: I Section: 3 Objective: 2 1
4. The statement "Sheila has a stain on her shirt" is an example of a(n)
   a. law. c. observation.
   b. hypothesis. d. prediction.
   Answer: B Difficulty: I Section: 3 Objective: 2
5. A hypothesis is often developed out of
   a. observations. c. laws.
   b. experiments. d. Both (a) and (b)
   Answer: B Difficulty: I Section: 3 Objective: 2
6. How many milliliters are in 3.5 kL?
   a. 3,500 mL c. 3,500, 000 mL
   b. 0.0035 mL d. 35,000 mL
   Answer: B Difficulty: I Section: 3 Objective: 2
7. A map of Seattle is an example of a
   a. law. c. model.
   b. theory. d. unit.
   Answer: B Difficulty: I Section: 3 Objective: 2
8. A lab has the safety icons shown below. These icons mean that you should wear
   a. only safety goggles. c. safety goggles and a lab apron.
   b. only a lab apron. d. safety goggles, a lab apron, and gloves.
   Answer: B Difficulty: I Section: 3 Objective: 2
9. The law of conservation of mass says the tot al mass before a chemical change is
   a. more than the total mass after the change.
   b. less than the total mass after the change.
   c. the same as the total mass after the change.
   d. not the same as the total mass after the change.
   Answer: B Difficulty: I Section: 3 Objective: 2
10. In which of the following areas might you find a geochemist at work?
    a. studying the chemistry of rocks c. studying fishes
    b. studying forestry d. studying the atmosphere
    Answer: B Difficulty: I Section: 3 Objective: 2

One-Stop Planner® CD-ROM

This CD-ROM includes all of the resources shown here and the following time-saving tools:

- *Lab Materials QuickList Software*
- *Customizable lesson plans*
- *Holt Calendar Planner*
- *The powerful ExamView® Test Generator*

For a preview of available worksheets covering math and science skills, see pages T12–T19. All of these resources are also on the One-Stop Planner®.

Meeting Individual Needs

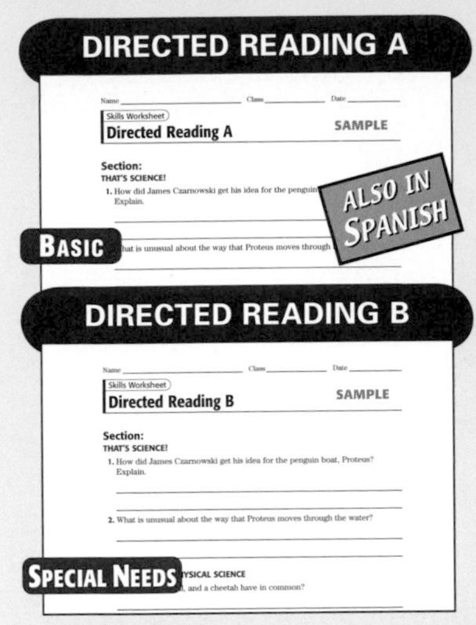

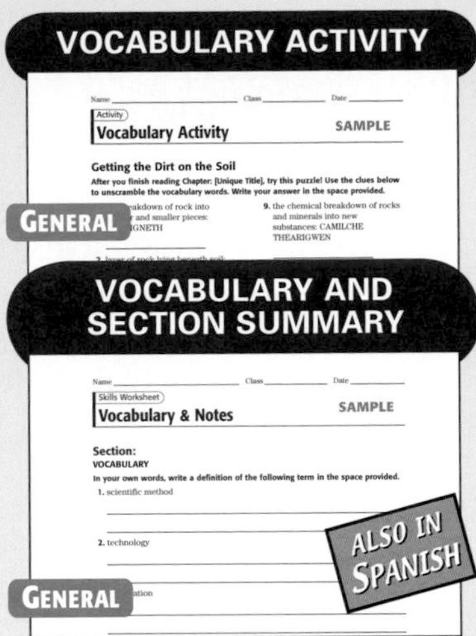

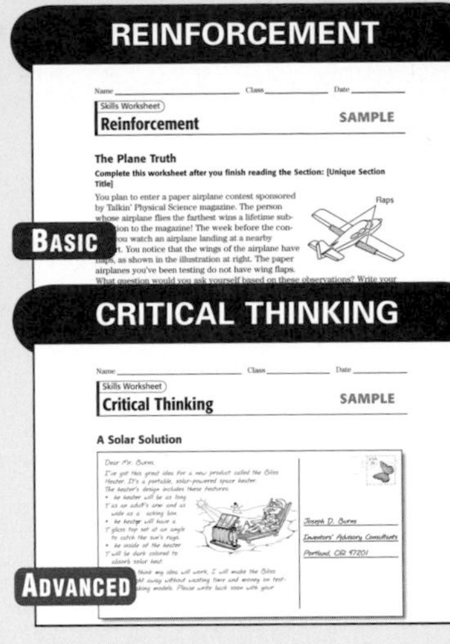

Labs and Activities

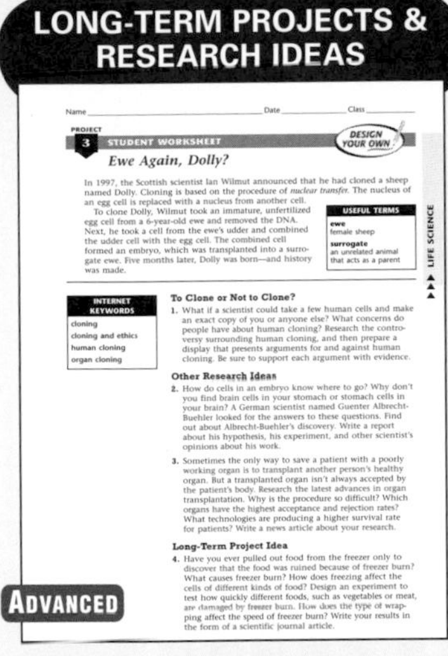

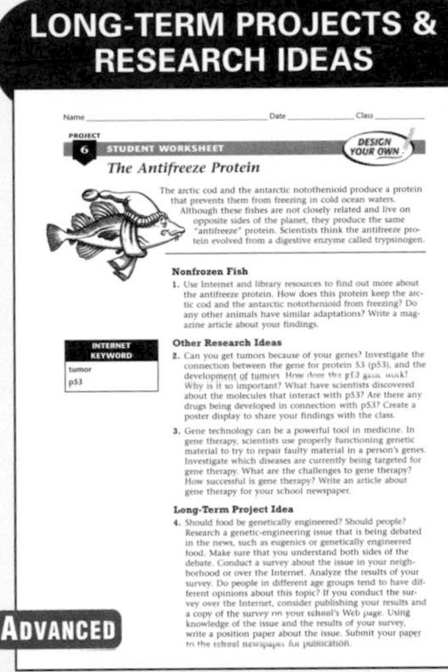

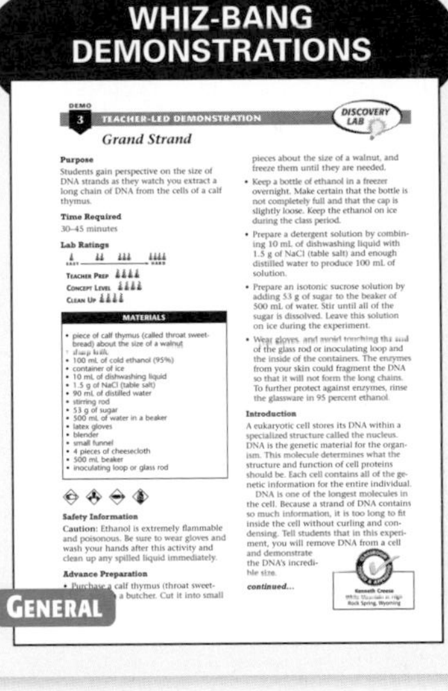

Review and Assessments

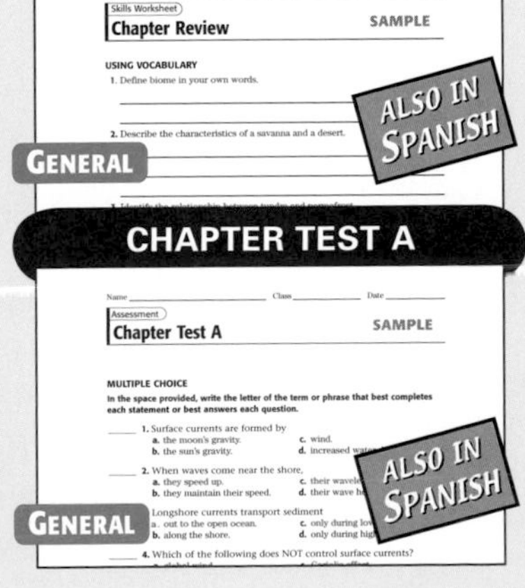

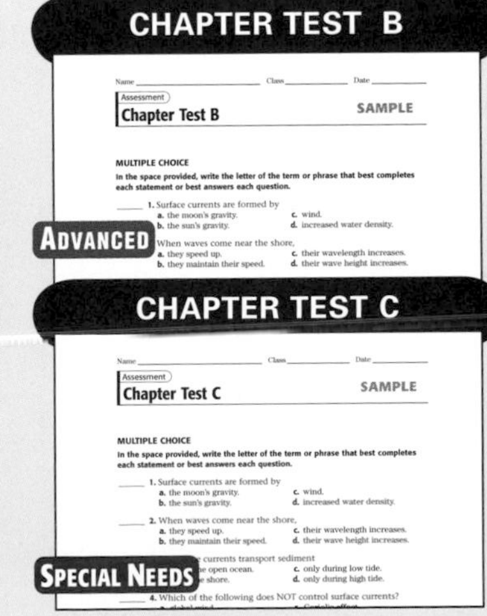

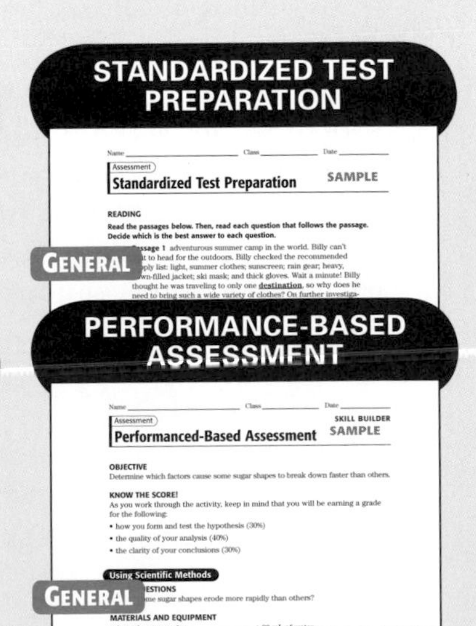

4 Chapter Enrichment

This Chapter Enrichment provides relevant and interesting information to expand and enhance your presentation of the chapter material.

Section 1

What Does DNA Look Like?

Discovering DNA

- In 1869, long before the time of Watson and Crick, a 22-year-old Swiss scientist isolated DNA from a cell nucleus. Unfortunately, he had no idea of its function, much less of its role in inheritance. It was not until 75 years later, in 1944, that an American geneticist named Oswald T. Avery found evidence that DNA is the carrier of genetic information.

Section 2

How DNA Works

Cracking the Genetic Code

- In the 1960s, scientists cracked the genetic code—the translation between codons (sequences of three bases) and amino acids. They have found that the genetic code is similar in all living organisms. If a codon aligns with a particular amino acid in humans, the same codon aligns with the same amino acid in bacteria. This similarity suggests that all life-forms have a common evolutionary ancestor.

Is That a Fact!

◆ Human DNA consists of about 3 billion base pairs. If you could print a book with all the genetic information carried in just one human cell, it would be 500,000 pages long.

Amino Acids

- All known organisms produce proteins using only 20 amino acids as building blocks (some use a rare 21st amino acid). The human body can manufacture 10 of these amino acids. The other 10 must be obtained from proteins in the diet and for this reason are called the *essential* amino acids. Foods that contain all the essential amino acids at once include eggs, milk, seafood, and meat. However, all amino acids can be obtained from a varied diet.

Protein Synthesis

- It took many years for scientists to determine how protein is synthesized in the cell. The discovery that DNA's nucleotide sequence corresponds to a certain amino acid sequence was a key step in unlocking this mystery. This link was conclusively proven by Charles Yanofsky and Sydney Brenner in 1964.

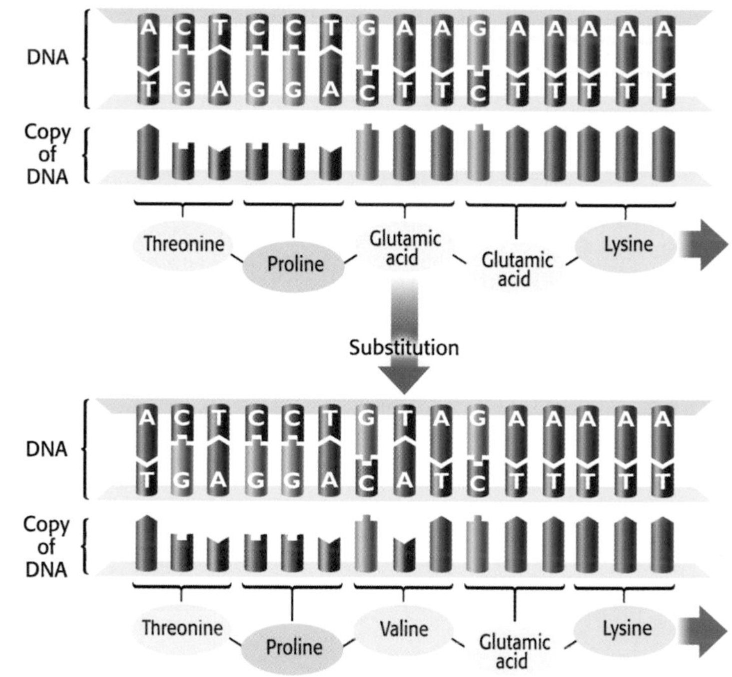

- The genetic sequences used to make proteins can be compared to sentences. Where each three-letter "word" in the genetic "sentence" starts and stops is very important for constructing a protein. For example, suppose the sentence to code for a particular protein read "PAT SAW THE FAT CAT." If you start just one base pair too late, the sentence would read "ATS AWT HEF ATC AT," which is meaningless.

Is That a Fact!

- If uncoiled, the DNA in the 46 chromosomes of a human body cell is about 2 m long. Within chromosomes, this DNA is so tightly coiled that if all 46 chromosomes were lined up end to end, they would span less than 0.5 cm.

Genetic Engineering

- Genetically engineered hybrid organisms are often called *chimeras*. The word *chimera* comes from Greek mythology, in which the Chimera was a fire-breathing monster, usually depicted as a composite of a lion, a goat, and a serpent.

- Scientists often disagree about the ethics of genetic engineering and about the safety risks involved. Dr. Maxine Frank Singer was one of the first scientists to warn the National Academy of Science of the potential hazards of genetic engineering. Because of the efforts of Dr. Singer and her colleagues, the National Institute of Health began to develop specific guidelines for genetic research as early as 1976. These guidelines, now regularly amended, continue to regulate the production and use of DNA and genetically engineered organisms.

The Human Genome Project

- The Human Genome Project (HGP) was started in 1990 as an international collaboration of scientists with the goal of mapping the entire sequence of DNA found in humans. In April 2003, in conjunction with the anniversary of the historic publication by Watson and Crick of DNA's molecular structure, the HGP announced that its work was mostly done. The HGP had completed mapping 99% of the human genetic code. Some mystery remained about the area of chromosomes called the *centromere*.
- Many potential benefits are predicted to result from the Human Genome Project, and some benefits have already been realized. Scientists working on the HGP have developed faster methods of determining the sequences within DNA samples. Also, scientists have improved methods of finding and tracking the functions of specific genes within cells. Such advances have made it easier to study the genetics of all kinds of organisms and to find the genetic indicators of specific kinds of cancer and other diseases.

DNA Fingerprints

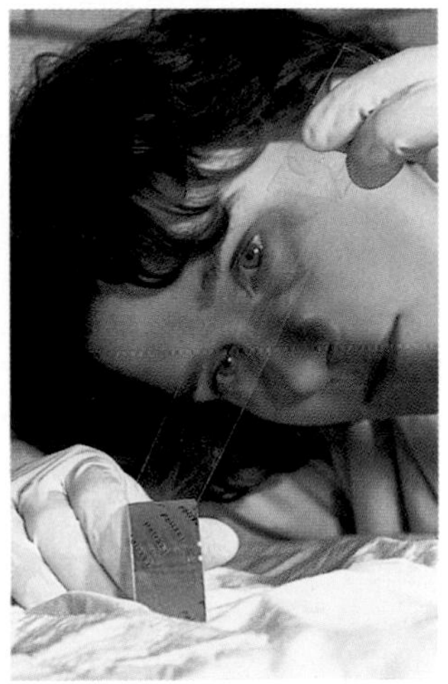

- DNA fingerprints are frequently used in criminal investigations. The DNA can come from hair, skin cells, blood, or other body fluids left at the crime scene by the perpetrator. Scientists use enzymes to make many copies of the DNA and then use other enzymes to cut the DNA into fragments. The fragments are separated by size and other characteristics on a specially treated plate. A photograph of the plate is taken, showing a unique set of dark bands. This set of bands is known as a *DNA fingerprint*. The fingerprint is then compared with the DNA fingerprint of the suspect to help determine innocence or guilt.

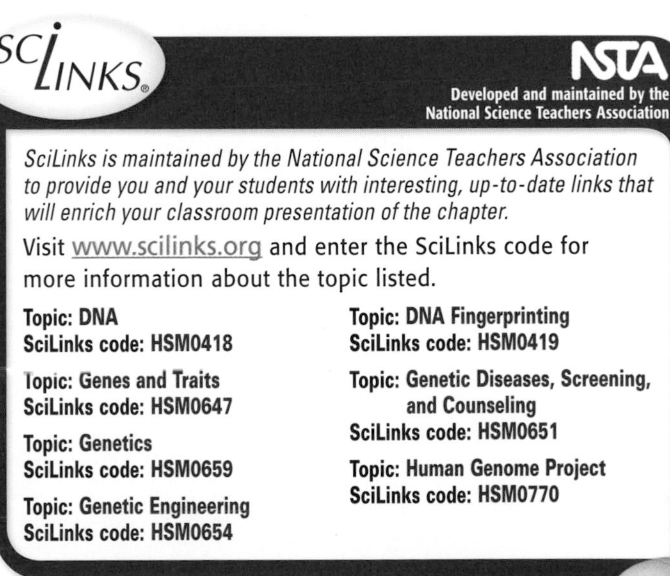

Overview
Tell students that this chapter is about DNA—the substance that makes up genes—and about how DNA works within cells to direct the growth and functioning of every organism.

Assessing Prior Knowledge
Students should be familiar with the following topics:

- cell structure
- mitosis and meiosis
- basic rules of heredity
- chromosomes

Identifying Misconceptions
The roles of DNA, RNA, and proteins in cells are very complex, and many puzzles remain. Students may tend to simplify their concept of the "rules" as they learn them. Students may remain unconvinced of the role of chance and probability in heredity. Also, students may have difficulty linking their knowledge of the functions of DNA at the cellular level to what they have learned and will learn about the functioning of tissues and organs within an entire organism.

4 Genes and DNA

About the PHOTO

These adult mice have no hair—not because their hair was shaved off but because these mice do not grow hair. In cells of these mice, the genes that normally cause hair to grow are not working. The genes were "turned off" by scientists who have learned to control the function of some genes. Scientists changed the genes of these mice to research medical problems such as cancer.

PRE-READING ACTIVITY

Graphic Organizer

Concept Map Before you read the chapter, create the graphic organizer entitled "Concept Map" described in the **Study Skills** section of the Appendix. As you read the chapter, fill in the concept map with details about DNA.

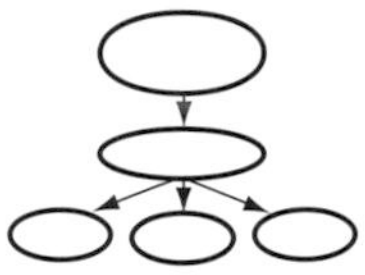

Standards Correlations

National Science Education Standards

The following codes indicate the National Science Education Standards that correlate to this chapter. The full text of the standards is at the front of the book.

Chapter Opener
UCP 5; SAI 1, 2; ST 2; LS 2e

Section 1 What Does DNA Look Like?
UCP 2, 5; SAI 1, 2; ST 1, 2; SPSP 5; HNS 1, 2, 3; LS 1a, 2d, 5a

Section 2 How DNA Works
UCP 1, 4, 5; SAI 1, 2; ST 2; SPSP 4, 5; LS 1c, 1e, 1f, 2b, 2c, 2d, 2e, 5b

Chapter Lab
UCP 2, 5; SAI 1, 2; HNS 1; LS 1a

Chapter Review
UCP 1, 2, 5; SAI 1, 2; ST 2; SPSP 4; HNS 2, 3; LS 1a, 1c, 1e, 1f, 2c, 2d, 2e, 5a, 5b

Science In Action
UCP 1, 2, 5; ST 2; SPSP 4 ,5; HNS 1, 2, 3; LS 1f

START-UP ACTIVITY

Fingerprint Your Friends

One way to identify people is by taking their fingerprints. Does it really work? Are everyone's fingerprints unique? Try this activity to find out.

Procedure

1. Rub the tip of a **pencil** back and forth across a **piece of tracing paper.** Make a large, dark mark.
2. Rub the tip of one of your fingers on the pencil mark. Then place a small **piece of transparent tape** over the darkened area on your finger.
3. Remove the tape, and stick it on **a piece of white paper.** Repeat steps 1–3 for the rest of your fingers.
4. Look at the fingerprints with a **magnifying lens.** What patterns do you see? Is the pattern the same on every finger?

Analysis

1. Compare your fingerprints with those of your classmates. Do any two people in your class have the same prints? Try to explain your findings.

START-UP ACTIVITY

MATERIALS

For Each Group

- magnifying lens
- paper, tracing (1 sheet)
- paper, white (1 sheet for each student)
- pencil or piece of charcoal
- tape, transparent

Safety Caution: Remind students to review all safety cautions and icons before beginning this lab activity. Charcoal is nontoxic, but it can stain clothes.

Teacher's Notes: The loop pattern is found in about 65% of the population, the whorl in about 30%, and the arch in about 5%.

Answers

1. The number of fingerprint types will vary for each class. No two students should have the same fingerprint (those of identical twins may be similar but still unique). Accept any reasonable explanation that incorporates variation in inherited traits among populations.

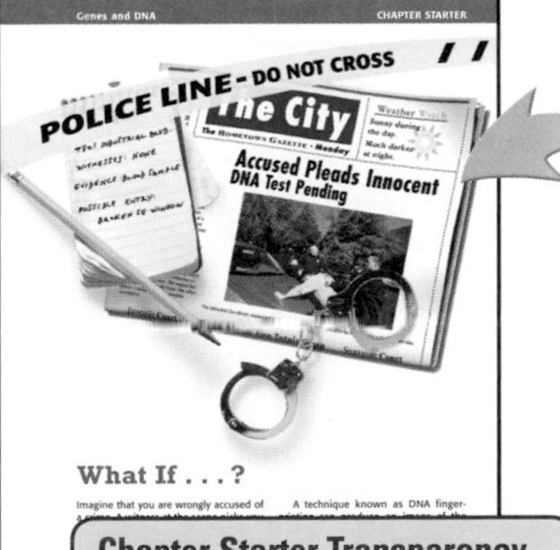

Chapter Starter Transparency
Use this transparency to help students begin thinking about genes and DNA.

CHAPTER RESOURCES

Technology

Transparencies 
- Chapter Starter Transparency

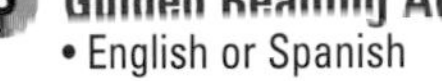
- English or Spanish

Classroom Videos
- Brain Food Video Quiz

Workbooks

Science Puzzlers, Twisters & Teasers
- Genes and DNA 

Focus

Overview

This section introduces students to the structure and function of DNA and to the process of DNA replication.

Bellringer

To test prior knowledge, have students answer the following questions:

1. Give an example of the difference between traits and characteristics. (Sample answer: Eye color is a characteristic, while having blue eyes is a trait.)
2. Where are genes found in cells? (in chromosomes; in cells that have nuclei, chromosomes are within the nucleus)

Motivate

ACTIVITY GENERAL

Modeling Code Create a code by pairing each letter of the alphabet with a numeral. For example, the numeral 1 could represent the letter *a*. Have students encode a brief message. Then, have students exchange and decode the message. Explain that a code is simply another way to represent information and that there are many types of codes. The genetic code is based on sequences of the four nucleotide bases of DNA. English Language Learners
LS Logical

SECTION 1

What Does DNA Look Like?

For many years, the structure of a DNA molecule was a puzzle to scientists. In the 1950s, two scientists deduced the structure while experimenting with chemical models. They later won a Nobel Prize for helping solve this puzzle!

READING WARM-UP

Objectives

- List three important events that led to understanding the structure of DNA.
- Describe the basic structure of a DNA molecule.
- Explain how DNA molecules can be copied.

Terms to Learn

DNA
nucleotide

READING STRATEGY

Prediction Guide Before reading this section, write the title of each heading in this section. Next, under each heading, write what you think you will learn.

Inherited characteristics are determined by genes, and genes are passed from one generation to the next. Genes are parts of chromosomes, which are structures in the nucleus of most cells. Chromosomes are made of protein and DNA. **DNA** stands for *deoxyribonucleic acid* (dee AHKS ee RIE boh noo KLEE ik AS id). DNA is the genetic material—the material that determines inherited characteristics. But what does DNA look like?

The Pieces of the Puzzle

Scientists knew that the material that makes up genes must be able to do two things. First, it must be able to give instructions for building and maintaining cells. Second, it must be able to be copied each time a cell divides, so that each cell contains identical genes. Scientists thought that these things could be done only by complex molecules, such as proteins. They were surprised to learn how much the DNA molecule could do.

Nucleotides: The Subunits of DNA

DNA is made of subunits called nucleotides. A **nucleotide** consists of a sugar, a phosphate, and a base. The nucleotides are identical except for the base. The four bases are *adenine, thymine, guanine,* and *cytosine.* Each base has a different shape. Scientists often refer to a base by the first letter of the base, *A, T, G,* and *C.* **Figure 1** shows models of the four nucleotides.

DNA **d**eoxyribo**n**ucleic **a**cid, a molecule that is present in all living cells and that contains the information that determines the traits that a living thing inherits and needs to live

nucleotide in a nucleic-acid chain, a subunit that consists of a sugar, a phosphate, and a nitrogenous base

Figure 1 The Four Nucleotides of DNA

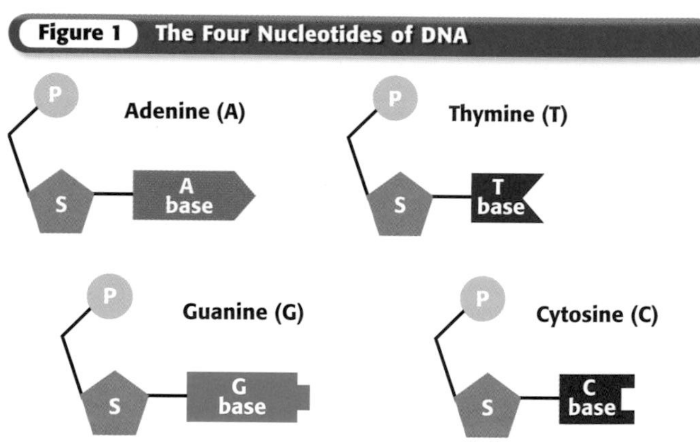

CHAPTER RESOURCES

Chapter Resource File

- Lesson Plan
- Directed Reading A BASIC
- Directed Reading B SPECIAL NEEDS

Technology

Transparencies
- Bellringer

Workbooks

Math Skills for Science
- A Shortcut for Multiplying Large Numbers GENERAL

Science Skills
- Science Drawing SPECIAL NEEDS

Science Bloopers

Missed Opportunity James Watson disliked Rosalind Franklin so much that at a 1951 lecture in which she gave information about the size and possible shape of the DNA molecule, Watson refused to take notes. It took Watson and Crick two more years to discover some of the same information on their own. When they did, their famous deduction of the shape of the DNA molecule came within two weeks!

Chargaff's Rules

In the 1950s, a biochemist named Erwin Chargaff found that the amount of adenine in DNA always equals the amount of thymine. And he found that the amount of guanine always equals the amount of cytosine. His findings are known as *Chargaff's rules*. At the time of his discovery, no one knew the importance of these findings. But Chargaff's rules later helped scientists understand the structure of DNA.

Reading Check **Summarize Chargaff's rules.** (*See the Appendix for answers to Reading Checks.*)

Franklin's Discovery

More clues about the structure of DNA came from scientists in Britain. There, chemist Rosalind Franklin, shown in **Figure 2,** was able to make images of DNA molecules. She used a process known as *X-ray diffraction* to make these images. In this process, X rays are aimed at the DNA molecule. When an X ray hits a part of the molecule, the ray bounces off. The pattern made by the bouncing rays is captured on film. Franklin's images suggested that DNA has a spiral shape.

Watson and Crick's Model

At about the same time, two other scientists were also trying to solve the mystery of DNA's structure. They were James Watson and Francis Crick, shown in **Figure 3.** After seeing Franklin's X-ray images, Watson and Crick concluded that DNA must look like a long, twisted ladder. They were then able to build a model of DNA by using simple materials from their laboratory. Their model perfectly fit with both Chargaff's and Franklin's findings. The model eventually helped explain how DNA is copied and how it functions in the cell.

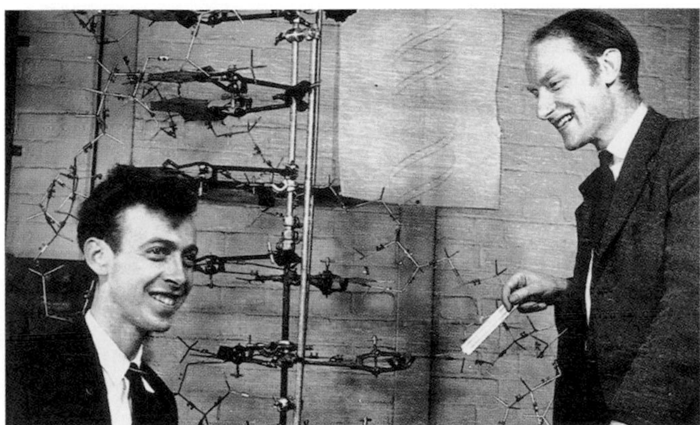

CONNECTION TO Chemistry

WRITING SKILL **Linus Pauling** Many scientists contributed to the discovery of DNA's structure. In fact, some scientists competed to be the first to make the discovery. One of these competitors was a chemist named Linus Pauling. Research and write a paragraph about how Pauling's work helped Watson and Crick.

Figure 2 *Rosalind Franklin used X-ray diffraction to make images of DNA that helped reveal the structure of DNA.*

Figure 3 *This photo shows James Watson (left) and Francis Crick (right) with their model of DNA.*

Teach

READING STRATEGY — BASIC

Mnemonics Have students create a mnemonic device that will remind them of the names of the bases and the way the bases form pairs. Examples such as "**A**toms are **T**iny" or "**A**dam is **T**errific" might help remind students that **a**denine pairs with **t**hymine. "**C**athy is **G**reat" might remind them that **c**ytosine pairs with **g**uanine.
LS Verbal/Logical

Answer to Reading Check

Guanine and cytosine always occur in equal amounts in DNA, as do adenine and thymine.

INCLUSION *Strategies*

- ***Learning Disabled***
- ***Attention Deficit Disorder***
- ***Behavior Control Issues***

Give students a chance to move around while they learn. Have students group themselves by eye color: all blues together, etc. Then, have students within each eye color group line up from lightest shade to darkest shade. Assign a spokesman within each group to explain how they decided the order for their line up. Also, ask each team to tell the number of unique eye colors within the team.
LS Kinesthetic English Language Learners

SCIENTISTS AT ODDS

Rosalind Franklin In 1951, Rosalind Franklin began working with Maurice Wilkins in London, but she and Wilkins never got along. Both Wilkins and James Watson belittled Franklin's work, but Francis Crick respected it. Franklin's study suggesting a helical structure to DNA was instrumental in Watson and Crick's final deduction of DNA's shape. If Franklin had not died in 1958, she might have received the Nobel Prize awarded to Wilkins, Watson, and Crick in 1962.

Answer to Connection to Chemistry

Linus Pauling was an innovator in the use of models to deduce chemical behavior. Whereas some scientists belittled the practice of "playing" with chemical models, Pauling inspired other scientists, such as Watson and Crick, to try this strategy. Watson and Crick's deduction of DNA's ladder structure was partly brought about by manipulating models of nucleotides.

Group ACTIVITY — GENERAL

A Place in History Have students imagine that they have just discovered the structure of DNA and must present their findings to a group of scientists. Have small groups of students use a model of DNA, a poster, or another visual aid to briefly describe the structure of DNA to their classmates. **LS Verbal**

Close

Reteaching — BASIC

DNA's Complementary Strands To help students understand how the term *complementary* relates to the structure of DNA, point out that the term means "completing." Using **Figure 4** and **Figure 5,** explain that complementary base pairs join to *complete* each rung on the spiral-staircase structure of DNA. Then, point out that complementary strands of DNA join to complete one DNA molecule. **LS Visual/Verbal**

Quiz — GENERAL

1. When is DNA copied? (every time a cell divides)
2. Name the four types of nucleotides. (adenine, thymine, guanine, and cytosine)

Alternative Assessment — GENERAL

PORTFOLIO **Custom Code** Have students create an alternative code that functions like DNA in the following ways:

- The code is based on four letters or symbols.
- Coded information can be split up and then reassembled.

Have students draw and explain their coding system. **LS Logical**

Making a Model of DNA

1. Gather assorted simple materials that you could use to build a basic model of DNA. You might use **clay, string, toothpicks, paper, tape, plastic foam,** or **pieces of food.**
2. Work with a partner or a small team to build your model. Use your book and other resources to check the details of your model.
3. Show your model to your classmates. Give your classmates feedback about the scientific aspects of their models.

DNA's Double Structure

The shape of DNA is shown in **Figure 4.** As you can see, a strand of DNA looks like a twisted ladder. This shape is known as a *double helix* (DUB uhl HEE LIKS). The two sides of the ladder are made of alternating sugar parts and phosphate parts. The rungs of the ladder are made of a pair of bases. Adenine on one side of a rung always pairs with thymine on the other side. Guanine always pairs with cytosine.

Notice how the double helix structure matches Chargaff's observations. When Chargaff separated the parts of a sample of DNA, he found that the matching bases were always present in equal amounts. To model how the bases pair, Watson and Crick tried to match Chargaff's observations. They also used information from chemists about the size and shape of each of the nucleotides. As it turned out, the width of the DNA ladder matches the combined width of the matching bases. Only the correct pairs of bases fit within the ladder's width.

Making Copies of DNA

The pairing of bases allows the cell to *replicate,* or make copies of, DNA. Each base always bonds with only one other base. Thus, pairs of bases are *complementary* to each other, and both sides of a DNA molecule are complementary. For example, the sequence CGAC will bond to the sequence GTCG.

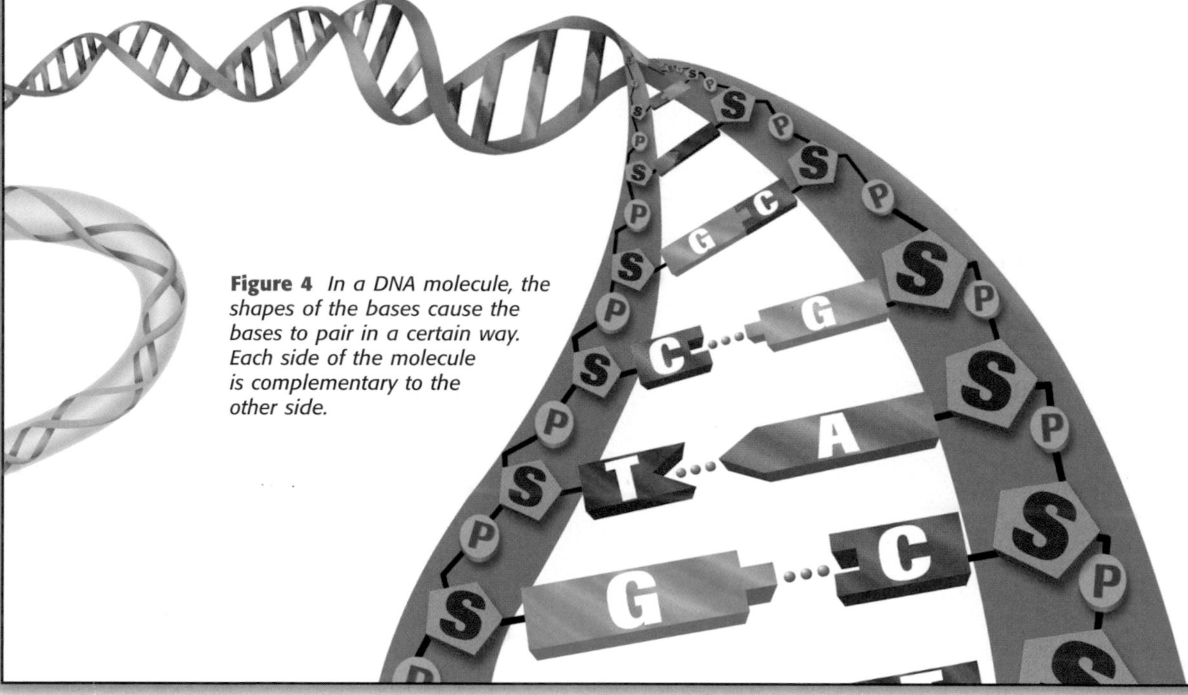

Figure 4 *In a DNA molecule, the shapes of the bases cause the bases to pair in a certain way. Each side of the molecule is complementary to the other side.*

MATERIALS

FOR EACH GROUP

- variety of materials, such as clay, string, toothpicks, paper, tape, plastic foam, beads or buttons and pipe cleaners or wire.
- food or candy items could be another option

Teacher's Note: Display student models within the school. Have students reevaluate or improve upon them later.

Safety Caution: Advise students to keep the area around them uncluttered. Students should exercise caution with sharp objects. Any food items used should not be eaten and should be disposed of.

Answers

2. Student models should resemble **Figure 4** in basic structure but may vary in size, color, and construction.
3. Students should suggest ways to make each model more accurate.

How Copies Are Made

During replication, as shown in **Figure 5,** a DNA molecule is split down the middle, where the bases meet. The bases on each side of the molecule are used as a pattern for a new strand. As the bases on the original molecule are exposed, complementary nucleotides are added to each side of the ladder. Two DNA molecules are formed. Half of each of the molecules is old DNA, and half is new DNA.

When Copies Are Made

DNA is copied every time a cell divides. Each new cell gets a complete copy of all the DNA. The job of unwinding, copying, and re-winding the DNA is done by proteins within the cell. So, DNA is usually found with several kinds of proteins. Other proteins help with the process of carrying out the instructions written in the code of the DNA.

Reading Check How often is DNA copied?

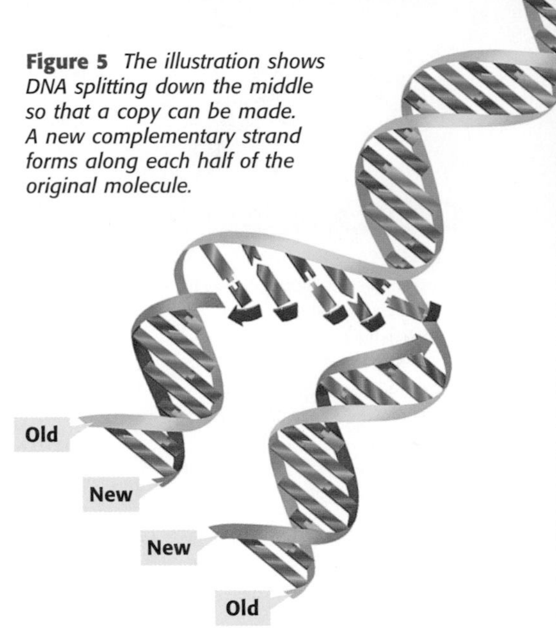

Figure 5 *The illustration shows DNA splitting down the middle so that a copy can be made. A new complementary strand forms along each half of the original molecule.*

SECTION Review

Summary

- DNA is the material that makes up genes. It carries coded information that is copied in each new cell.
- The DNA molecule looks like a twisted ladder. The two halves are long strings of nucleotides. The rungs are complementary pairs of bases.
- Because each base has a complementary base, DNA can be replicated accurately.

Using Key Terms

1. Use the term *DNA* in a sentence.
2. In your own words, write a definition for the term *nucleotide*.

Understanding Key Ideas

3. List three important events that led to understanding the structure of DNA.
4. Which of the following is NOT part of a nucleotide?
   a. base
   b. sugar
   c. fat
   d. phosphate

Math Skills

5. If a sample of DNA contained 20% cytosine, what percentage of guanine would be in this sample? What percentage of adenine would be in the sample? Explain.

Critical Thinking

6. **Making Inferences** Explain what is meant by the statement "DNA unites all organisms."
7. **Applying Concepts** What would the complementary strand of DNA be for the sequence of bases below?

   C T T A G G C T T A C C A

8. **Analyzing Processes** How are copies of DNA made? Draw a picture as part of your answer.

Answers to Section Review

1. Sample answer: DNA is the material that makes up genes and is found in all cells.
2. Sample answer: A nucleotide is a subunit of a DNA molecule and consists of a sugar, a phosphate, and a base; there are four kinds of bases and thus four kinds of nucleotides.
3. Sample answers: Scientists thought only complex materials such as proteins could make up genes; Erwin Chargaff discovered the rules of nucleotide base pairing; Rosalind Franklin made images of DNA molecules; Watson and Crick made a correct model of DNA's structure.
4. c
5. 20% guanine, because it should be equal to the amount of cytosine; 30% adenine, because the remaining 60% of the DNA should be made up of equal amounts of adenine and thymine
6. Sample answer: DNA is found in the cells of all organisms.
7. GAATCCGAATGGT
8. DNA copies are made by splitting the molecule down the middle and then adding new nucleotides to each side. (Students' drawings should resemble **Figure 5.**)

Answer to Reading Check

every time a cell divides

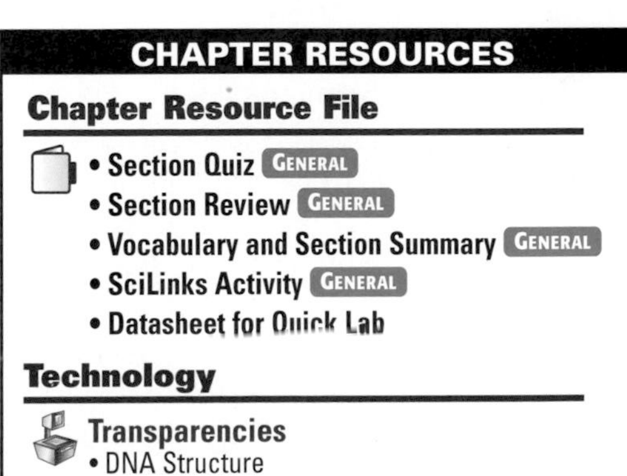

CHAPTER RESOURCES

Chapter Resource File

- Section Quiz GENERAL
- Section Review GENERAL
- Vocabulary and Section Summary GENERAL
- SciLinks Activity GENERAL
- Datasheet for Quick Lab

Technology

Transparencies
- DNA Structure

SECTION 2

Focus

Overview

This section shows how DNA is a part of chromosomes, how DNA is used as a template for making proteins, and how errors in DNA can lead to mutations and genetic disorders.

Bellringer

Have students unscramble the following words and use them both in one sentence:

tpsoneir (proteins)

neesg (genes)

(Sample answer: Genes contain instructions for making proteins.)

Motivate

Demonstration — GENERAL

A Tight Fit To illustrate the way that DNA is *supercoiled* within chromosomes and cells, hold up a long rubber band or thick piece of string. Begin to twist each end in opposite directions until coils form. Continue twisting until the band is highly compacted. Then, challenge students to fit 2 m of fine thread into a thimble or an empty gelatin capsule. English Language Learners

LS Kinesthetic

Answer to Reading Check

a string of nucleotides that give the cell information about how to make a specific trait

SECTION 2

How DNA Works

Almost every cell in your body contains 1.5 m of DNA. How does all of the DNA fit in a cell? And how does the DNA hold a code that affects your traits?

READING WARM-UP

Objectives

- Explain the relationship between DNA, genes, and proteins.
- Outline the basic steps in making a protein.
- Describe three types of mutations, and provide an example of a gene mutation.
- Describe two examples of uses of genetic knowledge.

Terms to Learn

RNA
ribosome
mutation

READING STRATEGY

Reading Organizer As you read this section, make a flowchart of the steps of how DNA codes for proteins.

DNA is found in the cells of all organisms, including bacteria, mosquitoes, and humans. Each organism has a unique set of DNA. But DNA functions the same way in all organisms.

Unraveling DNA

DNA is often wound around proteins, coiled into strands, and then bundled up even more. In a cell that lacks a nucleus, each strand of DNA forms a loose loop within the cell. In a cell that has a nucleus, the strands of DNA and proteins are bundled into chromosomes, as shown in **Figure 1**.

The structure of DNA allows DNA to hold information. The order of the bases on one side of the molecule is a code that carries information. A *gene* consists of a string of nucleotides that give the cell information about how to make a specific trait. There is an enormous amount of DNA, so there can be a large variety of genes. Humans have at least 30,000 genes.

Reading Check What makes up a gene? *(See the Appendix for answers to Reading Checks.)*

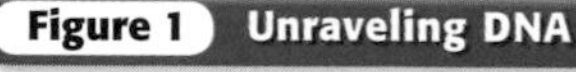

Figure 1 Unraveling DNA

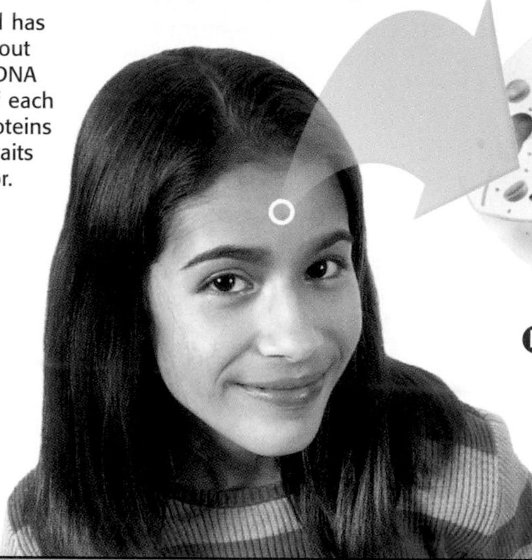

a A typical skin cell has a diameter of about 0.0025 cm. The DNA in the nucleus of each cell codes for proteins that determine traits such as skin color.

b The DNA in the nucleus is part of a material called *chromatin*. Long strands of chromatin are usually bundled loosely within the nucleus.

WEIRD SCIENCE

In 2003, the Human Genome Project had successfully mapped 99% of the 3 billion base pairs that make up a set of human DNA. But the project has raised new questions as well. For example, only about 3% of those base pairs are used in making proteins; the other 97% are regulatory sequences and nonfunctioning genes. Additionally, scientists originally expected to find over 50,000 human genes because human cells produce at least that many proteins. Instead, it seems that there are only 30,000 to 35,000 human genes, and each gene codes for multiple proteins. In this and other ways, human genes appear to be unique among organisms.

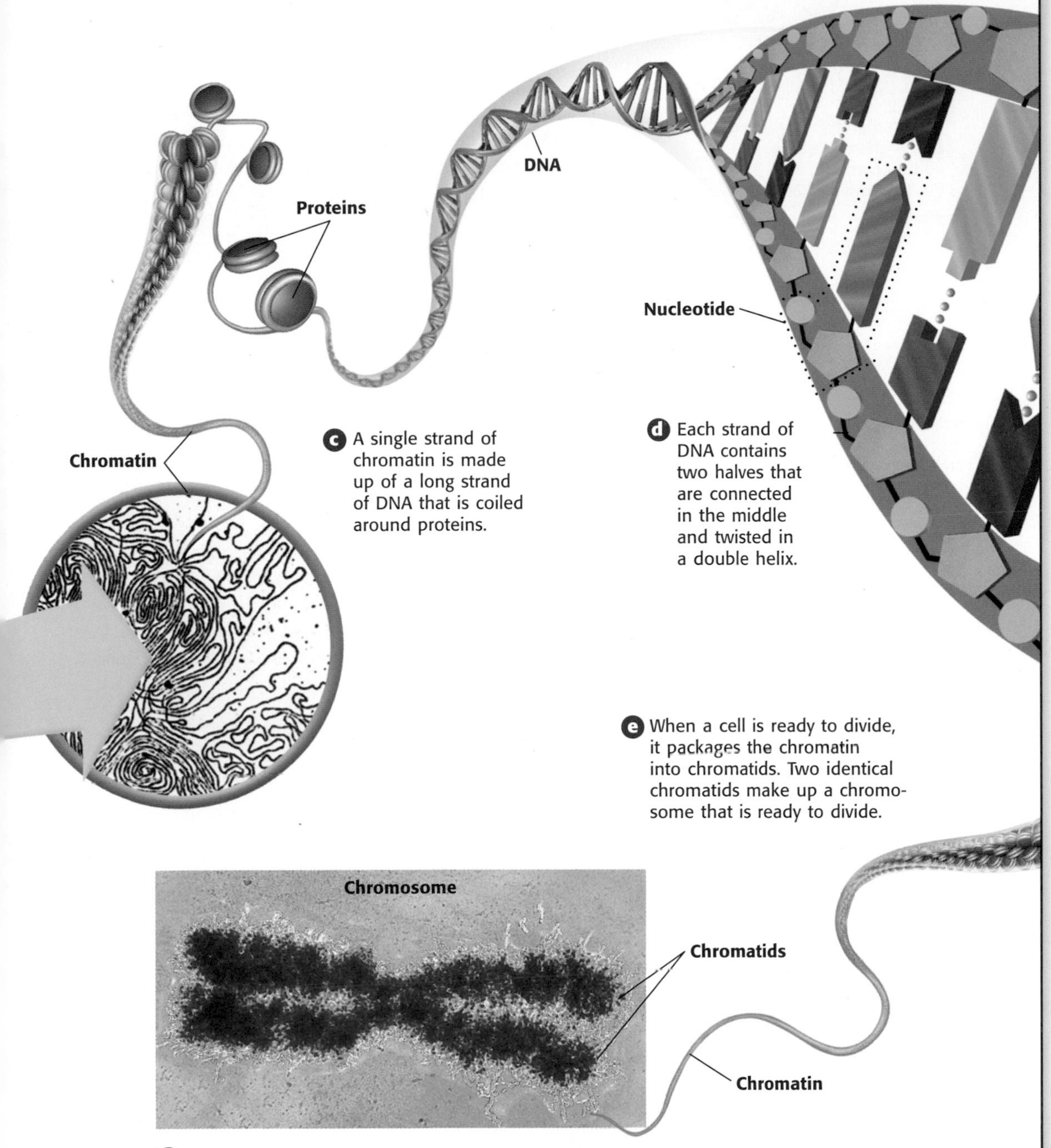

c A single strand of chromatin is made up of a long strand of DNA that is coiled around proteins.

d Each strand of DNA contains two halves that are connected in the middle and twisted in a double helix.

e When a cell is ready to divide, it packages the chromatin into chromatids. Two identical chromatids make up a chromosome that is ready to divide.

f Just before division, each human cell contains 46 chromosomes. These chromosomes contain two identical copies of all of the cell's genetic material.

Teach

Using the Figure—BASIC

Unraveling DNA Have students carefully study **Figure 1.** Remind them that each chromosome is a pair of chromatids, and each chromatid is one long strand of DNA. This DNA strand is usually somewhat wound up around proteins in the form of chromatin. The chromatin may be tightly bundled (and visible) or loose (and not visible) within the nucleus. Most of the time, the chromatid is loose. Ask students the following questions:

- Where is the DNA in your cells? (in the nucleus)
- How does so much DNA fit into the nucleus? (It is coiled up tightly around proteins.)
- What is the name for strands of DNA wound around proteins? (chromatin)
- When do chromosomes become visible in cells? (when the cell is about to divide)
- What are chromatids? (two identical copies of a chromosome that is about to divide)

LS Visual/Auditory

Hereditary Hearing

Researchers are trying to find out if a gene is responsible for "perfect pitch," the ability to determine any musical note upon hearing it. It is a rare ability—possessed by one in every 2,000 people—found most often among musicians. People with perfect pitch can easily tell the musical note of a dial tone, the hum of a refrigerator, or of any sound they hear. The researchers think that people with perfect pitch may inherit the ability, but an early education in music may also be necessary.

MISCONCEPTION ALERT

DNA from the Dead Can ancient DNA be used to produce dinosaurs? In some science fiction, scientists make dinosaurs by combining fragments of ancient DNA with DNA from modern organisms. In reality, less-ancient fragments of DNA have indeed been found. However, a fragment of DNA is not enough information to make an entire organism. And identifying the owner of a given DNA fragment is difficult.

CHAPTER RESOURCES

Technology

Transparencies
- Unraveling DNA

Teach, *continued*

CONNECTION ACTIVITY
Chemistry — ADVANCED

Proteins' Roles PORTFOLIO Have the students pair up. Then, have each pair of students select an important human protein to research. Suggest proteins such as insulin, hemoglobin, dopamine, somatostatin, erythropoetin (EPO), Alpha-1 antitrypsin (AAT), and Factor V (blood-clotting protein). Tell students to summarize the role the protein plays in the human body and to find out a disease that is caused by a mutation in the gene that codes for this protein. Have each pair of students present their findings on small placards that can be displayed in the classroom or school. LS **Logical/Verbal**

Group ACTIVITY — GENERAL

Skit Have groups of students write and perform a short skit to demonstrate the formation of a protein. For example, students could play the roles of a ribosome, an amino acid, a transfer enzyme, and a DNA copy.
LS **Kinesthetic/Interpersonal** English Language Learners

INTERNET ACTIVITY

For another activity related to this chapter, go to **go.hrw.com** and type in the keyword **HL5DNAW.**

Genes and Proteins

The DNA code is read like a book—from one end to the other and in one direction. The bases form the alphabet of the code. Groups of three bases are the codes for specific amino acids. For example, the three bases CCA form the code for the amino acid proline. The bases AGC form the code for the amino acid serine. A long string of amino acids forms a protein. Thus, each gene is usually a set of instructions for making a protein.

Proteins and Traits

How are proteins related to traits? Proteins are found throughout cells and cause most of the differences that you can see among organisms. Proteins act as chemical triggers and messengers for many of the processes within cells. Proteins help determine how tall you grow, what colors you can see, and whether your hair is curly or straight. Proteins exist in an almost limitless variety. A single organism may have thousands of genes that code for thousands of proteins.

RNA ribonucleic acid, a molecule that is present in all living cells and that plays a role in protein production

Help from RNA

Another type of molecule that helps make proteins is called **RNA,** or *ribonucleic acid* (RIE boh noo KLEE ik AS id). RNA is so similar to DNA that RNA can serve as a temporary copy of a DNA sequence. Several forms of RNA help in the process of changing the DNA code into proteins, as shown in **Figure 2.**

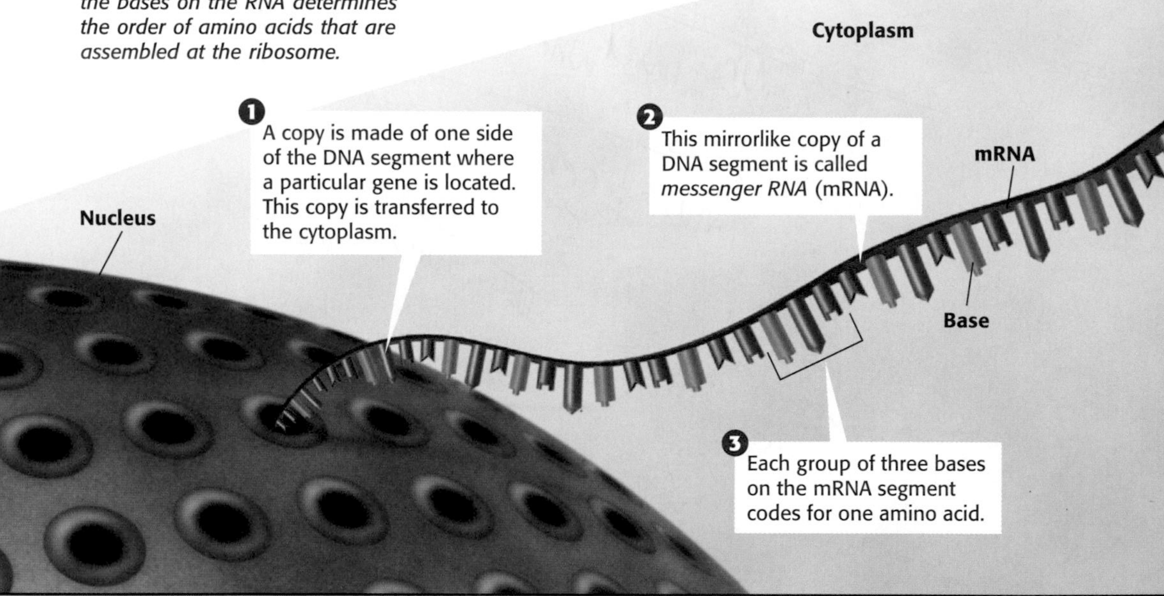

Figure 2 *Proteins are built in the cytoplasm by using RNA copies of a segment of DNA. The order of the bases on the RNA determines the order of amino acids that are assembled at the ribosome.*

Homework — GENERAL

Research Have students collect information on the use of amino acids to gain muscle. Suggest that they look at ads for amino acid supplements in health-food or fitness magazines or at labels of amino acid powdered drinks at the supermarket. Use these materials to discuss issues such as the expense of such supplements and how they might be used by the body. Discuss how amino acids might be acquired in a balanced diet. LS **Logical**

Q: What happens when an amateur-tein gets paid?

A: It becomes a pro-tein.

The Making of a Protein

The first step in making a protein is to copy one side of the segment of DNA containing a gene. A mirrorlike copy of the DNA segment is made out of RNA. This copy of the DNA segment is called *messenger RNA* (mRNA). It moves out of the nucleus and into the cytoplasm of the cell.

In the cytoplasm, the messenger RNA is fed through a protein assembly line. The "factory" that runs this assembly line is known as a ribosome. A **ribosome** is a cell organelle composed of RNA and protein. The messenger RNA is fed through the ribosome three bases at a time. Then, molecules of *transfer RNA* (tRNA) translate the RNA message. Each transfer RNA molecule picks up a specific amino acid from the cytoplasm. Inside the ribosome, bases on the transfer RNA match up with bases on the messenger RNA like pieces of a puzzle. The transfer RNA molecules then release their amino acids. The amino acids become linked in a growing chain. As the entire segment of messenger RNA passes through the ribosome, the growing chain of amino acids folds up into a new protein molecule.

Reading Check **What do the transfer RNA molecules transfer?**

Code Combinations

A given sequence of three bases codes for one amino acid. For example, AGT is one possible sequence. How many different sequences of the four DNA base types are possible? (Hint: Make a list.)

ribosome a cell organelle composed of RNA and protein; the site of protein synthesis

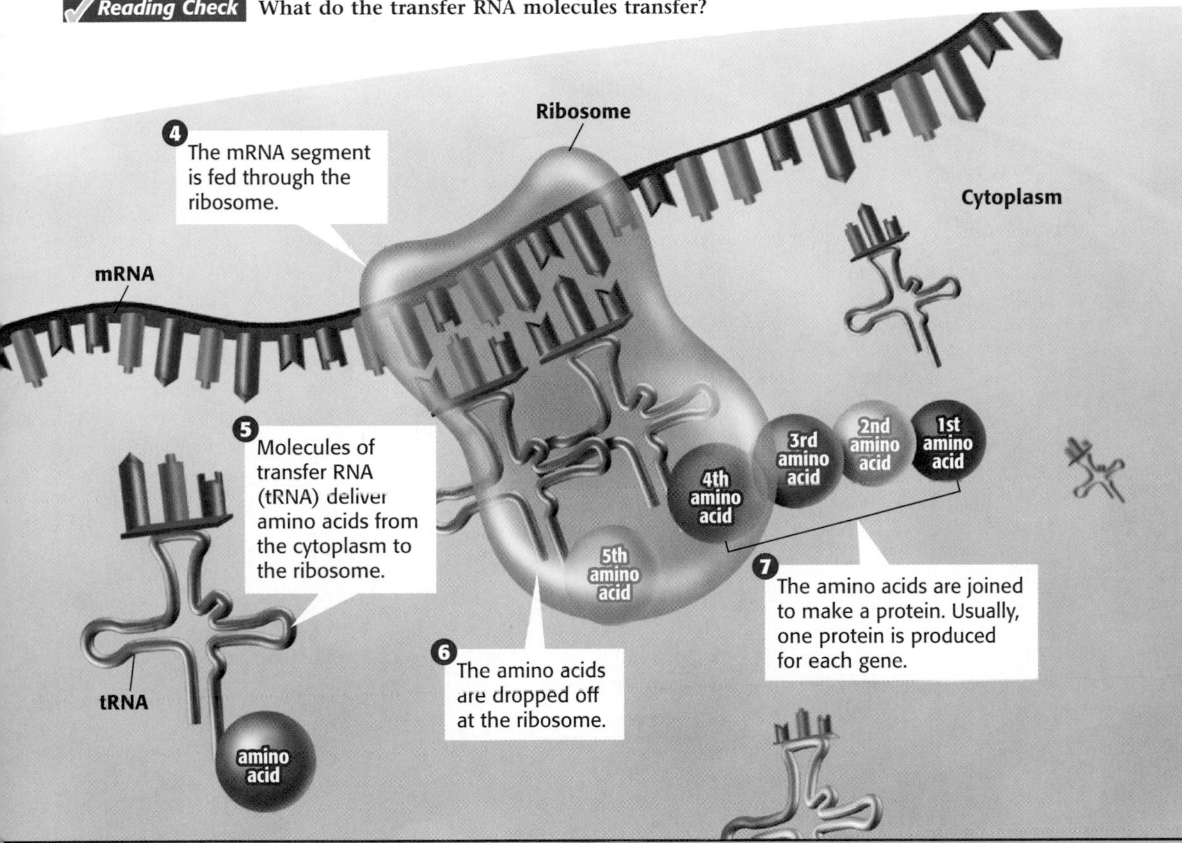

WEIRD SCIENCE

There is a gene located on the X chromosome that causes thick hair to grow on the upper body and face, including the ears, nose, cheeks, forehead, and even eyelids of people who have the gene. This condition is sometimes called *werewolf syndrome* because people who have the condition resemble fictional werewolves. This condition affects only appearance, however, not behavior.

CHAPTER RESOURCES

Technology

Transparencies

- The Making of a Protein: A
- The Making of a Protein: B

Answer to Math Practice

With four possible nucleotides in three possible positions, there are $4 \times 4 \times 4$, or 64, possible combinations.

Answer to Reading Check

They transfer amino acids to the ribosome.

CONNECTION ACTIVITY
Math — GENERAL

Redundant Code Mathematics has a lot to do with how DNA codes for amino acids. Each combination of three nucleotides that codes for one amino acid is called a *codon*. Yet cells use only 20 different amino acids to build proteins. Thus, most amino acids have several, redundant corresponding codons. This redundancy is another reason that mutations in genes do not always result in changes in proteins. To physically model the possible base combinations that make up codons, organize the class into small groups. Give each group four pieces of paper, with one of the following four letters printed on each piece: *A, T, C,* or *G.* Ask students to come up with as many different three-letter "words" as possible by using the four different bases. (There are 4^3, or 64, possible three-letter "words"—or codons. For example, the four possible combinations that would start with the base A are AAA, AAT, AAG, and AAC.) Check that students realize that the order of letters in each combination also matters. For example, *ATA* is not the same "word" as *AAT*. **LS Kinesthetic/Logical**

Teach, *continued*

ACTIVITY — BASIC

Complementary Code Write a sequence of DNA bases, such as AACTACGGT, on the board. Ask students to write the complementary base sequence by using base-pairing rules. (TTGATGCCA) Then, ask students to give examples of deletions, insertions, and substitutions. English Language Learners
LS Visual/Verbal

MISCONCEPTION ALERT

Mutants Among Us? Students may think mutations occur rarely, because organisms that are visibly "mutated" appear infrequently. However, scientists estimate that mistakes are made during DNA replication in approximately one out of every 1,000 base pairs. With cellular proofing mechanisms, the final error rate is lower—somewhere between one in a million and one in a billion. Still, we inherit hundreds of mutations from our parents' gametes. Many mutations have no apparent effect. For example, a mutation may occur in a cell that does not produce a particular protein or in a "junk" region of DNA that does not code for anything.

Answer to Reading Check

a physical or chemical agent that can cause a mutation in DNA

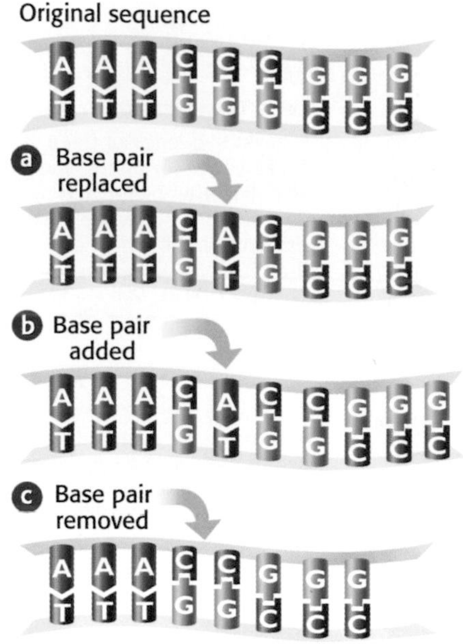

Figure 3 *The original base sequence on the top has been changed to illustrate (a) a substitution, (b) an insertion, and (c) a deletion.*

Changes in Genes

Imagine that you have been invited to ride on a new roller coaster at the state fair. Before you climb into the front car, you are told that some of the metal parts on the coaster have been replaced by parts made of a different substance. Would you still want to ride this roller coaster? Perhaps a strong metal was used as a substitute. Or perhaps a material that is not strong enough was used. Imagine what would happen if cardboard were used instead of metal!

Mutations

Substitutions like the ones in the roller coaster can accidentally happen in DNA. Changes in the number, type, or order of bases on a piece of DNA are known as **mutations.** Sometimes, a base is left out. This kind of change is known as a *deletion*. Or an extra base might be added. This kind of change is known as an *insertion*. The most common change happens when the wrong base is used. This kind of change is known as a *substitution*. **Figure 3** illustrates these three types of mutations.

Do Mutations Matter?

There are three possible consequences to changes in DNA: an improved trait, no change, or a harmful trait. Fortunately, cells make some proteins that can detect errors in DNA. When an error is found, it is usually fixed. But occasionally the repairs are not accurate, and the mistakes become part of the genetic message. If the mutation occurs in the sex cells, the changed gene can be passed from one generation to the next.

mutation a change in the nucleotide-base sequence of a gene or DNA molecule

How Do Mutations Happen?

Mutations happen regularly because of random errors when DNA is copied. In addition, damage to DNA can be caused by abnormal things that happen to cells. Any physical or chemical agent that can cause a mutation in DNA is called a *mutagen*. Examples of mutagens include high-energy radiation from X rays and ultraviolet radiation. Ultraviolet radiation is one type of energy in sunlight. It is responsible for suntans and sunburns. Other mutagens include asbestos and the chemicals in cigarette smoke.

Reading Check What is a mutagen?

CHAPTER RESOURCES

Technology

Transparencies

- ***LINK TO EARTH SCIENCE*** The Formation of Smog
- How Sickle Cell Anemia Results from a Mutation

CONNECTION to Earth Science — GENERAL

Pollution and Ozone Ozone is a gas made of three oxygen atoms. High in the atmosphere, ozone absorbs dangerous ultraviolet radiation (the high-energy light that can cause DNA mutations). When produced near the surface of the Earth, however, ozone is a pollutant that affects plant growth and makes breathing more difficult. Use the teaching transparency entitled "The Formation of Smog" to illustrate the process of ozone production. LS Visual

An Example of a Substitution

A mutation, such as a substitution, can be harmful because it may cause a gene to produce the wrong protein. Consider the DNA sequence GAA. When copied as mRNA, this sequence gives the instructions to place the amino acid glutamic acid into the growing protein. If a mistake happens and the original DNA sequence is changed to GTA, the sequence will code for the amino acid valine instead.

This simple change in an amino acid can cause the disease *sickle cell anemia.* Sickle cell anemia affects red blood cells. When valine is substituted for glutamic acid in a blood protein, as shown in **Figure 4,** the red blood cells are changed into a sickle shape.

The sickle cells are not as good at carrying oxygen as normal red blood cells are. Sickle cells are also likely to get stuck in blood vessels and cause painful and dangerous clots.

Reading Check What causes sickle cell anemia?

An Error in the Message

The sentence below is the result of an error similar to a DNA mutation. The original sentence was made up of three-letter words, but an error was made in this copy. Explain the idea of mutations to your parent. Then, work together to find the mutation, and write the sentence correctly.

THE IGB ADC ATA TET HEB IGR EDR AT.

ACTIVITY

Figure 4 How Sickle Cell Anemia Results from a Mutation

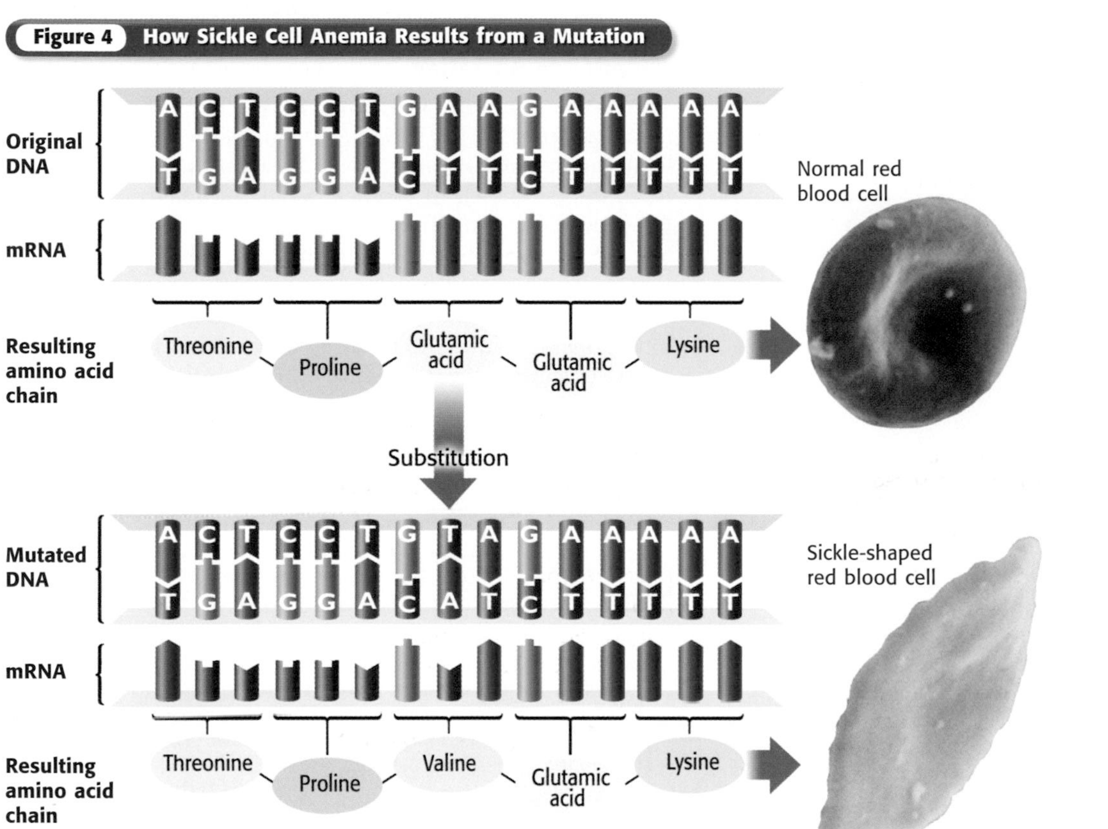

Answer to School-to-Home Activity

The mutation is a deletion. THE BIG BAD CAT ATE THE BIG RED RAT.

INCLUSION Strategies

- ***Behavior Control Issues***
- ***Attention Deficit Disorder***
- ***Visually Impaired***

Many students benefit from small-group work and learn well when actively involved. Divide all but four students into groups of three. Assign a DNA combination to each team (AT, CG, TA, or GC). Have students identify their DNA pairs by taping construction paper to their shirts. Ask students to line up to create a "human" DNA chain. Assign each of the remaining four students one of the four combinations. Have the four "extras" move around to create the three types of mutations: deletions, insertions, and substitutions. **LS Intrapersonal**

Homework — GENERAL

Writing **Genetic Diseases** Have students select a genetic disease about which to conduct research and write a report. Suggest diseases such as hemophilia, diabetes, Familial ALS (Amyotrophic Lateral Sclerosis, or Lou Gehrig's Disease), SCID (Severe Combined Immunodeficiency Syndrome, or "Plastic Bubble" syndrome), Huntington's disease, and neurofibromatosis ("elephantitis"). Suggest that their reports focus on historical occurrence of the disease, famous persons that had or have the disease, and treatments that have been tried. **LS Verbal** PORTFOLIO

CONNECTION to *Real Life* — GENERAL

Misunderstood Disease A person who carries a single allele for sickle cell anemia is said to have *sickle cell trait.* Only persons with two of these alleles usually develop anemia. In the past, many people did not understand that sickle cell trait and sickle cell anemia are not contagious. In some areas, children with the trait were banned from public schools. **LS Verbal**

Answer to Reading Check

Sickle cell anemia is caused by a mutation in a single nucleotide of DNA, which then causes the wrong amino acid to be assembled in a protein used in blood cells.

Close

Reteaching — BASIC

Mutations Write a sequence of DNA on the board, and invite students to come up to the board and change the sequence. Then, ask the class to discuss the possible consequences of such a mutation in DNA. (It might cause a different amino acid to be substituted in a protein. This substitution could result in a genetic disorder, an improvement, or no change at all.) Ask how the mutation could be corrected. (Special proteins may find and repair the error.) **LS Auditory**

Quiz — GENERAL

1. What is the function of the ribosome? (In the ribosome, the DNA code is translated into proteins.)
2. List some causes of DNA mutations. (Answers may include UV radiation, cigarette smoke, or X rays.)

Alternative Assessment — GENERAL

Writing **DNA How-To** Have students prepare an instruction manual for their DNA. The manual should include instructions for copying their DNA and translating it into proteins. It should also include information about protecting their DNA from mutations by avoiding mutagens and correcting any mutations that occur. **LS Intrapersonal**

English Language Learners 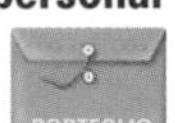

Uses of Genetic Knowledge

In the years since Watson and Crick made their model, scientists have learned a lot about genetics. This knowledge is often used in ways that benefit humans. But some uses of genetic knowledge also cause ethical and scientific debates.

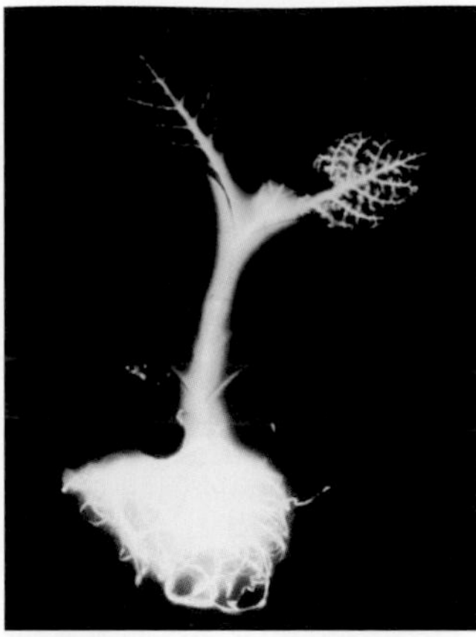

Figure 5 *This genetically engineered tobacco plant contains firefly genes.*

Genetic Engineering

Scientists can manipulate individual genes within organisms. This kind of manipulation is called *genetic engineering*. In some cases, genes may be transferred from one type of organism to another. An example of a genetically engineered plant is shown in **Figure 5.** Scientists added a gene from fireflies to this plant. The gene produces a protein that causes the plant to glow.

Scientists may use genetic engineering to create new products, such as drugs, foods, or fabrics. For example, bacteria may be used to make the proteins found in spider's silk. Or cows may be used to produce human proteins. In some cases, this practice could produce a protein that is needed by a person who has a genetic disease. However, some scientists worry about the dangers of creating genetically engineered organisms.

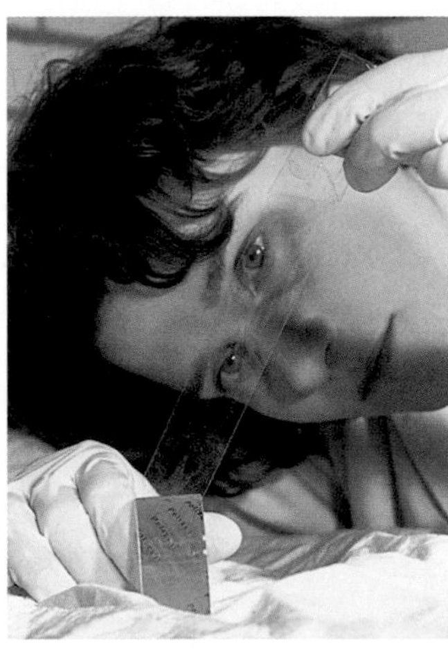

Figure 6 *This scientist is gathering dead skin cells from a crime scene. DNA from the cells could be used as evidence of a criminal's identity.*

Genetic Identification

Your DNA is unique, so it can be used like a fingerprint to identify you. *DNA fingerprinting* identifies the unique patterns in an individual's DNA. DNA samples are now used as evidence in crimes, as shown in **Figure 6.** Similarities between people's DNA can reveal other information, too. For example, DNA can be used to identify family relations or hereditary diseases.

Identical twins have truly identical DNA. Scientists are now able to create something like a twin, called a clone. A *clone* is a new organism that has an exact copy of another organism's genes. Clones of several types of organisms, including some mammals, have been developed by scientists. However, the possibility of cloning humans is still being debated among both scientists and politicians.

Reading Check What is a clone?

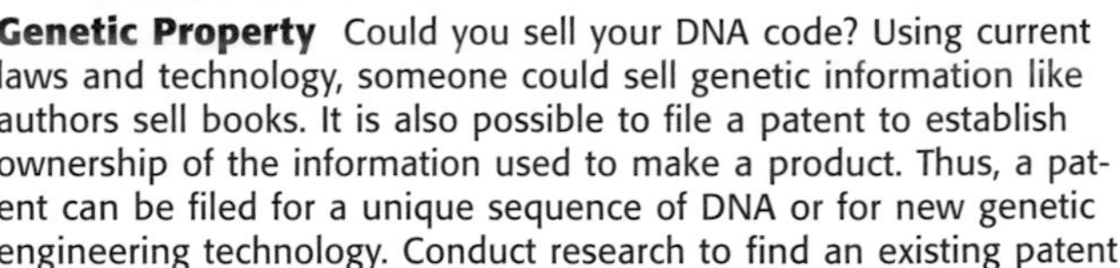

Genetic Property Could you sell your DNA code? Using current laws and technology, someone could sell genetic information like authors sell books. It is also possible to file a patent to establish ownership of the information used to make a product. Thus, a patent can be filed for a unique sequence of DNA or for new genetic engineering technology. Conduct research to find an existing patent on a genetic sequence or genetic engineering technology.

Answer to Connection to Social Studies

Students should be able to find examples of patents for specially bred or genetically engineered plants and seeds or for procedures that rely on genetic technologies to produce drugs or treat genetic diseases.

Answer to Reading Check

a near-identical copy of another organism, created with the original organism's genes

CONNECTION ACTIVITY Social Studies — ADVANCED

Ethics Debate Have interested students stage a debate about what kinds of regulations should be placed on the practices of genetic manipulation. Suggest students consider issues such as tranferring genes between different species, DNA fingerprinting, cloning, and genetic patents. **LS Interpersonal**

SECTION Review

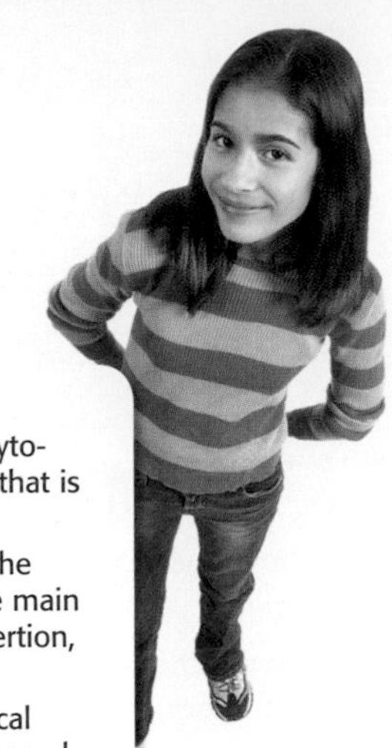

Summary

- A gene is a set of instructions for assembling a protein. DNA is the molecular carrier of these genetic instructions.
- Every organism has DNA in its cells. Humans have 1.5 m of DNA in each cell. This DNA makes up over 30,000 genes.
- Within a gene, each group of three bases codes for one amino acid. A sequence of amino acids is linked to make a protein.
- Proteins are fundamental to the function of cells and the expression of traits.
- Proteins are assembled within the cytoplasm through a multi-step process that is assisted by several forms of RNA.
- Genes can become mutated when the order of the bases is changed. Three main types of mutations are possible: insertion, deletion, and substitution.
- Genetic knowledge has many practical uses. Some applications of genetic knowledge are controversial.

Using Key Terms

1. Use each of the following terms in the same sentence: *ribosome* and *RNA*.
2. In your own words, write a definition for the term *mutation*.

Understanding Key Ideas

3. Explain the relationship between genes and proteins.
4. List three possible types of mutations.
5. Which type of mutation causes sickle cell anemia?
   a. substitution
   b. insertion
   c. deletion
   d. mutagen

Math Skills

6. A set of 23 chromosomes in a human cell contains 3.2 billion pairs of DNA bases in sequence. On average, about how many pairs of bases are in each chromosome?

Critical Thinking

7. **Applying Concepts** In which cell type might a mutation be passed from generation to generation? Explain.
8. **Making Comparisons** How is genetic engineering different from natural reproduction?

Interpreting Graphics

The illustration below shows a sequence of bases on one strand of a DNA molecule. Use the illustration below to answer the questions that follow.

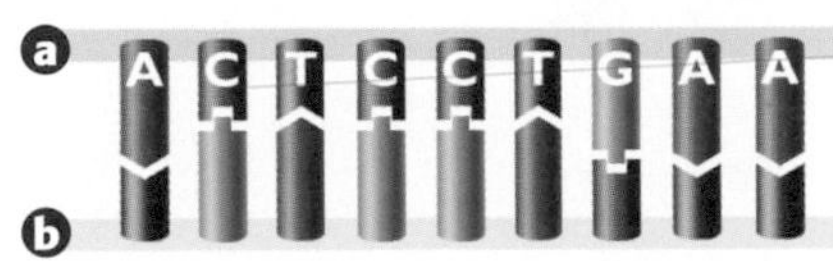

9. How many amino acids are coded for by the sequence on one side (A) of this DNA strand?
10. What is the order of bases on the complementary side of the strand (B), from left to right?
11. If a G were inserted as the first base on the top side (A), what would the order of bases be on the complementary side (B)?

For a variety of links related to this chapter, go to www.scilinks.org

Topic: Genetic Engineering
SciLinks code: HSM0654

Answers to Section Review

1. Sample answer: A ribosome is made of RNA and protein.
2. Sample answer: A mutation is a mistake in the DNA code.
3. Genes are sequences of DNA that code for particular proteins.
4. substitution, insertion, deletion
5. a
6. about 139 million (3,200,000,000 ÷ 23 = 139,130,435)
7. A sex cell (germ cell, sperm cell, or egg cell), because these cells contain the genes from which a new organism is formed.
8. Sample answer: Genetic engineering is deliberately controlled by humans and may involve processes that are rare or impossible in nature.
9. 3
10. TGAGGACTT
11. CTGAGGACTT

WEIRD SCIENCE

Gene therapy is an experimental field of medical research in which defective genes are replaced with healthy genes. One way to insert healthy genes involves using a delivery system called a gene gun to inject microscopic gold bullets coated with genetic material.

CHAPTER RESOURCES

Chapter Resource File

- Section Quiz GENERAL
- Section Review GENERAL
- Vocabulary and Section Summary GENERAL
- Reinforcement Worksheet BASIC
- Critical Thinking ADVANCED

Technology

Interactive Explorations CD-ROM
- DNA Pawprints GENERAL

Base-Pair Basics

Teacher's Notes

Time Required
One 45-minute class period

Lab Ratings

EASY → HARD

Teacher Prep 1
Student Set-Up 2
Concept Level 1
Clean Up 1

MATERIALS
The materials listed on the student page are enough for a group of 4–5 students. You may want to provide additional materials for the Applying Your Data section.

Safety Caution
Remind students to review all safety cautions and icons before beginning this lab activity. Students should always exercise care when using scissors.

Model-Making Lab

Base-Pair Basics

You have learned that DNA is shaped something like a twisted ladder. The side rails of the ladder are made of sugar parts and phosphate parts. The two side rails are connected to each other by parts called *bases*. The bases join in pairs to form the rungs of the ladder. Within DNA, each base can pair with only one other base. Each of these pairs is called a *base pair*. When DNA replicates, enzymes separate the base pairs, which breaks the rungs of the ladder in half. Then, each half of the DNA ladder can be used as a template for building a new half. In this activity, you will construct a paper model of DNA and use it to model the replication process.

OBJECTIVES

Construct a model of a DNA strand.

Model the process of DNA replication.

MATERIALS

- bag, large paper
- paper, colored (4 colors)
- paper, white
- scissors

SAFETY

Procedure

1. Trace the models of nucleotides below onto white paper. Label the pieces "A" (**a**denine), "T" (**t**hymine), "C" (**c**ytosine), and "G" (**g**uanine). Draw the pieces again on colored paper. Use a different color for each type of base. Draw the pieces as large as you want, and draw as many of the white pieces and as many of the colored pieces as time will allow.
2. Carefully cut out all of the pieces.
3. Put all of the colored pieces in the classroom into a large paper bag. Spread all of the white pieces in the classroom onto a large table.
4. Remove nine colored pieces from the bag. Arrange the colored pieces in any order in a straight column so that the letters *A, T, C,* and *G* are right side up. Be sure to fit the sugar notches to the phosphate tabs. Draw this arrangement.
5. Find the white bases that correctly pair with the nine colored bases. Remember the base-pairing rules, and pair the bases according to those rules.
6. Pair the pieces by fitting tabs to notches. The letters on the white pieces should be upside down. You now have a model of a double-stranded piece of DNA. The strand contains nine pairs of complementary nucleotides. Draw your model.

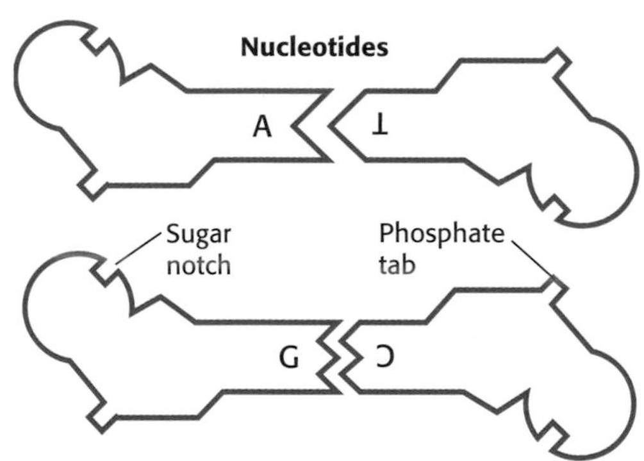

Lab Notes
You may wish to enlarge the nucleotide template for your students so that the models will be easier to cut out. Explain to students that the white pieces and the colored pieces represent the complementary sides of DNA strands. Also, suggest that students refer to the figure depicting DNA replication in their text. Remind students that this is a model of the parts of a DNA molecule and that the parts of real DNA molecules are three-dimensional and have a more complex shape.

CHAPTER RESOURCES

Chapter Resource File
- Datasheet for Chapter Lab
- Lab Notes and Answers

Technology

 **Classroom Videos**
- Lab Video

Workbooks

 **Whiz-Bang Demonstrations**
- Grand Strand GENERAL

Long-Term Projects & Research Ideas
- The Antifreeze Protein ADVANCED
- Ewe Again, Dolly? ADVANCED

Analyze the Results

1. **Identifying Patterns** Now, separate the two halves of your DNA strand along the middle of the base pair rungs of the ladder. Keep the side rails together by keeping the sugar notches fitted to the phosphate tabs. Draw this arrangement.

2. **Recognizing Patterns** Look at the drawing made in the previous step. Along each strand in the drawing, write the letters of the bases that complement the bases in that strand.

3. **Examining Data** Find all of the bases that you need to complete replication. Find white pieces to pair with the bases on the left, and find colored pieces to pair with the bases on the right. Be sure that the tabs and notches fit and the sides are straight. You have now replicated your model of DNA. Are the two models identical? Draw your results.

Draw Conclusions

4. **Interpreting Information** State the correct base-pairing rules. How do these rules make DNA replication possible?

5. **Evaluating Models** What happens when you attempt to pair thymine with guanine? Do they fit together? Are the sides straight? Do all of the tabs and notches fit? Explain.

Applying Your Data

Construct a 3-D model of a DNA molecule that shows DNA's twisted-ladder structure. Use your imagination and creativity to select materials. You may want to use licorice, gum balls, and toothpicks or pipe cleaners and paper clips.

1. Display your model in your classroom.
2. Take a vote to decide which models are the most accurate and the most creative.

Debra Sampson
Booker T. Washington Middle School
Elgin, Texas

Analyze the Results

1. Student drawings should show an "unzipped" DNA strand.
2. Student responses should always show A matched with T and C matched with G.
3. The two new molecules should exactly match each other and match the original molecule.

Draw Conclusions

4. G and C always pair, and A and T always pair. These pairings allow the two halves of a DNA molecule to be separated and replicated and ensure that identical new molecules can be formed.
5. The joining areas of guanine and thymine don't match up. They don't fit together well. The sides of the DNA molecule would not be straight and the parts would not line up if the bases were forced together in this way.

Applying Your Data

1. Student models should be more accurate than any models of DNA that they have previously constructed. Check for the correct "right-handed" orientation of the double-helix spiral, representation of the four base types, correct matching of the base-pairs and subunits, and overall uniformity of the helix.
2. Before voting, have students brainstorm their criteria for "accurate" and "creative." Take a separate vote for each category.

Chapter Review

Assignment Guide	
Section	Questions
1	4, 5, 10, 13, 16, 17, 21–23
2	1–3, 6–9, 11, 12, 14, 15, 18–20

ANSWERS

Using Key Terms

1. A mutagen is a substance that can cause a mutation in DNA.
2. nucelotides
3. ribosome

Understanding Key Ideas

4. d
5. b
6. b
7. b
8. a
9. b

Chapter Review

USING KEY TERMS

1. Use the following terms in the same sentence: *mutation* and *mutagen*.

The statements below are false. For each statement, replace the underlined term to make a true statement.

2. The information in DNA is coded in the order of <u>amino acids</u> along one side of the DNA molecule.

3. The "factory" that assembles proteins based on the DNA code is called a <u>gene</u>.

UNDERSTANDING KEY IDEAS

Multiple Choice

4. James Watson and Francis Crick
   a. took X-ray pictures of DNA.
   b. discovered that genes are in chromosomes.
   c. bred pea plants to study heredity.
   d. made models to figure out DNA's shape.

5. In a DNA molecule, which of the following bases pair together?
   a. adenine and cytosine
   b. thymine and adenine
   c. thymine and guanine
   d. cytosine and thymine

6. A gene can be all of the following EXCEPT
   a. a set of instructions for a trait.
   b. a complete chromosome.
   c. instructions for making a protein.
   d. a portion of a strand of DNA.

7. Which of the following statements about DNA is NOT true?
   a. DNA is found in all organisms.
   b. DNA is made up of five subunits.
   c. DNA has a structure like a twisted ladder.
   d. Mistakes can be made when DNA is copied.

8. Within the cell, where are proteins assembled?
   a. the cytoplasm
   b. the nucleus
   c. the amino acids
   d. the chromosomes

9. Changes in the type or order of the bases in DNA are called
   a. nucleotides.
   b. mutations.
   c. RNA.
   d. genes.

Short Answer

10. What would be the complementary strand of DNA for the following sequence of bases?

    C T T A G G C T T A C C A

11. If the DNA sequence TGAGCCATGA is changed to TGAGCACATGA, what kind of mutation has occurred?

12. Explain how the DNA in genes relates to the traits of an organism.

13. Why is DNA frequently found associated with proteins inside of cells?

14. What is the difference between DNA and RNA?

10. GAATCCGAATGGT
11. an insertion
12. The DNA in genes codes for specific proteins, and proteins control cells and result in traits.
13. because proteins do much of the work of copying and handling the DNA
14. DNA is deoxyribonucleic acid, and exact copies of a set of DNA are found in each cell of an organism. RNA is ribonucleic acid, which is similar to DNA but is used to carry copies of DNA code around the cell and to build proteins based on this code.

15 How does breeding differ from genetic engineering?

CRITICAL THINKING

16 **Concept Mapping** Use the following terms to create a concept map: *bases, adenine, thymine, nucleotides, guanine, DNA,* and *cytosine.*

17 **Analyzing Processes** Draw and label a picture that explains how DNA is copied.

18 **Analyzing Processes** Draw and label a picture that explains how proteins are made.

19 **Applying Concepts** The following DNA sequence codes for how many amino acids?

T C A G C C A C C T A T G G A

20 **Making Inferences** Why does the government make laws about the use of chemicals that are known to be mutagens?

INTERPRETING GRAPHICS

The illustration below shows the process of replication of a DNA strand. Use this illustration to answer the questions that follow.

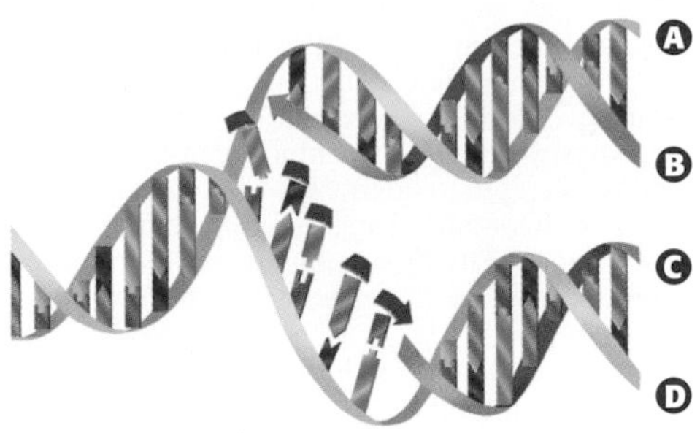

21 Which strands are part of the original molecule?

a. A and B
b. A and C
c. A and D
d. None of the above

22 Which strands are new?

a. A and B
b. B and C
c. C and D
d. None of the above

23 Which strands are complementary?

a. A and C
b. B and C
c. All of the strands
d. None of the strands

15. Sample answer: Pet breeding is usually simply selecting pet organisms with desirable traits and allowing them to reproduce naturally. Genetic engineering uses technology and may include a variety of ways of manipulating the genes of organisms, including mixing the genes of different species and manipulating cells and DNA in a laboratory.

Critical Thinking

16. 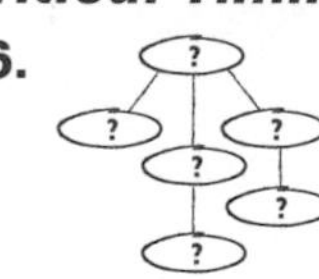 An answer to this exercise can be found at the end of this book.

17. Student drawings should resemble the diagram of replication in their student text and should have appropriate labels.

18. Student drawings should resemble the diagram of protein assembly in their student text and should have appropriate labels.

19. This sequence codes for five amino acids.

20. Sample answer: The government is trying to protect people from the risk of mutagens causing harmful mutations in people's cells—mutations could cause a disease such as cancer.

Interpreting Graphics

21. c
22. b
23. b

CHAPTER RESOURCES

Chapter Resource File

- Chapter Review GENERAL
- Chapter Test A GENERAL
- Chapter Test B ADVANCED
- Chapter Test C SPECIAL NEEDS
- Vocabulary Activity GENERAL

Workbooks

Study Guide
- Assessment resources are also available in Spanish.

Standardized Test Preparation

Teacher's Note

To provide practice under more realistic testing conditions, give students 20 minutes to answer all of the questions in this Standardized Test Preparation.

MISCONCEPTION ALERT

Answers to the standardized test preparation can help you identify student misconceptions and misunderstandings.

Passage 1

1. B
2. H

TEST DOCTOR

Question 2: This question asks for the main idea of the second paragraph. Main ideas are often introduced or summarized in the first sentence of a paragraph, and sometimes summarized or rephrased in the last sentence of a paragraph. The first sentence of the second paragraph is closest in meaning to answer H. Students who chose answer I may have looked for clues in the last sentence of the paragraph but missed the ontradiction indicated by the use of "however." For this type of question, advise students to reread the first and last sentences and check for contradictions before choosing an answer.

Standardized Test Preparation

READING

Read each of the passages below. Then, answer the questions that follow each passage.

Passage 1 The tension in the courtroom was so thick that you could cut it with a knife. The prosecuting attorney presented this evidence: "DNA analysis indicates that blood found on the defendant's shoes matches the blood of the victim. The odds of this match happening by chance are 1 in 20 million." The jury members were stunned by these figures. Can there be any doubt that the defendant is guilty?

DNA is increasingly used as evidence in court cases. Traditional fingerprinting has been used for more than 100 years, and it has been an extremely important identification tool. Recently, DNA fingerprinting, also called *DNA profiling,* has started to replace traditional techniques. DNA profiling has been used to clear thousands of wrongly accused or convicted individuals. However, there is some controversy over whether DNA evidence should be used to prove a suspect's guilt.

1. What does the first sentence in this passage describe?

A the air pollution in a particular place
B the feeling that a person might experience during an event
C the motion of an object
D the reason that a person was probably guilty of a crime

2. Which of the following best describes the main idea of the second paragraph of this passage?

F A defendant was proven guilty by DNA analysis.
G Court battles involving DNA fingerprinting are very exciting.
H The technique of DNA profiling is increasingly used in court cases.
I The technique of DNA profiling is controversial.

Passage 2 Most of the biochemicals found in living things are proteins. In fact, other than water, proteins are the most abundant molecules in your cells. Proteins have many functions, including regulating chemical activities, transporting and storing materials, and providing structural support.

Every protein is composed of small "building blocks" called *amino acids*. Amino acids are molecules that are composed of carbon, hydrogen, oxygen, and nitrogen atoms. Some amino acids also include sulfur atoms. Amino acids chemically bond to form proteins of many shapes and sizes.

The function of a protein depends on the shape of the bonded amino acids. If even a single amino acid is missing or out of place, the protein may not function correctly or may not function. Foods such as meat, fish, cheese, and beans contain proteins, which are broken down into amino acids as the foods are digested. Your body can then use these amino acids to make new proteins.

1. In the passage, what does *biochemical* mean?

A a chemical found in nonliving things
B a chemical found in living things
C a pair of chemicals
D a protein

2. According to the passage, which of the following statements is true?

F Amino acids contain carbon dioxide.
G Amino acids contain proteins.
H Proteins are made of living things.
I Proteins are made of amino acids.

Passage 2

1. B
2. I

TEST DOCTOR

Question 1: From the first sentence, one can infer that biochemicals are something found in living things and that proteins are one type of biochemical. Hence, the most likely meaning is answer B. Answer A is contradictory to the sentence. Answer C wrongly assumes that the "bi" in "biochemical" means "two." Answer D is a reasonable guess, but answer B best reflects the inference from the first sentence. Remind students to carefully read all answers and compare each with the question and the passage before deciding.

INTERPRETING GRAPHICS

The diagram below shows an original sequence of DNA and three possible mutations. Use the diagram to answer the questions that follow.

Original sequence

Mutation A

Mutation B

Mutation C

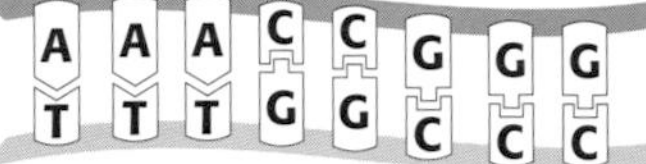

1. In which mutation was an original base pair replaced?

A Mutation A
B Mutation B
C Mutation C
D There is not enough information to determine the answer.

2. In which mutation was a new base pair added?

F Mutation A
G Mutation B
H Mutation C
I There is not enough information to determine the answer.

3. In which mutation was an original base pair removed?

A Mutation A
B Mutation B
C Mutation C
D There is not enough information to determine the answer.

MATH

Read each question below, and choose the best answer.

1. Mary was making a design on top of her desk with marbles. She put 3 marbles in the first row, 7 marbles in the second row, 15 marbles in the third row, and 31 marbles in the fourth row. If Mary continues this pattern, how many marbles will she put in the seventh row?

A 46
B 63
C 127
D 255

2. Bobby walked 3 1/2 km on Saturday, 2 1/3 km on Sunday, and 1 km on Monday. How many kilometers did Bobby walk on those 3 days?

F 5 1/6
G 5 5/6
H 6 1/6
I 6 5/6

3. Marie bought a new aquarium for her goldfish. The aquarium is 60 cm long, 20 cm wide, and 30 cm high. Which equation could be used to find the volume of water needed to fill the aquarium to 25 cm deep?

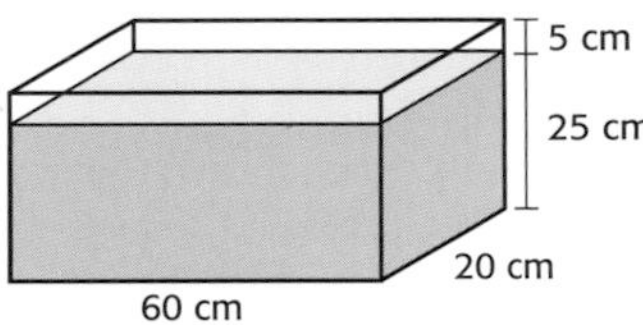

A $V = 30 \times 60 \times 20$
B $V = 25 \times 60 \times 20$
C $V = 30 \times 60 \times 20 - 5$
D $V = 30 \times 60 \times 25$

4. How is the product of $6 \times 6 \times 6 \times 4 \times 4 \times 4$ expressed in scientific notation?

F $6^4 \times 3^6$
G $6^3 \times 4^3$
H $3^6 \times 3^4$
I 24^6

Standardized Test Preparation

INTERPRETING GRAPHICS

1. B
2. F
3. C

Question 1: The student must recognize that, in Mutation B, the only change from the original sequence is a different base-pair in the middle of the sequence—a replacement. Mutation A is an insertion, and Mutation C is a deletion.

MATH

1. D
2. I
3. B
4. G

Question 1: The problem requires students to predict the next three values in a patterned sequence of numbers. The pattern is as follows:

3, 7, 15, 31, . . .

The logic of the pattern is to multiply each number by 2 and then add 1 to get the next number. Thus,

1	3
2	$(3 \times 2) + 1 = 7$
3	$(7 \times 2) + 1 = 15$
4	$(15 \times 2) + 1 = 31$
5	$(31 \times 2) + 1 = 63$
6	$(63 \times 2) + 1 = 127$
7	$(127 \times 2) + 1 = 255$

Question 3: The problem asks for the equation to find the volume of water in the aquarium, which is a rectangular box. The equation for the volume of a rectangular box is *length* × *width* × *height* (in any order). The problem and the diagram indicate that the depth of water needed is only 25 cm, so the value to use for *height* is 25 cm. Answer B uses the correct values in the order *height* × *length* × *width*.

CHAPTER RESOURCES

Chapter Resource File

• Standardized Test Preparation GENERAL

State Resources

For specific resources for your state, visit **go.hrw.com** and type in the keyword **HSMSTR.**

Scientific Debate

Background

The U.S. Food and Drug Administration began approving genetically modified organisms (GMOs) for consumer use in the 1990s. Some consumer groups have protested and boycotted such foods. Several countries around the world have banned the creation, sale, or importation of GMOs. Some consumer groups have asked that all GMO foods be clearly labeled. The majority of GMO foods being sold in the United States are made with corn or soybeans that contain bacterial genes.

Scientists have mixed opinions about GMOs. However, most scientists recognize that the potential to create new and unknown types of organisms should be undertaken with careful scientific scrutiny, should involve ethical considerations, and should be regulated by governments.

Science Fiction

Activity — ADVANCED

Further Reading If students liked this story, encourage them to read more of McKillip's stories, such as the following:

- *Fool's Run,* Warner, 1987
- *Something Rich and Strange,* Bantam, 1994
- *Winter Rose,* Ace, 1996

Scientific Debate

Supersquash or Frankenfruit?

Some food that you buy may have been developed in a new way. Food producers may use genetic engineering to make food crops easier to grow or sell, more nutritious, or resistant to pests and disease. More than half of the packaged foods sold in the United States are likely to contain ingredients from genetically modified organisms.

The U.S. government has stated that research shows that these foods are safe. But some scientists are concerned that genes introduced into crop plants could cause new environmental or health problems. For example, people who are allergic to peanuts might also be allergic to tomato plants that contain peanut genes.

Math Activity

Write a survey about genetically altered foods. Ask your teacher to approve your questions. Ask at least 15 people to answer your survey. Create graphs to summarize your results.

Science Fiction

"Moby James" by Patricia A. McKillip

Rob Trask and his family live on a space station. Rob thinks that his real brother was sent back to Earth. The person who claims to be his brother, James, is really either some sort of mutated plant or a mutant pair of dirty sweat socks.

Now, Rob has another problem—his class is reading Herman Melville's novel *Moby Dick*. As he reads the novel, Rob becomes convinced that his brother is a great white mutant whale—Moby James. To see how Rob solves his problems, read "Moby James" in the *Holt Anthology of Science Fiction*.

Language Arts Activity

WRITING SKILL Read "Moby James" by Patricia A. McKillip. Then, write your own short science-fiction story about a mutant organism. Be sure to incorporate some science into your science fiction.

Answer to Math Activity

Check that student surveys ask questions for which answers can be easily tallied, such as "Do you think that genetically modified foods should be labeled in the store?" Check that students have kept records and summarized their results accurately. Give them feedback about how well their graphs communicate the data they gathered.

Answer to Language Arts Activity

Instead of collecting and grading students' stories, you may want to have them read their stories to each other or to a family member, and then ask for feedback about how much science is included in their fiction.

People in Science

Lydia Villa-Komaroff

Genetic Researcher When Lydia Villa-Komaroff was young, science represented "a kind of refuge" for her. She grew up in a very large family that lived in a very small house. "I always wanted to find things out. I was one of those kids who took things apart."

In college, Villa-Komaroff became very interested in the process of embryonic development—how a simple egg grows into a complex animal. This interest led her to study genes and the way that genes code for proteins. For example, insulin is a protein that is normally produced by the human body. Often, people who suffer from diabetes lack the insulin gene, so their bodies can't make insulin. These people may need to inject insulin into their blood as a drug treatment.

Before the research by Villa-Komaroff's team was done, insulin was difficult to produce. Villa-Komaroff's team isolated the human gene that codes for insulin. Then, the scientists inserted the normal human insulin gene into the DNA of bacteria. This inserted gene caused the bacteria to produce insulin. This technique was a new and more efficient way to produce insulin. Now, most of the insulin used for diabetes treatment is made in this way. Many genetic researchers dream of making breakthroughs like the one that Villa-Komaroff made in her work with insulin.

Social Studies ACTIVITY

WRITING SKILL Do some research about several women, such as Marie Curie, Barbara McClintock, or Maxine Frank Singer, who have done important scientific research. Write a short biography about one of these women.

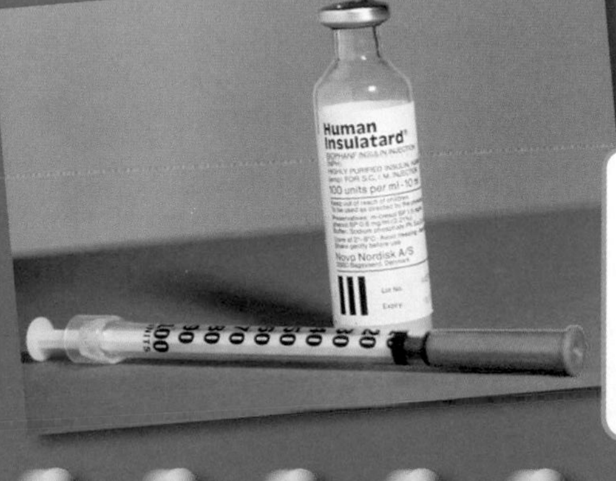

To learn more about these Science in Action topics, visit **go.hrw.com** and type in the keyword **HL5DNAF.**

Check out Current Science® articles related to this chapter by visiting go.hrw.com. Just type in the keyword HL5CS06.

Answer to Social Studies Activity

Suggest that students do research in the library or on the Internet for information. Additional women scientists to consider are as follows:

- Jewel Plummer Cobb
- Ruth Fulton Benedict
- Emma Perry Carr
- Rosalyn Yalow

Check student biographies for accuracy, and comment on any interesting facts.

People in Science

Background

Lydia Villa-Komaroff grew up in Santa Fe, New Mexico, in a household that loved to tell family stories. One favorite was the story of Villa-Komaroff's grandfather, Encarnacion Villa, and his brush with the Mexican revolutionary Pancho Villa. Encarnacion was going to be killed by Pancho Villa's soldiers when he refused to join their fight. But when Pancho Villa heard the captive's name, he ordered his release but told him he must have many sons. Pancho Villa probably could not imagine that a granddaughter of his former captive would someday become the third Mexican-American woman to earn a Ph.D. in the United States and would go on to make many important contributions to science.

When Lydia Villa-Komaroff and her colleagues inserted the human gene that directs the production of insulin into the DNA of bacteria, they were using recombinant DNA technology. In recombinant DNA technology, researchers identify which segment of DNA is the gene that directs the production of the desired substance, cut this section out of the DNA with special enzymes, and make copies, or clones. The researchers then take one of these clones and insert it, again using special enzymes, into the correct spot on the host DNA. The researchers look for a location on the host DNA that will ensure that the host organism will read the DNA and produce the substance.

The Evolution of Living Things
Chapter Planning Guide

Compression guide: To shorten instruction because of time limitations, omit Section 3.

OBJECTIVES	LABS, DEMONSTRATIONS, AND ACTIVITIES	TECHNOLOGY RESOURCES
PACING • 90 min pp. 106–115 **Chapter Opener**	**SE Start-up Activity,** p. 107 ◆ GENERAL	**OSP Parent Letter** ■ GENERAL **CD Student Edition on CD-ROM** **CD Guided Reading Audio CD** ■ **TR Chapter Starter Transparency*** **VID Brain Food Video Quiz**
Section 1 Change over Time • Identify two kinds of evidence that show that organisms have evolved. • Describe one pathway through which a modern whale could have evolved from an ancient mammal. • Explain how comparing organisms can provide evidence that they have ancestors in common.	**SE Connection to Geology** Sedimentary Rock, p. 111 ◆ GENERAL **TE Connection Activity** Math, p. 111 ADVANCED **TE Connection Activity** Art, p. 112 ◆ ADVANCED **TE Connection Activity** Geography, p. 113 ADVANCED	**CRF Lesson Plans*** **TR Bellringer Transparency*** **TR** *LINK TO EARTH SCIENCE* The Rock Cycle* **TR** Evidence of Whale Evolution: A* **TR** Evidence of Whale Evolution: B* **TR** Comparing Skeletal Structures*
PACING • 90 min pp. 116–121 **Section 2 How Does Evolution Happen?** • List four sources of Charles Darwin's ideas about evolution. • Describe the four parts of Darwin's theory of evolution by natural selection. • Relate genetics to evolution.	**TE Demonstration** Form and Function, p. 117 ◆ GENERAL **TE Connection Activity** Social Studies, p. 117 ADVANCED **TE Connection Activity** Geography, p. 118 GENERAL **SE Quick Lab** Population Growth Versus Food Supply, p. 119 ◆ GENERAL **CRF Datasheet for Quick Lab*** **TE Activity** Natural Selection, p. 120 BASIC **SE Skills Practice Lab** Survival of the Chocolates, p. 126 GENERAL **CRF Datasheet for Chapter Lab***	**CRF Lesson Plans*** **TR Bellringer Transparency*** **TR** Four Parts of Natural Selection* **CRF SciLinks Activity*** GENERAL **VID Lab Videos for Life Science**
PACING • 45 min pp. 122–125 **Section 3 Natural Selection in Action** • Give three examples of natural selection in action. • Outline the process of speciation.	**TE Connection Activity** Real World, p. 123 GENERAL **TE Group Activity** Amazing Adaptations, p. 123 ADVANCED **SE Science in Action** Math, Social Studies, and Language Arts Activities, p. 132–133 GENERAL **LB Whiz-Bang Demonstrations** Adaptation Behooves You* ◆ GENERAL **LB Long-Term Projects & Research Ideas** Evolution's Explosion* ADVANCED	**CRF Lesson Plans*** **TR Bellringer Transparency*** **TR** Evolution of the Galápagos Finches* **TE Internet Activity,** p. 123 GENERAL

PACING • 90 min

CHAPTER REVIEW, ASSESSMENT, AND STANDARDIZED TEST PREPARATION

CRF Vocabulary Activity* GENERAL
SE Chapter Review, pp. 128–129 GENERAL
CRF Chapter Review* ■ GENERAL
CRF Chapter Tests A* ■ GENERAL, **B*** ADVANCED, **C*** SPECIAL NEEDS
SE Standardized Test Preparation, pp. 130–131 GENERAL
CRF Standardized Test Preparation* GENERAL
CRF Performance-Based Assessment* GENERAL
OSP Test Generator GENERAL
CRF Test Item Listing* GENERAL

Online and Technology Resources

Visit **go.hrw.com** for a variety of free resources related to this textbook. Enter the keyword **HL5EVO.**

Students can access interactive problem-solving help and active visual concept development with the *Holt Science and Technology* Online Edition available at **www.hrw.com.**

Guided Reading Audio CD

A direct reading of each chapter using instructional visuals as guideposts. For auditory learners, reluctant readers, and Spanish-speaking students. Available in English and Spanish.

KEY

SE Student Edition	**CRF** Chapter Resource File	**SS** Science Skills Worksheets	* Also on One-Stop Planner
TE Teacher Edition	**OSP** One-Stop Planner	**MS** Math Skills for Science Worksheets	◆ Requires advance prep
	LB Lab Bank	**CD** CD or CD-ROM	■ Also available in Spanish
	TR Transparencies	**VID** Classroom Video/DVD	

SKILLS DEVELOPMENT RESOURCES	SECTION REVIEW AND ASSESSMENT	STANDARDS CORRELATIONS
SE Pre-Reading Activity, p. 106 GENERAL **OSP Science Puzzlers, Twisters & Teasers*** GENERAL		National Science Education Standards UCP 2, 5; SAI 1, 2; LS 1a, 5a
CRF Directed Reading A* ■ BASIC, **B*** SPECIAL NEEDS **CRF Vocabulary and Section Summary*** ■ GENERAL **SE Reading Strategy** Paired Summarizing, p. 108 GENERAL **TE Connection to Earth Science** Rock Layers, p. 110 GENERAL **SE Math Practice** The Weight of Whales, p. 113 GENERAL **TE Inclusion Strategies,** p. 113	**SE Reading Checks,** pp. 108, 110, 112, 114 GENERAL **TE Reteaching,** p. 114 BASIC **TE Quiz,** p. 114 GENERAL **TE Alternative Assessment,** p. 114 GENERAL **TE Homework,** p. 115 GENERAL **SE Section Review,*** p. 115 ■ GENERAL **CRF Section Quiz*** ■ GENERAL	UCP 2, 4, 5; SAI 2; HNS 2; LS 2e, 3a, 3d, 4a, 5a, 5b, 5c
CRF Directed Reading A* ■ BASIC, **B*** SPECIAL NEEDS **CRF Vocabulary and Section Summary*** ■ GENERAL **SE Reading Strategy** Brainstorming, p. 116 GENERAL **TE Connection to Geography** Galápagos, p. 117 GENERAL **TE Reading Strategy** Prediction Guide, p. 118 GENERAL **MS Math Skills for Science** Multiplying Whole Numbers* GENERAL **CRF Reinforcement Worksheet** Bicentennial Celebration* BASIC	**SE Reading Checks,** pp. 117, 119, 120 GENERAL **TE Homework,** p. 116 GENERAL **TE Homework,** p. 118 ADVANCED **TE Reteaching,** p. 120 BASIC **TE Quiz,** p. 120 GENERAL **TE Alternative Assessment,** p. 120 ADVANCED **SE Section Review,*** p. 121 ■ GENERAL **CRF Section Quiz*** ■ GENERAL	UCP 1, 2, 4, 5; SAI 1, 2; SPSP 2, 5; HNS 1, 2, 3; LS 2a, 2b, 2d, 2e, 3d, 5a, 5b; *Chapter Lab:* UCP 2, 4; SAI 1, 2
CRF Directed Reading A* ■ BASIC, **B*** SPECIAL NEEDS **CRF Vocabulary and Section Summary*** ■ GENERAL **SE Reading Strategy** Prediction Guide, p. 122 GENERAL **TE Inclusion Strategies,** p. 124 ◆ **CRF Critical Thinking** Taking the Earth's Pulse* ADVANCED	**SE Reading Checks,** pp. 123, 124 GENERAL **TE Reteaching,** p. 124 BASIC **TE Quiz,** p. 124 GENERAL **TE Alternative Assessment,** p. 124 ADVANCED **SE Section Review,*** p. 125 ■ GENERAL **CRF Section Quiz*** ■ GENERAL	UCP 1, 3, 4; SPSP 4, 5; LS 2a, 2e, 3d, 4d, 5b

This convenient CD-ROM includes:
- **Lab Materials QuickList Software**
- **Holt Calendar Planner**
- **Customizable Lesson Plans**
- **Printable Worksheets**
- **ExamView® Test Generator**

cnnstudentnews.com

Find the latest news, lesson plans, and activities related to important scientific events.

www.scilinks.org

Maintained by the **National Science Teachers Association.** See Chapter Enrichment pages for a complete list of topics.

Check out ***Current Science*** articles and activities by visiting the HRW Web site at **go.hrw.com.** Just type in the keyword **HL5CS07T.**

Classroom Videos
- **Lab Videos** demonstrate the chapter lab.
- **Brain Food Video Quizzes** help students review the chapter material.

5 Chapter Resources

Visual Resources

CHAPTER STARTER TRANSPARENCY

The Evolution of Living Things — CHAPTER STARTER

What If . . . ?

The time is 50 million years ago. The place is a swamp in North America. Imagine yourself trekking through the steamy swamp, sidestepping snakes and spiders. Suddenly, out of the trees dashes a 182 kg giant with a huge head, a thick neck, and long, muscular legs.

What is this beast? A velociraptor? A giant sloth? A prehistoric bear? None of the above. It's a *Diatryma*, a kind of flightless bird that was common during the Cenozoic era of prehistory, 57 to 35 million years ago! *Diatryma* stood over 2 m tall and had an enormous beak and sharp claws.

Scientists know about *Diatryma* from many fossils dug up in Wyoming, New Mexico, and New Jersey. *Diatryma* was probably forced out of existence by large mammals. Though the monster bird is long gone, smaller versions of it live in poultry coops around the world. *Diatryma*'s fossils indicate that it was a distant cousin of the present-day chicken!

BELLRINGER TRANSPARENCIES

The Evolution of Living Things — BELLRINGER TRANSPARENCY

Section: Change Over Time

The cockroach originated on Earth over 250 million years ago and is thriving today all over the world. A giant deer that was 2 m tall first appeared less than 1 million years ago and became extinct around 11,000 years ago. Why do you think one animal thrived and the other one perished?

Record your answer in your **science journal.**

Section: How Does Evolution Happen?

The following are characteristics that almost all humans have in common: upright walking, hair, fingerprints, binocular vision, speech. List the advantages and disadvantages of each characteristic. Do you think the advantages are greater than the disadvantages? Why or why not?

Record your responses in your **science journal.**

TEACHING TRANSPARENCIES

The Evolution of Living Things — TEACHING TRANSPARENCY

Evidence of Whale Evolution: A

A *Pakicetus* (PAK uh SEE tuhs) Scientists think that whales evolved from land-dwelling mammals that could run on four legs. One of these ancestors may have been *Pakicetus*, which lived about 50 million years ago. The fossil skeleton and an artist's illustration of *Pakicetus* are shown here. *Pakicetus* was about the size of a wolf.

B *Ambulocetus* (AM byoo loh SEE tuhs) This mammal lived in coastal waters about 49 million years ago. It could swim by kicking its legs and using its tail for balance. It could also waddle on land by using its short legs. *Ambulocetus* was about the size of a dolphin.

The Evolution of Living Things — TEACHING TRANSPARENCY

Evidence of Whale Evolution: B

C *Dorudon* (DOH roo DAHN) This mammal lived in the oceans about 40 million years ago. It resembled a giant dolphin and propelled itself with its massive tail. *Dorudon* had tiny hind limbs that it could not use for walking or swimming.

D Modern toothed whale — Modern whales' forelimbs are flippers. Modern whales do not have hind limbs, but they do have tiny hip bones. Modern whales range in size from 1.4 m porpoises to 33 m blue whales.

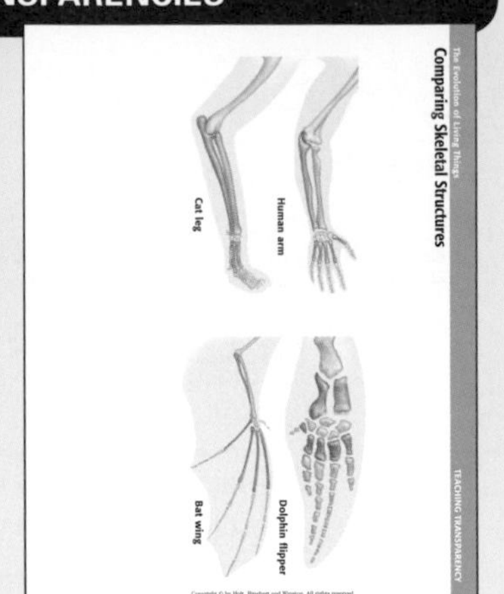

TEACHING TRANSPARENCIES

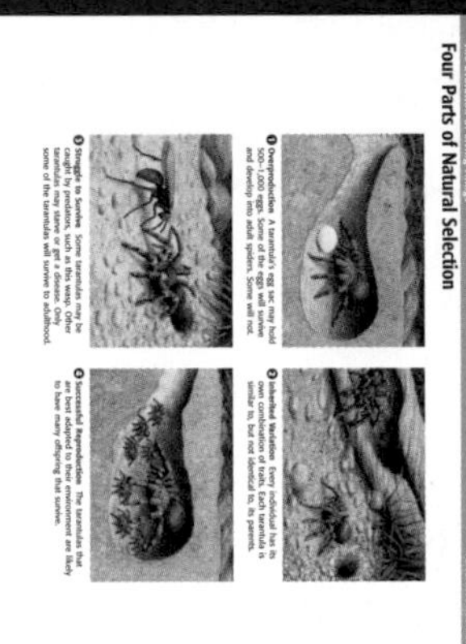

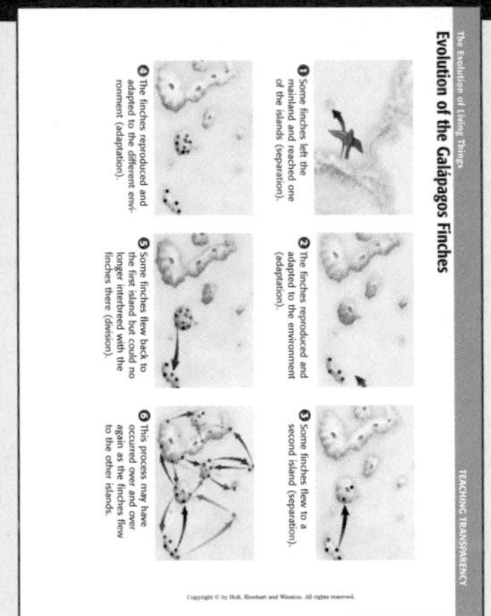

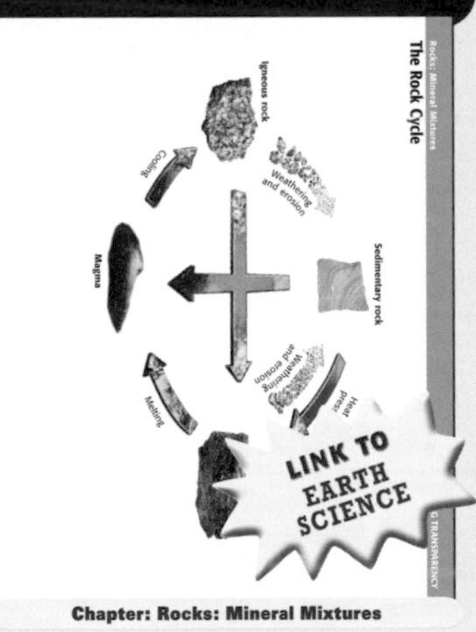

Chapter: Rocks: Mineral Mixtures

CONCEPT MAPPING TRANSPARENCY

The Evolution of Living Things — CONCEPT MAPPING TRANSPARENCY

Use the following terms to complete the concept map below: evolution, evidence, extinct species, living species, common ancestors, DNA, time, fossil record, body structures

of

exists in the

of

of

that changed over

that descended from

Planning Resources

LESSON PLANS

Lesson Plan — SAMPLE

Section: Waves

Pacing

Regular Schedule: with lab(s):2 days without lab(s)2 days
Block Schedule: with lab(s):1 1/2 days without lab(s)1 day

Objectives

1. Relate the seven properties of life to a living organism.
2. Describe seven themes that can help you to organize what you learn about biology.
3. Identify the tiny structures that make up all living organisms.
4. Differentiate between reproduction and heredity and between metabolism and homeostasis.

National Science Education Standards Covered

LSInter6: Cells have particular structures that underlie their functions.

LSMat1: Most cell functions involve chemical reactions.

LSBeh1: Cells store and use information to guide their functions.

UCP1: Cell functions are regulated.

SI1: Cells can differentiate and form complete multicellular organisms.

PS1: Species evolve over time.

ESS1: The great diversity of organisms is the result of more than 3.5 billion years of evolution.

ESS2: Natural selection and its evolutionary consequences provide a scientific explanation for the fossil record of ancient life forms as well as for the striking molecular similarities observed among the diverse species of living organisms.

ST1: The millions of different species of plants, animals, and microorganisms that live on Earth today are related by descent from common ancestors.

ST2: The energy for life primarily comes from the sun.

SPSP1: The complexity and organization of organisms accommodates the need for obtaining, transforming, transporting, releasing, and eliminating the matter and energy used to sustain the organism.

SPSP6: As matter and energy flows through different levels of organization of living systems—cells, organs, communities—and between living systems and the physical environment, chemical elements are recombined in different ways.

HNS1: Organisms have behavioral responses to internal changes and to external stimuli.

PARENT LETTER

SAMPLE

Dear Parent,

Your son's or daughter's science class will soon begin exploring the chapter entitled "The World of Physical Science." In this chapter, students will learn about how the scientific method applies to the world of physical science and the role of physical science in the world. By the end of the chapter, students should demonstrate a clear understanding of the chapter's main ideas and be able to discuss the following topics:

1. physical science as the study of energy and matter (Section 1)
2. the role of physical science in the world around them (Section 1)
3. careers that rely on physical science (Section 1)
4. the steps used in the scientific method (Section 2)
5. examples of technology (Section 2)
6. how the scientific method is used to answer questions and solve problems (Section 2)
7. how our knowledge of science changes over time (Section 2)
8. how models represent real objects or systems (Section 3)
9. examples of different ways models are used in science (Section 3)
10. the importance of the International System of Units (Section 4)
11. the appropriate units to use for particular measurements (Section 4)
12. how area and density are derived quantities (Section 4)

Questions to Ask Along the Way

You can help your son or daughter learn about these topics by asking interesting questions such as the following:

- What are some surprising careers that use physical science?
- What is a characteristic of a good hypothesis?
- When is it a good idea to use a model?
- Why do Americans measure things in terms of inches and yards [...] and meters ?

TEST ITEM LISTING

TEST ITEM LISTING
The World of Science — SAMPLE

MULTIPLE CHOICE

1. A limitation of models is that
   a. they are large enough to see.
   b. they do not act exactly like the things that they model.
   c. they are smaller than the things that they model.
   d. they model unfamiliar things.
   Answer: B Difficulty: I Section: 3 Objective: 2
2. The length 10 m is equal to
   a. 100 cm. c. 10,000 mm.
   b. 1,000 cm. d. Both (b) and (c)
   Answer: B Difficulty: I Section: 3 Objective: 2
3. To be valid, a hypothesis must be
   a. testable. c. made into a law.
   b. supported by evidence. d. Both (a) and (b)
   Answer: B Difficulty: I Section: 3 Objective: 2 1
4. The statement "Sheila has a stain on her shirt" is an example of a(n)
   a. law. c. observation.
   b. hypothesis. d. prediction.
   Answer: B Difficulty: I Section: 3 Objective: 2
5. A hypothesis is often developed out of
   a. observations. c. laws.
   b. experiments. d. Both (a) and (b)
   Answer: B Difficulty: I Section: 3 Objective: 2
6. How many milliliters are in 3.5 kL?
   a. 3,500 mL c. 3,500, 000 mL
   b. 0.0035 mL d. 35,000 mL
   Answer: B Difficulty: I Section: 3 Objective: 2
7. A map of Seattle is an example of a
   a. law. c. model.
   b. theory. d. unit.
   Answer: B Difficulty: I Section: 3 Objective: 2
8. A lab has the safety icons shown below. These icons mean that you should wear
   a. only safety goggles. c. safety goggles and a lab apron.
   b. only a lab apron. d. safety goggles, a lab apron, and gloves.
   Answer: B Difficulty: I Section: 3 Objective: 2
9. The law of conservation of mass says the tot al mass before a chemical change is
   a. more than the total mass after the change.
   b. less than the total mass after the change.
   c. the same as the total mass after the change.
   d. not the same as the total mass after the change.
   Answer: B Difficulty: I Section: 3 Objective: 2
10. In which of the following areas might you find a geochemist at work?
    a. studying the chemistry of rocks c. studying fishes
    b. studying forestry d. studying the atmosphere
    Answer: B Difficulty: I Section: 3 Objective: 2

One-Stop Planner® CD-ROM

This CD-ROM includes all of the resources shown here and the following time-saving tools:

- *Lab Materials QuickList Software*
- *Customizable lesson plans*
- *Holt Calendar Planner*
- *The powerful ExamView® Test Generator*

For a preview of available worksheets covering math and science skills, see pages T12–T19. All of these resources are also on the One-Stop Planner®.

Meeting Individual Needs

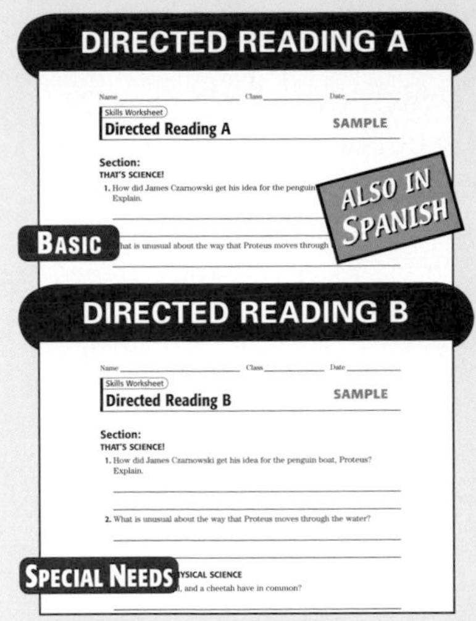

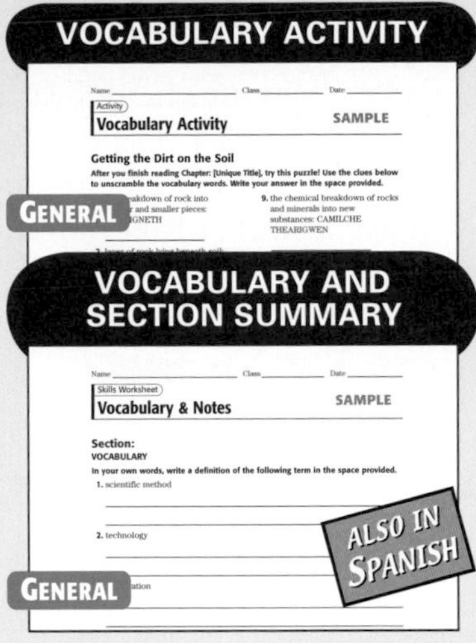

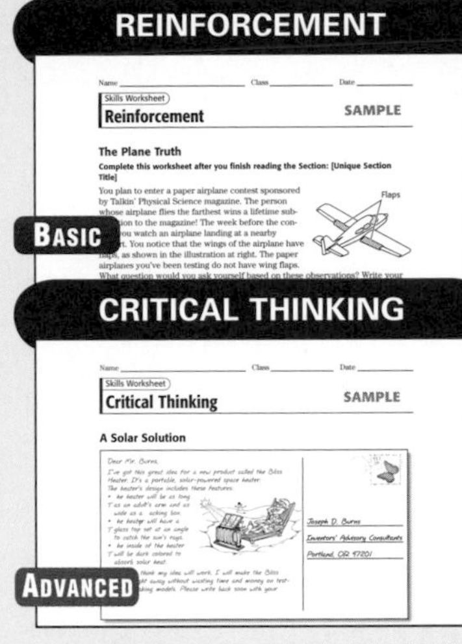

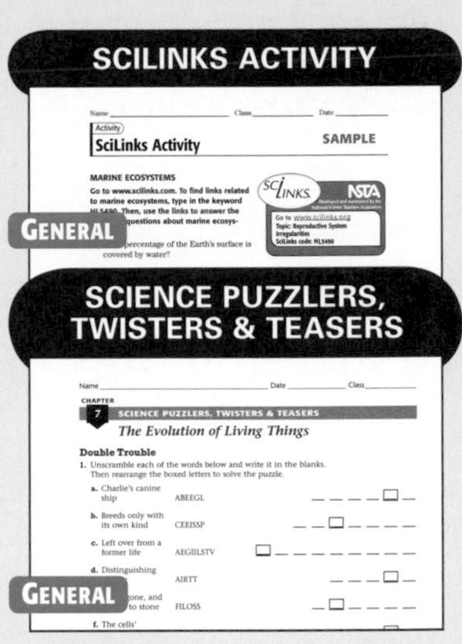

Labs and Activities

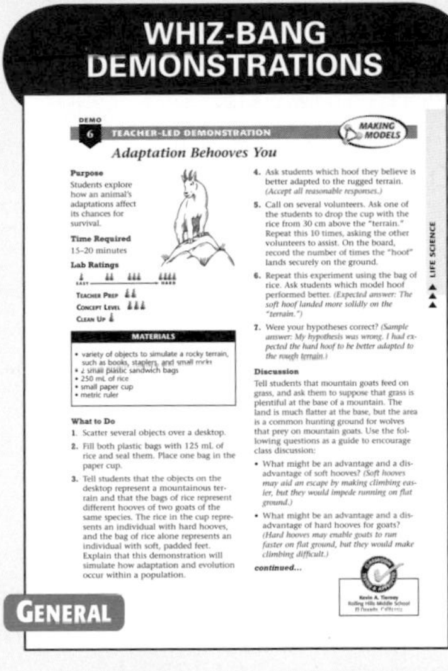

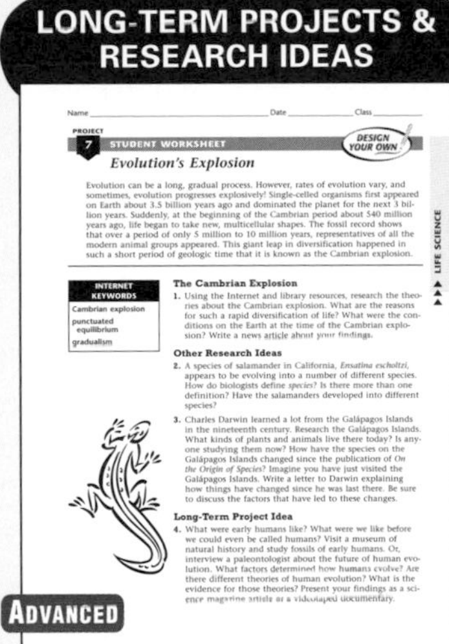

Review and Assessments

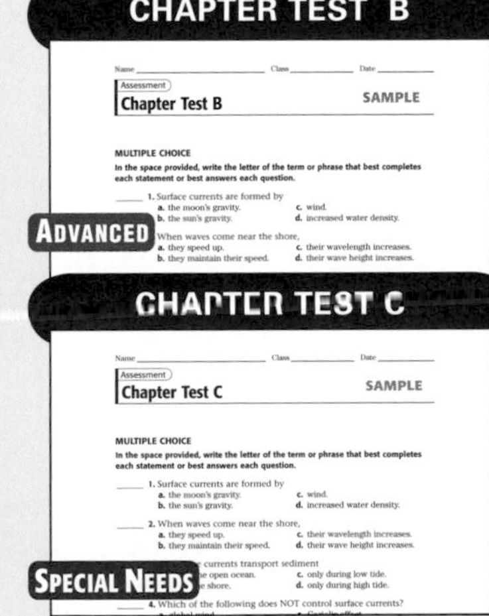

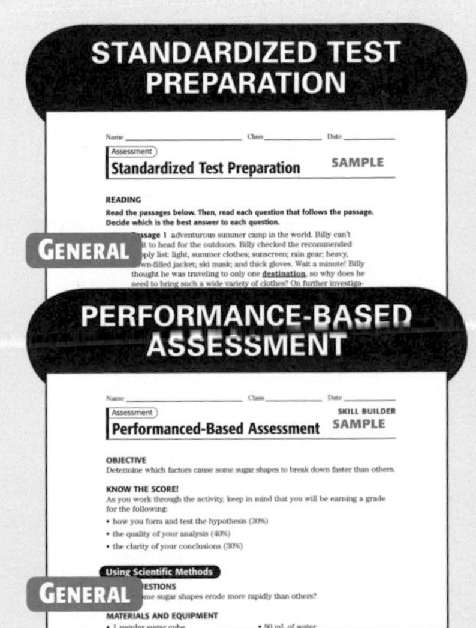

5 Chapter Enrichment

This Chapter Enrichment provides relevant and interesting information to expand and enhance your presentation of the chapter material.

Section 1

Change over Time

Evolution of Whales and Other Mammals

- Scientists think that all mammals evolved from a shrewlike ancestor. This ancestor survived the mass extinction that wiped out the dinosaurs about 65 million years ago. This hypothesis is supported by fossils formed during and after the time of the dinosaurs.

- The first ocean-dwelling mammals appeared in the fossil record about 50 million years ago. Scientists think that these mammals were the ancestors of whales and shared an ancestor with the artiodactyl group (even-toed, hoofed mammals). However, other types of aquatic mammals, such as dugongs, manatees, and sea lions, probably evolved separately and from later branches of the mammal lineage.

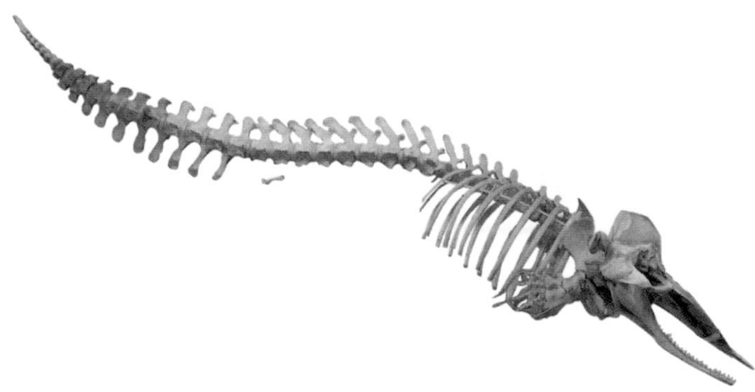

- In many ways, whales are more similar to their hoofed mammal relatives than they are to fish. Similarities include internal structures, behavior, and DNA. Also, whales swim by moving their tails up and down, in a motion similar to a gallop and to the swimming of an otter, whereas fish move their tails sideways.

Homologous Structures

- *Homologous structures* are anatomical features that have similar evolutionary and embryological origins and exhibit similar anatomical patterns. For example, bird wings, human arms, whale flippers, and deer forelimbs are all homologous. However, bird wings and butterfly wings are *analogous structures* because they function similarly but are anatomically dissimilar.

- Cellular components and biochemicals may also be homologous. For example, hemoglobin molecules from different vertebrate species have similar amino-acid sequences. But hemocyanin, which transports oxygen in crabs, has a very different sequence and is therefore analogous to hemoglobin; that is, the two molecules have a similar function but different structure.

Is That a Fact!

- ◆ The California halibut belongs to the family Bothidae, also known as the *left-eyed flounders.* Despite the name, about 40% of California halibut adults have both eyes on the right side of their body.

Convergent Evolution

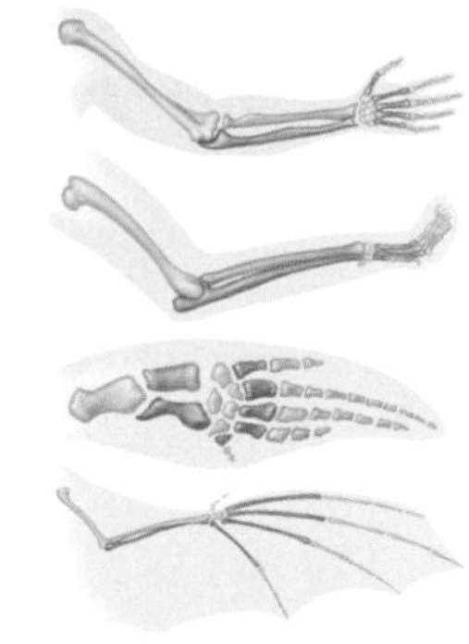

- When scientists study the fossils, skeletons, and DNA of species thought to be related, the scientists sometimes find that the organisms are not related at all. For example, the jerboa and the kangaroo rat look almost identical, but scientists have concluded that they have different ancestors. Such cases illustrate *convergent evolution,* where different species developed similar adaptations to similar environmental conditions and roles.

Frozen Fossils

- In some cases, scientists can obtain DNA from ancient tissues that have not completely decomposed or fossilized. Two Japanese geneticists are hoping to create a mammoth-elephant hybrid by using tissue from a Siberian mammoth that died and was frozen thousands of years ago. However, the chances of finding intact DNA are small, and the genetic structures of mammoths and elephants are not fully compatible.

Is That a Fact!

- ◆ The human appendix is a *vestigial organ,* or an organ that performs little or no apparent function but that is thought to have had a function in ancestors. The appendix is a narrow tube attached to the large intestine. In chimpanzees, gorillas, and orangutans, the appendix is an intestinal sac that helps them digest tough plant material.

For background information about teaching strategies and issues, refer to the ***Professional Reference for Teachers.***

Section 2

How Does Evolution Happen?

Alfred Russel Wallace

- Alfred Wallace (1823–1913) was born in England. He came from a poor family and had no formal scientific education. Though originally interested in botany, he began to study insects with the encouragement of British naturalist Henry Walter Bates, whom Wallace met when he was about 20 years old. Bates and Wallace explored the Amazon from 1848 to 1852 and found much evidence to support a theory of evolution by natural selection.

- From 1854 to 1862, Wallace traveled in the Malay Archipelago to find more evidence of evolution. In 1855, he published a preliminary essay, "On the Law Which Has Regulated the Introduction of New Species." Meanwhile, nearly 20 years after Charles Darwin's voyage on the HMS *Beagle,* Darwin was still mulling over his data. In 1858, Wallace mailed an essay to Darwin that explained Wallace's theory that natural selection pressures species to change.

- In July 1858, Wallace's essay was presented along with a paper by Darwin at a meeting of the Linnean Society in London. In the following year, after nearly two decades of delay (because of his doubts and repeated analysis), Darwin published *On the Origin of Species by Means of Natural Selection.*

Charles Lyell

- Charles Lyell (1797–1875), the eldest of 10 children, was born in Scotland and raised in England. His father was a naturalist who traveled with him to collect butterflies and aquatic insects, informal research that Lyell continued throughout college. Lyell's research led him to the belief that natural processes occurring over millions of years have shaped the Earth's features. This idea was known as uniformitarianism. Lyell's work influenced Darwin's formulation of his theory of natural selection.

Section 3

Natural Selection in Action

Adaptive Coloration

- Penguins, puffins, killer whales, and blue sharks are just some of the ocean animals that have white bellies and black or dark blue dorsal surfaces. This type of coloration is called *countershading.* When seen from below, the white underside helps the animal blend into the lighter sky above the water. When viewed from above, the dark coloration makes the animal difficult to see against the ocean depths.

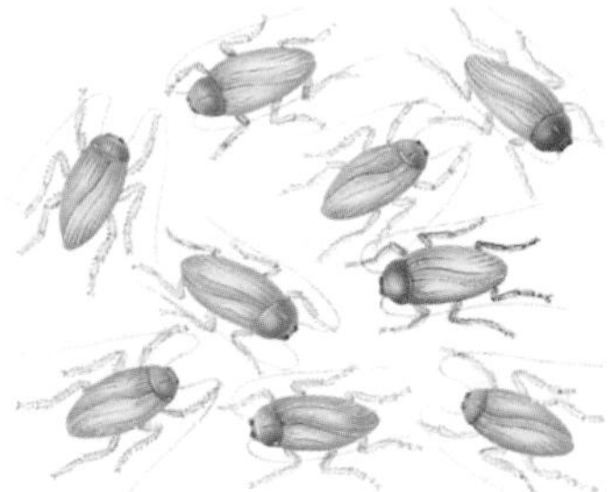

Sexual Selection

- *Sexual selection* is the term for the selection of traits brought about by a specific pattern of mating. In many sexual organisms, members of one sex must compete with each other for access to mates. Biologists think this behavior results when one sex's investment in the next generation is greater than that of the other sex. At an extreme, the "choosiness" of one sex may drive the evolution of traits that confer no other apparent advantage to the opposite sex. The long tails and colorful plumage of many male birds are considered examples of such "runaway sexual selection."

Overview

Tell students that this chapter will introduce them to *evolution*—the process by which organisms on Earth change over time. Evolution helps explain the variations and adaptations that we see in organisms around us and in evidence of the past.

Assessing Prior Knowledge

Students should be familiar with the following topics:

- scientific methods and models
- heredity and genetics

Identifying Misconceptions

Students may have heard that evolution is "just a theory." But in academic biology, evolution (defined as the process by which species change over time) is accepted in the way that "cell theory" is now accepted. Furthermore, the theory of evolution by natural selection (integrated with modern genetic knowledge) is considered to be strongly supported and widely accepted. Specific models, mechanisms, rates, and other aspects of evolution continue to be investigated and debated among scientists, but few biologists doubt that evolution happens.

The Evolution of Living Things

About the 

What happened to this fish's face? This flounder wasn't born this way, but it did develop naturally. When young, a flounder looks and swims as most fish do. But as it becomes an adult, one of its eyes moves to the other side of its head, and the flounder begins to swim sideways. An adult flounder is adapted to swim and hide along the sandy bottom of coastal areas.

PRE-READING ACTIVITY

Graphic Organizer

Concept Map Before you read the chapter, create the graphic organizer entitled "Concept Map" described in the **Study Skills** section of the Appendix. As you read the chapter, fill in the concept map with details about evolution and natural selection.

Standards Correlations

National Science Education Standards

The following codes indicate the National Science Education Standards that correlate to this chapter. The full text of the standards is at the front of the book.

Chapter Opener
UCP 2, 5; SAI 1, 2; LS 1a, 5a

Section 1 Change over Time
UCP 2, 4, 5; SAI 2; HNS 2; LS 2e, 3a, 3d, 4a, 5a, 5b, 5c

Section 2 How Does Evolution Happen?
UCP 1, 2, 4, 5; SAI 1, 2; SPSP 2, 5; HNS 1, 2, 3; LS 2a, 2b, 2d, 2e, 3d, 5a, 5b

Section 3 Natural Selection in Action
UCP 1, 3, 4; SPSP 4, 5; LS 2a, 2e, 3d, 4d, 5b

Chapter Lab
UCP 2, 4; SAI 1, 2

Chapter Review
SAI 2; SPSP 4; HNS 1, 2, 3; LS 2e, 3d, 4a, 5a, 5b, 5c

Science in Action
SPSP 2, 4, 5; HNS 1, 3

START-UP ACTIVITY

Out of Sight, Out of Mind

In this activity, you will see how traits can affect the success of an organism in a particular environment.

Procedure

1. Count out **25 colored marshmallows** and **25 white marshmallows.**
2. Ask your partner to look away while you spread the marshmallows out on a **white cloth.** Do not make a pattern with the marshmallows. Now, ask your partner to turn around and pick the first marshmallow that he or she sees.
3. Repeat step 2 ten times.

Analysis

1. How many white marshmallows did your partner pick? How many colored marshmallows did he or she pick?
2. What did the marshmallows and the cloth represent in your investigation? What effect did the color of the cloth have?
3. When an organism blends into its environment, the organism is *camouflaged*. How does this activity model camouflaged organisms in the wild? What are some weaknesses of this model?

START-UP ACTIVITY

MATERIALS

For Each Pair

- cloth, white, approximately 20 cm × 20 cm
- marshmallows, colored (all same color), miniature (25)
- marshmallows, white, miniature (25)

Teacher's Note: To avoid the use of marshmallows in this activity, try the following alternative: Instead of using marshmallows, punch 100 holes from the classified advertisement section of a newspaper. Instead of using cloth, use the newspaper for the background. Spread the holes on the paper all at once, and have the "hunter" pick up as many as he or she can in 15 s. Tally the results.

Answers

1. Answers may vary, but students are likely to pick up more colored marshmallows than white ones.
2. Sample answer: The marshmallows represent organisms that could be eaten; the cloth represents the area where they live.
3. Sample answer: Many organisms in the wild blend into their surroundings by having colors or patterns that make them hard to see. This might help them hide from things trying to eat them. A weakness of this model is that it's very simple—a real "wild" environment would be more than two colors and would contain a variety of organisms.

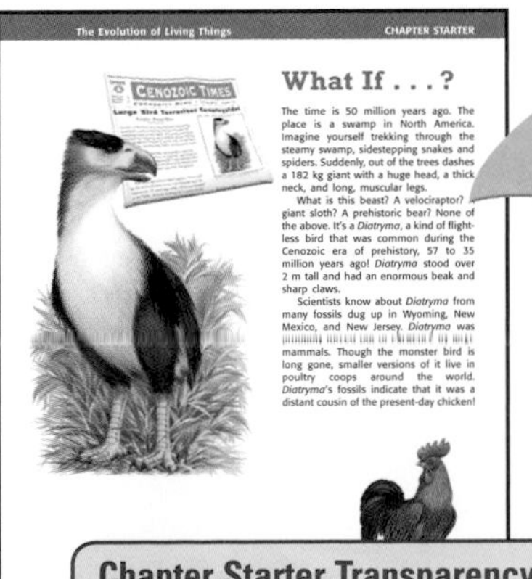

The Evolution of Living Things — CHAPTER STARTER

What If . . . ?

The time is 50 million years ago. The place is a swamp in North America. Imagine yourself trekking through the steamy swamp, sidestepping snakes and spiders. Suddenly, out of the trees dashes a 182 kg giant with a huge head, a thick neck, and long, muscular legs.

What is this beast? A velociraptor? A giant sloth? A prehistoric bear? None of the above. It's a *Diatryma*, a kind of flightless bird that was common during the Cenozoic era of prehistory, 57 to 35 million years ago! *Diatryma* stood over 2 m tall and had an enormous beak and sharp claws.

Scientists know about *Diatryma* from many fossils dug up in Wyoming, New Mexico, and New Jersey. *Diatryma* was [illegible] mammals. Though the monster bird is long gone, smaller versions of it live in poultry coops around the world. *Diatryma*'s fossils indicate that it was a distant cousin of the present-day chicken!

Chapter Starter Transparency
Use this transparency to help students begin thinking about changes in species over time.

CHAPTER RESOURCES

Technology

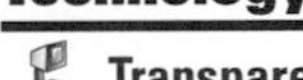

Transparencies
- Chapter Starter Transparency

Student Edition on CD-ROM

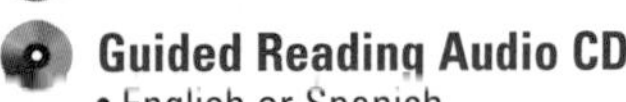

Guided Reading Audio CD
- English or Spanish

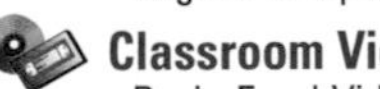

Classroom Videos
- Brain Food Video Quiz

Workbooks

Science Puzzlers, Twisters & Teasers
- The Evolution of Living Things GENERAL

SECTION 1

Focus

Overview

This section introduces the concept of evolution as change over time in populations of organisms. Students will survey evidence used to understand evolution and determine ancestry, including the fossil record and comparisons of organisms' physical and genetic traits.

Bellringer

Have students respond to the following prompt: "The cockroach originated on Earth more than 250 million years ago and is thriving today all over the world. A giant deer (over 2 m tall!) evolved less than 1 million years ago and became extinct around 11,000 years ago. Why do you think one animal thrived and the other perished?" (Accept all reasonable answers.)

Motivate

Discussion — GENERAL

Adaptation Ask students if a polar bear could live comfortably in Hawaii. Ask if a fish could survive in a forest. Why or why not? Discuss various characteristics of animals, such as physical adaptations, that make the animals well suited for a specific environment. LS Verbal

SECTION 1

Change over Time

READING WARM-UP

Objectives

- Identify two kinds of evidence that show that organisms have evolved.
- Describe one pathway through which a modern whale could have evolved from an ancient mammal.
- Explain how comparing organisms can provide evidence that they have ancestors in common.

Terms to Learn

adaptation
species
evolution
fossil
fossil record

READING STRATEGY

Paired Summarizing Read this section silently. In pairs, take turns summarizing the material. Stop to discuss ideas that seem confusing.

If someone asked you to describe a frog, you might say that a frog has long hind legs, has bulging eyes, and croaks. But what color skin would you say that a frog has?

Once you start to think about frogs, you realize that frogs differ in many ways. These differences set one kind of frog apart from another. The frogs in **Figures 1, 2,** and **3** look different from each other, yet they may live in the same areas.

Differences Among Organisms

As you can see, each frog has a different characteristic that might help the frog survive. A characteristic that helps an organism survive and reproduce in its environment is called an **adaptation.** Adaptations may be physical, such as a long neck or striped fur. Or adaptations may be behaviors that help an organism find food, protect itself, or reproduce.

Living things that have the same characteristics may be members of the same species. A **species** is a group of organisms that can mate with one another to produce fertile offspring. For example, all strawberry poison arrow frogs are members of the same species and can mate with each other to produce more strawberry poison arrow frogs. Groups of individuals of the same species living in the same place make up a *population.*

Reading Check How can you tell that organisms are members of the same species? (*See the Appendix for answers to Reading Checks.*)

Figure 1 The red-eyed tree frog hides among a tree's leaves during the day and comes out at night.

Figure 2 The bright coloring of the strawberry poison arrow frog warns predators that the frog is poisonous.

Figure 3 The smoky jungle frog blends into the forest floor.

CHAPTER RESOURCES

Chapter Resource File

- Lesson Plan
- Directed Reading A BASIC
- Directed Reading B SPECIAL NEEDS

Technology

Transparencies
- Bellringer

Answer to Reading Check

if they mate with each other and produce more of the same type of organism

Do Species Change over Time?

In a single square mile of rain forest, there may be dozens of species of frogs. Across the Earth, there are millions of different species of organisms. The species that live on Earth today range from single-celled bacteria, which lack cell nuclei, to multicellular fungi, plants, and animals. Have these species always existed on Earth?

Scientists think that Earth has changed a great deal during its history, and that living things have changed, too. Scientists estimate that the planet is 4.6 billion years old. Since life first appeared on Earth, many species have died out, and many new species have appeared. **Figure 4** shows some of the species that have existed during Earth's history.

Scientists observe that species have changed over time. They also observe that the inherited characteristics in populations change over time. Scientists think that as populations change over time, new species form. Thus, newer species descend from older species. The process in which populations gradually change over time is called **evolution.** Scientists continue to develop theories to explain exactly how evolution happens.

adaptation a characteristic that improves an individual's ability to survive and reproduce in a particular environment

species a group of organisms that are closely related and can mate to produce fertile offspring

evolution the process in which inherited characteristics within a population change over generations such that new species sometimes arise

Figure 4 *This diagram shows some of the many kinds of organisms that have lived on Earth since the planet formed 4.6 billion years ago.*

Is That a Fact!

There are more than 100,000 living mollusk species, and at least 35,000 extinct forms are known from the fossil record. As a group, mollusks are very successful—there have been mollusks on Earth for nearly 600 million years.

WEIRD SCIENCE

The gastric brooding frogs of Australia, now extinct, incubated their tadpoles in their stomachs and gave birth to their young through their mouths!

Teach

Using the Figure – GENERAL

Species and Change Ask students to compare the frogs in **Figures 1, 2,** and **3** and discuss the unique adaptations of each. Ask students: "If you see a frog that looks like one of these frogs, can you assume that they are both the same species? (no, not necessarily) How can you tell? (by whether they breed with each other or not.) Next, have students look at **Figure 4.** Ask them how they think changes in the planet could have affected the appearance and disappearance of so many types of organisms over time. (Sample answer: Temperature fluctuations due to ice ages and other climatic changes would have affected which plants and animals could survive. Climate changes caused changes in the vegetation and in the availability of food for animals. Until the planet was able to support a lot of vegetation, there wouldn't have been much food for animals to eat.) **LS Visual/Verbal**

Discussion — BASIC

Populations Versus Species Understanding evolution requires understanding that changes in populations can lead to changes in species. Thus, it is important that students distinguish between populations and species. Reinforce these meanings with the following examples: All domestic cats are the same species. Those that live in one city may be a population, but cats that live in different cities are probably not part of the same population. All of the same kind of lizard that live on an island are a population and may be the only population of a single species. All humans are the same species, with many populations. (Note, however, that even professional biologists are sometimes unsure about the designation of closely related groups of organisms.) **LS Verbal**

Teach, continued

Cultural Awareness — GENERAL

Historic Paleontologist Mary Anning (1799–1847) made some of the most important fossil discoveries of her time. She was born in Lyme Regis in southern Great Britain, an area with many fossils. Her father, a cabinet-maker and amateur fossil collector, died when Anning was 11 years old, leaving the family in debt. Anning's fossil-finding skills provided the family with needed income. Even before she reached her teens, Anning had discovered part of the first *Ichthyosaurus* to be recognized by scientists in London.

In the early 1820s, a professional fossil collector sold his private collection and gave the proceeds to the Anning family. He recognized that they had contributed many specimens for scientific investigation. Soon after, Mary Anning took charge of the family fossil business. However, many of Anning's finds ended up uncredited. Many scientists could not accept that a person of her financial and educational background could have acquired such expertise. Have students research to find out one of Anning's significant fossil finds (For example, she discovered the first plesiosaur fossil.) LS Verbal

Answer to Reading Check

by their estimated ages and physical similarities

Figure 5 *The fossil on the left is of a trilobite, an ancient aquatic animal. The fossils on the right are of seed ferns.*

Evidence of Changes over Time

Evidence that evolution has happened is buried within Earth. Earth's crust is arranged in layers. These layers are made up of different kinds of rock and soil stacked on top of each other. These layers form when *sediments*, particles of sand, dust, or soil, are carried by wind and water and are deposited in an orderly fashion. Older layers are deposited before newer layers and are buried deeper within Earth.

Fossils

Sometimes, the remains or imprints of once-living organisms are found in the layers of rock. These remains are called **fossils.** Examples of fossils are shown in **Figure 5.** Fossils can be complete organisms, parts of organisms, or just a set of footprints. Fossils usually form when a dead organism is covered by a layer of sediment. Over time, more sediment settles on top of the organism. Minerals in the sediment may seep into the organism and gradually replace the organism with stone. If the organism rots away completely after being covered, it may leave an imprint of itself in the rock.

fossil the remains or physical evidence of an organism preserved by geological processes

fossil record a historical sequence of life indicated by fossils found in layers of the Earth's crust

The Fossil Record

By studying fossils, scientists have made a timeline of life that is known as the **fossil record.** The fossil record organizes fossils by their estimated ages and physical similarities. Fossils found in newer layers of Earth's crust tend to be similar to present-day organisms. This similarity indicates that the fossilized organisms were close relatives of present-day organisms. Fossils from older layers are less similar to present-day organisms than fossils from newer layers are. The older fossils are of earlier life-forms, which may not exist anymore.

✓ *Reading Check* How does the fossil record organize fossils?

CHAPTER RESOURCES

Technology

Transparencies

- ***LINK TO EARTH SCIENCE*** The Rock Cycle

CONNECTION to *Earth Science* — GENERAL

Rock Layers Using sedimentary layers as reference points, scientists can find the relative age of a fossil. Use the teaching transparency entitled "The Rock Cycle" to illustrate the ways that sedimentary rock is continually formed on Earth. Tell students that a common way for fossils to form is for an organism to be buried under sediment that becomes sedimentary rock. Point out that a layer of sedimentary rock can form only on top of older rock. LS Visual

Evidence of Ancestry

The fossil record provides evidence about the order in which species have existed. Scientists observe that all living organisms have characteristics in common and inherit characteristics in similar ways. So, scientists think that all living species descended from common ancestors. Evidence of common ancestors can be found in fossils and in living organisms.

Drawing Connections

Scientists examine the fossil record to figure out the relationships between extinct and living organisms. They draw models, such as the one shown in **Figure 6,** that illustrate their hypotheses. The short horizontal line at the top left in the diagram represents a species that lived in the past. Each branch in the diagram represents a group of organisms that descended from that species.

As shown in **Figure 6,** scientists think that whales and some types of hoofed mammals have a common ancestor. This ancestor was probably a mammal that lived on land between 50 million and 70 million years ago. During this time period, the dinosaurs died out and a variety of mammals appeared in the fossil record. The first ocean-dwelling mammals appeared about 50 million years ago. Scientists think that all mammal species alive today evolved from common ancestors.

Scientists have named and described hundreds of thousands of living and ancient species. Scientists use information about these species to sketch out a "tree of life" that includes all known organisms. But scientists know that their information is incomplete. For example, parts of Earth's history lack a fossil record. In fact, fossils are rare because specific conditions are necessary for fossils to form.

CONNECTION TO Geology

Sedimentary Rock Fossils are most often found in sedimentary rock. *Sedimentary rock* usually forms when rock is broken into sediment by wind, water, and other means. The wind and water move the sediment around and deposit it. Over time, layers of sediment pile up. Lower layers are compressed and changed into rock. Find out if your area has any sedimentary rocks that contain fossils. Mark the location of such rocks on a copy of a local map.

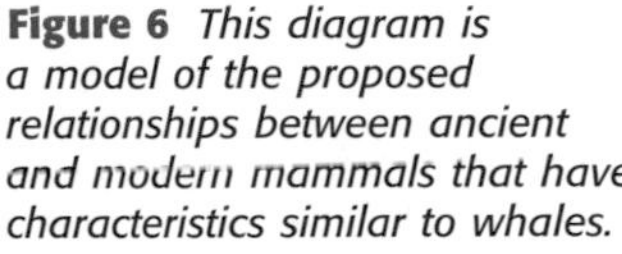

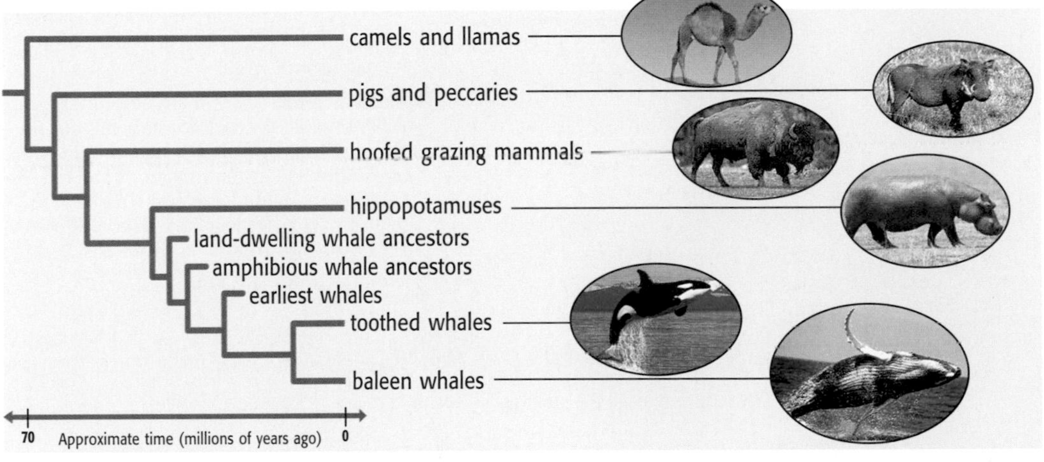

Figure 6 *This diagram is a model of the proposed relationships between ancient and modern mammals that have characteristics similar to whales.*

CONNECTION ACTIVITY
Math — ADVANCED

Species Countdown Present the following scenario to students: Imagine that you are a scientific time traveler assigned to count the species in your home state 1,000 years from now. You know that, at present, your home state contains 2,300 animal species and 4,500 plant species. But many species are becoming endangered or extinct. Only 55% of the state's species are expected to remain in existence after 1,000 years. How many species should you expect to count during your visit to the future? (Tip: Round your answer to two significant figures.) (Animals: $2,300 \times 0.55$ = approximately 1,300; Plants: $4,500 \times 0.55$ = approximately 2,500) **LS Logical**

MISCONCEPTION ALERT

Adaptation on Purpose? Students commonly confuse evolutionary adaptation with intentional change. For example, a student might say that an organism "learns" to adapt or "grows" an adaptation to a changing environment. Explain to students that the evolution of adaptations is something that happens to populations over many generations. Evolution never happens within a single individual. Furthermore, new traits arise by chance and become adaptations in a population only if those traits are both advantageous to individuals and inherited by their offspring. Students may also confuse evolutionary adaptation with acclimation, such as adjustment to seasonal changes. Adaptations that enable organisms to adjust to changing conditions can evolve, but only if those abilities are passed on to future generations.

How Many Toes? Hoofed mammals, or *ungulates,* are classified into two major groups. The even-toed *artiodactyls* have either two or four toes on each foot. Examples are pigs, deer, and cows. The odd-toed *perissodactyls* carry their weight on a middle toe. Examples are horses, zebras, tapirs, and rhinoceroses. All ungulates have similar foot and ankle bones.

Q: How did the dinosaurs listen to music?

A: on their fossil records

Teach, *continued*

Using the Figure—GENERAL

Evidence of Whale Evolution Have students examine each of the skeletons in **Figure 7** carefully. Ask them to describe one similarity and one difference between each successive species. (Sample answer: *Pakicetus* and *Ambulocetus* have similar limbs and feet but those of *Ambulocetus* are shorter.) **LS Visual/Logical** English Language Learners

CONNECTION ACTIVITY

Art—ADVANCED

Scientific Illustration The role of a scientific illustrator is to create accurate pictures of organisms and things that scientists study. In the case of long-extinct species, such as dinosaurs, artists must sometimes fill in where science leaves off. Have students look for and compare several examples of illustrations of a specific extinct organism. Have students try to identify ways in which artistic interpretation is used. For comparison, show students examples of similar illustrations from hundreds of years ago, when much less was known about many fossil organisms. **LS Visual** English Language Learners

Answer to Reading Check

a four-legged land mammal

Examining Organisms

Examining an organism carefully can give scientists clues about its ancestors. For example, whales seem similar to fish. But unlike fish, whales breathe air, give birth to live young, and produce milk. These traits show that whales are *mammals*. Thus, scientists think that whales evolved from ancient mammals.

Case Study: Evolution of the Whale

Scientists think that the ancient ancestor of whales was probably a mammal that lived on land and that could run on four legs. A more recent ancestor was probably a mammal that spent time both on land and in water. Comparisons of modern whales and a large number of fossils have supported this hypothesis. **Figure 7** illustrates some of this evidence.

Reading Check **What kind of organism do scientists think was an ancient ancestor of whales?**

Figure 7 **Evidence of Whale Evolution**

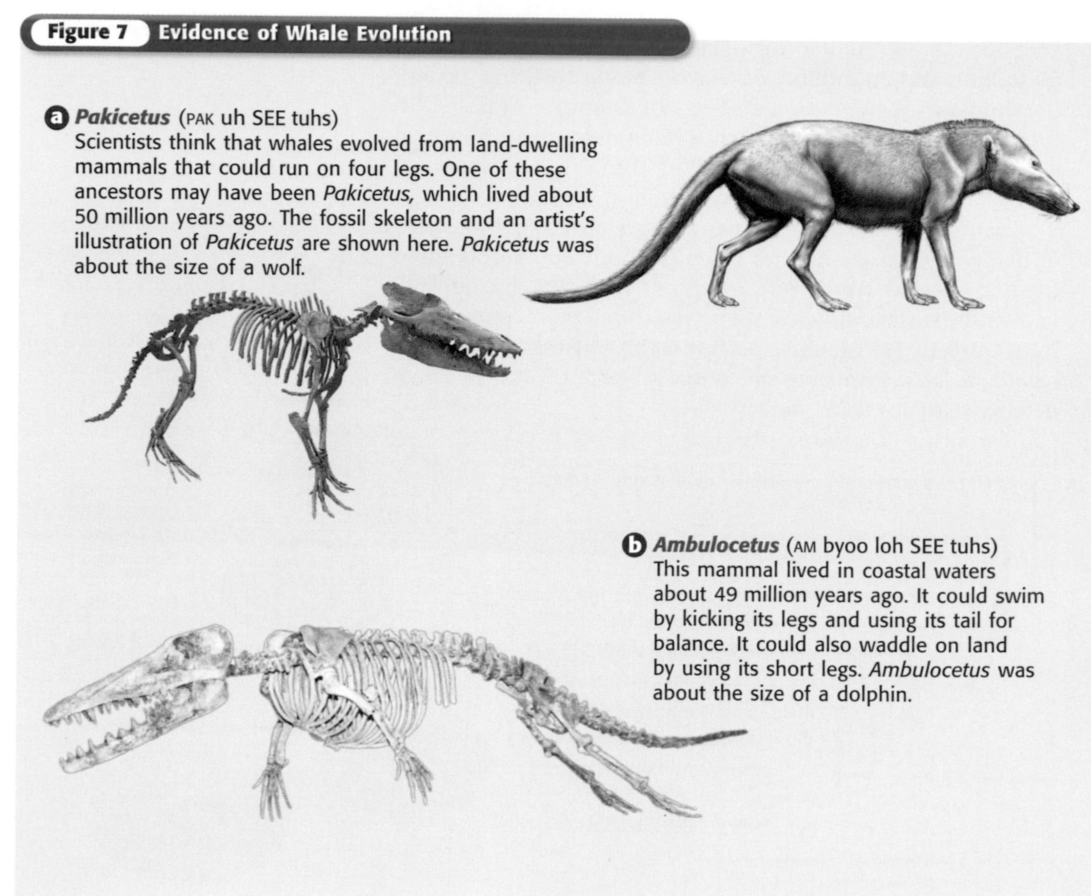

a ***Pakicetus*** (PAK uh SEE tuhs) Scientists think that whales evolved from land-dwelling mammals that could run on four legs. One of these ancestors may have been *Pakicetus,* which lived about 50 million years ago. The fossil skeleton and an artist's illustration of *Pakicetus* are shown here. *Pakicetus* was about the size of a wolf.

b ***Ambulocetus*** (AM byoo loh SEE tuhs) This mammal lived in coastal waters about 49 million years ago. It could swim by kicking its legs and using its tail for balance. It could also waddle on land by using its short legs. *Ambulocetus* was about the size of a dolphin.

CHAPTER RESOURCES

Technology

Transparencies
- Evidence of Whale Evolution: A
- Evidence of Whale Evolution: B

WEIRD SCIENCE

In 1938, some fishermen near the coast of South Africa caught an unusual fish called a *coelacanth*. Until that time, this type of fish was known only from fossils and was thought to have been extinct.

Walking Whales

The organisms in **Figure 7** form a sequence between ancient four-legged mammals and modern whales. Several pieces of evidence indicate that these species are related by ancestry. Each species shared some traits with an earlier species. However, some species had new traits that were shared with later species. Yet, each species had traits that allowed it to survive in a particular time and place in Earth's history.

Further evidence can be found inside the bodies of living whales. For example, although modern whales do not have hind limbs, inside their bodies are tiny hip bones, as shown in **Figure 7.** Scientists think that these hip bones were inherited from the whales' four-legged ancestors. Scientists often look at this kind of evidence when they want to determine the relationships between organisms.

The Weight of Whales

Whales are the largest animals ever known on Earth. One reason whales can grow so large is that they live in water, which supports their weight in a way that their bones could not. The blue whale—the largest type of whale in existence—is about 24 m long and has a mass of about 99,800 kg. Convert these measurements into feet and pounds, and round to whole numbers.

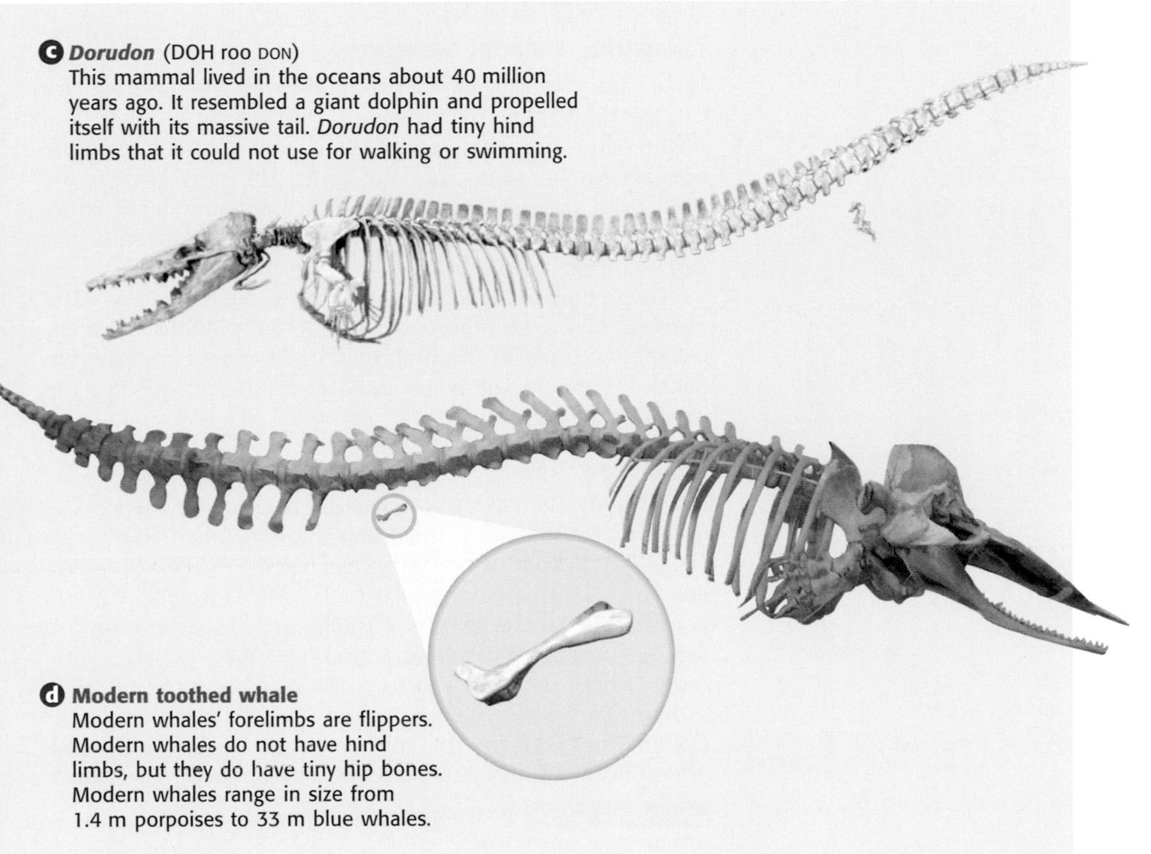

c ***Dorudon*** (DOH roo DON)
This mammal lived in the oceans about 40 million years ago. It resembled a giant dolphin and propelled itself with its massive tail. *Dorudon* had tiny hind limbs that it could not use for walking or swimming.

d **Modern toothed whale**
Modern whales' forelimbs are flippers. Modern whales do not have hind limbs, but they do have tiny hip bones. Modern whales range in size from 1.4 m porpoises to 33 m blue whales.

Answer to Math Practice

length: 24 m × 3.3 ft/m = 79 ft
mass: 99,800 kg × 2.21 lb/kg = 221,000 lb

Note: Final answers should be rounded according to the least number of significant figures used in the starting calculations.

CONNECTION ACTIVITY
Geography — ADVANCED

Inland Whales Explain to students that in 1849, workers constructing a railroad near the town of Charlotte, Vermont, discovered bones that were later identified as those of a beluga whale. Ask students to locate Charlotte on a map and explain why this discovery is so unusual. (Charlotte is more than 150 miles from the nearest ocean.) Ask students what these bones tell us about the history of the land around Charlotte. (It used to be part of an ocean.) Explain that the Champlain Sea, an extension of the ocean, existed for 2,500 years after the last glaciers retreated 12,500 years ago. Recently, many fossils of whale ancestors have been found near the Himalaya Mountains in Pakistan. **LS Visual/Verbal**

INCLUSION *Strategies*

- ***Gifted and Talented***

Some students benefit from exploring a topic in greater depth. Ask these students to research current theories about the evolution of whales from ancient land mammals. Tell them to include the following at each stage: a picture (skeletons or artists' conception), how it walked or moved, what the animal probably ate, what type of an environment it probably lived in, and how it breathed. **LS Visual/Verbal**

Is That a Fact!

Baby blue whales can weigh about 9 tons (9,000 kg or 20,000 lb) at birth and can grow to 190 tons. Blue whales are baleen whales and are now the largest of all whales. The largest among the toothed whales is the sperm whale. The largest sperm whale on record weighed more than 50 tons. Toothed whales include orcas, dolphins, porpoises, narwhals, belugas, beaked whales, and bottle-nosed whales.

A Vestigial Tail Scientists think that the tailbone in humans is a *vestigial structure*. In other words, it is an inherited remnant of the tails of humans' mammal ancestors.

Close

Reteaching — BASIC

Concept Map While prompting students for input, create a large concept map with key ideas from this chapter. **LS Visual**

Quiz — GENERAL

1. Use the words *adaptations, population,* and *evolution* together in a sentence. (Sample answer: Evolution is the process by which a population accumulates inherited adaptations over time.)
2. Why are scientists unsure about some parts of the "tree of life" on Earth? (Sample answer: There are many species to consider and incomplete information. Some parts of Earth's history lack a fossil record because fossils are formed only rarely.)

Alternative Assessment — ADVANCED

Writing **Panda Pedigrees** Scientists study animal skeletons and DNA to determine evolutionary relationships and development because merely looking at the outward appearance of a species can be misleading. The giant panda and red panda illustrate this problem. The two pandas seem similar, but scientists now believe that the red panda is more closely related to raccoons than to the giant panda. Have students investigate and write a report based on recent studies on the classification of these two pandas. **LS Verbal/Logical**

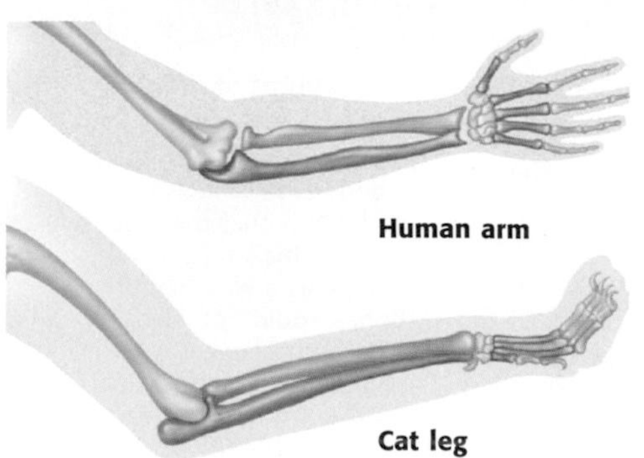

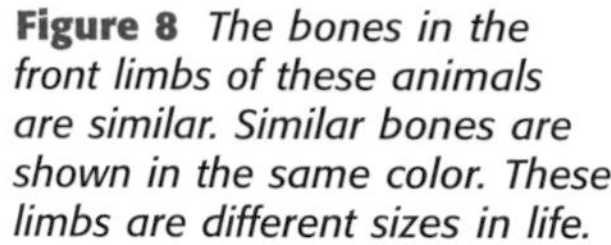

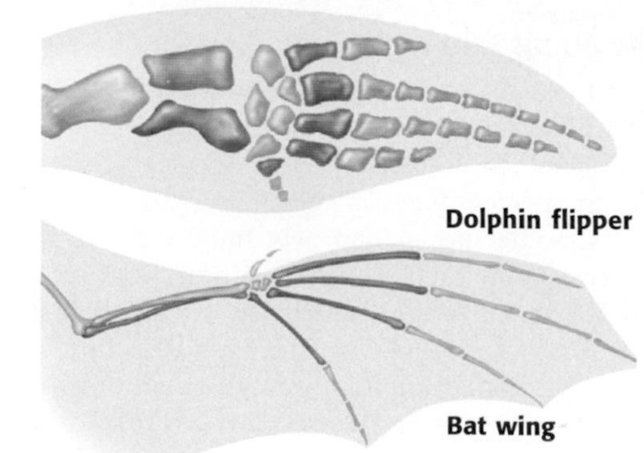

Figure 8 *The bones in the front limbs of these animals are similar. Similar bones are shown in the same color. These limbs are different sizes in life.*

Comparing Organisms

Evidence that groups of organisms have common ancestry can be found by comparing the groups' DNA. Because every organism inherits DNA, every organism inherits the traits determined by DNA. Organisms contain evidence that populations and species undergo changes in traits and DNA over time.

Comparing Skeletal Structures

What does your arm have in common with the front leg of a cat, the front flipper of a dolphin, or the wing of a bat? You might notice that these structures do not look alike and are not used in the same way. But under the surface, there are similarities. Look at **Figure 8.** The structure and order of bones of a human arm are similar to those of the front limbs of a cat, a dolphin, and a bat.

These similarities suggest that cats, dolphins, bats, and humans had a common ancestor. Over millions of years, changes occurred in the limb bones of the ancestor's descendants. Eventually, the bones performed different functions in each type of animal.

Comparing DNA

Interestingly, the DNA of a house cat is similar to the DNA of a tiger. Scientists have learned that traits are inherited through DNA's genetic code. So, scientists can test the following hypothesis: If species that have similar traits evolved from a common ancestor, the species will have similar genetic information. In fact, scientists find that species that have many traits in common do have similarities in their DNA. For example, the DNA of house cats is more similar to the DNA of tigers than to the DNA of dogs. The fact that all existing species have DNA supports the theory that all species share a common ancestor.

Reading Check **If two species have similar DNA, what hypothesis is supported?**

Answer to Reading Check
that they have common ancestry

Is That a Fact!

In the late 1990s, analysis of genetic and hereditary molecular material from a variety of mammals showed that whales share more genetic similarities to hippopotamuses than to any other living mammal group.

SECTION Review

Summary

- Evolution is the process in which inherited characteristics within a population change over generations, sometimes giving rise to new species. Scientists continue to develop theories to explain how evolution happens.
- Evidence that organisms evolve can be found by comparing living organisms to each other and to the fossil record. Such comparisons provide evidence of common ancestry.
- Scientists think that modern whales evolved from an ancient, land-dwelling mammal ancestor. Fossil organisms that support this hypothesis have been found.
- Evidence of common ancestry among living organisms is provided by comparing DNA and inherited traits. Species that have a common ancestor will have traits and DNA that are more similar to each other than to those of distantly related species.

Using Key Terms

Complete each of the following sentences by choosing the correct term from the word bank.

adaptation | species
fossil | evolution

1. Members of the same ___ can mate with one another to produce offspring.

2. A(n) ___ helps an organism survive.

3. When populations change over time, ___ has occurred.

Understanding Key Ideas

4. A human's arm, a cat's front leg, a dolphin's front flipper, and a bat's wing

a. have similar kinds of bones.
b. are used in similar ways.
c. are very similar to insect wings and jellyfish tentacles.
d. have nothing in common.

5. How does the fossil record show that species have changed over time?

6. What evidence do fossils provide about the ancestors of whales?

Critical Thinking

7. Making Comparisons Other than the examples provided in the text, how are whales different from fishes?

8. Forming Hypotheses Is a person's DNA likely to be more similar to the DNA of his or her biological parents or to the DNA of one of his or her cousins? Explain your answer.

Interpreting Graphics

9. The photograph below shows the layers of sedimentary rock exposed during the construction of a road. Imagine that a species that lived 200 million years ago is found in layer **b**. Would the species' ancestor, which lived 250 million years ago, most likely be found in layer **a** or in layer **c**? Explain your answer.

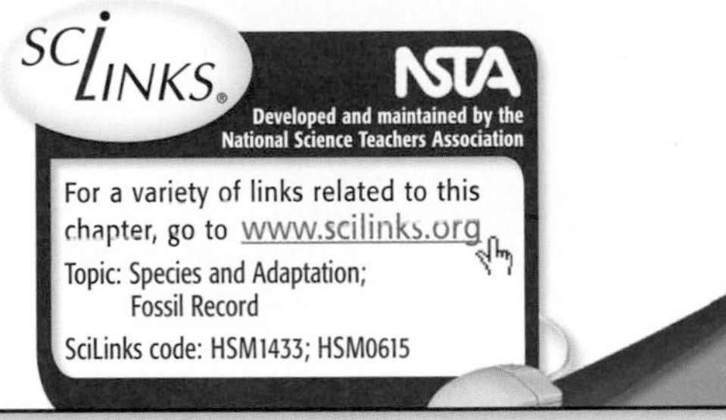

Answers to Section Review

1. species

2. adaptation

3. evolution

4. a

5. Sample answer: Fossils of types of organisms that no longer exist are found. These fossils form a sequence of change and adaptation of populations of organisms over time.

6. Sample answer: There are fossils of four-legged land mammals that share some characteristics with modern whales and other hoofed mammals. Also, there is a sequence of fossil organisms that have characteristics in between those of the ancient fossils and modern whales.

7. Sample answer: Whales breathe air with lungs (not gills) and breathe through a spout that is like a nasal passage. Whales swim in an up-and-down "galloping" motion, as otters do (not side to side as fish do).

8. Sample answer: A person's DNA is likely to be most similar to that of his or her biological parents, because the parents are most closely related to the person by birth.

9. layer c; That layer is under layer b, so layer c was probably deposited earlier than layer b and is thus older.

Homework — GENERAL

Writing **Horse Evolution Report** Have students research to find the four main ancestors of the horse known from the fossil record. (Eohippus, Mesohippus, Merychippus, Pliohippus) Ask students to make a poster that shows each ancestral horse in order of appearance and to write a paragraph about each one, explaining its unique physical characteristics. Students should conclude their reports with an explanation of the origin of wild horses in North America. LS **Verbal**

CHAPTER RESOURCES

Chapter Resource File

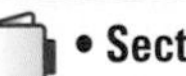

- **Section Quiz** GENERAL
- **Section Review** GENERAL
- **Vocabulary and Section Summary** GENERAL

Technology

Transparencies
- Comparing Skeletal Structures

SECTION 2

Focus

Overview

This section introduces students to Charles Darwin and his famous life history. Students will learn the observations and ideas that helped Darwin formulate his theory of natural selection. Finally, students will connect concepts of genetics to explanations of evolution.

Bellringer

Have students respond to the following prompt: "The following are characteristics that almost all humans have in common: upright walking, hair, fingerprints, binocular vision, and speech. List the advantages and disadvantages of each characteristic." (Accept all reasonable answers.)

Motivate

Discussion — GENERAL

Dinosaurs Ask students to describe a dinosaur. Ask them to explain why there are no dinosaurs alive today. Ask why they think dinosaurs became extinct. (Sample answer: Dinosaurs were well adapted to their environment and lived over 150 million years on Earth. But a catastrophic event changed the environment faster than the dinosaurs could adapt, and they became extinct.) LS **Verbal**

SECTION 2

How Does Evolution Happen?

READING WARM-UP

Objectives

- List four sources of Charles Darwin's ideas about evolution.
- Describe the four parts of Darwin's theory of evolution by natural selection.
- Relate genetics to evolution.

Terms to Learn

trait
selective breeding
natural selection

READING STRATEGY

Brainstorming The key idea of this section is natural selection. Brainstorm words and phrases related to natural selection.

Imagine that you are a scientist in the 1800s. Fossils of some very strange animals have been found. And some familiar fossils have been found where you would least expect them. How did seashells end up on the tops of mountains?

In the 1800s, geologists began to realize that the Earth is much older than anyone had previously thought. Evidence showed that gradual processes had changed the Earth's surface over millions of years. Some scientists saw evidence of evolution in the fossil record. However, no one had been able to explain *how* evolution happens—until Charles Darwin.

Charles Darwin

In 1831, 21-year-old Charles Darwin, shown in **Figure 1,** graduated from college. Like many young people just out of college, Darwin didn't know what he wanted to do with his life. His father wanted him to become a doctor, but seeing blood made Darwin sick. Although he eventually earned a degree in theology, Darwin was most interested in the study of plants and animals.

So, Darwin signed on for a five-year voyage around the world. He served as the *naturalist*—a scientist who studies nature—on the British ship the HMS *Beagle,* similar to the ship in **Figure 2.** During the trip, Darwin made observations that helped him form a theory about how evolution happens.

Figure 1 *Charles Darwin wanted to understand the natural world.*

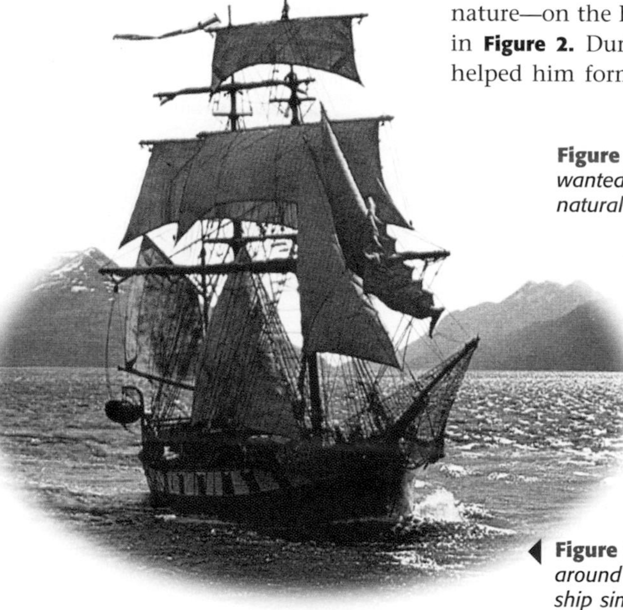

Figure 2 *Darwin sailed around the world on a ship similar to this one.*

CHAPTER RESOURCES

Chapter Resource File

- Lesson Plan
- Directed Reading A  BASIC
- Directed Reading B SPECIAL NEEDS

Technology

 **Transparencies**
- Bellringer

Workbooks

 **Math Skills for Science**
- Multiplying Whole Numbers BASIC

Homework — GENERAL

PORTFOLIO **Poster Project** Have students research the natural history and current status of a specific sea turtle species to find examples for each of the four steps of natural selection. Have them construct a display to present their findings. Require them to include information about the turtle's reproductive habits, physical adaptations, and factors in its environment that affect its success. LS **Verbal/Visual**

Figure 3 *The course of the HMS* Beagle *is shown by the red line. The journey began and ended in England.*

Darwin's Excellent Adventure

The *Beagle*'s journey is charted in **Figure 3.** Along the way, Darwin collected thousands of plant and animal samples. He kept careful notes of his observations. One interesting place that the ship visited was the Galápagos Islands. These islands are found 965 km (600 mi) west of Ecuador, a country in South America.

Reading Check **Where are the Galápagos Islands?** *(See the Appendix for answers to Reading Checks.)*

Darwin's Finches

Darwin noticed that the animals and plants on the Galápagos Islands were a lot like those in Ecuador. However, they were not exactly the same. The finches of the Galápagos Islands, for example, were a little different from the finches in Ecuador. And the finches on each island differed from the finches on the other islands. As **Figure 4** shows, the beak of each finch is adapted to the way the bird usually gets food.

Figure 4 **Some Finches of the Galápagos Islands**

The **large ground finch** has a wide, strong beak that it uses to crack open big, hard seeds. This beak works like a nutcracker.

The **cactus finch** has a tough beak that it uses for eating cactus parts and insects. This beak works like a pair of needle-nose pliers.

The **warbler finch** has a small, narrow beak that it uses to catch small insects. This beak works like a pair of tweezers.

Is That a Fact!

The giant tortoises of the Galápagos Islands weigh up to 270 kg and can live for more than 150 years.

Answer to Reading Check

965 km (600 mi) west of Ecuador

Teach

Demonstration — GENERAL

Form and Function Present and identify to students the following pieces of clothing: sneaker, dress pump, loafer, necktie, scarf, anklet, knee sock, baseball cap, and ski cap. Discuss the ways that all these items are related and how some are more closely related than others. (Related items are best suited for one particular function.) Explain that this relationship of similarities and differences is what scientists such as Charles Darwin observe in studying organisms. **LS Kinesthetic**

CONNECTION ACTIVITY Social Studies — ADVANCED

Writing **Darwin's Voyage** Have interested students research in greater detail Darwin's voyage and similar long-distance travel by explorers in the 1800s. Topics for reports include the types of ships used for travel in that era, the kinds of food eaten by the explorers, and the sophistication and thoroughness of maps in the 1800s. **LS Verbal**

CONNECTION to Geography — GENERAL

Galápagos While displaying a world map or atlas, explain the following: The Galápagos Islands are officially part of the country of Ecuador, though they are 1,000 km west of the mainland. They are a group of 19 volcanically formed islands. Though they have a land area of only 8,000 km^2, they are dispersed over almost 60,000 km^2 of the Pacific Ocean. Biologists now know that unique new species are likely to arise under these conditions. **LS Visual/Logical**

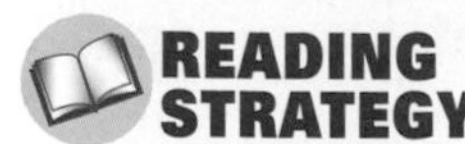

Teach, *continued*

READING STRATEGY — GENERAL

Prediction Guide Before students read this page, have them answer the following questions:

- Why did the finches Darwin saw on the Galápagos Islands look similar to those he saw in South America?
- Why did they look a little different?

Have students share and evaluate their answers with a partner after they read the page.
LS **Verbal/Intrapersonal**

Homework — ADVANCED

Island Biogeography Report *Biogeography* is the study of where animals and plants are found and how they came to live in their particular location. Biogeography uses information from the fossil record and integrates ideas from biology, geology, paleontology, and chemistry. Encourage interested students to write a report about island biogeography. Have them include information about how it is used to design and manage terrestrial wildlife refuges. LS **Verbal/ Intrapersonal**

PORTFOLIO

Darwin's Thinking

After returning to England, Darwin puzzled over the animals of the Galápagos Islands. He tried to explain why the animals seemed so similar to each other yet had so many different adaptations. For example, Darwin hypothesized that the island finches were descended from South American finches. The first finches on the islands may have been blown from South America by a storm. Over many generations, the finches may have adapted to different ways of life on the islands.

During the course of his travels, Darwin came up with many new ideas. Before sharing these ideas, he spent several years analyzing his evidence. He also gathered ideas from many other people.

Ideas About Breeding

trait a genetically determined characteristic

selective breeding the human practice of breeding animals or plants that have certain desired characteristics

In Darwin's time, farmers and breeders had produced many kinds of farm animals and plants. These plants and animals had traits that were desired by the farmers and breeders. **Traits** are specific characteristics that can be passed from parent to offspring through genes. The process in which humans select which plants or animals to reproduce based on certain desired traits is called **selective breeding.** Most pets, such as the dogs in **Figure 5,** have been bred for various desired traits.

You can see the results of selective breeding in many kinds of organisms. For example, people have bred horses that are particularly fast or strong. And farmers have bred crops that produce large fruit or that grow in specific climates.

Figure 5 *Over the past 12,000 years, dogs have been selectively bred to produce more than 150 breeds.*

CONNECTION ACTIVITY
Geography — GENERAL

Bird Barrier Locate the Rocky Mountains on a map. Explain to students that bird-identification guides for North America usually classify birds into those that are east of the Rocky Mountains and those that are west of the Rocky Mountains. Brainstorm reasons why ornithologists might use this system. (Sample answer: because the Rockies present a large geographical barrier.) LS **Visual/Logical**

Is That a Fact!

As a result of selective breeding, the smallest horse is the Falabella, which is only about 76 cm tall. The largest is the Shire, originally bred in England. It can grow more than 1.73 m high at the shoulder and weigh as much as 910 kg.

Population Growth Versus Food Supply

1. Get an **egg carton** and a **bag of rice.** Use a **marker** to label one row of the carton "Food supply." Then, label the second row "Human population."
2. In the row labeled "Food supply," place one grain of rice in the first cup. Place two grains of rice in the second cup, and place three grains of rice in the third cup. In each subsequent cup, place one more grain than you placed in the previous cup. Imagine that each grain represents enough food for one person's lifetime.
3. In the row labeled "Human population," place one grain of rice in the first cup. Place two grains in the second cup, and place four grains in the third cup. In each subsequent cup, place twice as many grains as you placed in the previous cup. This rice represents people.
4. How many units of food are in the sixth cup? How many "people" are in the sixth cup? If this pattern continued, what would happen?
5. Describe how the patterns in the food supply and in the human population differ. Explain how the patterns relate to Malthus's hypothesis.

Ideas About Population

During Darwin's time, Thomas Malthus wrote a famous book entitled *An Essay on the Principle of Population.* Malthus noted that humans have the potential to reproduce rapidly. He warned that food supplies could not support unlimited population growth. **Figure 6** illustrates this relationship. However, Malthus pointed out that human populations are limited by choices that humans make or by problems such as starvation and disease.

After reading Malthus's work, Darwin realized that any species can produce many offspring. He also knew that the populations of all species are limited by starvation, disease, competition, and predation. Only a limited number of individuals survive to reproduce. Thus, there is something special about the survivors. Darwin reasoned that the offspring of the survivors inherit traits that help the offspring survive in their environment.

Malthus's Description of Unlimited Population Growth

Figure 6 *Malthus thought that the human population could increase more quickly than the food supply, with the result that there would not be enough food for everyone.*

Ideas About Earth's History

Darwin had begun to think that species could evolve over time. But most geologists at the time did not think that Earth was old enough to allow for slow changes. Darwin learned new ideas from *Principles of Geology,* a book by Charles Lyell. This book presented evidence that Earth had formed by natural processes over a long period of time. It became clear to Darwin that Earth was much older than anyone had imagined.

Reading Check What did Darwin learn from Charles Lyell?

Is That a Fact!

The full title of Thomas Malthus's famous essay was "An Essay on the Principle of Population, as it Affects the Future Improvement of Society with Remarks on the Speculations of Mr. Godwin, M. Condorcet, and Other Writers."

Science Bloopers

Malthus's work was important in influencing ecological scientists and prompting social planners to consider the potential problems of rapid population growth. However, Malthus was wrong in his projections of the growth of food supplies. The use of machinery, fossil fuels, and chemicals since Malthus's time enabled food production to increase more rapidly than he thought was possible.

MATERIALS

For Each Student

- egg carton, 12-egg size, empty (2)
- marker
- rice (about 1 cup) (or lentils or small pebbles)

Answers

4. There are 32 "people" and 6 units of "food." If this pattern continued, there would be a lot more people than food and not enough food to keep the people alive.
5. The human population is growing much faster than the food supply. This pattern is similar to that in Malthus's prediction.

Using the Figure—Advanced

Two Kinds of Growth Have students examine **Figure 6.** Ask students to describe the behavior of the graph for food supply. (It rises steadily in a straight line.) Explain that the line graphing food supply represents *linear growth* and increases by *adding* a certain amount in each time interval. Next, ask students to describe the behavior of the graph for human population. (It rises quickly and in a curved line.) Explain that the line graphing human population represents *exponential growth* and increases by *multiplying* by a certain percentage in each time interval. Ask students to suggest alternative titles for this graph. (Sample answer: "Human Population Growth Vs. Food Supply.") **LS Visual/Logical**

Answer to Reading Check

that Earth had been formed by natural processes over a long period of time

Close

Reteaching — BASIC

Terms Have students list the key terms and any unfamiliar terms from this chapter. For each term, they should write a definition and then write sample sentences using the term. LS **Verbal**

Quiz — GENERAL

1. Who was Charles Lyell? (He was a British geologist.)
2. What did Darwin learn from Lyell's data about the age of Earth? (Darwin learned from Lyell that Earth was old enough for slow changes to happen in a population.)

Alternative Assessment — ADVANCED

Darwin's Journal Charles Darwin's journals contain notes and records from his travels. Ask students to imagine that they are traveling with Darwin and keeping their own journals. Their notes and drawings should reflect what they see, the questions that arise from their observations, and the hypotheses that they form. Encourage students to write journal entries about other animals on the Galápagos Islands besides the finches, such as the Galápagos tortoise and marine iguanas. LS **Verbal/Intrapersonal**

Answer to Reading Check

On the Origin of Species by Means of Natural Selection

natural selection the process by which individuals that are better adapted to their environment survive and reproduce more successfully than less well adapted individuals do; a theory to explain the mechanism of evolution

Darwin's Theory of Natural Selection

After he returned from his voyage on the HMS *Beagle*, Darwin privately struggled with his ideas for about 20 years. Then, in 1858, Darwin received a letter from a fellow naturalist named Alfred Russel Wallace. Wallace had arrived at the same ideas about evolution that Darwin had. Darwin grew more and more motivated to present his ideas. In 1859, Darwin published a famous book called *On the Origin of Species by Means of Natural Selection*. In his book, Darwin proposed the theory that evolution happens through a process that he called **natural selection.** This process, explained in **Figure 7,** has four parts.

Reading Check What is the title of Darwin's famous book?

Figure 7 Four Parts of Natural Selection

1 Overproduction A tarantula's egg sac may hold 500–1,000 eggs. Some of the eggs will survive and develop into adult spiders. Some will not.

2 Inherited Variation Every individual has its own combination of traits. Each tarantula is similar to, but not identical to, its parents.

3 Struggle to Survive Some tarantulas may be caught by predators, such as this wasp. Other tarantulas may starve or get a disease. Only some of the tarantulas will survive to adulthood.

4 Successful Reproduction The tarantulas that are best adapted to their environment are likely to have many offspring that survive.

ACTIVITY — BASIC

Natural Selection Have students carefully study **Figure 7** and begin to create their own table, concept map, or other graphic organizer about the four parts of natural selection. For each part of the figure, call on several students to restate the meaning in their own words, and then ask students to write their own version of the explanation on their graphic organizer. Finally, for each part, ask students to describe an additional example of the same process with another organism besides a tarantula. LS **Visual**

Science Bloopers

In 1809, French naturalist Jean Baptiste Lamarck's theory of evolution by *acquired characteristics* stated that if an animal changed a body part through use or nonuse, that change would be inherited by its offspring. For example, larger leg muscles as a result of extensive running would be passed on to the next generation. However, genetic studies in the 1930s and 1940s disproved this mechanism for inheriting traits.

Genetics and Evolution

Darwin lacked evidence for parts of his theory. For example, he knew that organisms inherit traits, but not *how* they inherit traits. He knew that there is great variation among organisms, but not *how* that variation occurs. Today, scientists have found most of the evidence that Darwin lacked. They know that variation happens as a result of differences in genes. Changes in genes may happen whenever organisms produce offspring. Some genes make an organism more likely to survive to reproduce. The process called *selection* happens when only organisms that carry these genes can survive to reproduce. New fossil discoveries and new information about genes add to scientists' understanding of natural selection and evolution.

SECTION Review

Summary

- Darwin explained that evolution occurs through natural selection. His theory has four parts:
  1. Each species produces more offspring than will survive to reproduce.
  2. Individuals within a population have slightly different traits.
  3. Individuals within a population compete with each other for limited resources.
  4. Individuals that are better equipped to live in an environment are more likely to survive to reproduce.
- Modern genetics helps explain the theory of natural selection.

Using Key Terms

1. In your own words, write a definition for the term *trait*.
2. Use the following terms in the same sentence: *selective breeding* and *natural selection*.

Understanding Key Ideas

3. Modern scientific explanations of evolution
   a. have replaced Darwin's theory.
   b. rely on genetics instead of natural selection.
   c. fail to explain how traits are inherited.
   d. combine the principles of natural selection and genetic inheritance.
4. Describe the observations that Darwin made about the species on the Galápagos Islands.
5. Summarize the ideas that Darwin developed from books by Malthus and Lyell.
6. Describe the four parts of Darwin's theory of evolution by natural selection.
7. What knowledge did Darwin lack that modern scientists now use to explain evolution?

Math Skills

8. In a sample of 80 beetles, 50 beetles had 4 spots each, and the rest had 6 spots each. What was the average number of spots per beetle?

Critical Thinking

9. **Making Comparisons** In selective breeding, humans influence the course of evolution. What determines the course of evolution in natural selection?
10. **Predicting Consequences** Suppose that an island in the Pacific Ocean was just formed by a volcano. Over the next million years, how might species evolve on this island?

Answers to Section Review

1. Sample answer: A trait is a specific characteristic that is inherited from ancestors.
2. Sample answer: Selective breeding happens when humans choose which organisms will reproduce, and natural selection happens when the environment "chooses."
3. d
4. Darwin observed that many of the species of the Galápagos Islands were similar to those of South America, but they had unique adaptations.
5. From Malthus, Darwin developed the idea that populations are limited by food and other problems. From Lyell, Darwin developed tha idea that conditions on Earth could change slowly over long periods of time.
6. Sample answer: Overproduction means that every organism can produce more offspring than will likely survive; genetic variation means that all offspring will have some differences; struggle to survive means the offspring have to compete with each other and with other organisms around them; and successful reproduction means those that are best adapted will probably have more offpsring like themselves.
7. Sample answer: Darwin lacked the knowledge of genetics that helps explain how organisms inherit traits and why there is so much variety in organisms.
8. $\{(50 \times 4) + [(80 - 50) \times 6]\} \div 80 = 4.75$ spots average per beetle
9. Sample answer: natural forces and other organisms
10. Sample answer: Species on the island might evolve in a way similar to those of the Galápagos Islands.

SCIENTISTS AT ODDS

Rate Debate Evolutionary scientists do not yet agree on how often new species arise. *Gradualism*, the theory that Darwin supported, holds that changes in species occur slowly and steadily over thousands of years. In the 1970s, Stephen Jay Gould and others proposed the theory of *punctuated equilibrium*, which holds that species can remain unchanged for millions of years until dramatic environmental changes prompt speciation.

CHAPTER RESOURCES

Chapter Resource File

- Section Quiz GENERAL
- Section Review GENERAL
- Vocabulary and Section Summary GENERAL
- Reinforcement Worksheet BASIC
- SciLinks Activity GENERAL
- Datasheet for Quick Lab

Technology

Transparencies
- Four Parts of Natural Selection

SECTION 3

Focus

Overview

In this section, students will see examples of natural selection at work. They will relate a species' generation time to its ability to adapt. Students will also examine the process of speciation.

Bellringer

Have students respond to the following prompt: "Write the four parts of natural selection, and create a mnemonic device to remember each part by using the first letter of the words." (Sample answer: Overproduction, genetic Variation, Struggle to survive, successful Reproduction; Olga's Vacation Seemed Relaxing.)

Motivate

Debate — GENERAL

People and Nature During the past several hundred years, a rapidly expanding human population has caused some species to become extinct either from habitat destruction or overhunting. Have students takes sides and debate the following issue: If people are as much a part of the environment as trees and birds, are people's actions just another natural process?

LS Verbal/Intrapersonal

SECTION 3

READING WARM-UP

Objectives

- Give three examples of natural selection in action.
- Outline the process of speciation.

Terms to Learn

generation time
speciation

READING STRATEGY

Prediction Guide Before reading this section, write the title of each heading in this section. Next, under each heading, write what you think you will learn.

Natural Selection in Action

Have you ever had to take an antibiotic? Antibiotics are supposed to kill bacteria. But sometimes, bacteria are not killed by the medicine. Do you know why?

A population of bacteria might develop an adaptation through natural selection. Most bacteria are killed by the chemicals in antibiotics. But in some cases, a few bacteria are naturally *resistant* to the chemicals, so they are not killed. These survivors are then able to pass this adaptation to their offspring. This situation is an example of how natural selection works.

Changes in Populations

The theory of natural selection explains how a population changes in response to its environment. If natural selection is always taking place, a population will tend to be well adapted to its environment. But not all individuals are the same. The individuals that are likely to survive and reproduce are those that are best adapted at the time.

Adaptation to Hunting

Changes in populations are sometimes observed when a new force affects the survival of individuals. In Uganda, scientists think that hunting is affecting the elephant population. In 1930, about 99% of the male elephants in one area had tusks. Only 1% of the elephants were born without tusks. Today, as few as 85% of the male elephants in that area have tusks. What happened?

A male African elephant that has tusks is shown in **Figure 1.** The ivory of an elephant's tusks is very valuable. People hunt the elephants for their tusks. As a result, fewer of the elephants that have tusks survive to reproduce, and more of the tuskless elephants survive. When the tuskless elephants reproduce, they pass the tuskless trait to their offspring.

Figure 1 *The ivory tusks of African elephants are very valuable. Some elephants are born without tusks.*

CHAPTER RESOURCES

Chapter Resource File

- Lesson Plan
- Directed Reading A BASIC
- Directed Reading B SPECIAL NEEDS

Technology

Transparencies
- Bellringer

MISCONCEPTION ALERT

Tuskless Elephants It is important to understand that there were always some tuskless elephants in the wild populations. These animals were naturally tuskless—they were born without tusks and never developed tusks. Because the tuskless elephants are not hunted, they are more likely to pass their traits to future generations.

Figure 2 Natural Selection of Insecticide Resistance

1 An insecticide will kill most insects, but a few may survive. These survivors have genes that make them resistant to the insecticide.

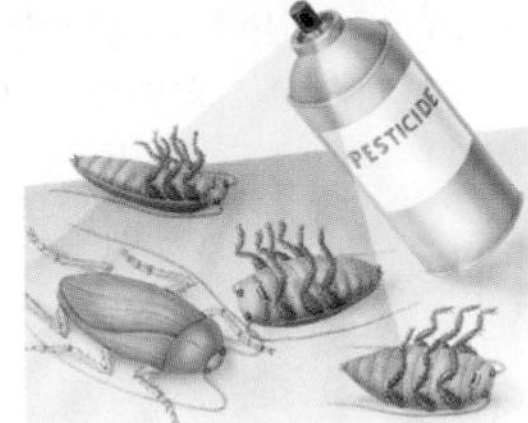

2 The survivors then reproduce, passing the insecticide-resistance genes to their offspring.

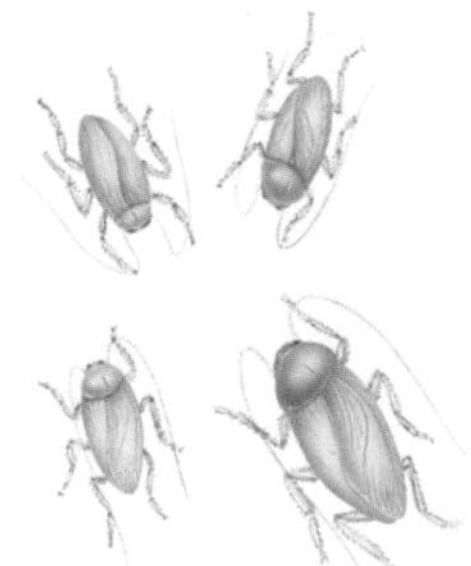

3 In time, the replacement population of insects is made up mostly of individuals that have the insecticide-resistance genes.

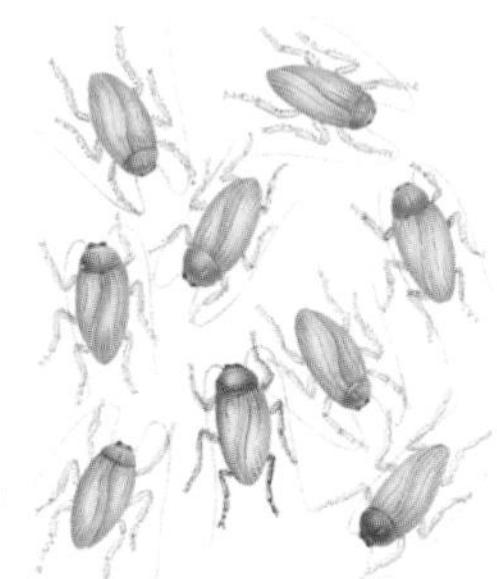

4 When the same kind of insecticide is used on the insects, only a few are killed because most of them are resistant to that insecticide.

Insecticide Resistance

People have always wanted to control the insect populations around their homes and farms. Many insecticides are used to kill insects. But some chemicals that used to work well do not work as well anymore. Some individual insects within the population are resistant to certain insecticides. **Figure 2** shows how a population of insects might become resistant to common insecticides.

More than 500 kinds of insects are now resistant to certain insecticides. Insects can quickly develop resistance because they often produce many offspring and have short generation times. **Generation time** is the average time between one generation of offspring and the next.

generation time the period between the birth of one generation and the birth of the next generation

Reading Check **Why do insects quickly develop resistance to insecticides?** *(See the Appendix for answers to Reading Checks.)*

Competition for Mates

In the process of evolution, survival is simply not enough. Natural selection is at work when individuals reproduce. In organisms that reproduce sexually, finding a mate is part of the struggle to reproduce. Many species have so much competition for mates that interesting adaptations result. For example, the females of many bird species prefer to mate with males that have certain types of colorful feathers.

Teach

CONNECTION ACTIVITY Real World — GENERAL

Natural Pest Control The use of natural predators against insect pests can provide an alternative to chemical insecticides. Ladybugs, for example, which are actually beetles, are used to combat infestations of aphids, whiteflies, fruitworms, mites, broccoli worms, and tomato hornworms. Ladybugs can be purchased in large numbers and released on crops. Many species of bats are voracious hunters of mosquitoes and other small insects. Bat houses are becoming increasingly popular. Encourage students to research methods for encouraging ladybugs, bats, and other insect predators.

LS Kinesthetic/Intrapersonal English Language Learners

Answer to Reading Check

because they often produce many offspring and have short generation times

Group ACTIVITY — ADVANCED

Amazing Adaptations Have groups of students collaborate to create depictions of hypothetical organisms with interesting adaptations for bizarre imaginary environments or conditions, or perhaps other planets. Students should create captions to explain each organism's unique traits. As an alternative, have students research and make posters about interesting cultural practices of people who live in extreme environments such as the Arctic or desert.

LS Visual/Kinesthetic English Language Learners
Co-op Learning

WEIRD SCIENCE

In several species, typical courtship behaviors involve one sex offering food to the other sex. For example, males of the fairy tern fly around the female and offer her a mackerel. In many spider species, males offer an insect carcass to the female. In some spider species, such as the Australian redback spider, the male himself may become a meal!

INTERNET ACTIVITY Essay — GENERAL

For an internet activity related to this chapter, have students go to **go.hrw.com** and type in the keyword **HL5EVOW.**

Close

Reteaching — Basic

Selection Have students place 20 black beans and 20 red beans on a piece of black paper. Call the display *Generation 1.* Tell students that the beans are fish and ask which would most likely be eaten first by the bean shark. Then tell them to add 5 black beans and take away 5 red ones. Call this *Generation 2.* Have students predict and model *Generation 3.* LS **Visual**

Quiz — General

Concept Mapping Construct a concept map that shows how a population of mosquitoes can develop resistance to a pesticide.

Alternative Assessment — Advanced

PORTFOLIO **Species Report** Have each student research and give an oral presentation on how the three steps of speciation (separation, adaptation, and division) worked in providing a particular animal with a distinctive feature. Species of the Galapágos Islands are good examples. LS **Verbal**

Answer to Reading Check

Sample answer: A newly formed canyon, mountain range, or lake could divide the members of a population.

Forming a New Species

Sometimes, drastic changes that can form a new species take place. In the animal kingdom, a *species* is a group of organisms that can mate with each other to produce fertile offspring. A new species may form after a group becomes separated from the original population. This group forms a new population. Over time, the two populations adapt to their different environments. Eventually, the populations can become so different that they can't mate anymore. Each population may then be considered a new species. The formation of a new species as a result of evolution is called **speciation** (SPEE shee AY shuhn). **Figure 3** shows how new species of Galápagos finches may have formed. Speciation may happen in other ways as well.

speciation the formation of new species as a result of evolution

Separation

Speciation often begins when a part of a population becomes separated from the rest. The process of separation can happen in several ways. For example, a newly formed canyon, mountain range, or lake can divide the members of a population.

Reading Check How can parts of a population become separated?

Figure 3 The Evolution of Galápagos Finch Species

1 Some finches left the mainland and reached one of the islands (separation).

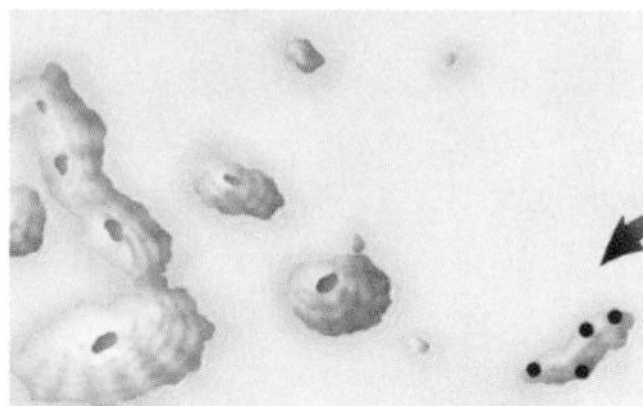

2 The finches reproduced and adapted to the environment (adaptation).

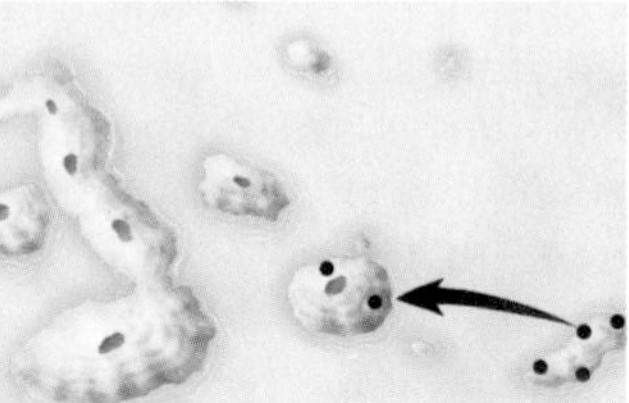

3 Some finches flew to a second island (separation).

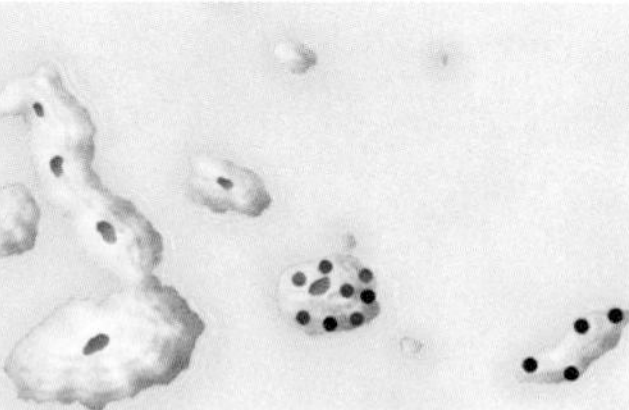

4 The finches reproduced and adapted to the different environment (adaptation).

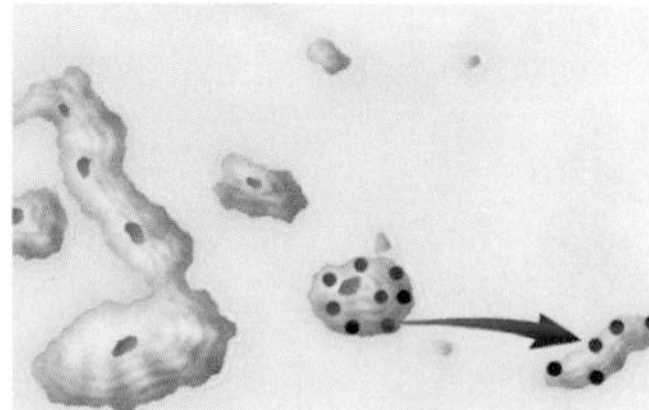

5 Some finches flew back to the first island but could no longer interbreed with the finches there (division).

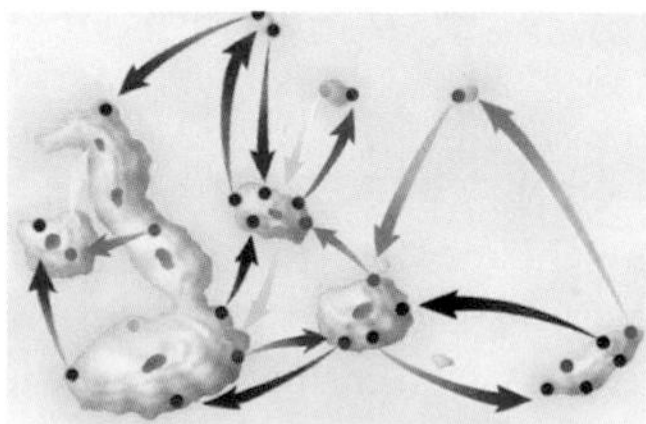

6 This process may have occurred over and over again as the finches flew to the other islands.

INCLUSION *Strategies*

- ***Learning Disabled***
- ***Attention Deficit Disorder***
- ***Behavior Control Issues***

Guide students through a simulation. Trace nine rabbit outlines on a single page, and place an X on five of them. For each student, make one copy on each of three colors of paper. Have each student label his or her rabbits. Prompt students through the following stages:

1. Overproduction: Cut out the rabbits.
2. Genetic variation: Spread the rabbits on your desk. Notice the colors.
3. Struggle to survive: For 1 min, try to take rabbits from each other. Yours are "safe" if you are touching them.
4. Successful reproduction: Look at your surviving rabbits. Suppose only those that have an X will reproduce.

As a class, tally and discuss the results.
LS **Visual**
Co-op Learning
English Language Learners

Adaptation

Populations constantly undergo natural selection. After two groups have separated, natural selection may act on each group in different ways. Over many generations, the separated groups may evolve different sets of traits. If the environmental conditions for each group differ, the adaptations in the groups will also differ.

Division

Over many generations, two separated groups of a population may become very different. Even if a geographical barrier is removed, the groups may not be able to interbreed anymore. At this point, the two groups are no longer the same species.

Figure 4 shows another way that populations may stop interbreeding. Leopard frogs and pickerel frogs probably had the same ancestor species. Then, at some point, some of these frogs began to mate at different times during the year.

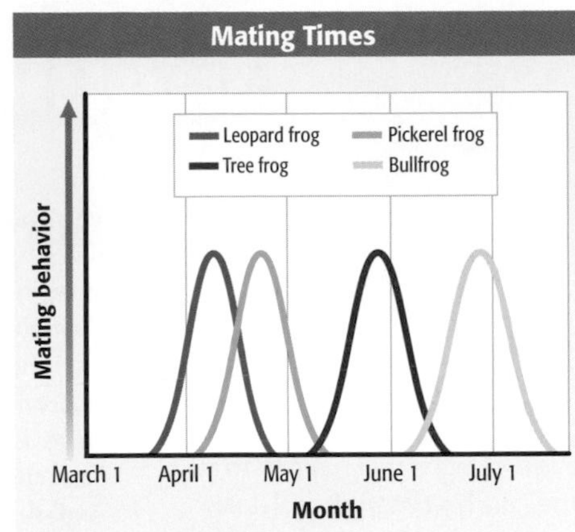

Figure 4 *The leopard frog and the pickerel frog are similar species. However, leopard frogs do not search for mates at the same time of year that pickerel frogs do.*

SECTION Review

Summary

- Natural selection explains how populations adapt to changes in their environment. A variety of examples of such adaptations can be found.
- Natural selection also explains how one species may evolve into another. Speciation occurs as populations undergo separation, adaptation, and division.

Using Key Terms

1. In your own words, write a definition for the term *speciation*.

Understanding Key Ideas

2. Two populations have evolved into two species when
   - **a.** the populations are separated.
   - **b.** the populations look different.
   - **c.** the populations can no longer interbreed.
   - **d.** the populations adapt.
3. Explain why the number of tuskless elephants in Uganda may be increasing.

Math Skills

4. A female cockroach can produce 80 offspring at a time. If half of the offspring produced by a certain female are female and each female produces 80 offspring, how many cockroaches are there in the third generation?

Critical Thinking

5. **Forming Hypotheses** Most kinds of cactus have leaves that grow in the form of spines. The stems or trunks become thick, juicy pads or barrels. Explain how these cactus parts might have evolved.
6. **Making Comparisons** Suggest an organism other than an insect that might evolve an adaptation to human activities.

For a variety of links related to this chapter, go to www.scilinks.org

Topic: Species and Adaptation
SciLinks code: HSM1433

Answers to Section Review

1. Sample answer: Speciation is one species evolving into two.
2. c
3. Sample answer: The number of tuskless elephants is increasing because hunters are killing off the elephants with tusks before they can breed.
4. 128,000 cockroaches
   Calculations:
   First generation:
   80 cockroaches
   Second generation:
   $(80 \div 2) \times 80 =$
   3,200 cockroaches
   Third generation:
   $(3{,}200 \div 2) \times 80 =$
   128,000 cockroaches
5. Sample answer: Cactuses evolved from plants with adaptations to dry conditions, such as spiny leavers that keep animals from eating the plant or thick stems that store water.
6. Sample answer: Rodents might adapt to eating our garbage.

Is That a Fact!

Some species that have adapted to live in total darkness no longer even have eyes! Just as whales have evolved into legless forms, these species have completely adapted to life without light, and some have evolved forms lacking eyes altogether. There are blind cave fish, eels, salamanders, worms, shrimp, crayfish, spiders, beetles, and crickets.

CHAPTER RESOURCES

Chapter Resource File

- Section Quiz GENERAL
- Section Review GENERAL
- Vocabulary and Section Summary GENERAL
- Critical Thinking ADVANCED

Technology

Transparencies
- Evolution of the Galápagos Finches

Inquiry Lab

Survival of the Chocolates

Teacher's Notes

Time Required

One or two 45-minute class periods

Lab Ratings

Teacher Prep 1

Student Set-Up 2

Concept Level 2

Clean Up 1

Safety Caution

Safety concerns will vary with each design.

Preparation Notes

Be prepared for a variety of experimental designs. For example, students may wish to test which color will crack easiest under physical stress or which color will dissolve more quickly in water. This lab is an opportunity to reinforce scientific methods and practice designing experiments. Encourage students to brainstorm a variety of possible hypotheses and ways of testing the hypotheses. Have students identify scientific methods in their experiments.

Using Scientific Methods

Inquiry Lab

Survival of the Chocolates

OBJECTIVES

Form a hypothesis about the fate of the candy-coated chocolates.

Predict what will happen to the candy-coated chocolates.

Design and conduct an experiment to test your hypothesis.

MATERIALS

- chocolates, candy-coated, small, in a variety of colors (about 100)
- items to be determined by the students and approved by the teacher

SAFETY

Imagine a world populated with candy, and hold that delicious thought in your head for just a moment. Try to apply the idea of natural selection to a population of candy-coated chocolates. According to the theory of natural selection, individuals who have favorable adaptations are more likely to survive. In the "species" of candy-coated chocolates you will study in this experiment, the characteristics of individual chocolates may help them "survive." For example, shell strength (the strength of the candy coating) could be an adaptive advantage. Plan an experiment to find out which characteristics of the chocolates are favorable "adaptations."

Ask a Question

1. What might "survival" mean for a candy-coated chocolate? What are some ways you can test which chocolates are the "strongest" or "most fit" for their environment? Also, write down any other questions that you could ask about the "survival" of the chocolates.

Form a Hypothesis

2. Form a hypothesis, and make a prediction. For example, if you chose to study candy color, your prediction might be similar to this: If the ___ colored shell is the strongest, then fewer of the chocolates with this color of shell will ___ when ___.

Karma Houston-Hughes
Kyrene Middle School
Tempe, Arizona

CHAPTER RESOURCES

Chapter Resource File

- Datasheet for Chapter Lab
- Lab Notes and Answers

Technology

Classroom Videos
- Lab Video

Test the Hypothesis

3. Design a procedure to determine which type of candy-coated chocolate is most likely to survive. In your plan, be sure to include materials and tools you may need to complete this procedure.
4. Check your experimental design with your teacher before you begin. Your teacher will supply the candy and assist you in gathering materials and tools.
5. Record your results in a data table. Be sure to organize your data in a clear and understandable way.

Analyze the Results

1. **Describing Events** Write a report that describes your experiment. Be sure to include tables and graphs of the data you collected.

Draw Conclusions

2. **Evaluating Data** In your report, explain how your data either support or do not support your hypothesis. Include possible errors and ways to improve your procedure.

Applying Your Data

Can you think of another characteristic of the chocolates that can be tested to determine which type is best adapted to survive? Explain your idea, and describe how you might test it.

Form a Hypothesis

2. Answers may vary. The example statement is only an example for format. Students may wish to investigate a characteristic other than candy shell hardness. Help them make a prediction about their own experiment. Check that all students have formed testable hypotheses.

Test the Hypothesis

4. Answers may vary. Check that students have planned a controlled experiment and that each factor is accounted for. Also, check that they have planned for all materials they will need.

5. Answers may vary. Students should conduct their own experiment and record all procedures, observations, and results. Students should use data tables to record results where appropriate.

Analyze the Results

1. Reports may vary but should describe all parts of the experiment and present the results with tables, diagrams, or graphs as appropriate.

Draw Conclusions

2. Reports may vary but should include a conclusion that the hypothesis was supported or not. Check that student conclusions are directly related to the hypothesis and were logically drawn from the experimental results.

CHAPTER RESOURCES

Workbooks

Whiz-Bang Demonstrations
- Adaptation Behooves You GENERAL

Long-Term Projects & Research Ideas
- Evolution's Explosion ADVANCED

Chapter Review

Assignment Guide

Section	Questions
1	2–4, 6, 8, 9, 12, 13, 21, 22
2	5, 10, 11, 14, 15, 17, 23
3	1, 7, 16, 18, 19, 20

ANSWERS

Using Key Terms

1. speciation
2. evolution
3. species
4. fossil record
5. selective breeding
6. adaptation
7. generation time

Understanding Key Ideas

8. a
9. b
10. b
11. c

Chapter Review

USING KEY TERMS

Complete each of the following sentences by choosing the correct term from the word bank.

adaptation
evolution
generation time
species
speciation
fossil record
selective breeding
natural selection

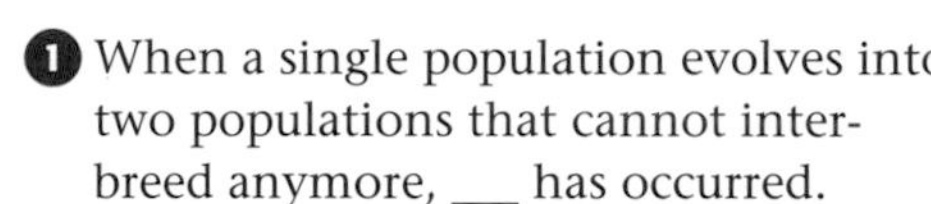

1. When a single population evolves into two populations that cannot interbreed anymore, ___ has occurred.
2. Darwin's theory of ___ explained the process by which organisms become well-adapted to their environment.
3. A group of organisms that can mate with each other to produce offspring is known as a(n) ___.
4. The ___ provides information about organisms that have lived in the past.
5. In ___, humans select organisms with desirable traits that will be passed from one generation to another.
6. A(n) ___ helps an organism survive better in its environment.
7. Populations of insects and bacteria can evolve quickly because they usually have a short ___.

UNDERSTANDING KEY IDEAS

Multiple Choice

8. Fossils are commonly found in
   a. sedimentary rock.
   b. all kinds of rock.
   c. granite.
   d. loose sand.
9. The fact that all organisms have DNA as their genetic material is evidence that
   a. all organisms undergo natural selection.
   b. all organisms may have descended from a common ancestor.
   c. selective breeding takes place every day.
   d. genetic resistance rarely occurs.
10. Charles Darwin puzzled over differences in the ___ of the different species of Galápagos finches.
    a. webbed feet
    b. beaks
    c. bone structure of the wings
    d. eye color
11. Darwin observed variations among individuals within a population, but he did not realize that these variations were caused by
    a. interbreeding.
    b. differences in food.
    c. differences in genes.
    d. selective breeding.

12. Sample answer: Living organisms can be compared in terms of body structures with other living organisms and with organisms from the fossil record. Also, the DNA of living organisms can be compared.

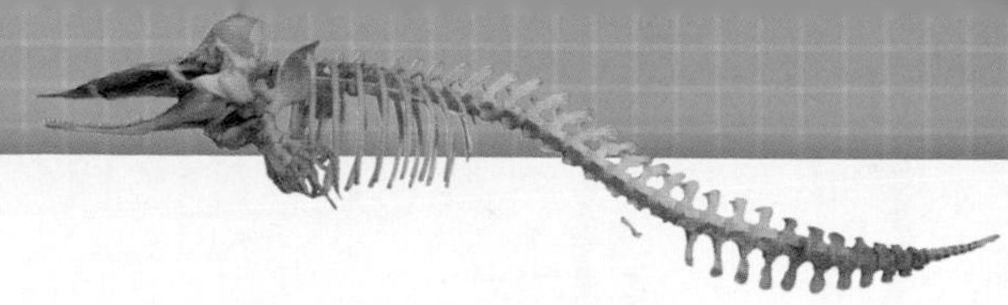

Short Answer

12. Identify two ways that organisms can be compared to provide evidence of evolution from a common ancestor.

13. Describe evidence that supports the hypothesis that whales evolved from land-dwelling mammals.

14. Why are some animals more likely to survive to adulthood than other animals are?

15. Explain how genetics is related to evolution.

16. Outline an example of the process of speciation.

CRITICAL THINKING

17. **Concept Mapping** Use the following terms to create a concept map: *struggle to survive, theory, genetic variation, Darwin, overpopulation, natural selection,* and *successful reproduction.*

18. **Making Inferences** How could natural selection affect the songs that birds sing?

19. **Forming Hypotheses** In Australia, many animals look like mammals from other parts of the world. But most of the mammals in Australia are marsupials, which carry their young in pouches after birth. Few kinds of marsupials are found anywhere else in the world. What is a possible explanation for the presence of so many of these unique mammals in Australia?

20. **Analyzing Relationships** Geologists have evidence that the continents were once a single giant continent. This giant landform eventually split apart, and the individual continents moved to their current positions. What role might this drifting of continents have played in evolution?

INTERPRETING GRAPHICS

The graphs below show information about the infants that are born and the infants that have died in a population. The weight of each infant was measured at birth. Use the graphs to answer the questions that follow.

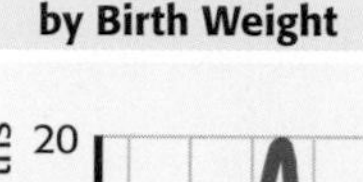

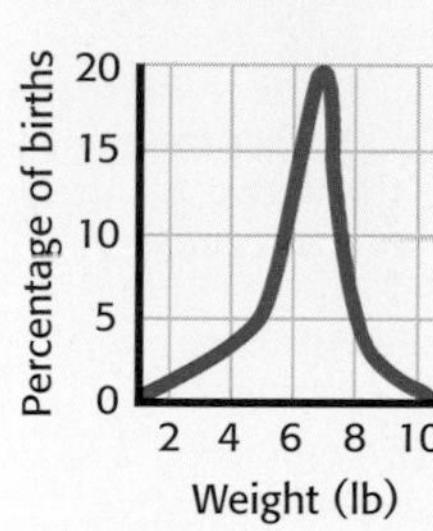

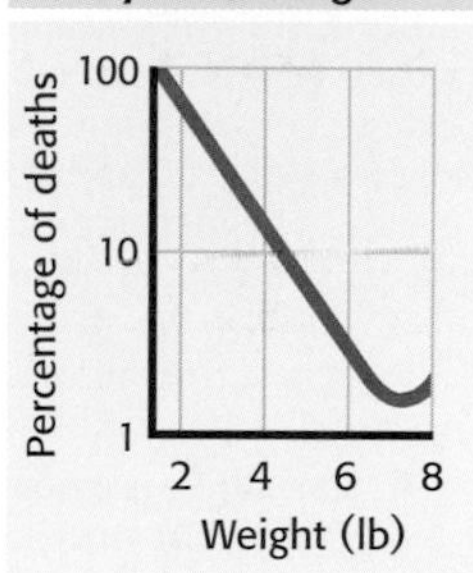

21. What is the most common birth weight?

22. At which birth weight is an infant most likely to survive?

23. How do the principles of natural selection help explain why there are more deaths among babies whose birth weights are low than among babies whose birth weights are average?

13. Sample answer: Whales share many internal similarities with hoofed land mammals. Ancient fossils of four-legged land mammals exist from times when whales did not exist, but some of these fossils shared characteristics with modern whales and other hoofed mammals. A sequence of fossil organisms shows how the characteristics of modern whales could have evolved from those of ancient land mammals.

14. Sample answer: Those animals that are better adapted to the conditions of their environment, including competition with other organisms, are more likely to survive to adulthood.

15. Sample answer: Genetics provides an explanation of what happens inside cells as organisms evolve.

16. Answers may vary. Student answers may resemble the description of the speciation of the Galápagos finches given in the student text.

Critical Thinking

17. 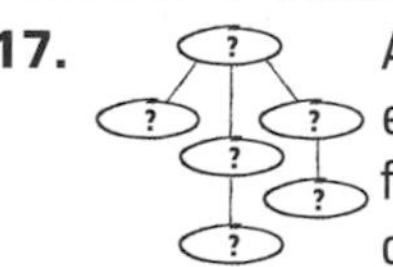

An answer to this exercise can be found at the end of this book.

18. Sample answer: A bird's song could be an advantage if the song helps the bird find mates or food, but the song could be a disadvantage if it attracts predators. So, natural selection would mean that birds whose songs were not an advantage might not survive, and songs would evolve that gave birds some kind of advantage.

19. Sample answer: Australia is an island, so the marsupials there could have evolved separately from other mammals around the world.

20. Sample answer: As the continents drifted apart, populations of species would have been separated and may have had to adapt to new environmental conditions. The separated populations would likely have evolved into separate species over time.

Interpreting Graphics

21. about 7 lb

22. about 7 lb

23. Sample answer: The infants who are best adapted to survive birth are those that weigh about 7 lb at birth.

CHAPTER RESOURCES

Chapter Resource File

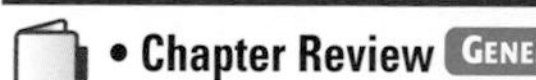

- Chapter Review GENERAL
- Chapter Test A GENERAL
- Chapter Test B ADVANCED
- Chapter Test C SPECIAL NEEDS
- Vocabulary Activity GENERAL

Workbooks

Study Guide

- Assessment resources are also available in Spanish.

Teacher's Note

To provide practice under more realistic testing conditions, give students 20 minutes to answer all of the questions in this Standardized Test Preparation.

MISCONCEPTION ALERT

Answers to the standardized test preparation can help you identify student misconceptions and misunderstandings.

READING

Passage 1

1. A
2. H

Question 1: When a test question asks for a main idea, students should re-read the first and last sentences of each paragraph, and then try to summarize the main idea for themselves. In this case, the combination of the first and last sentences of the passage closely matches answer A. If students are still not sure about the best answer from among the options given, they should skim the passage to be sure that the majority of the sentences relate to the possible main idea.

Standardized Test Preparation

READING

Read each of the passages below. Then, answer the questions that follow each passage.

Passage 1 When the Grand Canyon was forming, a single population of tassel-eared squirrels may have been separated into two groups. Today, descendants of the two groups live on opposite sides of the canyon. The two groups share many characteristics, but they do not look the same. For example, both groups have tasseled ears, but each group has a unique fur color pattern. An important difference between the groups is that the Abert squirrels live on the south rim of the canyon, and the Kaibab squirrels live on the north rim.

The environments on the two sides of the Grand Canyon are different. The north rim is about 370 m higher than the south rim. Almost twice as much precipitation falls on the north rim than on the south rim every year. Over many generations, the two groups of squirrels have adapted to their new environments. Over time, the groups became very different. Many scientists think that the two types of squirrels are no longer the same species. The development of these two squirrel groups is an example of speciation in progress.

1. Which of the following statements **best** describes the main idea of this passage?
   - **A** Speciation is evident in two groups of squirrels in the Grand Canyon area.
   - **B** Two groups of squirrels in the Grand Canyon area are closely related.
   - **C** Two species can form from one species. This process is called *speciation*.
   - **D** There are two groups of squirrels because the Grand Canyon has two sides.

2. Which of the following statements about the two types of squirrels is true?
   - **F** They look the same.
   - **G** They live in similar environments.
   - **H** They have tasseled ears.
   - **I** They can interbreed with each other.

Passage 2 You know from experience that individuals in a population are not exactly the same. If you look around the room, you will see a lot of differences among your classmates. You may have even noticed that no two dogs or two cats are exactly the same. No two individuals have exactly the same adaptations. For example, one cat may be better at catching mice, and another is better at running away from dogs. Observations such as these form the basis of the theory of natural selection. Because adaptations help organisms survive to reproduce, the individuals that are better adapted to their environment are more likely to pass their traits to future generations.

1. In the passage, what does *population* mean?
   - **A** a school
   - **B** some cats and dogs
   - **C** a group of the same type of organism
   - **D** a group of individuals that are the same

2. In this passage, which of the following are given as examples of adaptations?
   - **F** differences among classmates
   - **G** differences among cats
   - **H** differences between cats and dogs
   - **I** differences among environments

3. Which of the following statements about the individuals in a population that survive to reproduce is true?
   - **A** They have the same adaptations.
   - **B** They are likely to pass on adaptations to the next generation.
   - **C** They form the basis of the theory of natural selection.
   - **D** They are always better hunters.

Passage 2

1. C
2. G
3. B

Question 1: The question asks for an example of a concept that can be found in the passage. A good strategy for the reader is to skim the paragraph for the use of the word *adaptation* and then read the surrounding sentences. In this case, the sentence that begins "For example," clearly offers examples of adaptations. The remaining task is to find the answer (G) that best paraphrases or summarizes the types of examples in that sentence: differences among cats.

INTERPRETING GRAPHICS

The graph below shows average beak sizes of a group of finches on one island over several years. Use the graph to answer the questions that follow.

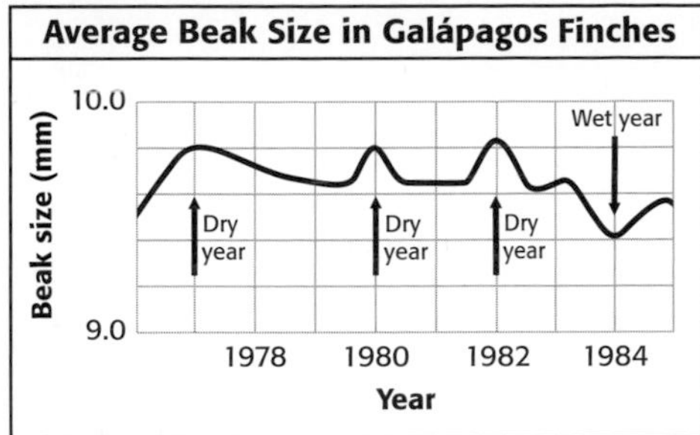

1. In which of the years studied was average beak size the largest?
   A 1977
   B 1980
   C 1982
   D 1984

2. If beak size in this group of birds is linked to the amount of rainfall, what can you infer about the year 1976 on this island?
   F The year 1976 was drier than 1977.
   G The year 1976 was drier than 1980.
   H The year 1976 was wetter than 1977.
   I The year 1976 was wetter than 1984.

3. During which year(s) was rainfall probably the lowest on the island?
   A 1978, 1980, and 1982
   B 1977, 1980, 1982, and 1984
   C 1982
   D 1984

4. Which of the following statements **best** summarizes this data?
   F Average beak size stayed about the same except during wet years.
   G Average beak size decreased during dry years and increased during wet years.
   H Average beak size increased during dry years and decreased during wet years.
   I Average beak size changed randomly.

MATH

Read each question below, and choose the best answer.

Average Beak Measurements of Birds of the Colores Islands

Island	Average beak length (mm)	Average beak width (mm)	Number of unique species
Verde	9.7	6.5	5
Azul	8.9	8.7	15
Rosa	5.2	8.0	10

1. What is the ratio of the number of species on Verde Island to the total number of species on all three of the Colores Islands?
   A 1:2
   B 1:5
   C 1:6
   D 5:15

2. What percentage of all bird species on the Colores Islands are on Rosa Island?
   F approximately 15%
   G approximately 30%
   H approximately 50%
   I approximately 80%

3. On which of the islands is the ratio of average beak length to average beak width closest to 1:1?
   A Verde Island
   B Azul Island
   C Rosa Island
   D There is not enough information to determine the answer.

4. On which island does the bird with the smallest beak length live?
   F Verde Island
   G Azul Island
   H Rosa Island
   I There is not enough information to determine the answer.

Standardized Test Preparation

INTERPRETING GRAPHICS

1. C
2. H
3. C
4. H

 TEST DOCTOR

Question 1: This question simply requires understanding the labels on the graph. Students who miss this question may have difficulty understanding titles and axis labels.

Question 2: This question asks about a year (1976) that is not directly labeled, so the student needs to count two spaces backward from 1978 in order to determine which part of the data is from 1976. Then, the student needs to infer from the trends in the graph that 1976 was wetter than 1977.

MATH

1. C
2. G
3. B
4. I

 TEST DOCTOR

Question 1: The question asks for the ratio of the number in one group to the number in all groups combined. Thus, the ratio is Verde:all or 5:30, which simplifies to 1:6.

Question 3: Because the question asks for a ratio that is close to 1:1, the best strategy is to scan the columns of beak length and beak width, looking for the pair of numbers that are closest in value (Azul Island).

Question 4: For this question, students must recall that an *average* (or mean) does not necessarily indicate the *range* of values that may have been sampled. Thus, there is no way of knowing from the data given in the table what the smallest or largest beak sizes were for each island's set of measurements.

CHAPTER RESOURCES

Chapter Resource File

 • **Standardized Test Preparation** GENERAL

State Resources

 For specific resources for your state, visit **go.hrw.com** and type in the keyword **HSMSTR.**

Science in Action

Science, Technology, and Society

Teaching Strategy- GENERAL

To help students understand changes in agriculture, create a table that contrasts traditional and industrial farming practices. Under "Traditional," list phrases such as *smaller scale, few machines, manual labor, more plant varieties,* and *mainly for sustenance.* Under "Industrial," list phrases such as *larger scale, more mechanized, fewer plant varieties,* and *primarily for profit.* Point out that a shrinking number of farmers practice some form of traditional agriculture. Scientists at seed banks try to obtain samples of traditionally farmed seeds.

Science Fiction

Background

About the Author Scott Sanders (1945–) writes many different kinds of stories. Early in life, he chose to become a writer rather than a scientist, although he still has an interest science. Sanders has written about folklore, physics, the naturalist John James Audubon, and settlers of Indiana. Much of his work is nonfiction. His writing has been published in books and periodicals, such as the *Chicago Sun-Times, Harper's,* and *Omni.*

HOLT ANTHOLOGY OF Science Fiction

HOLT, RINEHART AND WINSTON

Science, Technology, and Society

Seed Banks

All over the world, scientists are making deposits in a special kind of bank. These banks are not for money, but for seeds. Why should seeds be saved? Saving seeds saves plants that may someday save human lives. These plants could provide food or medicine in the future. Throughout human history, many medicines have been developed from plants. And scientists keep searching for new chemicals among the incredible variety of plants in the world. But time is running out. Many plant species are becoming extinct before they have even been studied.

Math ACTIVITY

Many drugs were originally developed from plants. Suppose that 100 plants are used for medicines this year, but 5% of plant species become extinct each year. How many of the medicinal plants would be left after 1 year? after 10 years? Round your answers to whole numbers.

Science Fiction

"The Anatomy Lesson" by Scott Sanders

Do you know the feeling you get when you have an important test? A medical student faces a similar situation in this story. The student needs to learn the bones of the human body for an anatomy exam the next day. The student goes to the anatomy library to study. The librarian lets him check out a box of bones that are supposed to be from a human skeleton. But something is wrong. There are too many bones. They are the wrong shape. They don't fit together correctly. Somebody must be playing a joke! Find out what's going on and why the student and the librarian will never be the same after "The Anatomy Lesson." You can read it in the *Holt Anthology of Science Fiction.*

Language Arts ACTIVITY

WRITING SKILL Before you read this story, predict what you think will happen. Write a paragraph that "gives away" the ending that you predict. After you have read the story, listen to some of the predictions made by your classmates. Discuss your opinions about the possible endings.

Answer to Math Activity

year	number of plants
0	100
1	95
5	77
10	60

Answer to Language Arts Activity

Student paragraphs may vary. Have students compare their predictions to the story's ending and to other students' predictions.

People in Science

Raymond Pierotti

Canine Evolution Raymond Pierotti thinks that it's natural that he became an evolutionary biologist. He grew up exploring the desert around his home in New Mexico. He was fascinated by the abundant wildlife surviving in the bleak landscape. "One of my earliest memories is getting coyotes to sing with me from my backyard," he says.

Pierotti now studies the evolutionary relationships between wolves, coyotes, and domestic dogs. Some of his ideas come from the traditions of the Comanches. According to the Comanche creation story, humans came from wolves. Although Pierotti doesn't believe that humans evolved from wolves, he sees the creation story as a suggestion that humans and wolves have evolved together. "Wolves are very similar to humans in many ways," says Pierotti. "They live in family groups and hunt together. It is possible that wolves actually taught humans how to hunt in packs, and there are ancient stories of wolves and humans hunting together and sharing the food. I think it was this relationship that inspired the Comanche creation stories."

Social Studies ACTIVITY

WRITING SKILL Research a story of creation that comes from a Greek, Roman, or Native American civilization. Write a paragraph summarizing the myth, and share it with a classmate.

To learn more about these Science in Action topics, visit **go.hrw.com** and type in the keyword **HL5EVOF.**

Current Science

Check out Current Science® articles related to this chapter by visiting go.hrw.com. Just type in the keyword HL5CS07.

People in Science

ACTIVITY — GENERAL

Have every student write or present a report on a breed of dog. The report should focus on the origin and evolution of the breed, with particular attention paid to the culture that bred it and why those characteristics were chosen. The report could also explore whether these breeds make good household pets and why. Students can easily find information on dog breeds on the Internet by searching for either the name of a breed or for "dog breeds" and visiting any of several sites that collect information on different breeds.

Answer to Social Studies Activity

Student summaries may vary. Have students share their summaries with each other or with the entire class, and then discuss similarities.

The History of Life on Earth
Chapter Planning Guide

Compression guide:
To shorten instruction because of time limitations, omit Section 3.

OBJECTIVES	LABS, DEMONSTRATIONS, AND ACTIVITIES	TECHNOLOGY RESOURCES
PACING • 90 min pp. 134–141 **Chapter Opener**	**SE Start-up Activity,** p. 135 ◆ GENERAL	**OSP Parent Letter** ■ GENERAL **CD Student Edition on CD-ROM** **CD Guided Reading Audio CD** ■ **TR Chapter Starter Transparency*** **VID Brain Food Video Quiz**
Section 1 Evidence of the Past • Explain how fossils can be formed and how their age can be estimated. • Describe the geologic time scale and the way that scientists use it. • Compare two ways that conditions for life on Earth have changed over time.	**TE Activity** Newspaper Layers, p. 137 ◆ GENERAL **SE Connection to Social Studies** A Place in Time, p. 138 ◆ GENERAL **TE Group Activity** Detailed Geologic Timeline, p. 138 ADVANCED **SE Quick Lab** Making a Geologic Timeline, p. 139 ◆ GENERAL **CRF Datasheet for Quick Lab*** **TE Activity** Rock Collectors, p. 139 GENERAL **SE Skills Practice Lab** The Half-Life of Pennies, p. 189 ◆ GENERAL **CRF Datasheet for LabBook***	**CRF Lesson Plans*** **TR Bellringer Transparency*** TR Using Half-Lives to Date Fossils* TR The Geologic Time Scale* TR Moving Continents and Tectonic Plates* TR ***LINK TO EARTH SCIENCE*** The South American Plate*
PACING • 45 min pp. 142–147 **Section 2 Eras of the Geologic Time Scale** • Outline the major developments that allowed life to exist on Earth. • Describe the types of organisms that arose during the four major divisions of the geologic time scale.	**TE Connection Activity** Earth Science, p. 143 ◆ GENERAL **TE Activity** Using Maps, p. 143 ◆ GENERAL **TE Group Activity** Ancient Plants, p. 144 GENERAL **TE Connection Activity** Real World, p. 144 ◆ GENERAL **LB Long-Term Projects & Research Ideas** A Horse is a Horse* ADVANCED	**CRF Lesson Plans*** **TR Bellringer Transparency*** **SE Internet Activity,** p. 143 GENERAL **CRF SciLinks Activity*** GENERAL **CD Interactive Explorations CD-ROM** Rock On! GENERAL
PACING • 90 min pp. 148–153 **Section 3 Humans and Other Primates** • Describe two characteristics that all primates share. • Describe three major groups of hominids.	**TE Activity** Exploring Vision, p. 149 GENERAL **TE Activity** Primate Characteristics, p. 150 ◆ BASIC **TE Group Activity** Comparing Hominids, p. 150 ◆ GENERAL **SE School-to-Home Activity** Thumb Through This, p. 151 GENERAL **TE Connection Activity** Art, p. 151 ◆ GENERAL **TE Activity** Classifying Primates, p. 153 ADVANCED **SE Inquiry Lab** Mystery Footprints, p. 154 ◆ GENERAL **CRF Datasheet for LabBook*** **SE Science in Action** Math, Social Studies, and Language Arts Activities, pp. 160–161 GENERAL	**CRF Lesson Plans*** **TR Bellringer Transparency*** TR Comparison of Primate Skeletons* **VID Lab Videos for Life Science**

PACING • 90 min

CHAPTER REVIEW, ASSESSMENT, AND STANDARDIZED TEST PREPARATION

CRF Vocabulary Activity* GENERAL
SE Chapter Review, pp. 156–157 GENERAL
CRF Chapter Review* ■ GENERAL
CRF Chapter Tests A* ■ GENERAL, **B*** ADVANCED, **C*** SPECIAL NEEDS
SE Standardized Test Preparation, pp. 158–159 GENERAL
CRF Standardized Test Preparation* GENERAL
CRF Performance-Based Assessment* GENERAL
OSP Test Generator GENERAL
CRF Test Item Listing* GENERAL

Online and Technology Resources

Visit **go.hrw.com** for a variety of free resources related to this textbook. Enter the keyword **HL5HIS.**

Students can access interactive problem-solving help and active visual concept development with the *Holt Science and Technology* Online Edition available at **www.hrw.com.**

Guided Reading Audio CD

A direct reading of each chapter using instructional visuals as guideposts. For auditory learners, reluctant readers, and Spanish-speaking students. Available in English and Spanish.

KEY

SE Student Edition	CRF Chapter Resource File	SS Science Skills Worksheets	* Also on One-Stop Planner
TE Teacher Edition	OSP One-Stop Planner	MS Math Skills for Science Worksheets	◆ Requires advance prep
	LB Lab Bank	CD CD or CD-ROM	■ Also available in Spanish
	TR Transparencies	VID Classroom Video/DVD	

SKILLS DEVELOPMENT RESOURCES	SECTION REVIEW AND ASSESSMENT	STANDARDS CORRELATIONS
SE **Pre-Reading Activity,** p. 134 GENERAL OSP **Science Puzzlers, Twisters & Teasers** GENERAL		National Science Education Standards UCP 2, 3, SAI 1, 2; SPSP 5; HNS 3
CRF **Directed Reading A*** ■ BASIC, **B*** SPECIAL NEEDS CRF **Vocabulary and Section Summary*** ■ GENERAL SE **Reading Strategy** Reading Organizer, p. 136 GENERAL SE **Math Practice** Fractions of Fractions, p. 137 GENERAL SE **Connection to Geology** Mid-Atlantic Ridge, p. 141 GENERAL TE **Inclusion Strategies,** p. 140 MS **Math Skills for Science** Radioactive Decay and the Half-Life* GENERAL MS **Math Skills for Science** Geologic Time Scale* GENERAL CRF **Reinforcement Worksheet** Earth Timeline* BASIC	SE **Reading Checks,** pp. 137, 139, 140 GENERAL TE **Homework,** p. 137 GENERAL TE **Homework,** p. 139 GENERAL TE **Reteaching,** p. 140 BASIC TE **Quiz,** p. 140 GENERAL TE **Alternative Assessment,** p. 141 GENERAL SE **Section Review,*** p. 141 ■ GENERAL CRF **Section Quiz*** ■ GENERAL	UCP 1, 2, 4; SAI 1, 2; HNS 1, 2, 3; LS 1a, 3d, 5b, 5c; *LabBook:* UCP 1, 3; SAI 1; LS 5c
CRF **Directed Reading A*** ■ BASIC, **B*** SPECIAL NEEDS CRF **Vocabulary and Section Summary*** ■ GENERAL SE **Reading Strategy** Mnemonics, p. 142 GENERAL SE **Connection to Oceanography** Prehistoric Marine Organisms, p. 144 GENERAL TE **Inclusion Strategies,** p. 144 SE **Math Focus** Relative Scale, p. 146 GENERAL TE **Connection to Math** Another Time Scale, p. 146 ADVANCED MS **Math Skills for Science** Subtraction Review* GENERAL CRF **Reinforcement Worksheet** Condensed History* GENERAL	SE **Reading Checks,** pp. 142, 145, 146 GENERAL TE **Reteaching,** p. 146 BASIC TE **Quiz,** p. 146 GENERAL TE **Alternative Assessment,** p. 146 GENERAL SE **Section Review,*** p. 147 ■ GENERAL CRF **Section Quiz*** ■	UCP 1, 2, 3, 4; SAI 1; LS 1a, 1b, 3d, 5a, 5b, 5c
CRF **Directed Reading A*** BASIC, **B*** SPECIAL NEEDS CRF **Vocabulary and Section Summary*** ■ GENERAL SE **Reading Strategy** Discussion, p. 148 GENERAL CRF **Critical Thinking** Fossil Revelations* ADVANCED	SE **Reading Checks,** pp. 149, 150, 153 GENERAL TE **Reteaching,** p. 152 BASIC TE **Quiz,** p. 152 GENERAL TE **Alternative Assessment,** p. 152 GENERAL TE **Homework,** p. 152 GENERAL SE **Section Review,*** p. 153 ■ GENERAL CRF **Section Quiz*** ■ GENERAL	UCP 2, 4, 5; SAI 1, 2; ST 1, 2; HNS 2, 3; LS 1a, 3d, 5a, 5b, 5c; *Chapter Lab:* UCP 2, 5; SAI 1; HNS 2

This convenient CD-ROM includes:
- **Lab Materials QuickList Software**
- **Holt Calendar Planner**
- **Customizable Lesson Plans**
- **Printable Worksheets**
- **ExamView® Test Generator**

CNN Student News

cnnstudentnews.com

Find the latest news, lesson plans, and activities related to important scientific events.

www.scilinks.org

Maintained by the **National Science Teachers Association.** See Chapter Enrichment pages for a complete list of topics.

Current Science®

Check out ***Current Science*** articles and activities by visiting the HRW Web site at **go.hrw.com.** Just type in the keyword **HL5CG0T.**

Classroom Videos
- **Lab Videos** demonstrate the chapter lab.
- **Brain Food Video Quizzes** help students review the chapter material.

Chapter Resources

Visual Resources

CHAPTER STARTER TRANSPARENCY

The History of Life on Earth — CHAPTER STARTER

Imagine . . .

One day you and your friends learn about a secret underground passage that leads into an old abandoned mansion, and you set out to find it. As you walk around a field in search of the passage, you stumble across a large hole in the ground. Could this be it?

One by one, you and your friends squeeze down into the hole. You slide down the sloping tunnel and finally land. Dusting yourself off, you turn on your flashlight. Instead of finding a passage to an abandoned mansion, you and your friends find yourselves in an immense cavern. Painted high on the cavern's walls are pictures of bulls, cows, horses, and stags. You get the feeling these images have been here for a very long time. Who made these paintings?

This adventure actually occurred in southern France in the late 1940s. Four teenage boys went hunting for a passageway to the old manor of Lascaux. Instead of finding a passageway, they stumbled onto a 17,000-year-old connection with our ancestors, the Cro-Magnons. Three of the adventurers are shown below talking to their teacher.

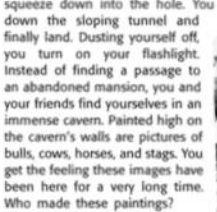

BELLRINGER TRANSPARENCIES

The History of Life on Earth — BELLRINGER TRANSPARENCY

Section: Evidence of the Past

Imagine that you haven't cleaned your room for 30 years and you finally decide to sort through the 2 m pile of stuff on your floor. What might you find on the top of the pile? in the middle? on the bottom?

Write your responses in your **science journal.**

Section: Eras of the Geologic Time Scale

Suppose that electrical energy had never been developed. How would your life differ from what it is like now? What do you do every day that requires electricity?

Write your answers in your **science journal.**

TEACHING TRANSPARENCIES

The History of Life on Earth — TEACHING TRANSPARENCY

Using Half-Lives to Date Fossils

A The unstable atoms in this sample of rock have a half-life of 1.3 billion years. The sample contained 4 mg of unstable atoms when it formed.

B After 1.3 billion years, (one half-life for this type of unstable atom), 2 mg of the unstable atoms have decayed to become stable atoms, and 2 mg of unstable atoms remain.

C After 2.6 billion years (two half-lives for this sample), the rock sample contains 3 mg of stable decay atoms and 1 mg of unstable atoms.

D After three half-lives, only 0.5 mg of unstable atoms remain in the rock sample. This is equal to one-eighth of the original amount.

The History of Life on Earth — TEACHING TRANSPARENCY

Moving Continents and Tectonic Plates

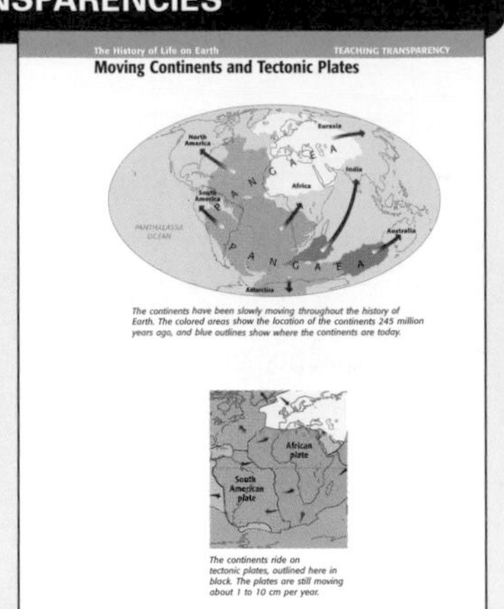

The continents have been slowly moving throughout the history of Earth. The colored areas show the location of the continents 245 million years ago, and blue outlines show where the continents are today.

The continents ride on tectonic plates, outlined here in black. The plates are still moving about 1 to 10 cm per year.

TEACHING TRANSPARENCIES

The History of Life on Earth — TEACHING TRANSPARENCY

The Geologic Time Scale

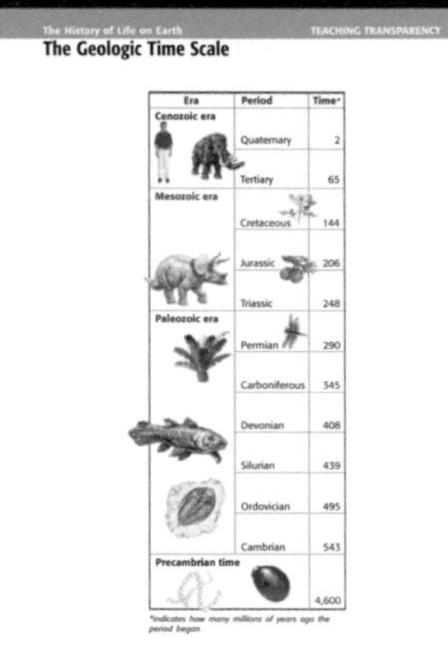

The History of Life on Earth — TEACHING TRANSPARENCY

Comparison of Primate Skeletons

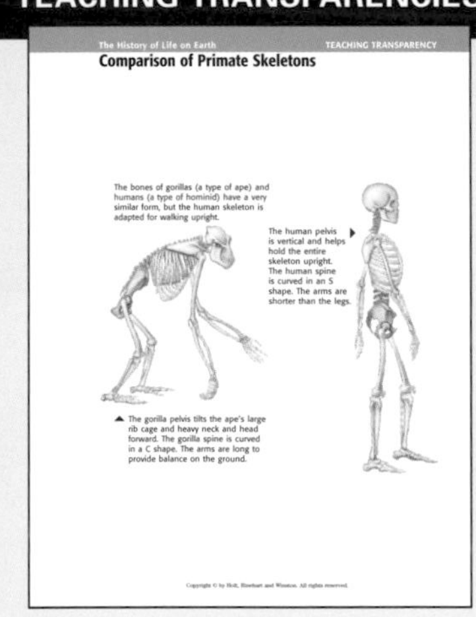

The Tectonic Plates; Close-up of a Tectonic Plate

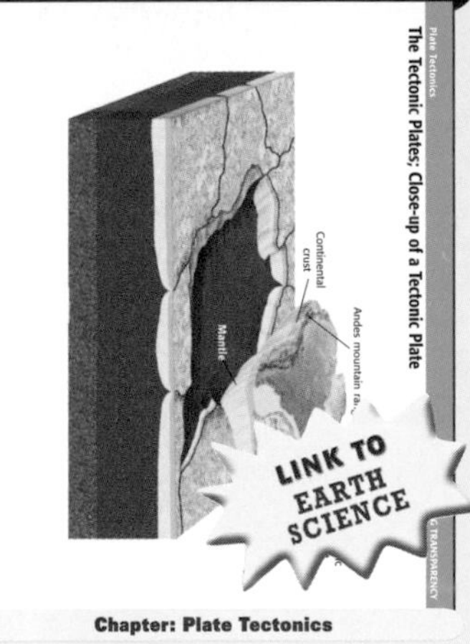

Chapter: Plate Tectonics

CONCEPT MAPPING TRANSPARENCY

The History of Life on Earth — CONCEPT MAPPING

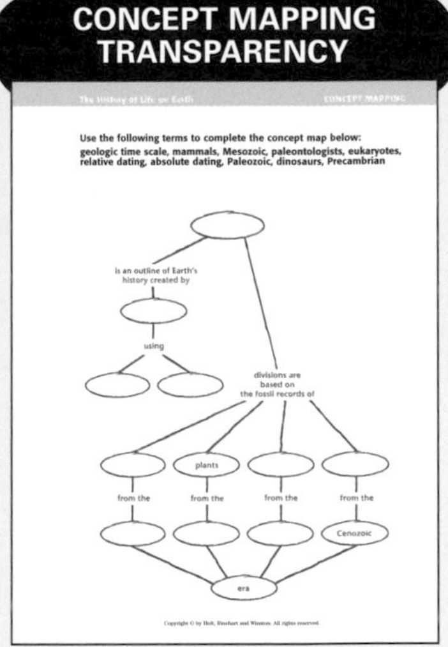

Planning Resources

LESSON PLANS

Lesson Plan SAMPLE

Section: Waves

Pacing

Regular Schedule: with lab(s):2 days without lab(s)2 days

Block Schedule: with lab(s):1 1/2 days without lab(s)1 day

Objectives

1. Relate the seven properties of life to a living organism.
2. Describe seven themes that can help you to organize what you learn about biology.
3. Identify the tiny structures that make up all living organisms.
4. Differentiate between reproduction and heredity and between metabolism and homeostasis.

National Science Education Standards Covered

LSInter6: Cells have particular structures that underlie their functions.

LSMat1: Most cell functions involve chemical reactions.

LSBeh1: Cells store and use information to guide their functions.

UCP1: Cell functions are regulated.

SI1: Cells can differentiate and form complete multicellular organisms.

PS1: Species evolve over time.

ESS1: The great diversity of organisms is the result of more than 3.5 billion years of evolution.

ESS2: Natural selection and its evolutionary consequences provide a scientific explanation for the fossil record of ancient life forms as well as for the striking molecular similarities observed among the diverse species of living organisms.

ST1: The millions of different species of plants, animals, and microorganisms that live on Earth today are related by descent from common ancestors.

ST2: The energy for life primarily comes from the sun.

SPSP1: The complexity and organization of organisms accommodates the need for obtaining, transforming, transporting, releasing, and eliminating the matter and energy used to sustain the organism.

SPSP6: As matter and energy flows through different levels of organization of living systems—cells, organs, communities—and between living systems and the physical environment, chemical elements are recombined in different ways.

HNS1: Organisms have behavioral responses to internal changes and to external stimuli.

PARENT LETTER

SAMPLE

Dear Parent,

Your son's or daughter's science class will soon begin exploring the chapter entitled "The World of Physical Science." In this chapter, students will learn about how the scientific method applies to the world of physical science and the role of physical science in the world. By the end of the chapter, students should demonstrate a clear understanding of the chapter's main ideas and be able to discuss the following topics:

1. physical science as the study of energy and matter (Section 1)
2. the role of physical science in the world around them (Section 1)
3. careers that rely on physical science (Section 1)
4. the steps used in the scientific method (Section 2)
5. examples of technology (Section 2)
6. how the scientific method is used to answer questions and solve problems (Section 2)
7. how our knowledge of science changes over time (Section 2)
8. how models represent real objects or systems (Section 3)
9. examples of different ways models are used in science (Section 3)
10. the importance of the International System of Units (Section 4)
11. the appropriate units to use for particular measurements (Section 4)
12. how area and density are derived quantities (Section 4)

Questions to Ask Along the Way

You can help your son or daughter learn about these topics by asking interesting questions such as the following:

- What are some surprising careers that use physical science?
- What is a characteristic of a good hypothesis?
- When is it a good idea to use a model?
- Why do Americans measure things in terms of inches and yards and meters ?

ALSO IN SPANISH

TEST ITEM LISTING

TEST ITEM LISTING

The World of Science SAMPLE

MULTIPLE CHOICE

1. A limitation of models is that
   a. they are large enough to see.
   b. they do not act exactly like the things that they model.
   c. they are smaller than the things that they model.
   d. they model unfamiliar things.
   Answer: B Difficulty: I Section: 3 Objective: 2
2. The length 10 m is equal to
   a. 100 cm. c. 10,000 mm.
   b. 1,000 cm. d. Both (b) and (c)
   Answer: B Difficulty: I Section: 3 Objective: 2
3. To be valid, a hypothesis must be
   a. testable. c. made into a law.
   b. supported by evidence. d. Both (a) and (b)
   Answer: B Difficulty: I Section: 3 Objective: 2 1
4. The statement "Sheila has a stain on her shirt" is an example of a(n)
   a. law. c. observation.
   b. hypothesis. d. prediction.
   Answer: B Difficulty: I Section: 3 Objective: 2
5. A hypothesis is often developed out of
   a. observations. c. laws.
   b. experiments. d. Both (a) and (b)
   Answer: B Difficulty: I Section: 3 Objective: 2
6. How many milliliters are in 3.5 kL?
   a. 3,500 mL c. 3,500, 000 mL
   b. 0.0035 mL d. 35,000 mL
   Answer: B Difficulty: I Section: 3 Objective: 2
7. A map of Seattle is an example of a
   a. law. c. model.
   b. theory. d. unit.
   Answer: B Difficulty: I Section: 3 Objective: 2
8. A lab has the safety icons shown below. These icons mean that you should wear
   a. only safety goggles. c. safety goggles and a lab apron.
   b. only a lab apron. d. safety goggles, a lab apron, and gloves.
   Answer: B Difficulty: I Section: 3 Objective: 2
9. The law of conservation of mass says the tot al mass before a chemical change is
   a. more than the total mass after the change.
   b. less than the total mass after the change.
   c. the same as the total mass after the change.
   d. not the same as the total mass after the change.
   Answer: B Difficulty: I Section: 3 Objective: 2
10. In which of the following areas might you find a geochemist at work?
    a. studying the chemistry of rocks c. studying fishes
    b. studying forestry d. studying the atmosphere
    Answer: B Difficulty: I Section: 3 Objective: 2

One-Stop Planner® CD-ROM

This CD-ROM includes all of the resources shown here and the following time-saving tools:

- *Lab Materials QuickList Software*
- *Customizable lesson plans*
- *Holt Calendar Planner*
- *The powerful ExamView® Test Generator*

For a preview of available worksheets covering math and science skills, see pages T12–T19. All of these resources are also on the One-Stop Planner®.

Meeting Individual Needs

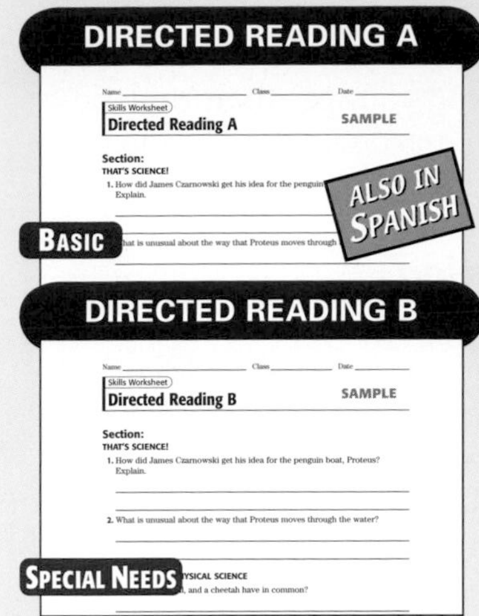

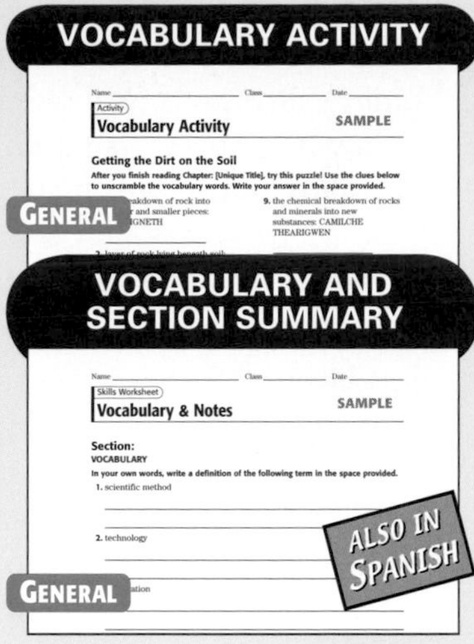

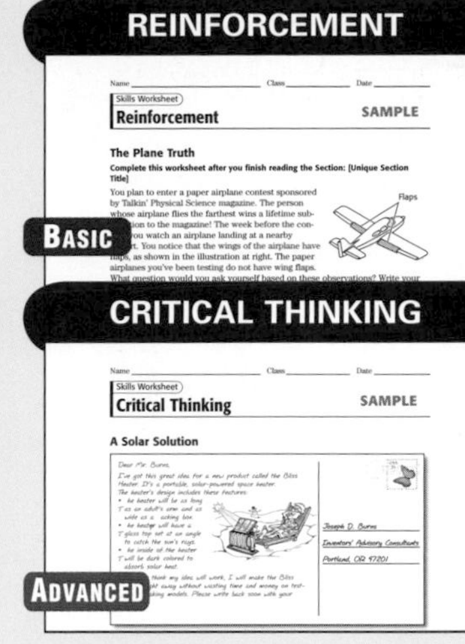

Labs and Activities

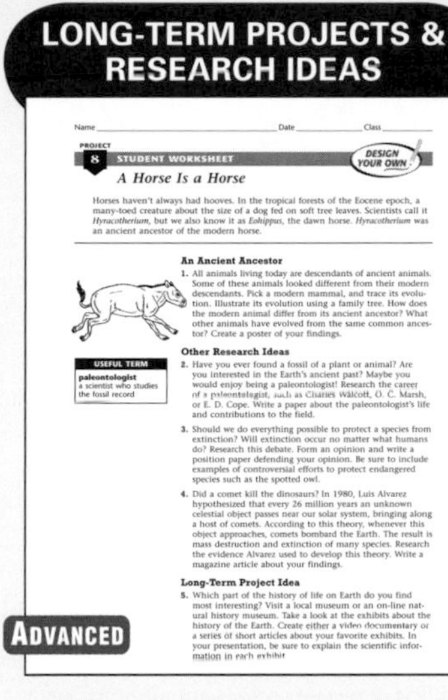

Review and Assessments

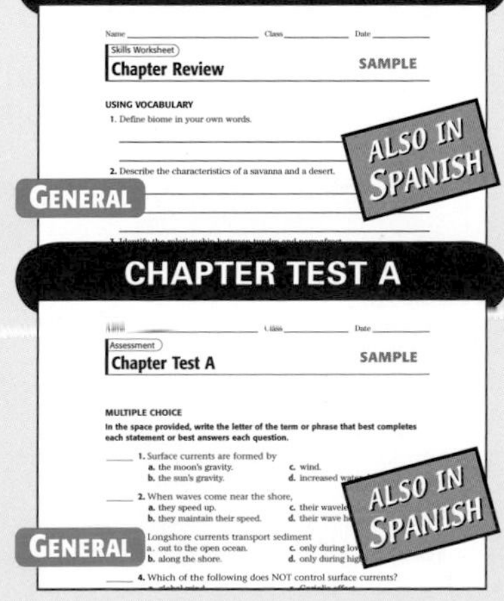

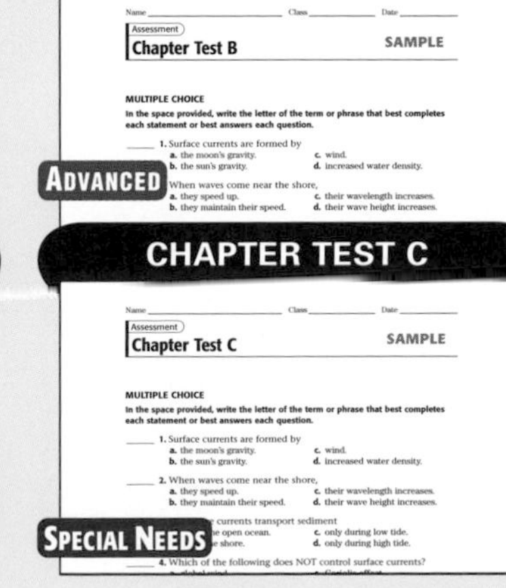

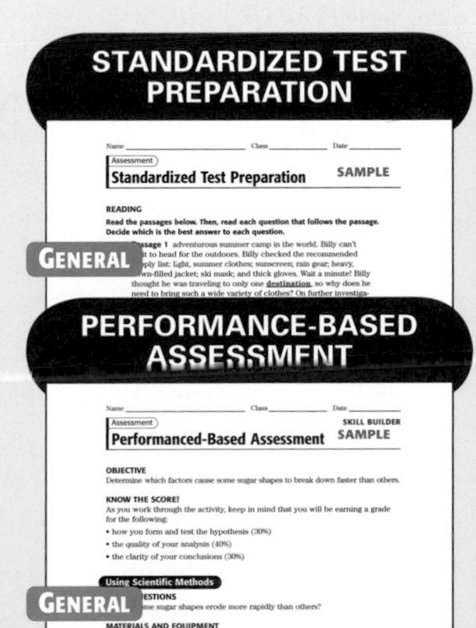

This Chapter Enrichment provides relevant and interesting information to expand and enhance your presentation of the chapter material.

Section 1

Evidence of the Past

Paul Sereno

- Paul Sereno has traveled around the world to study and document dinosaur fossils. He teaches at the University of Chicago and also involves his students in searching museum collections and combing deserts for new fossils. Sereno's teams have made many important finds. One of the first was in 1988 in Argentina, where his team unearthed the skeletons of a primitive 12-foot-long dinosaur called *Herrerasaurus.* The fossils in that area shed light on how and when the Age of Reptiles began. Sereno has continued to map the dinosaur family tree by studying fossils in the Sahara, in Niger, and in Morocco.

Fossils

- Fossils may be mere traces of organisms. Preserved footprints, feces, gnaw marks, and dug-out holes can all be considered fossils. Also, traces or remains of organisms can be preserved in materials other than sedimentary rock,such as amber, tar, or lava.

- Despite what many people think, fossils are not difficult to find. Nearly every state in the United States contains an abundance of fossils. However, scientists think that only a tiny fraction of the countless organisms that lived on Earth has been preserved as fossils. Many organisms have lived and died without leaving evidence of their existence in the fossil record.

Is That a Fact!

◆ The oldest fossils known are structures called *stromatolites* that are more than 3.5 billion years old. These structures are bands of sedimentary rock that are very similar to layered mats formed today by colonies of bacteria and cyanobacteria.

Law of Superposition

- The law of superposition states that in a series of undisturbed sedimentary rock layers, each layer is older than the one above it and younger than the one below it. This law is based on an observation made by Nicolaus Steno, a Danish physician, in 1669.

Methods of Absolute Dating

- Radioisotope dating is the most widely used method for dating a fossil. This method analyzes samples of igneous rock found within the same rock formation as the fossil. The method differs depending on the type of chemical isotope analyzed. The older the rocks are, the less accurate the dating. Isotopes with shorter half-lives provide a more accurate range of possible ages for younger rocks and fossils. Radiocarbon dating by accelerator mass spectrometry (AMS) has become a preferred method to date with high accuracy fossils less than 60,000 years old.

Is That a Fact!

◆ The beginning of the Paleozoic era is sometimes called the *Cambrian explosion.* Within the first 100 million years of this period, a large variety of multicellular organisms appeared for the first time on Earth, including most of the major groups of animals.

Modern Mass Extinction

- Because of natural selection, there will always be some extinctions of species within any given time period. Mass extinctions, however, are periods of acceleration of the average rate of extinction. Many scientists think that our planet has entered another era of mass extinction and that human activities are prompting these extinctions. Species all over the Earth are threatened by habitat destruction, pollution, and invasive nonnative species. During the last 200 years, more than 50 species of birds, more than 75 species of mammals, and hundreds of other species have become extinct.

For background information about teaching strategies and issues, refer to the ***Professional Reference for Teachers.***

Section 2

Eras of the Geologic Time Scale

Experiment About the Origin of Life

- In 1953, American scientist Stanley Miller devised a famous experiment to simulate life-forming conditions on the early Earth. He mixed together hydrogen, ammonia, and methane (to represent the air) and water (to represent the oceans) in a flask. When he applied electricity to the mixture, amino acids were produced. His experiment demonstrated that the building blocks of life could be created on Earth through simple chemistry. Scientists have since found amino acids in meteorites, confirming that conditions favorable for their formation also exist elsewhere.

Dinosaur Whodunit

- Scientists continue to debate various hypotheses about the cause of the mass extinction that wiped out the dinosaurs at the end of the Cretaceous period of the Mesozoic era. The prime suspect for many scientists is an asteroid that created the 185 km wide Chicxulub crater in the Yucatán area of the Gulf of Mexico. Seismology studies support this hypothesis. However, a sample of rock from the core of this crater contains evidence that the Chicxulub asteroid did not result in sudden climate change. An alternative hypothesis is that a series of asteroid impacts eventually caused climate changes that led to the mass extinction.

Is That a Fact!

◆ Dinosaurs are not the biggest animals ever to have lived on Earth. Blue whales are bigger than the largest known dinosaur.

Section 3

Humans and Other Primates

Clues to Migration Route

- Scientists think that people passed through the Nile Valley of Egypt when they migrated from Africa, perhaps as early as 100,000 years ago. The first evidence supporting this idea was an adundance of fossil tools and other artifacts in the Nile Valley area. Then, in 1994, the team of Belgian archaeologist Pierre Vermeersch found an ancient—but clearly human—skeleton in the area. The skeleton appears to have been a child that was ritually buried over 80,000 years ago. The skull and teeth show similarities to those of equally old human remains from East Africa and the Middle East. These similarities show a link between the African and the Middle Eastern populations.

Dawn of Language

- Scientists Matt Cartmill and Richard Kay examined fossil hominid skulls and measured the hole through which the hypoglossal nerve passes in its course from the brain to the tongue. The hypoglossal nerve enables precise control over the tongue movements needed for speech. A large hole suggests a larger nerve. Chimpanzees have much smaller holes in their skulls than do modern humans. Because australopithecine skulls have small holes, like the skulls of chimpanzees, Cartmill and Kay think that australopithecines were unable to form words.

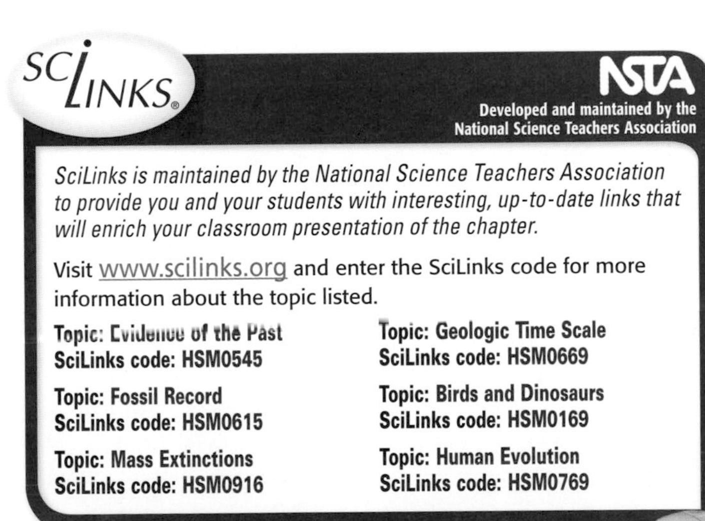

Overview

In this chapter, students will learn about the evidence of the history of life on earth. Students will study the geologic time scale and theories about the evolution of hominids.

Assessing Prior Knowledge

Students should be familiar with the following topics:

- cells
- the basic chemistry of life
- classification
- evolution

Identifying Misconceptions

As students learn the material in this chapter, they may have misconceptions about the length of time living organisms have been on Earth and the length of time needed for geologic processes. For example, mass extinctions are "sudden" on a geologic time scale but may take thousands of years. Furthermore, students may have misconceptions about how long humans have been on Earth. Students may also be unaware of the large amount of evidence scientists have gathered in order to determine the time and order of events in Earth's history.

6

The History of Life on Earth

About the PHOTO

What is 23,000 years old and 9 ft tall? The partial remains of the woolly mammoth in this picture! The mammoth was found in the frozen ground in Siberia in 1999. Scientists think that several types of woolly mammoths roamed the northern hemisphere until about 4,000 years ago.

PRE-READING ACTIVITY

FOLDNOTES **Layered Book** Before you read the chapter, create the Foldnote entitled "Layered Book" described in the **Study Skills** section of the Appendix. Label the tabs of the layered book with "Precambrian time," "Paleozoic era," "Mesozoic era," and "Cenozoic era." As you read the chapter, write information you learn about each category under the appropriate tab.

Standards Correlations

National Science Education Standards

The following codes indicate the National Science Education Standards that correlate to this chapter. The full text of the standards is at the front of the book.

Chapter Opener
UCP 2, 3; SAI 1, 2; SPSP 5; HNS 3

Section 1 Evidence of the Past
UCP 1, 2, 4; SAI 1, 2; HNS 1, 2, 3; LS 1a, 3d, 5b, 5c; *LabBook:* UCP 1, 3; SAI 1; LS 5c

Section 2 Eras of the Geologic Time Scale
UCP 1, 2, 3, 4; SAI 1; LS 1a, 1b, 3d, 5a, 5b, 5c

Section 3 Humans and Other Primates
UCP 2, 4, 5; SAI 1, 2; ST 1, 2; HNS 2, 3; LS 1a, 3d, 5a, 5b, 5c

Chapter Lab
UCP 2, 5; SAI 1; HNS 2

Chapter Review
LS 1a, 1b, 3d, 5a, 5b, 5c

Science in Action
UCP 2; ST 2; SPSP 5; HNS 1, 3

Start-Up Activity

Making a Fossil

In this activity, you will make a model of a fossil.

Procedure

1. Get a **paper plate,** some **modeling clay,** and a **leaf** or a **shell** from your teacher.
2. Flatten some of the modeling clay on the paper plate. Push the leaf or shell into the clay. Be sure that your leaf or shell has made a mark in the clay. Remove the leaf or shell carefully.
3. Ask your teacher to cover the clay with **plaster of Paris.** Allow the plaster to dry overnight.
4. Carefully remove the paper plate and the clay from the plaster the next day.

Analysis

1. Consider the following objects—a clam, a seed, a jellyfish, a crab, a leaf, and a mushroom. Which of the objects do you think would make good fossils? Explain your answers.
2. In nature, fossils form only under certain conditions. For example, fossils may form when a dead organism is covered by tiny bits of sand or dirt for a long period of time. The presence of oxygen can prevent fossils from forming. Considering these facts, what are some limitations of your model of how a fossil is formed?

Start-Up Activity

MATERIALS

For Each Student
- clay, modeling
- leaf or shell
- plaster of Paris
- plate, paper

Teacher's Note: Coating the leaves or shells with a light vegetable oil will help prevent the clay and plaster from sticking.

Answers

1. Sample answer: The crab, the clam, and perhaps the seed will make the best fossils because they have hard parts that would decay slowly and leave clear impressions. The softer objects—the jellyfish, the leaf, and the mushroom—are less likely to make impressions in the sediment and more likely to decay quickly, so they are less likely to form fossils.
2. Sample answer: Like all models, this model cannot encompass all real possibilities. This model does not exactly duplicate the way most fossils are formed. For example, this model does not show how fossils of softer organisms can be formed. Also, in this model, oxygen is present, but the plaster forms into rock quickly, so the objects are not given time to decay. Finally, in this model, the objects do not move on their own, but moving organisms might leave fossils of traces such as footprints.

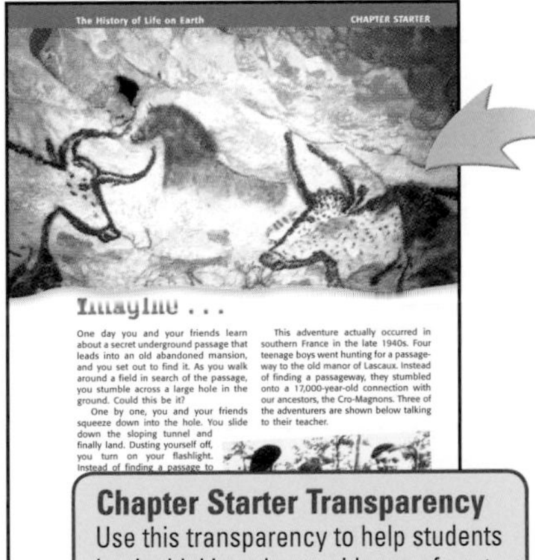

The History of Life on Earth CHAPTER STARTER

Imagine . . .

One day you and your friends learn about a secret underground passage that leads into an old abandoned mansion, and you set out to find it. As you walk around a field in search of the passage, you stumble across a large hole in the ground. Could this be it?

One by one, you and your friends squeeze down into the hole. You slide down the sloping tunnel and finally land. Dusting yourself off, you turn on your flashlight. Instead of finding a passage to

This adventure actually occurred in southern France in the late 1940s. Four teenage boys went hunting for a passageway to the old manor of Lascaux. Instead of finding a passageway, they stumbled onto a 17,000-year-old connection with our ancestors, the Cro-Magnons. Three of the adventurers are shown below talking to their teacher.

Chapter Starter Transparency
Use this transparency to help students begin thinking about evidence of past life on Earth.

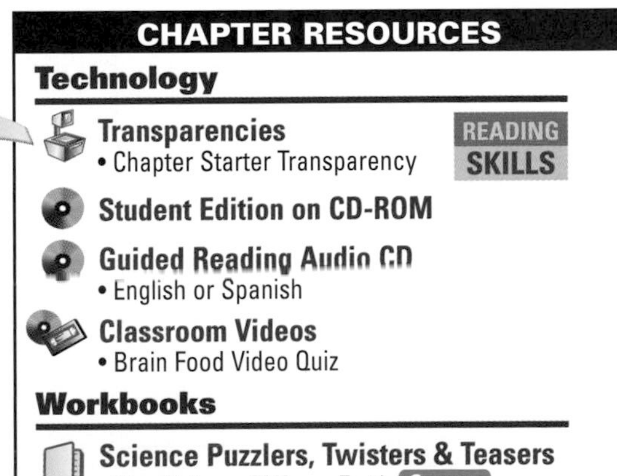

CHAPTER RESOURCES

Technology

- **Transparencies** READING SKILLS
  - Chapter Starter Transparency
- **Student Edition on CD-ROM**
- **Guided Reading Audio CD**
  - English or Spanish
- **Classroom Videos**
  - Brain Food Video Quiz

Workbooks

- **Science Puzzlers, Twisters & Teasers**
  - The History of Life on Earth GENERAL

SECTION 1

Focus

Overview

This section introduces students to fossils and how they provide clues to Earth's past. Students learn how fossils most commonly form in sedimentary rock. They explore the methods scientists use to determine the age of fossils. They learn how scientists place events in the Earth's history in the correct order and how mass extinctions mark the boundaries of eras. Finally, they learn how plate tectonics has moved continents slowly and affected life over time.

Bellringer

Ask students to imagine that they haven't cleaned their room for 30 years. After 30 years, they finally decide to sort through the 2 m pile of stuff on their floor. Ask students, "What might you find on the top of the pile? in the middle? on the bottom?" (The items on the bottom are most likely to be those that were left on the floor at an earlier time than were the items above them.)

SECTION 1

Evidence of the Past

In 1995, scientist Paul Sereno found a dinosaur skull that was 1.5 m long in the Sahara, a desert in Africa. The dinosaur may have been the largest land predator that has ever existed!

READING WARM-UP

Objectives

- Explain how fossils can be formed and how their age can be estimated.
- Describe the geologic time scale and the way that scientists use it.
- Compare two ways that conditions for life on Earth have changed over time.

Terms to Learn

fossil
relative dating
absolute dating
geologic time scale
extinct
plate tectonics

READING STRATEGY

Reading Organizer As you read this section, make a concept map by using the terms above.

Scientists such as Paul Sereno look for clues to help them reconstruct what happened in the past. These scientists, called *paleontologists* (PAY lee uhn TAHL uh jists), use fossils to reconstruct the history of life before humans existed. Fossils show us that life on Earth has changed a great deal. They also provide us clues about how those changes happened.

Fossils

Fossils are traces or imprints of living things—such as animals, plants, bacteria, and fungi—that are preserved in rock. Fossils sometimes form when a dead organism is covered by a layer of sediment. The sediment may later be pressed together to form sedimentary rock. **Figure 1** shows one way that fossils can form in sedimentary rock.

fossil the remains or physical evidence of an organism preserved by geological processes

Figure 1 One Way Fossils Can Form

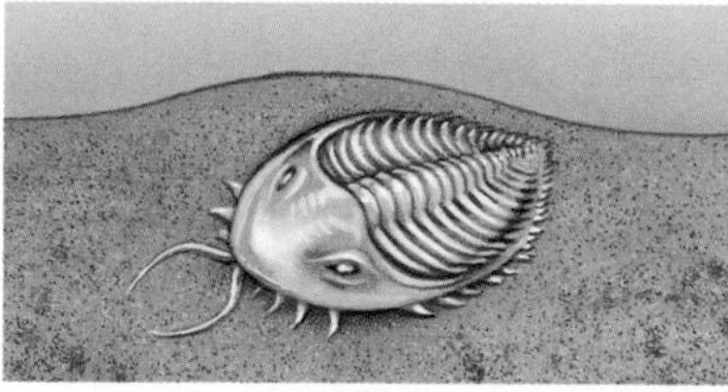

1 Fossils can form in several ways. The most common way is when an organism dies and becomes buried in sediment.

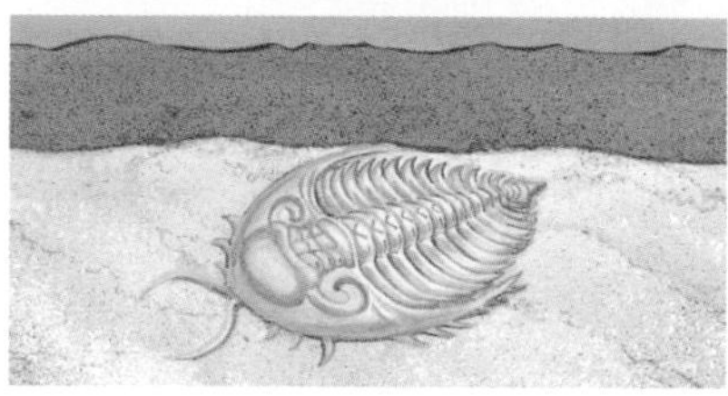

2 The organism gradually decomposes and leaves a hollow impression, or *mold*, in the sediment.

3 Over time, the mold fills with sediment, which forms a *cast* of the organism.

CHAPTER RESOURCES

Chapter Resource File

- Lesson Plan
- Directed Reading A BASIC
- Directed Reading B SPECIAL NEEDS

Technology

Transparencies
- Bellringer
- Using Half-Lives to Date Fossils

SCIENTISTS AT ODDS

Fossil Finds In the 1870s, two American scientists, Edward Drinker Cope and Othniel Charles Marsh, studied dinosaur fossils. They became bitter rivals and often argued. In 1878, Marsh and Cope were both excavating fossils in Wyoming. They had separate excavations and didn't want to share their findings. Both groups found more fossils than they could carry. To prevent the other group from taking their fossils, each group smashed all the fossils that couldn't be carried away.

Figure 2 Using Half-Lives to Date Fossils

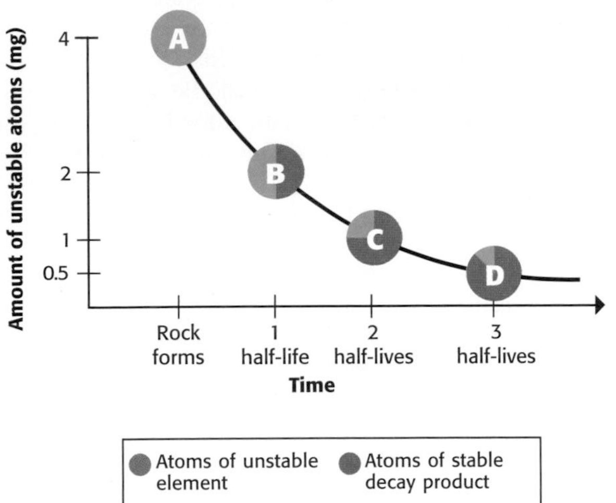

A The unstable atoms in this sample of rock have a half-life of 1.3 billion years. The sample contained 4 mg of unstable atoms when it formed.

B After 1.3 billion years, (one half-life for this type of unstable atom), 2 mg of the unstable atoms have decayed to become stable atoms, and 2 mg of unstable atoms remain.

C After 2.6 billion years (two half-lives for this sample), the rock sample contains 3 mg of stable decay atoms and 1 mg of unstable atoms.

D After three half-lives, only 0.5 mg of unstable atoms remain in the rock sample. This is equal to one-eighth of the original amount.

The Age of Fossils

Sedimentary rock has many layers. The oldest layers are usually on the bottom. The newest layers are usually on the top. The layers can tell a scientist the relative age of fossils. Fossils found in the bottom layers are usually older than the fossils in the top layers. So, scientists can determine whether a fossil is older or younger than other fossils based on its position in sedimentary rock. Estimating the age of rocks and fossils in this way is called **relative dating.**

In addition, scientists can determine the age of a fossil more precisely. **Absolute dating** is a method that measures the age of fossils or rocks in years. In one type of absolute dating, scientists examine atoms. *Atoms* are the particles that make up all matter. Atoms, in turn, are made of smaller particles. Some atoms are unstable and will decay by releasing energy, particles, or both. When an atom decays it becomes a different, and more stable, kind of atom. Each kind of unstable atom decays at its own rate. As shown in **Figure 2,** the time it takes for half of the unstable atoms in a sample to decay is the *half-life* of that type of unstable atom. By measuring the ratio of unstable atoms to stable atoms, scientists can determine the approximate age of a sample of rock.

Reading Check Which type of fossil dating is more precise? *(See the Appendix for answers to Reading Checks.)*

relative dating any method of determining whether an event or object is older or younger than other events or objects

absolute dating any method of measuring the age of an object or event in years

Fractions of Fractions

Find the answer to each of the following problems. Be sure to show your work. You may want to draw pictures.

1. 1/2 × 1/2 × 1/2 × 1/2

2. 1/2 × 1/8

3. 1/4 × 1/4

Motivate

ACTIVITY — GENERAL

Newspaper Layers Obtain a stack of old daily newspapers. Tell students that the dailies represent layers of sedimentary rock and that the pictures on the front pages represent fossils. Scramble the newspapers to make sure that they are not ordered according to date. Then, have students stack the newspapers from oldest to newest to simulate the layering of fossils over time. **LS Kinesthetic**

Teach

CONNECTION to Chemistry — ADVANCED

Half-Lives The unstable atoms referred to in **Figure 2** are unstable because they have an unusual number of neutrons. Atoms of the same element can occur as *isotopes* in which the number of neutrons per atom differs. For example, most carbon atoms have 12 protons, 12 neutrons, and 12 electrons. But the isotope carbon-14 has 14 neutrons. Scientists have experimentally determined the half-lives of various isotopes in order to determine the accuracy of age estimates based on a given isotope. **LS Logical**

Answer to Reading Check

absolute dating

Homework — GENERAL

Calculating To help students grasp the concept of half-life, assign these problems.

1. Thorium-232 has a half-life of 14.1 billion years. How much of an 8 mg sample will be unchanged after 1 half-life? (4 mg) 2 half lives? (2 mg) 4 half-lives? (0.5 mg)
2. Carbon-14 has a half-life of 5,730 years. How much of the original sample will be left after 11,460 years? (one-fourth of the original sample) after 17,190 years? (one-eighth of the original sample)

LS Logical

Answers to Math Practice

1. 1/16

2. 1/16

3. 1/16

Teach, continued

Group ACTIVITY — ADVANCED

Detailed Geologic Timeline
Table 1 shows that geologic time is divided into *eras,* which are broken into smaller divisions called *periods*. But periods can be divided into *epochs,* and scientists continue to add many more details to this table. Challenge students to construct a giant geologic timeline that identifies all of these divisions. Direct them to scale the size of the divisions relative to time. Allow them to do research in the library or on the Internet for information. **LS Visual**

Research — GENERAL

Extinct Species Students could research plant and animal species that have become extinct within the last 200 years. Many of the extinctions were caused by human activities. Extinct birds include the dodo, great auk, Labrador duck, moa, and passenger pigeon. Extinct mammals include the Steller's sea cow and the quagga. **LS Verbal**

Discussion — GENERAL

Abbreviations In geological and paleontological literature, students may encounter the abbreviations *MYA* and *BYA*. Explain to students that *MYA* means "million years ago." Likewise, *BYA* means "billion years ago." Ask students why they think geologists use this form of dating. **LS Verbal**

Table 1 Geologic Time Scale

Era	Period	Time*
Cenozoic era	Quaternary	2
	Tertiary	65
Mesozoic era	Cretaceous	144
	Jurassic	206
	Triassic	248
Paleozoic era	Permian	290
	Carboniferous	345
	Devonian	408
	Silurian	439
	Ordovician	495
	Cambrian	543
Precambrian time		4,600

*indicates how many millions of years ago the period began

The Geologic Time Scale

Think about important events that have happened during your lifetime. You usually recall each event in terms of the day, month, or year in which it happened. These divisions of time make it easier to recall when you were born, when you kicked the winning soccer goal, or when you started the fifth grade. Scientists also use a type of calendar to divide the Earth's long history. The span of time from the formation of the Earth to now is very long. Therefore, the calendar is divided into very long units of time.

The calendar scientists use to outline the history of life on Earth is called the **geologic time scale,** shown in **Table 1.** After a fossil is dated, a paleontologist can place the fossil in chronological order with other fossils. This ordering forms a picture of the past that shows how organisms have changed over time.

Divisions in the Geologic Time Scale

Paleontologists have divided the geologic time scale into large blocks of time. Each block may be divided into smaller blocks of time as scientists continue to find more fossil information.

The divisions known as *eras* are characterized by the type of organism that dominated the Earth at the time. For instance, the Mesozoic era—dominated by dinosaurs and other reptiles—is referred to as the *Age of Reptiles.* Eras began with a change in the type of organism that was most dominant.

Paleontologists sometimes adjust and add details to the geologic time scale. For example, the early history of the Earth has been poorly understood. There is little evidence that life existed billions of years ago. So, the earliest part of the geologic time scale is not named as an era. But more evidence of life before the Paleozoic era is being gathered. Scientists have proposed using this evidence to name new eras before the Paleozoic era.

CONNECTION TO Social Studies

A Place in Time Most of the periods of the Paleozoic era were named by geologists for places where rocks from that period are found. Research the name of each period of the Paleozoic era listed in **Table 1.** On a copy of a world map, label the locations related to each name.

ACTIVITY

Private Fossil Collectors In 1997, the most complete skeleton ever found of a *Tyrannosaurus rex* was auctioned. The winning bid was $8.36 million, made by the Field Museum of Natural History in Chicago. Scientists were relieved that a museum won the bid; their fear was that a private collector would buy the skeleton fossil and not allow scientists to study it.

Answer to Connection to Social Studies

Cambrian: for Cambria, the Latin name for Wales in Great Britain; Ordovician: after the Ordivices, a Celtic tribe; Silurian: after the Silures, a Celtic tribe; Devonian: for Devonshire, England; Carboniferous: for coal-bearing rocks in England; Permian: for the Russian province of Perm

Figure 3 *Scientists think that a meteorite hit Earth about 65 million years ago and caused major climate changes.*

Mass Extinctions

Some of the important divisions in the geologic time scale mark times when rapid changes happened on Earth. During these times, many species died out completely, or became **extinct**. When a species is extinct, it does not reappear. At certain points in the Earth's history, a large number of species disappeared from the fossil record. These periods when many species suddenly become extinct are called *mass extinctions*.

Scientists are not sure what caused each of the mass extinctions. Most scientists think that the extinction of the dinosaurs happened because of extreme changes in the climate on Earth. These changes could have resulted from a giant meteorite hitting the Earth, as shown in **Figure 3.** Or, forces within the Earth could have caused many volcanoes and earthquakes.

geologic time scale the standard method used to divide the Earth's long natural history into manageable parts

extinct describes a species that has died out completely

Reading Check What are mass extinctions?

Making a Geologic Timeline

1. Use a **metric ruler** to mark 10 cm sections on a **strip of paper** that is 46 cm long.
2. Label each 10 cm section in order from top to bottom as follows: 1 bya (billion years ago), 2 bya, etc. The timeline begins at 4.6 bya.
3. Divide each 10 cm section into 10 equal subsections. Divide the top 1 cm into 10 subsections. Calculate the number of years that are represented by 1 mm on this scale.
4. On your timeline, label the following events:
   a. Earth forms. (4.6 billion years ago)
   b. First animals appear. (600 million years ago)
   c. Dinosaurs appear. (251 million years ago)
   d. Dinosaurs are extinct. (65 million years ago)
   e. Humans appear. (100,000 years ago)
5. Label other events from the chapter.
6. Describe what most of the timeline looks like.
7. Compare the length of time dinosaurs existed with the length of time humans have existed.

WEIRD SCIENCE

One species that became extinct during the time of the dinosaurs was the insect having the largest wingspan on record. The insect belonged to the order Protodonata, and it measured an astonishing 76 cm (30 in.) from wingtip to wingtip. Its body was 46 cm (18 in.) long. It died out about 200 million years ago. Fossils of this insect have been found in Kansas.

CHAPTER RESOURCES

Workbooks

Math Skills for Science
- Geologic Time Scale GENERAL
- Radioactive Decay and the Half-Life GENERAL

Technology

Transparencies
- Geologic Time Scale

Activity — GENERAL

Rock Collectors If any students have rock collections, ask them to show these collections to the class. Allow students time to observe the different collections and to ask questions about them. English Language Learners
LS Interpersonal

Answer to Reading Check

periods of sudden extinction of many species

MATERIALS

For Each Group
- paper, adding machine, 46 cm strip
- ruler, metric

Teacher's Note: A fun variation on this activity uses a full roll of perforated toilet paper instead of adding-machine paper. At the scale of 20 million years per perforated sheet, 4.6 billion years would be represented by 230 sheets. This activity variation would best be conducted outdoors or in a long hallway.

Answers

6. Most of the timeline should be blank and should be taken up by Precambrian time.
7. Dinosaurs existed for a much longer time than humans have.

Homework — GENERAL

PORTFOLIO **Poster Project** Have students research what their local area was like millions of years ago. They can develop a written report and a poster describing the climate, living things, and landforms at different points in time. LS Visual

Close

Reteaching — BASIC

Illustrating Drift Have students cut out each continent from a copy of an area-proportionate world map. Have them try to fit the cut-out continents together into one landmass, using **Figure 4** as a model. Then, ask students to model the movements of the continents into their current locations. Finally, ask them to predict where the continents might be in the future if they keep moving in a similar way. (Predictions might include a wider Atlantic Ocean and a shift northward of Africa, Australia, and South America.) **LS Visual/Kinesthetic**

Quiz — GENERAL

1. How are fossils most commonly formed? (An organism is buried in sediments that harden into rock.)
2. What can scientists learn about Earth's past from fossils? (how life on Earth has changed over time)
3. Why would fossils found at the top of a canyon probably be younger than those found at the bottom of the canyon? (The upper layers were deposited more recently.)

Answer to Reading Check

the idea that the Earth's continents once formed a single landmass surrounded by oceans

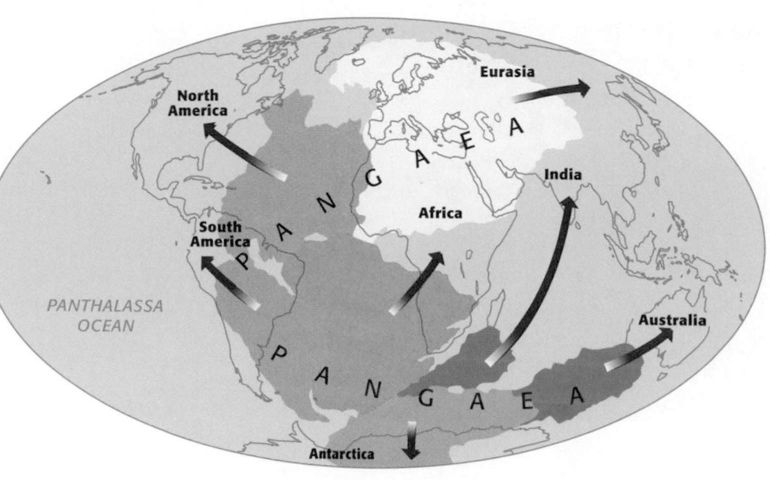

Figure 4 *The continents have been slowly moving throughout the history of Earth. The colored areas show the location of the continents 245 million years ago, and blue outlines show where the continents are today.*

The Changing Earth

Did you know that fossils of tropical plants have been found in Antarctica? Antarctica, now frozen, must have once had a warm climate to support these plants. The fossils provide evidence that Antarctica was once located near the equator!

Pangaea

Have you ever noticed that the continents look like pieces of a puzzle? German scientist Alfred Wegener had a similar thought in the early 1900s. He proposed that long ago the continents formed one landmass surrounded by a gigantic ocean. Wegener called that single landmass *Pangaea* (pan JEE uh), which means "all Earth." **Figure 4** shows how the continents may have formed from Pangaea.

Reading Check What idea did Alfred Wegener propose?

Do the Continents Move?

plate tectonics the theory that explains how large pieces of the Earth's outermost layer, called *tectonic plates,* move and change shape

In the mid-1960s, J. Tuzo Wilson of Canada came up with the idea that the continents were not moving by themselves. Wilson thought that huge pieces of the Earth's crust were pushed around by forces within the planet. Each huge piece of crust is called a *tectonic plate.* Wilson's theory of how these huge pieces of crust move around the Earth is called **plate tectonics.**

According to Wilson, the outer crust of the Earth is broken into seven large, rigid plates and several smaller ones. The continents and oceans ride on top of these plates. The motion of the plates causes the continents to move. For example, the plates that carry South America and Africa are slowly moving apart, as shown in **Figure 5.**

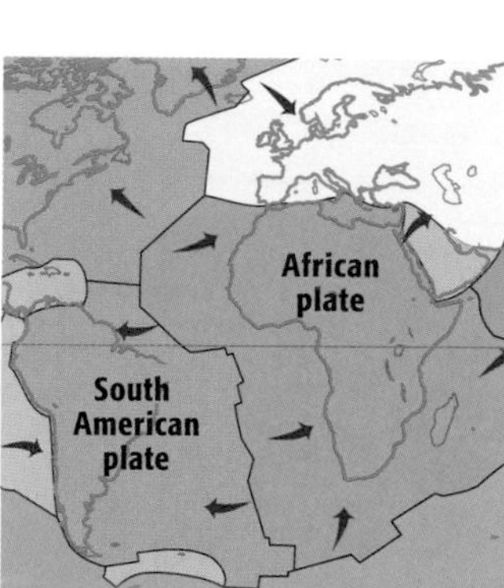

Figure 5 *The continents ride on tectonic plates, outlined here in black. The plates are still moving about 1 to 10 cm per year.*

Cultural Awareness — GENERAL

Iceland Tell the class that the Mid-Atlantic Ridge rises above water only in Iceland. Point to Iceland (in the Arctic Circle) on a map. Tell students that Iceland is a place of both extreme cold and extreme heat. Ask students to do some research on what life is like in Iceland. Have them focus on how geologic processes, such as glaciers, geysers, and volcanic activity, affect the lifestyle of the people who live there. **LS Visual**

INCLUSION Strategies

- *Visually Impaired*
- *Learning Disabled*
- *Developmentally Delayed*

Assist students in understanding the concepts in **Figure 4** by creating a cardboard replica of each of the continents. Let students experiment with fitting the pieces together. Then, have them approximate the shape of Pangaea, as shown in **Figure 4.** English Language Learners **LS Kinesthetic**

Adaptation to Slow Changes

When conditions on the Earth change, organisms may become extinct. A rapid change, such as a meteorite impact, may cause a mass extinction. But slow changes, such as moving continents, allow time for adaptation.

Anywhere on Earth, you are able to see living things that are well adapted to the location where they live. Yet in the same location, you may find evidence of organisms that lived there in the past that were very different. For example, the animals currently living in Antarctica are able to survive very cold temperatures. But under the frozen surface of Antarctica are the remains of tropical forests. Conditions on Earth have changed many times in history, and life has changed, too.

Mid-Atlantic Ridge In 1947, scientists examined rock from a ridge that runs down the middle of the Atlantic Ocean, between Africa and the Americas. They found that this rock was much younger than the rock on the continents. Explain what this finding indicates about the tectonic plates.

SECTION Review

Summary

- Fossils are formed most often in sedimentary rock. The age of a fossil can be determined using relative dating and absolute dating.
- The geologic time scale is a timeline that is used by scientists to outline the history of Earth and life on Earth.
- Conditions for life on Earth have changed many times. Rapid changes, such as a meteorite impact, might have caused mass extinctions. But many groups of organisms have adapted to changes such as the movement of tectonic plates.

Using Key Terms

1. Use the following terms in the same sentence: *fossil* and *extinct*.
2. In your own words, write a definition for the term *plate tectonics*.

Understanding Key Ideas

3. Explain how a fossil forms in sedimentary rock.
4. What kind of information does the geologic time scale show?
5. About how many years of Earth's history was Precambrian time?
6. What are two possible causes of mass extinctions?

Math Skills

7. The Earth formed 4.6 billion years ago. Modern humans have existed for about 100,000 years. Simple worms have existed for at least 500 million years. For what fraction of the history of Earth have humans existed? have worms existed?

Critical Thinking

8. **Identifying Relationships** Why are both absolute dating and relative dating used to determine the age of fossils?
9. **Making Inferences** Fossils of *Mesosaurus,* the small aquatic reptile shown below, have been found only in Africa and South America. Using what you know about plate tectonics, how would you explain this finding?

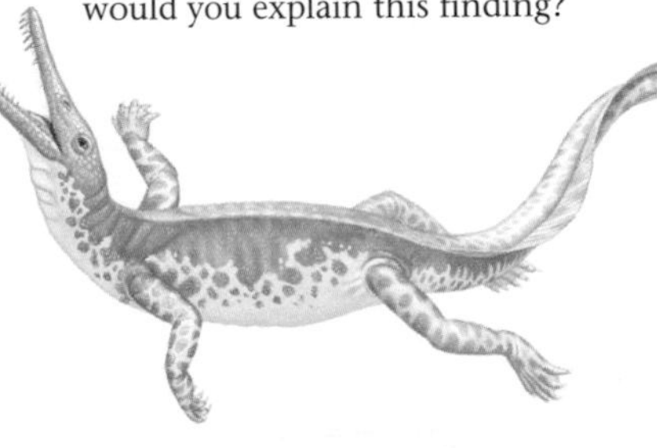

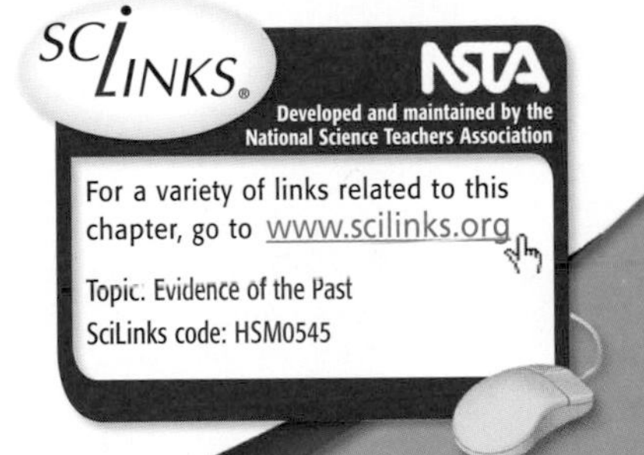

CONNECTION to Earth Science — GENERAL

The South American Plate Have students identify the boundaries of the South American tectonic plate on a map. Ask students, "What kinds of features mark the east side of the plate?" (a rift or canyon down the middle of the Atlantic Ocean) Ask, "Which direction is the plate drifting?" (to the west, away from Africa.) Ask, "What is happening on the west side of the plate?" (the plate is crashing into another plate and pushing up the Andes moutnains) LS **Visual**

CHAPTER RESOURCES

Chapter Resource File

- **Section Quiz** GENERAL
- **Section Review** GENERAL
- **Vocabulary and Section Summary** GENERAL
- **Reinforcement Worksheet** BASIC
- **Datasheet for Quick Lab**

Technology

Transparencies
- Moving Continents and Tectonic Plates
- **LINK TO EARTH SCIENCE** The South American Plate

Alternative Assessment — GENERAL

Geologic Time Scale Each student in a group receives three blank cards. The group looks at **Table 1** in this section and thinks of questions that could be answered using information from the table. Students write a question on one side of a card. They stack the cards with the blank sides up. In turn, each student draws a card, reads the question on it, and attempts to answer the question. Group members determine if the answer is correct by consulting the table. LS **Visual/Interpersonal**

Answer to Connection to Geology

This finding indicates that the plates are spreading apart as newer rock forms at the center.

Answers to Section Review

1. Sample answer: Scientists can learn about extinct organisms from fossils.
2. Sample answer: Plate tectonics is a theory about how the plates of the Earth's crust move around.
3. A fossil may form when a dead organism is covered by a layer of sediment. Then, these sediments are slowly pressed together to form sedimentary rock.
4. how organisms and Earth have changed over time
5. about 4 billion years
6. meteorite impacts and forces within the Earth
7. Humans have existed for 1/46,000 of the history of Earth. Worms have existed for 5/46 of Earth's history.
8. Sample answer: to compare the results of each method so there can be greater confindence in the conclusions
9. Sample answer: Perhaps *Mesosaurus* evolved after Pangaea broke up but before South America and Africa split apart.

SECTION 2

Focus

Overview

This section discusses current theories regarding the origin of life. Students are introduced to the four major divisions of geologic time in chronological order: Precambrian time, the Paleozoic era, the Mesozoic era, and the Cenozoic era. They learn about the organisms that characterize each era.

Bellringer

Ask students to respond to the following question:

> Suppose that electrical energy had never been developed. How would your life differ from what it is like now?

Discuss with students the consequences of great changes over time.

Answer to Reading Check

The early Earth was very different from Earth as it is today. There were violent events and a harsh atmosphere.

SECTION 2

Eras of the Geologic Time Scale

READING WARM-UP

Objectives

- Outline the major developments that allowed life to exist on Earth.
- Describe the types of organisms that arose during the four major divisions of the geologic time scale.

Terms to Learn

Precambrian time
Paleozoic era
Mesozoic era
Cenozoic era

READING STRATEGY

Mnemonics As you read this section, create a mnemonic device to help you remember the eras of geologic time.

The walls of the Grand Canyon are layered with different kinds and colors of rocks. The deeper down into the canyon you go, the older the layers of rocks. Try to imagine a time when the bottom layer was the only layer that existed.

Each layer of rock tells a story about what was happening on Earth when that layer was on top. The rocks and fossils in each layer tell the story. Scientists have compared the stories told by fossils and rocks all over the Earth. From these stories, scientists have divided geologic history into four major parts. These divisions are Precambrian time, the Paleozoic era, the Mesozoic era, and the Cenozoic era.

Precambrian Time

The layers at the bottom of the Grand Canyon are from the oldest part of the geologic time scale. **Precambrian time** (pree KAM bree UHN TIEM) is the time from the formation of Earth 4.6 billion years ago to about 543 million years ago. Life on Earth began during this time.

Precambrian time the period in the geologic time scale from the formation of the Earth to the beginning of the Paleozoic era, from about 4.6 billion to 543 million years ago

Scientists think that the early Earth was very different than it is today. The atmosphere was made of gases such as water vapor, carbon dioxide, and nitrogen. Also, the early Earth was a place of great turmoil, as illustrated in **Figure 1.** Volcanic eruptions, meteorite impacts, and violent storms were common. Intense radiation from the sun bombarded Earth's surface.

Reading Check Describe the early Earth. (*See the Appendix for answers to Reading Checks.*)

Figure 1 *This illustration shows the conditions under which the first life on Earth may have formed.*

CHAPTER RESOURCES

Chapter Resource File

- Lesson Plan
- Directed Reading A  BASIC
- Directed Reading B SPECIAL NEEDS

Technology

 **Transparencies**
- Bellringer

Is That a Fact!

Throughout Earth's history, the forces of erosion have been altering the planet's surface, making it almost impossible to find rocks older than 3.5 billion years. However, a number of rocks dating from about 3.5 billion to 3.9 billion years ago have been found in Canada and Greenland. The oldest was found in the Northwest Territories of Canada in 1989.

How Did Life Begin?

Scientists think that life developed from simple chemicals in the oceans and in the atmosphere. Energy from radiation and storms could have caused these chemicals to react. Some of these reactions formed the complex molecules that made life possible. Eventually, these molecules may have joined to form structures such as cells.

The early atmosphere of the Earth did not contain oxygen gas. The first organisms did not need oxygen to survive. These organisms were *prokaryotes* (proh KAR ee OHTS), or single-celled organisms that lack a nucleus.

INTERNET ACTIVITY

For another activity related to this chapter, go to **go.hrw.com** and type in the keyword **HL5HISW.**

Photosynthesis and Oxygen

There is evidence that *cyanobacteria,* a new kind of prokaryotic organism, appeared more than 3 billion years ago. Some cyanobacteria are shown in **Figure 2.** Cyanobacteria use sunlight to produce their own food. Along with doing other things, this process releases oxygen. The first cyanobacteria began to release oxygen gas into the oceans and air.

Eventually, some of the oxygen formed a new layer of gas in the upper atmosphere. This gas, called *ozone*, absorbs harmful radiation from the sun, as shown in **Figure 3.** Before ozone formed, life existed only in the oceans and underground. The new ozone layer reduced the radiation on Earth's surface.

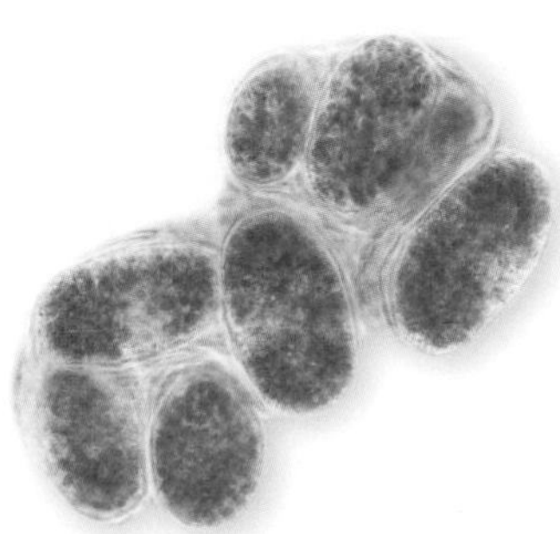

Figure 2 *Cyanobacteria are the simplest living organisms that use the sun's energy to produce their own food.*

Multicellular Organisms

After about 1 billion years, organisms that were larger and more complex than prokaryotes appeared in the fossil record. These organisms, known as *eukaryotes* (yoo KAR ee OHTS), contain a nucleus and other complex structures in their cells. Eventually, eukaryotic cells may have evolved into organisms that are composed of many cells.

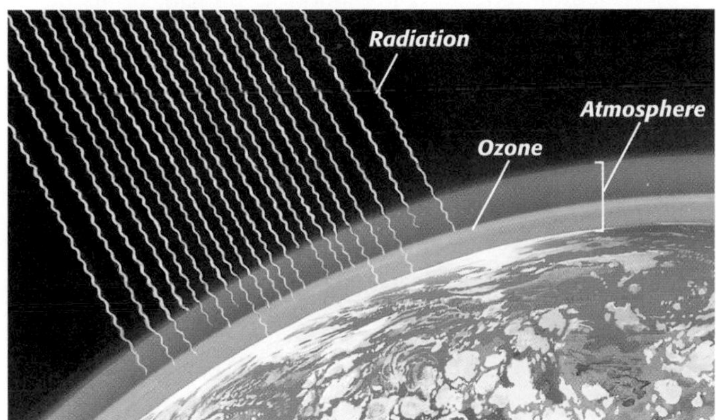

Figure 3 *Oxygen in the atmosphere formed a layer of ozone, which helps to absorb harmful radiation from the sun.*

Motivate

Discussion — GENERAL

Historical Perspective Have students pretend that they are in a time-travel machine. What scientifically significant events would they witness as they travel back in time to Earth's origin? List the events on the board in the order that they are suggested. Ask students how they can determine whether the events are in chronological order. (Students can consult the geologic time scale or other scientific materials.) LS **Verbal/Logical**

Teach

ACTIVITY — GENERAL

Using Maps Have students locate the following three earthquake and volcano zones on a world map. One zone extends nearly all the way around the edge of the Pacific Ocean. A second zone is located near the Mediterranean Sea and extends across Asia into India. The third zone extends through Iceland to the middle of the Atlantic Ocean. LS **Visual**

READING STRATEGY — GENERAL

Concept Mapping Have students construct a concept map that shows how life on Earth developed from nonliving matter. They should base their map on information presented on these pages. The primary subject headings should refer to the chemicals that already exist in the environment. The final part of the concept map should identify prokaryotes as the first lifeforms to appear on Earth. LS **Logical/Visual**

PORTFOLIO

CONNECTION ACTIVITY
Earth Science — GENERAL

Ancient Mountains Provide a large wall map of the world, and provide map pins or tacks in three colors. Have students locate the following mountain ranges on a map. Have students place pins on the map for each range to match the eras when each range was formed. Use the following list for reference:

Blue (Paleozoic):
- Caledonian (Scandinavia), Acadian (New York), Appalachian (eastern North America), Ural (Russia)

Green (Mesozoic):
- Palisades (New Jersey), Rockies (western North America)

Red (Cenozoic):
- Andes (South America), Alps (Europe), Himalayas (Asia)

LS **Visual/Kinesthetic**

Teach, *continued*

Group ACTIVITY — GENERAL

Ancient Plants Organize the class into five groups, and assign each group one of the following groups of paleozoic plants: club mosses, ferns, horsetails, ginkos, or conifers.

Have each group use encyclopedias or botany books to look up the plant group and prepare a poster about it. Have students include diagrams of key features of the plant group and pictures of fossils and living examples of the group. LS **Interpersonal/Visual**

Co-op Learning

CONNECTION ACTIVITY

Real World — GENERAL

Fossil Fuels The huge plants that grew in forests during the Paleozoic era later became coal. Ask students to research the locations of the world's coal deposits and to mark them on a world map. (Most of the known coal reserves are in Australia, China, Germany, Poland, Great Britain, India, Russia, South Africa, the United States, and Canada.)

LS **Visual**

Answer to Connection to Oceanography

Descriptions may vary.

Figure 4 *Organisms that first appeared in the Paleozoic era include reptiles, amphibians, fishes, worms, and ferns.*

The Paleozoic Era

The **Paleozoic era** (PAY lee OH ZOH ik ER uh) began about 543 million years ago and ended about 248 million years ago. Considering the length of Precambrian time, you can see that the Paleozoic era was relatively recent. Rocks from the Paleozoic era are rich in fossils of animals such as sponges, corals, snails, clams, squids, and trilobites. Fishes, the earliest animals with backbones, appeared during this era, and sharks became abundant. **Figure 4** shows an artist's depiction of life in the Paleozoic era.

The word *Paleozoic* comes from Greek words that mean "ancient life." When scientists first named this era, they thought it held the earliest forms of life. Scientists now think that earlier forms of life existed, but less is known about those life-forms. Before the Paleozoic era, most organisms lived in the oceans and left few fossils.

Life on Land

During the 300 million years of the Paleozoic era, plants, fungi, and air-breathing animals slowly colonized land. By the end of the era, forests of giant ferns, club mosses, horsetails, and conifers covered much of the Earth. All major plant groups except for flowering plants appeared during this era. These plants provided food and shelter for animals.

Fossils indicate that crawling insects were some of the first animals to live on land. They were followed by large salamander-like animals. Near the end of the Paleozoic era, reptiles and winged insects appeared.

The largest mass extinction known took place at the end of the Paleozoic era. By 248 million years ago, as many as 90% of all Paleozoic species had become extinct. The mass extinction wiped out entire groups of marine organisms, such as trilobites. The oceans were completely changed.

Paleozoic era the geologic era that followed Precambrian time and that lasted from 543 million to 248 million years ago

CONNECTION TO Oceanography

Prehistoric Marine Organisms Find a variety of pictures and descriptions of marine organisms from the Cambrian period of the Paleozoic era. Choose three organisms that you find interesting. Draw or write a description of each organism. Find out whether scientists think the organism is related to any living group of organisms, and add this information to your description.

INCLUSION *Strategies*

- ***Hearing Impaired*** • ***Learning Disabled***
- ***Visually Impaired***

Demonstrate the superposition of geologic layers. Have student teams each use a different color of modeling clay to create a 3 in diameter circle, representing a piece of land. Ask teams to use other pieces of clay to add organisms to the land. Choose one team's model to represent early organisms, and place it where all students can see it. Then, choose a different team's model, and place it on top of the first, squashing the bottom organisms. Continue in this manner until all circles have been stacked. Slice a cross section through the stack for all to see.

LS **Kinesthetic** English Language Learners

The Mesozoic Era

The **Mesozoic era** (MES oh ZOH ik ER uh) began about 248 million years ago and lasted about 183 million years. *Mesozoic* comes from Greek words that mean "middle life." Scientists think that the surviving reptiles evolved into many different species after the Paleozoic era. Therefore, the Mesozoic era is commonly called the *Age of Reptiles*.

Life in the Mesozoic Era

Dinosaurs are the most well known reptiles that evolved during the Mesozoic era. Dinosaurs dominated the Earth for about 150 million years. A great variety of dinosaurs lived on Earth. Some had unique adaptations, such as ducklike bills for feeding or large spines on their bodies for defense. In addition to dinosaurs roaming the land, giant marine lizards swam in the ocean. The first birds also appeared during the Mesozoic era. In fact, scientists think that some of the dinosaurs became the ancestors of birds.

The most important plants during the early part of the Mesozoic era were conifers, which formed large forests. Flowering plants appeared later in the Mesozoic era. Some of the organisms of the Mesozoic era are illustrated in **Figure 5.**

Figure 5 *The Mesozoic era was dominated by dinosaurs. The era ended with the mass extinction of many species.*

The Extinction of Dinosaurs

At the end of the Mesozoic era, 65 million years ago, dinosaurs and many other animal and plant species became extinct. What happened to the dinosaurs? According to one hypothesis, a large meteorite hit the Earth and generated giant dust clouds and enough heat to cause worldwide fires. The dust and smoke from these fires blocked out much of the sunlight and caused many plants to die out. Without enough plants to eat, the plant-eating dinosaurs died out. And the meat-eating dinosaurs that fed on the plant-eating dinosaurs died. Global temperatures may have dropped for many years. However, some mammals and birds survived.

Reading Check **What kind of event happened at the end of both the Paleozoic and Mesozoic eras?**

Mesozoic era the geologic era that lasted from 248 million to 65 million years ago; also called the *Age of Reptiles*

READING STRATEGY — GENERAL

Prediction Guide Before students read the text on this page and the next, ask them to guess whether the following events occurred in the Mesozoic era or in the Cenozoic era:

- Dinosaurs dominated Earth and then died out by the end of this era. (Mesozoic)
- Mammals appeared. (Mesozoic)
- Rock layers close to Earth's surface contain fossils from this era. (Cenozoic)
- The first birds appeared. (Mesozoic)
- Humans appeared. (Cenozoic)

LS Logical/Auditory

Research — GENERAL

Reptile Research Have students research reptiles and list their distinctive characteristics. (Reptiles are ectothermic, or "cold-blooded," they all have a distinctive three-chambered heart and at least one lung, most reptiles have scales but not feathers or fur, many reptiles have claws, all reptiles produce amniotic eggs, and most lay these eggs on land.) Tell students that all living reptiles fall into four main groups—turtles, lizards and snakes, crocodiles and related forms, and the tuatara. Have students investigate these orders and list examples of each.

LS Verbal/Logical

Making Models — GENERAL

Dinosaur Models Dinosaurs varied greatly in size and appearance. Have groups of students consult reference books to find information about the many kinds of dinosaurs. Then, have them use art materials to make models of different dinosaurs. Models should include flying and marine reptiles as well as land-dwelling reptiles.

LS Kinesthetic English Language Learners

Cultural Awareness — GENERAL

Dinosaur Food Recently, a fossil plesiosaur was found in a riverside cliff in Hokkaido, Japan, with shellfish remains in its stomach. Scientists had long suspected that plesiosaurs were predators, but they had no proof. Geologist Tamaki Sato thinks that the plesiosaur swallowed its tiny prey whole because the plesiosaur's long, sharp teeth were unsuitable for crushing hard outer shells.

Answer to Reading Check

a mass extinction

Close

Reteaching — Basic

Comparing Organisms Have students compare the characteristics of each of the Paleozoic, Mesozoic, and Cenozoic organisms described in this section with those of a living descendant (if one exists) of each of the organisms. Students can organize the information in the form of a chart. **LS Visual**

Quiz — General

On index cards, write the names of several types of organisms that appeared in each the four major divisions of geologic time mentioned in this section. Then, on paper strips, write the names of the geologic time divisions, and place the strips on a tabletop. Direct students to classify each organism named on a card by placing the card under the appropriate paper strip.

Alternative Assessment — General

Diorama Organize students into groups of three or four. Groups should use boxes with covers and art materials to make a diorama of one of the four major divisions of geologic time mentioned in this section.
LS Interpersonal/Kinesthetic

Answer to Reading Check

"recent life"

Figure 6 *Many types of mammals evolved during the Cenozoic era.*

The Cenozoic Era

The **Cenozoic era** (SEN uh ZOH ik ER uh) began about 65 million years ago and continues today. *Cenozoic* comes from Greek words that mean "recent life." Scientists have more information about the Cenozoic era than about any of the previous eras. Fossils from the Cenozoic era formed recently in geologic time, so they are found in rock layers closer to the Earth's surface. The closer the fossils are to the surface, the easier they are to find.

During the Cenozoic era, many kinds of mammals, birds, insects, and flowering plants appeared. Some organisms that appeared in the Cenozoic era are shown in **Figure 6.**

Reading Check What does *Cenozoic* mean?

The Age of Mammals

The Cenozoic era is sometimes called the *Age of Mammals*. Mammals have dominated the Cenozoic era the way reptiles dominated the Mesozoic era. Early Cenozoic mammals were small, forest dwellers. Larger mammals appeared later in the era. Some of these larger mammals had long legs for running, teeth that were specialized for eating different kinds of food, and large brains. Cenozoic mammals have included mastodons, saber-toothed cats, camels, giant ground sloths, and small horses.

Math Focus

Relative Scale It's hard to imagine 4.6 billion years. One way is to use a *relative scale*. For example, we can represent all of Earth's history by using the 12 h shown on a clock. The scale would begin at noon, representing 4.6 billion years ago, and end at midnight, representing the present. Because 12 h represent 4.6 billion years, 1 h represents about 383 million years. (Hint: 4.6 billion ÷ 12 = 383 million) So, what time on the clock represents the beginning of the Paleozoic era, 543 million years ago?

Step 1: Write the ratio.

$$\frac{x}{543{,}000{,}000 \text{ years}} = \frac{1 \text{ h}}{383{,}000{,}000 \text{ years}}$$

Step 2: Solve for x.

$$x = \frac{543{,}000{,}000 \text{ years} \times 1 \text{ h}}{383{,}000{,}000 \text{ years}} = 1.42 \text{ h}$$

Step 3: Convert the answer to the clock scale.

$$1.42 \text{ h} = 1 \text{ h} + (0.42 \times 60 \text{ min/h})$$

$$1.42 \text{ h} = 1 \text{ h } 25 \text{ min}$$

So, the Paleozoic era began 1 h 25 min before midnight, at about 10:35.

Now It's Your Turn

1. Use this method to calculate the relative times at which the Mesozoic and Cenozoic eras began.

Answer to Math Focus

Mesozoic:

12 h − [(348 ÷ 383) × 60] min = 11:21

Cenozoic:

12 h − [(65 ÷ 383) × 60] min = 11:50

Connection to Math — Advanced

Another Time Scale Have students calculate the length of the major divisions of geologic time relative to a 365-day calendar. They should state the month and day that each of the eras began. Use the Math Focus as an example, and provide a calendar for reference. Also, have students calculate the day that humans appeared (about 150,000 years ago). (Precambrian: Jan. 1; Paleozoic: Nov. 16; Mesozoic: Dec. 12; Cenozoic: Dec. 26; humans appeared: Dec. 31) **LS Logical**

The Cenozoic Era Today

We are currently living in the Cenozoic era. Modern humans appeared during this era. The environment and landscapes that we see around us today are part of this era.

However, the climate has changed many times during the Cenozoic era. Earth's history includes some periods called *ice ages,* during which the climate was very cold. During the ice ages, ice sheets and glaciers extended from the Earth's poles. To survive, many organisms migrated toward the equator. Other organisms adapted to the cold or became extinct.

When will the Cenozoic era end? No one knows. In the future, geologists might draw the line at a time when life on Earth again undergoes major changes.

Cenozoic era the most recent geologic era, beginning 65 million years ago; also called the *Age of Mammals*

SECTION Review

Summary

- The Earth is about 4.6 billion years old. Life formed from nonliving matter long ago.
- Precambrian time includes the formation of the Earth and the appearance of simple organisms.
- The first cells did not need oxygen. Later, photosynthetic cells evolved and released oxygen into the atmosphere.
- During the Paleozoic era, animals appeared in the oceans and on land, and plants grew on land.
- Dinosaurs dominated the Earth during the Mesozoic era.
- Mammals have dominated the Cenozoic era. This era continues today.

Using Key Terms

1. Use each of the following terms in a separate sentence: *Precambrian time, Paleozoic era, Mesozoic era,* and *Cenozoic era.*

Understanding Key Ideas

2. Unlike the atmosphere today, the atmosphere 3.5 billion years ago did not contain
   a. carbon dioxide.
   b. nitrogen.
   c. gases.
   d. ozone.
3. How do prokaryotic cells and eukaryotic cells differ?
4. Explain why cyanobacteria were important to the development of life on Earth.
5. Place in chronological order the following events on Earth:
   a. The first cells appeared that could make their own food from sunlight.
   b. The ozone layer formed.
   c. Simple chemicals reacted to form the molecules of life.
   d. Animals appeared.
   e. The first organisms appeared.
   f. Humans appeared.
   g. The Earth formed.

Math Skills

6. Calculate the total number of years that each of the geologic eras lasted, rounding to the nearest 100 million. Then, calculate each of these values as a percentage of the total 4.6 billion years of Earth's history. Round your answer to the units place.

Critical Thinking

7. **Making Inferences** Which chemicals probably made up the first cells on Earth?
8. **Forming Hypotheses** Think of your own hypothesis to explain the disappearance of the dinosaurs. Explain your hypothesis.

Answers to Section Review

1. Sample answer: Most of Earth's history was Precambrian time. The Paleozoic era was the time when life began to colonize land. The Mesozoic era is known as the Age of Reptiles. The Cenozoic era is known as the Age of Mammals.
2. d
3. Eukaryotes have a nucleus and are usually larger and more complex than prokaryotes.
4. Cyanobacteria were the first significant source of atmospheric oxygen on the planet.
5. g, c, e, a, b, d, f
6. Precambrian time: 4,057 million years, 88%; Paleozoic era: 295 million years, 6%; Mesozoic era: 183 million years, 4%; Cenozoic era: 65 million years, 1%
7. Sample answer: carbon dioxide, water, and nitrogen (or carbon, hydrogen, oxygen, and nitrogen)
8. Sample answers: A devastating virus attacked all reptiles on Earth, and only a few survived; a new kind of organism, such as a mammal, began to outcompete the dinosaurs; aliens bombed the Earth with climate-changing gases.

SCIENTISTS AT ODDS

Fire or Ice Scientists continue to debate whether another ice age is coming soon or whether the Earth is overheating. During Earth's history, the climate has changed many times, slowly switching between icy cold and lush warmth. Scientists call the warmer periods *interglacial* because each is usually followed by another ice age. However, in recent decades average temperatures on Earth seem to keep getting warmer.

CHAPTER RESOURCES

Chapter Resource File

- Section Quiz GENERAL
- Section Review GENERAL
- Vocaulary and Section Summary GENERAL
- Reinforcement Worksheet BASIC
- SciLinks Activity GENERAL

Technology

Interactive Explorations CD-ROM
- Rock On! GENERAL

Workbooks

Math Skills for Science
- Subtraction Review BASIC

SECTION 3

Focus

Overview

In this section, students will learn that scientists think humans share a common ancestor and common characteristics with other primates, such as apes and monkeys. This section describes the characteristics of hominids and examines trends in their evolution that could have led to modern humans.

Bellringer

Ask students to write an answer to the following question:

What makes you unique among your family members? (Responses might include food preferences, health condition, physical appearance, and talents.)

Point out that understanding human ancestry requires recognizing similarities and differences, such as those seen in families.

SECTION 3

Humans and Other Primates

Have you ever heard someone say that humans descended from monkeys or apes? Well, scientists would not exactly say that. The scientific theory is that humans, apes, and monkeys share a common ancestor. This common ancestor probably lived more than 45 million years ago.

Most scientists agree that there is enough evidence to support this theory. Many fossils of organisms have been found that show traits of both humans and apes. Also, comparisons of modern humans and apes support this theory.

READING WARM-UP

Objectives

- Describe two characteristics that all primates share.
- Describe three major groups of hominids.

Terms to Learn

primate
hominid
Homo sapiens

READING STRATEGY

Discussion Read this section silently. Write down questions that you have about this section. Discuss your questions in a small group.

Primates

What characteristics make us human? Humans are classified as primates. **Primates** are a group of mammals that includes humans, apes, monkeys, and lemurs. Primates have the characteristics illustrated in **Figure 1.**

primate a type of mammal characterized by opposable thumbs and binocular vision

The First Primates

The ancestors of primates may have co-existed with the dinosaurs. These ancestors were probably mouselike mammals that were active at night, lived in trees, and ate insects. The first primates did not exist until after the dinosaurs died out. About 45 million years ago, primates that had larger brains appeared. These were the first primates that had traits similar to monkeys, apes, and humans.

Figure 1 Characteristics of Primates

Both eyes are located at the front of the head, and they provide binocular, or three-dimensional, vision.

Almost all primates, such as these orangutans, have five flexible fingers—four fingers and an opposable thumb. This thumb enables primates to grip objects. Most primates besides humans also have opposable big toes.

CHAPTER RESOURCES

Chapter Resource File

- Lesson Plan
- Directed Reading A BASIC
- Directed Reading B SPECIAL NEEDS

Technology

Transparencies

- Bellringer
- Comparison of Primate Skeletons

Is That a Fact!

Although the skulls of a human and a chimpanzee appear similar, there are significant differences. The cranium of a human skull is domed, whereas the chimpanzee's cranium is flattened. Also, canine teeth in humans do not overlap as they do in chimpanzees. And because humans walk upright, the place where the spine connects to the skull is more centered under the skull in humans.

Apes and Chimpanzees

Scientists think that the chimpanzee, a type of ape, is the closest living relative of humans. This theory does not mean humans descended from chimpanzees. It means that humans and chimpanzees share a common ancestor. Sometime between 5 million and 30 million years ago, the ancestors of humans, chimpanzees, and other apes began to evolve along different lines.

Hominids

Humans are in a family separate from other primates. This family, called **hominids,** includes only humans and their human-like ancestors. The main characteristic that separates hominids from other primates is bipedalism. *Bipedalism* means "walking primarily upright on two feet." Evidence of bipedalism can be seen in a primate's skeletal structure. **Figure 2** shows a comparison of the skeletal features of apes and hominids.

hominid a type of primate characterized by bipedalism, relatively long lower limbs, and lack of a tail

Reading Check In which family are humans classified? *(See the Appendix for answers to Reading Checks.)*

Figure 2 **Comparison of Primate Skeletons**

The bones of gorillas (a type of ape) and humans (a type of hominid) have a very similar form, but the human skeleton is adapted for walking upright.

▲ The gorilla pelvis tilts the ape's large rib cage and heavy neck and head forward. The gorilla spine is curved in a C shape. The arms are long to provide balance on the ground.

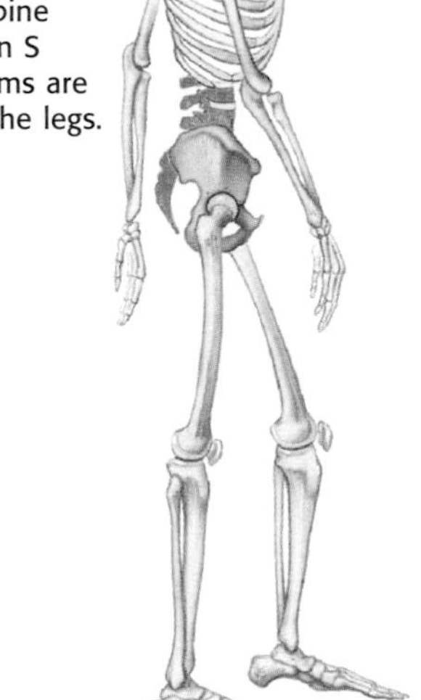

The human pelvis is vertical and helps hold the entire skeleton upright. The human spine is curved in an S shape. The arms are shorter than the legs. ▶

CONNECTION to Anthropology — ADVANCED

Bipedalism The tendency to walk fully upright distinguishes us from apes. But how do our physical features relate to our posture? Thomas Greiner, a physical anthropologist, developed a computer model that shows how muscle action relates to bone shape. Greiner concluded that in order to regularly walk upright, an ape would need larger gluteus maximus muscles and a larger ileum. These attributes are seen in human bodies.

Answer to Reading Check

the hominid family

Motivate

Discussion — GENERAL

Comparing Primates Display a picture of an ape and a picture of a human for students to compare. Have students identify characteristics that the two organisms have in common. (Answers will probably include references to common physical characteristics.) Then, ask students how the two animals are different from each other. (Answers will probably include references to differences in behavior or abilities.) **LS Visual/Verbal**

Teach

Using the Figure — GENERAL

Binocular Vison Discuss with students how binocular vision, illustrated in **Figure 1,** is important. How is it useful to humans? How is it different in most other animals? (Sample answer: Binocular vision enables humans to perceive depth and thus to judge distances, to hunt, to use tools, to drive vehicles, and to play sports. Except for some predators, most other animals do not have such abilities.) **LS Visual**

ACTIVITY — GENERAL

Exploring Vision Students can explore the utility of binocular vision by making a dot on a sheet of paper and placing the paper on a desktop. Have them stand about half a meter away from the desk and close their right eye. Then, have them try to touch the dot on the paper with the tip of a pencil. Have them repeat the action with their left eye closed and then with both eyes open. (Students should find that it is easier to touch the dot with a pencil when they have both eyes open than when they have one eye closed.) English Language Learners **LS Visual**

Teach, continued

Activity — BASIC

Primate Characteristics Help students identify the characteristics that distinguish primates from other mammal groups. Show them pictures of primate and nonprimate mammals. Ask students to describe how the primates differ from the other animals. (Sample answer: Primates generally have flatter faces than nonprimates. Their eyes are located at the front of the head rather than at the sides, their snouts are small, and their fingers are flexible.) LS **Visual/Logical**

Answer to Reading Check

Africa

Group Activity — GENERAL

Comparing Hominids Create and display a large table to compare the distinguishing characteristics of the primates and hominids discussed in this section. Column heads might include "Binocular vision," "Bipedalism," "Brain size," "Tool use," "Known locations," and "Estimated dates." Call on students to make additions to the table. LS **Logical/Verbal**

Figure 3 *This skull was found in the Sahel desert in Chad, Africa. The skull is estimated to be 6 million to 7 million years old.*

Hominids Through Time

Paleontologists are constantly filling in pieces of the hominid family picture. They have found many different fossils of ancient hominids and have named at least 18 types of hominids. However, scientists do not agree on the classification of every fossil. Fossils are classified as hominids when they share some of the characteristics of modern humans. But each type of hominid was unique in terms of size, the way it walked, the shape of its skull, and other characteristics.

The Earliest Hominids

The earliest hominids had traits that were more humanlike than apelike. These traits include the ability to walk upright as well as smaller teeth, flatter faces, and larger brains than earlier primates. The oldest hominid fossils have been found in Africa. So, scientists think hominid evolution began in Africa. **Figure 3** shows a fossil that may be from one of the earliest hominids. It is 6 million to 7 million years old.

Reading Check Where are the earliest hominid fossils found?

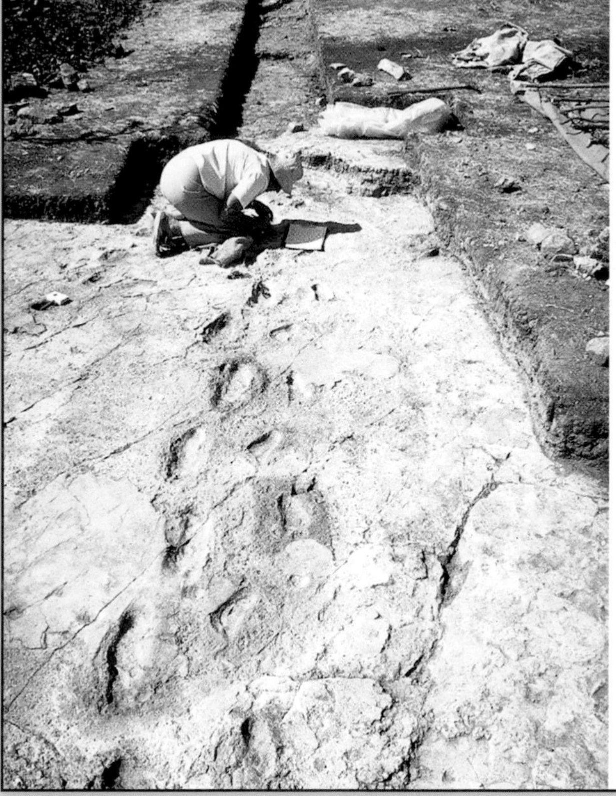

Australopithecines

Many early hominids are classified as *australopithecines* (AW struh LOH PITH uh SEENS). Members of this group were similar to apes but were different from apes in several ways. For example, their brains were slightly larger than the brains of apes. Some of them may have used stone tools. They climbed trees but also walked on two legs.

Fossil evidence of australopithecines has been found in several places in Africa. The fossilized footprints in **Figure 4** were probably made by a member of this group over 3 million years ago. Some skeletons of australopithecines have been found near what appear to be simple tools.

Figure 4 *Anthropologist Mary Leakey discovered these 3.6 million year old footprints in Tanzania, Africa.*

Scientists at Odds

In 1975, fossils of 13 hominids were found in Ethiopia by Donald Johansen, a contemporary of Mary Leakey. These fossils differed in body size and jaw shape. Some anthropologists think that the larger fossils represent the males and the smaller fossils represent the females of a particular species. Johansen and others believe that the differences indicate that the fossils are of two distinct species of australopithecines.

Is That a Fact!

One of the most famous skeletons of an australopithecine was found by Donald Johansen in Ethiopia 1974 and was nicknamed "Lucy." This nickname came from the Beatles' hit song "Lucy in the Sky with Diamonds," which was playing around the time when the fossil was discovered.

A Variety of Early Hominids

Many australopithecines and other types of hominids lived at the same time. Some australopithecines had slender bodies. They had humanlike jaws and teeth but had small, apelike skulls. They probably lived in forests and grasslands and ate a vegetarian diet. Scientists think that some of these types of hominids may have been the ancestors of modern humans.

Some early hominids had large bodies and massive teeth and jaws. They had a unique skull structure and relatively small brains. Most of these types of hominids lived in tropical forests and probably ate tough plant material, such as roots. Scientists do not think that these large-bodied hominids are the ancestors of modern humans.

Global Hominids

About 2.3 million years ago, a new group of hominids appeared. These hominids were similar to the slender australopithecines but were more humanlike. These new hominids had larger and more complex brains, rounder skulls, and flatter faces than early hominids. They showed advanced tool-making abilities and walked upright.

These new hominids were members of the group *Homo,* which includes modern humans. Fossil evidence indicates that several members of the *Homo* group existed at the same time and on several continents. Members of this group were probably scavengers that ate a variety of foods. Some of these hominids may have adapted to climate change by migrating and changing the way they lived.

An early member of this new group was *Homo habilis* (HOH moh HAB uh luhs), which lived about 2 million years ago. In another million years, a hominid called *Homo erectus* (HOH moh i REK tuhs) appeared. This type of hominid could grow as tall as modern humans do. A museum creation of a member of *Homo erectus* is shown in **Figure 5.** No one knows what early hominids looked like. Scientists construct models based on skulls and other evidence.

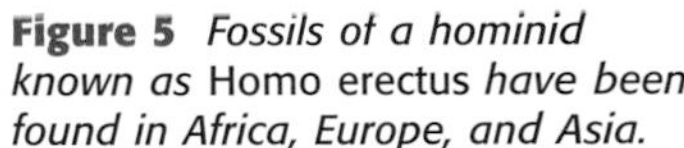

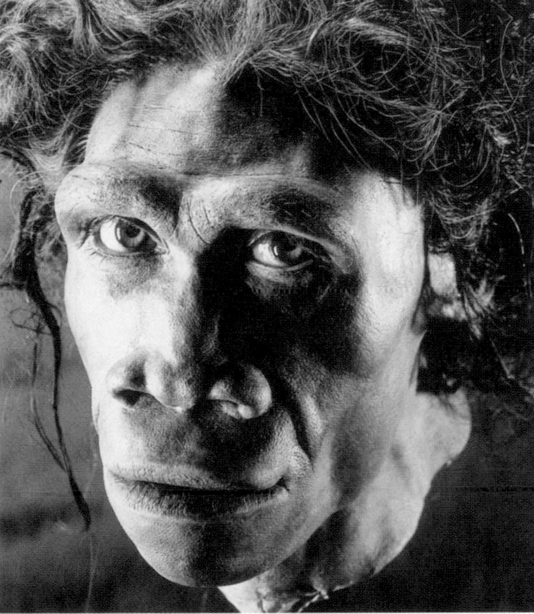

Figure 5 *Fossils of a hominid known as* Homo erectus *have been found in Africa, Europe, and Asia.*

Thumb Through This

1. Keep your thumbs from moving by attaching them to the sides of your hands with **tape.**
2. Attempt each of the following tasks: using a **pencil sharpener,** using **scissors,** tying your **shoelaces,** buttoning **buttons.**
3. After each attempt, answer the following questions:
   a. Is the task more difficult with an opposable thumb or without one?
   b. Do you think you would carry out this task on a regular basis if you did not have an opposable thumb?

ACTIVITY

Answers to School-to-Home Activity

3. a. Answers may vary. Most of the tasks will probably be more difficult without the use of the thumb.
   b. Answers may vary. Some people would not like to carry out such a difficult task on a regular basis.

Discussion — GENERAL

Analyzing Tools *Homo habilis* is thought to have made one of the oldest recognizable stone tools. The tool was a pebble with some sharp edges. Ask students how they think the tool was made. (Sample answer: Flakes were chipped off the pebble to sharpen it.) **LS Verbal/Logical**

CONNECTION ACTIVITY
ART — GENERAL

Sculpting Sculptors probably helped paleoanthropologists determine the physical appearance of the hominid shown in **Figure 5.** Sculptors can apply their knowledge of anatomy to reconstruct body features. Have students research how sculptors are called upon to reconstruct hominid faces and heads, based on skulls. Then, have interested students use clay to sculpt a model of the head of a hominid. English Language Learners **LS Kinesthetic**

MISCONCEPTION ALERT

Students often misunderstand the meaning of the scientific name of the genus *Homo*. This word comes from the greek word for "earth" or "ground" and was later used to mean "man" or "human being" in latin. In some other words, the word root *homo-* comes from the greek word meaning "same."

CONNECTION to
History — GENERAL

Insulting Apes? Many cartoons in the 19th century satirized the idea that humans are related to apes. In one such cartoon, Henry Bergh, the founder of the Society for the Prevention of Cruelty to Animals, chided Charles Darwin for insulting apes by suggesting that they are related to humans. Suggest that students look for examples of these historical cartoons. **LS Visual/Interpersonal**

Close

Reteaching — Basic

Timeline Have students make a timeline that shows the order of appearance of the primates discussed in the chapter. When students have finished their timelines, have them review the timeline of another student. LS **Visual**

Quiz — General

Among primates, what is distinctive about hominids? (The main characteristic that distinguishes hominids from other primates is walking upright on two legs as their main way of moving around.)

Alternative Assessment — Advanced

Hominid Poster Have students construct a poster with a detailed timeline of the appearance of different types of primates and hominds. Ask students to include pictures and information about the distinguishing characteristics of each group. Encourage students to conduct additional research to find the latest discoveries. LS **Visual/Logical**

Recent Hominids

As recently as 30,000 years ago, two types of hominids may have lived in the same areas at the same time. Both had the largest brains of any hominids and made advanced tools, clothing, and art. Scientists think that modern humans may have descended from one of these two types of hominids.

Neanderthals

One recent hominid is known as *Neanderthal* (nee AN duhr TAWL). Neanderthals lived in Europe and western Asia. They may have lived as early as 400,000 years ago. They hunted large animals, made fires, and wore clothing. They also may have cared for the sick and elderly and buried their dead with cultural rituals. About 30,000 years ago, Neanderthals disappeared. No one knows what caused their extinction.

Homo sapiens the species of hominids that includes modern humans and their closest ancestors and that first appeared about 100,000 to 150,000 years ago

Early and Modern Humans

Modern humans are classified as the species ***Homo sapiens*** (HOH moh SAY pee UHNZ). The earliest *Homo sapiens* existed in Africa 100,000 to 150,000 years ago. The group probably migrated out of Africa about 40,000 years ago. Compared with Neanderthals, *Homo sapiens* has a smaller and flatter face, and has a skull that is more rounded. Of all known hominids, only *Homo sapiens* still exists.

Homo sapiens seems to be the first to create art. Early humans produced sculptures, carvings, paintings, and clothing such as that shown in **Figure 6.** The preserved villages and burial grounds of early humans show that they had an organized and complex society.

Figure 6 *These photos show museum recreations of early* Homo sapiens.

Homework — General

Writing **Future Scientists** Have students write a page from an anthropologist's journal that will be written 100,000 years in the future.Tell students that the anthropologist is studying an archaeological site that contains the remains or traces of people from today. Suggest that students describe the scientist's thoughts and hypotheses about the site. LS **Visual**

Science Bloopers

A skull of a *Homo sapiens* who had dental problems was found in Zambia. There was a hole in one side of the skull and signs of a partially healed abscess. This skull was made famous by a writer who imagined that the hole was caused by a bullet shot from an interplanetary visitor's gun 120,000 years ago.

Drawing the Hominid Family Tree

Paleontologists review their hypotheses when they learn something new about a group of organisms and their related fossils. As more hominid fossils are discovered, there are more features to compare. Sometimes, scientists add details to the relationships they see between each group. Sometimes, new groups of hominids are recognized. Human evolution was once thought to be a line of descent from ancient primates to modern humans. But scientists now speak of a "tree" or even a "bush" to describe the evolution of various hominids in the fossil record.

Reading Check What is likely to happen when a new hominid fossil is discovered?

SECTION Review

Summary

- Humans, apes, and monkeys are primates. Almost all primates have opposable thumbs and binocular vision.
- Hominids, a subgroup of primates, include humans and their humanlike ancestors. The oldest known hominid fossils may be 7 million years old.
- Early hominids included australopithecines and the *Homo* group.
- Early *Homo sapiens* did not differ very much from present-day humans. *Homo sapiens* is the only type of hominid living today.

Using Key Terms

1. Use each of the following words in the same sentence: *primate, hominid,* and *Homo sapiens.*

Understanding Key Ideas

2. The unique characteristics of primates are
   a. bipedalism and thumbs.
   b. opposable thumbs.
   c. opposable thumbs and binocular vision.
   d. opposable toes and thumbs.
3. Describe the major evolutionary developments from early hominids to modern humans.
4. Compare members of the *Homo* group with australopithecines.

Critical Thinking

5. **Forming Hypotheses** Suggest some reasons why Neanderthals might have become extinct.
6. **Making Inferences** Imagine you are a scientist excavating an ancient campsite. What might you infer about the people who used the site if you found the charred bones of large animals and various stone blades among human fossils?

Interpreting Graphics

The figure below shows a possible ancestral relationships between humans and some modern apes. Use this figure to answer the questions that follow.

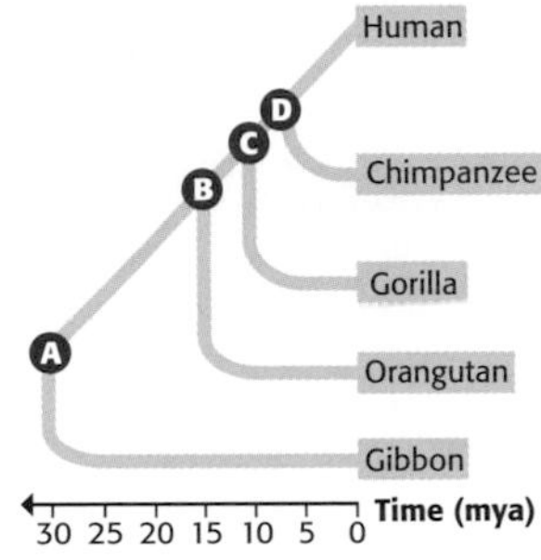

7. Which letter represents the ancestor of all the apes?
8. To which living ape are gorillas most closely related?

ACTIVITY — ADVANCED

Classifying Primates Have students find out the family or genus of the apes and hominids mentioned in this section.

- Apes: family Pongidae (great apes; includes orangutans, gorillas, chimpanzees)
- Hominids: family Hominidae
- Australopithecines: genus *Australopithecus* (slender) and genus *Paranthropus* (robust)
- Homo group: genus *Homo* (includes species *Homo habilis, Homo erectus, Homo neanderthalensis*, and *Homo sapiens*)

LS Verbal

CHAPTER RESOURCES

Chapter Resource File

- Section Quiz GENERAL
- Section Review GENERAL
- Vocabulary and Section Summary GENERAL
- Critical Thinking ADVANCED

Answer to Reading Check

Sample Answer: Paleontologists will review their ideas about the evolution of hominids.

Answers to Section Review

1. Sample answer: *Homo sapiens* is the only living type of hominid but not the only living type of primate.
2. c
3. walking upright, larger brains, changed diet, using tools, and culture
4. Sample answer: Compared to australopithecines, *Homo* were larger, had larger brains and different skull shapes, moved out of Africa and into other continents, and used tools.
5. Sample answer: They may have been killed off by larger animals, or they may have run out of food and starved to extinction.
6. Sample answer: It is possible that the people had hunted animals for food.
7. A
8. chimpanzee

Mystery Footprints

Teacher's Notes

Time Required

Two 45-minute class periods

Lab Ratings

Teacher Prep 3

Student Set-Up 2

Concept Level 2

Clean Up 2

Preparation Notes

To set up this lab, you will need to either find a sandy area outside or construct a long, shallow sandbox out of wood or cardboard. You may prefer to perform this activity outside because it is likely to be messy. Ask a boy and a girl (preferably students who are not in your science class) or two adults, one male and one female, to walk through the sand with their bare feet. The sand should be about 16 cm deep, and the area to be walked through should be long enough that three or four footprints can be seen in the sand. Slightly moistened sand will hold the best footprints. You may want to make the footprints more permanent by using plaster of Paris. If you do not have access to sand, look for a type of soil that will hold a footprint.

Using Scientific Methods

Mystery Footprints

OBJECTIVES

Form a hypothesis to explain observations of traces left by other organisms.

Design and **conduct** an experiment to test one of these hypotheses.

Analyze and **communicate** the results in a scientific way.

MATERIALS

- ruler, metric or meterstick
- sand, slightly damp
- large box, at least 1 m^2 or large enough to contain 3 or 4 footprints

SAFETY

Sometimes, scientists find clues preserved in rocks that are evidence of the activities of organisms that lived thousands of years ago. Evidence such as preserved footprints can provide important information about an organism. Imagine that your class has been asked by a group of paleontologists to help study some human footprints. These footprints were found embedded in rocks in an area just outside of town.

Ask a Question

1. Your teacher will give you some mystery footprints in sand. Examine the mystery footprints. Brainstorm what you might learn about the people who walked on this patch of sand.

Form a Hypothesis

2. As a class, formulate several testable hypotheses about the people who left the footprints. Form groups of three people, and choose one hypothesis for your group to investigate.

Test the Hypothesis

3. Draw a table for recording your data. For example, if you have two sets of mystery footprints, your table might look similar to the one below.

Mystery Footprints		
	Footprint set 1	Footprint set 2
Length		
Width		
Depth of toe	DO NOT WRITE IN BOOK	
Depth of heel		
Length of stride		

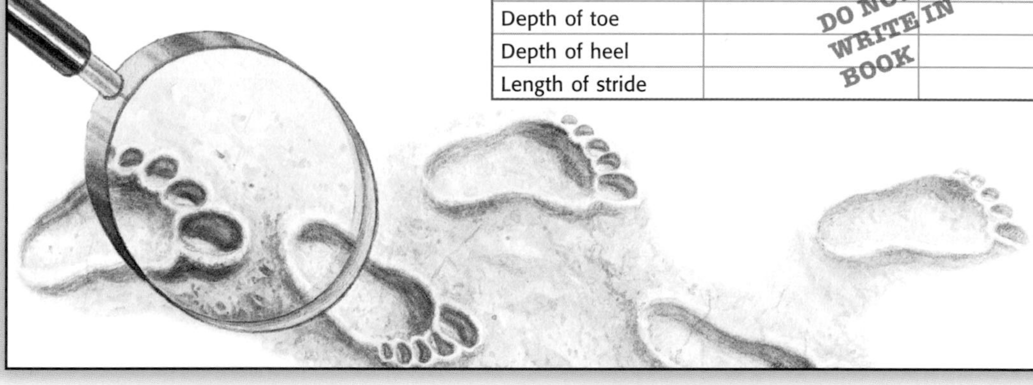

CLASSROOM TESTED & APPROVED

Maurine Marchani
Raymond Park Middle School
Indianapolis, Indiana

CHAPTER RESOURCES

Chapter Resource File

- Datasheet for Chapter Lab
- Lab Notes and Answers

Technology

Classroom Videos
- Lab Video

LabBook
- The Half-Life of Pennies

4. With the help of your group, you may first want to analyze your own footprints to help you draw conclusions about the mystery footprints. For example, use a meterstick to measure your stride when you are running. Is your stride different when you are walking? What part of your foot touches the ground first when you are running? When you are running, which part of your footprint is deeper?
5. Make a list of the kind of footprint each different activity produces. For example, you might write, "When I am running, my footprints are deep near the toe area and 110 cm apart."

Analyze the Results

1. **Classifying** Compare the data from your footprints with the data from the mystery footprints. How are the footprints alike? How are they different?
2. **Identifying Patterns** How many people do you think made the mystery footprints? Explain your interpretation.
3. **Analyzing Data** Can you tell if the mystery footprints were made by men, women, children, or a combination? Can you tell if they were standing still, walking, or running? Explain your interpretation.

Draw Conclusions

4. **Drawing Conclusions** Do your data support your hypothesis? Explain.
5. **Evaluating Methods** How could you improve your experiment?

Communicating Your Data

WRITING SKILL Summarize your group's conclusions in a report for the scientists who asked for your help. Begin by stating your hypothesis. Then, summarize the methods you used to study the footprints. Include the comparisons you made between your footprints and the mystery footprints. Add pictures if you wish. State your conclusions. Finally, offer some suggestions about how you could improve your investigation.

CHAPTER RESOURCES

Workbooks

Long-Term Projects & Research Ideas
- A Horse Is a Horse ADVANCED

Answers

The answers for this activity will depend on the footprints your students observe. Students should be able to compare their own activities with variations in the footprints they leave. Then, they should be able to apply what they've learned to the mystery footprints.

Safety Caution

Remind students to review all safety cautions and icons before beginning this lab activity. Supervise students during this activity. Provide students with ample space and a safe location to test walking and running on the sandy area. Remind each group of students to keep out of the way of others. Have students keep shoes, or at least socks, on while walking on the sandy area.

Lab Notes

Tell the students to imagine that a group of scientists wish to analyze human footprints found in the rocks near fossilized remains and that the scientists have contacted the class to help with the investigation. The scientists want to know how the students intend to gather information to make inferences about the humans who left the prints. Explain that a scientist should be able to make the same type of inferences about an organism from fresh tracks as from preserved tracks. Use the mystery footprints in the sand to help students design investigations for gathering data. From the data, students can learn to draw inferences.

A large proportion of research on evolution depends on making scientific inferences and checking for corroboration among different sources of data. To conclude the laboratory experience, lead the students in a discussion about the importance of large sets of data in helping scientists make inferences.

Chapter Review

Assignment Guide	
Section	**Questions**
1	5, 7, 11, 12, 20–22, 26, 27
2	1–4, 8, 9, 13–16, 19, 23, 24
3	6, 10, 17–19, 25

ANSWERS

Using Key Terms

1. Precambrian time
2. Cenozoic era
3. Mesozoic era
4. Paleozoic era
5. Absolute dating tries to determine a specific range of dates, while relative dating determines the order of events but not specific dates.
6. Hominids are a type of primate, but not all primates are hominids.

Understanding Key Ideas

7. d
8. a
9. d
10. d

Chapter Review

USING KEY TERMS

Complete each of the following sentences by choosing the correct term from the word bank.

Precambrian time	Paleozoic era
Mesozoic era	Cenozoic era

1. During ___, life is thought to have originated from nonliving matter.
2. The Age of Mammals refers to the ___.
3. The Age of Reptiles refers to the ___.
4. Plants colonized land during the ___.

For each pair of terms, explain how the meanings of the terms differ.

5. *relative dating* and *absolute dating*
6. *primates* and *hominids*

UNDERSTANDING KEY IDEAS

Multiple Choice

7. If the half-life of an unstable element is 5,000 years, what percentage of the parent material will be left after 10,000 years?
   a. 100%
   b. 75%
   c. 50%
   d. 25%

8. The first cells on Earth appeared in
   a. Precambrian time.
   b. the Paleozoic era.
   c. the Mesozoic era.
   d. the Cenozoic era.

9. In which era are we currently living?
   a. Precambrian time
   b. Paleozoic era
   c. Mesozoic era
   d. Cenozoic era

10. Scientists think that the closest living relatives of humans are
    a. lemurs.
    b. monkeys.
    c. gorillas.
    d. chimpanzees.

Short Answer

11. Describe how plant and animal remains can become fossils.
12. What information do fossils provide about the history of life?
13. List three important steps in the early development of life on Earth.
14. List two important groups of organisms that appeared during each of the three most recent geologic eras.
15. Describe the event that scientists think caused the mass extinction at the end of the Mesozoic era.

11. Sample answer: The remains of organisms can become fossils when geologic processes preserve them or their traces. An example is sediment covering an organism and the shape becoming hardened into rock.
12. Fossils tell us about the kinds of organisms that existed and the way they changed over time.
13. Simple chemicals reacted to form the molecules that make up life. Simple cells formed. Photosynthetic cells developed and began to produce oxygen.
14. Sample answer: Precambrian: prokaryotes and eukaryotes; Paleozoic: multicellular organisms, plants, insects, and amphibians; Mesozoic: dinosaurs and other reptiles, birds, and small mammals; Cenozoic: large mammals, including humans, and more-diverse birds and insects
15. A giant meteorite hit the Earth, disturbing ecosystems and causing climate change.

16 From which geologic era are fossils most commonly found?

17 Describe two characteristics that are shared by all primates.

18 Which hominid species is alive today?

CRITICAL THINKING

19 **Concept Mapping** Use the following terms to create a concept map: *Earth's history, humans, Paleozoic era, dinosaurs, Precambrian time, land plants, Mesozoic era, cyanobacteria,* and *Cenozoic era.*

20 **Applying Concepts** Can footprints be fossils? Explain your answer.

21 **Making Inferences** If you find rock layers containing fish fossils in a desert, what can you infer about the history of the desert?

22 **Applying Concepts** Explain how an environmental change can threaten the survival of a species. Give two examples.

23 **Analyzing Ideas** Why do scientists think the first cells did not need oxygen to survive?

24 **Identifying Relationships** How does the extinction that occurred at the end of the Mesozoic era relate to the Age of Mammals?

25 **Making Comparisons** Make a table listing the similarities and differences between australopithecines, early members of the group *Homo*, and modern members of the species *Homo sapiens*.

INTERPRETING GRAPHICS

The graph below shows data about fossilized teeth that were found within a series of rock layers. Use this graph to answer the questions that follow.

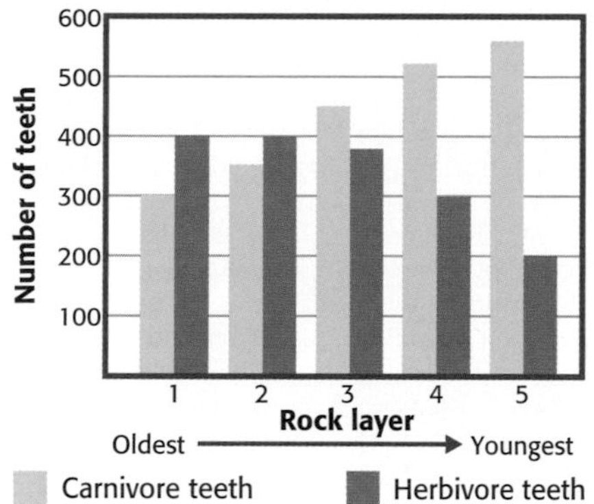

26 Which of the following statements best describes the information presented in the graph?

a. Over time, the number of carnivores decreased and the number of herbivores increased.

b. Over time, the number of carnivores increased and the number of herbivores increased.

c. Over time, the number of carnivores and herbivores remained the same.

d. Over time, the number of carnivores increased and the number of herbivores decreased.

27 At what point did carnivore teeth begin to outnumber herbivore teeth?

a. between layer 1 and layer 2

b. between layer 2 and layer 3

c. between layer 3 and layer 4

d. between layer 4 and layer 5

16. Fossils in the Cenozoic era are most common because they are closer to the surface of Earth and were formed more recently.

17. Primates have opposable thumbs that allow them to grasp objects, and and they have eyes positioned at the front of their heads that allow them to tell how far away something is.

18. *Homo sapiens*

Critical Thinking

19. 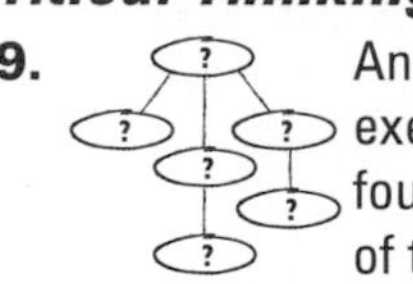

An answer to this exercise can be found at the end of this book.

20. yes; A footprint is a trace left by an organism.

21. that the rocks there may have once been underwater

22. Sample answer: If an environment changes suddenly, species may not be adapted to survive the change. Examples might be a sudden change from a hot to a cold climate or a sudden change from a wet to a dry environment.

23. There was no oxygen available when the first cells developed.

24. After the dinosaurs were wiped out (with the exception of birds), mammals were able to take their place in ecosystems.

25. Sample answer:

	Australo-pithecines	Early homo	*Homo sapiens*
Bipedalism	yes	yes	yes
Brains	medium	larger	largest
Tools	none	some	many
Art	none	none	a lot
Known locations	Africa	several continents	world-wide

Interpreting Graphics

26. d

27. b

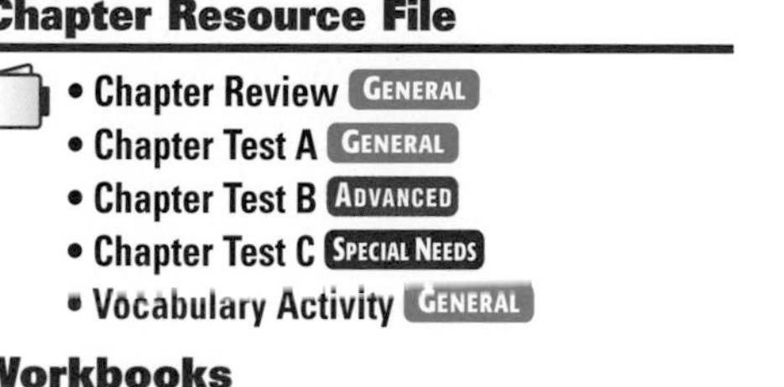

CHAPTER RESOURCES

Chapter Resource File

- Chapter Review GENERAL
- Chapter Test A GENERAL
- Chapter Test B ADVANCED
- Chapter Test C SPECIAL NEEDS
- Vocabulary Activity GENERAL

Workbooks

Study Guide

- Assessment resources are also available in Spanish.

Standardized Test Preparation

Teacher's Note

To provide practice under more realistic testing conditions, give students 20 minutes to answer all of the questions in this Standardized Test Preparation.

MISCONCEPTION ALERT

Answers to the standardized test preparation can help you identify student misconceptions and misunderstandings.

READING

Passage 1

1. D
2. F
3. B

Question 2: The answer is drawn by inference from the fourth sentence in the passage, ". . . its bladelike teeth meant certain death for its prey." Students who miss this question may have missed the implication or may have difficulty with long sentences.

Question 3: The answer is drawn from the fourth sentence in the passage. Answers A and C are false, given the facts in the passage. Students who chose answer D may have made an incorrect inference from the last sentence in the passage.

Standardized Test Preparation

READING

Read each of the passages below. Then, answer the questions that follow each passage. Decide which is the best answer to each question.

Passage 1 In 1995, paleontologist Paul Sereno and his team were working in an unexplored region of Morocco when they made an incredible find—an enormous dinosaur skull! The skull measured about 1.6 m in length, which is about the height of a refrigerator. Given the size of the skull, Sereno concluded that the skeleton of the entire animal must have been about 14 m long—about as big as a school bus, and even larger than *Tyrannosaurus rex.* This 90-million-year-old predator most likely chased other dinosaurs by running on large, powerful hind legs, and its bladelike teeth meant certain death for its prey. Sereno named his new discovery *Carcharodontosaurus saharicus*, which means "shark-toothed reptile from the Sahara."

1. Paul Sereno estimated the total size of this *Carcharodontosaurus* based on
- **A** the size of *Tyrannosaurus rex.*
- **B** the fact that it was a predator.
- **C** the fact that it had bladelike teeth.
- **D** the fact that its skull was 1.6 m long.

2. Which of the following is evidence that the *Carcharodontosaurus* was a predator?
- **F** It had bladelike teeth.
- **G** It had a large skeleton.
- **H** It was found with the bones of a smaller animal nearby.
- **I** It is 90 million years old.

3. Which of the following is a fact in the passage?
- **A** *Carcharodontosaurus* was the largest predator that ever existed.
- **B** *Carcharodontosaurus* had bladelike teeth.
- **C** *Carcharodontosaurus* was as large as *Tyrannosaurus rex.*
- **D** *Carcharodontosaurus* was a shark-like reptile.

Passage 2 In 1912, Alfred Wegener proposed a hypothesis called *continental drift.* At the time, many scientists laughed at his idea. Yet Wegener's idea jolted the very foundations of geology.

Wegener used rock, fossil, and glacial evidence from opposite sides of the Atlantic Ocean to support the idea that continents can "drift." For example, Wegener recognized similar rocks and rock structures in the Appalachian Mountains and the Scottish Highlands, as well as similarities between rock layers in South Africa and Brazil. He thought that these striking similarities could be explained only if these geologic features were once part of the same continent. Wegener proposed that because continents are less dense, they float on top of the denser rock of the ocean floor.

Although continental drift explained many of Wegener's observations, he could not find evidence to explain exactly how continents move. But by the 1960s, this evidence was found and continental drift was well understood. However, Wegener's contributions went unrecognized until years after his death.

1. Which of the following did Wegener use as evidence to support his hypothesis?
- **A** similarities between nearby rock layers
- **B** similarities between rock layers in different parts of the world
- **C** a hypothesis that continents float
- **D** an explanation of how continents move

2. Which of the following statements is supported by the above passage?
- **F** A hypothesis is never proven.
- **G** A new hypothesis may take many years to be accepted by scientists.
- **H** The hypothesis of continental drift was not supported by evidence.
- **I** Wegener's hypothesis was proven wrong.

Passage 2

1. B
2. G

Question 1: The answer is drawn from the second paragraph in the passage. Students who missed this question may have failed to distinguish between evidence that supports a hypothesis and other ideas presented in the passage. Answer A is an idea not present in the passage.

Question 2: The answer requires logical reasoning and an inference of the main idea of the passage. Answers H and I are logically inconsistent with the passage. Answer F is a true statement but not supported by the passage.

INTERPRETING GRAPHICS

The map below shows the areas where fossils of certain organisms have been found. Use the map below to answer the questions that follow.

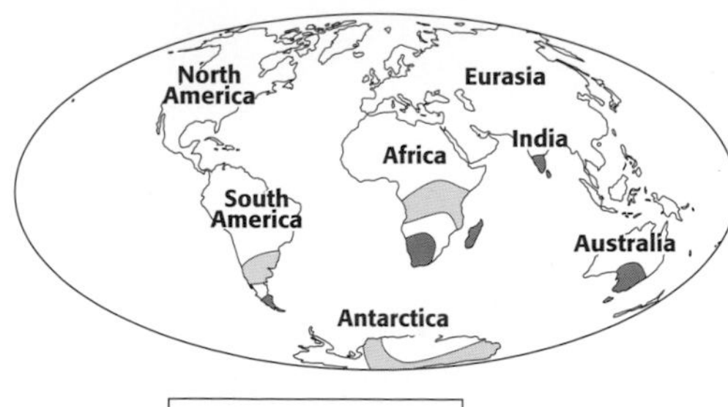

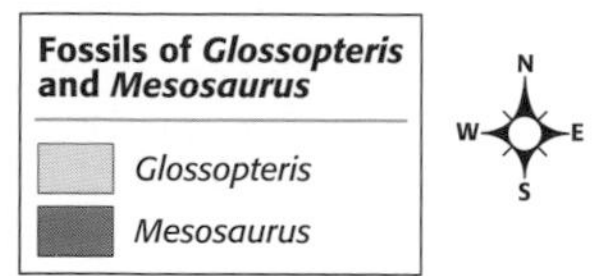

1. *Mesosaurus* was a small, aquatic reptile and *Glossopteris* was an ancient plant species. What do these two have in common?
   A Their fossils have been found on several continents.
   B Their fossils are found in exactly the same places.
   C Their fossils have been found only in North America.
   D Their fossils have only been found near oceans.

2. Which of the following statements is best supported by these findings?
   F All of the continents were once connected to each other.
   G South America was once connected to Africa.
   H *Glossopteris* is adapted to life at the South Pole.
   I *Mesosaurus* could swim.

3. The map provides evidence that the following continents were once connected to each other, with the exception of
   A North America.
   B Africa.
   C Antarctica.
   D South America.

MATH

Read each question below, and choose the best answer.

1. Four students are sharing a birthday cake. The first student takes half of the cake. The second student take half of what remains of the cake. Then, the third student takes half of what remains of the cake. What fraction of the cake is left for the fourth student?
   A 1/2
   B 1/4
   C 1/8
   D 1/16

2. One sixteenth is equal to what percentage?
   F 6.25%
   G 12.5%
   H 25%
   I 50%

3. What is one-half of one-fourth?
   A 1/2
   B 1/4
   C 1/8
   D 1/16

4. Half-life is the time it takes for one-half of the radioactive atoms in a rock sample to decay, or change into different atoms. Carbon-14 is a radioactive isotope with a half-life of 5,730 years. In a sample that is 11,460 years old, what percentage of carbon-14 from the original sample would remain?
   F 100%
   G 50%
   H 25%
   I 12.5%

5. If a sample contains an isotope with a half-life of 10,000 years, how old would the sample be if 1/8 of the original isotope remained in the sample?
   A 5,000 years
   B 10,000 years
   C 20,000 years
   D 30,000 years

CHAPTER RESOURCES

Chapter Resource File

• Standardized Test Preparation GENERAL

State Resources

For specific resources for your state, visit go.hrw.com and type in the keyword HSMSTR.

INTERPRETING GRAPHICS

1. A
2. G
3. A

TEST DOCTOR

Question 1: The answer requires careful viewing of the graphic to find information that supports the correct answer. Simply subsituting "the two kinds of shading" for "their fossils" in each statement makes the answer more apparent. Answer A is supported by observing the presence of shading on several continents. Answers B, C, and D are false, as indicated by observing locations where the shading contradicts the statement.

Question 2: The answer requires the student to locate continents in the graphic that have no shading to indicate the presence of either type of fossil. Answers B, C, and D are each false because the continents listed do have some shading on them.

MATH

1. C
2. F
3. C
4. H
5. D

TEST DOCTOR

Question 1: This question is equivalent to the following problem:

$$1 \times {}^1/_2 \times {}^1/_2 \times {}^1/_2$$

Students who miss this question may have difficulty conceptualizing fractions of fractions or may have multiplied by 1/2 an incorrect number of times.

Science in Action

Science, Technology, and Society

Background

The computer-generated skull image is reconstructed from the fossil skull of a Neanderthal child. The fossil find is named *Le Moustier 1*; it was excavated from a cave in Le Moustier, France. This site is also important because of the unique types of tools found there. The tools indicate a different form of toolmaking than is found in many other Neanderthal sites. One theory related to this finding is that Neanderthals learned new toolmaking techniques from *Homo sapiens* at some point when the two groups came in contact.

Scientific Debate

Discussion — GENERAL

Ask students to discuss what they might do if human fossils were discovered in their backyard. Then, discuss a scenario in which a company is prevented or delayed from conducting business when human fossils are discovered on their property. Brainstorm other similar scenarios or research and discuss real scenarios with which the students might be familiar. Ask students to think of ways that these conflicts might be resolved.

Science in Action

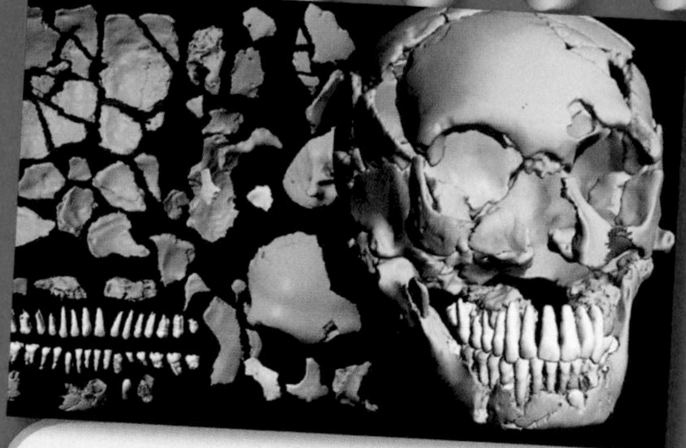

Residents of this neighborhood in Jerusalem, Israel, objected when anthropologists started to dig in the area.

Science, Technology, and Society

Using Computers to Examine Fossils

Paleontologists want to examine fossils without taking apart or damaging the fossils. Fortunately, they can now use a technology called *computerized axial tomography*, or *CAT scanning,* which provides views inside objects without touching the objects. A CAT scan is a series of cross-section pictures of an object. A computer can assemble these "slices" to create a three-dimensional picture of the entire object. Computer graphic programs can also be used to move pictures of fossil pieces around to see how the pieces fit together. The fossil skull above was reconstructed using CAT scans and computers.

Math ACTIVITY

The average volume of a Neanderthal adult's brain was about 1,400 cm^3, while that of an adult gorilla is about 400 cm^3. Calculate how much larger a Neanderthal brain was than a gorilla brain. Express your answer as a percentage.

Scientific Debate

Who Owns the Past?

Does a piece of land include all the layers below it? If you start digging, you may find evidence of past life. In areas that have been inhabited by human ancestors, you may find artifacts that they left behind. But who has the right to dig up these "leftovers" from the past? And who owns them?

In areas that contain many remains of the past, digging up land often leads to conflicts. Landowners may want to build on their own land. But when remains of ancient human cultures are found, living relatives of those cultures may lay claim to the remains. Paleontologists are often caught in the middle, because they want to study and preserve evidence of past life.

Social Studies ACTIVITY

WRITING SKILL Research an area where there is a debate over what to do with fossils or remains of human ancestors. Write a newspaper article about the issue. Be sure to present all sides of the debate.

Answer to Math Activity

350%

Sample calculation:

$1,400 \div 400 = 3.5$

$3.5 \times 100\% = 350\%$

Answer to Social Studies Activity

Student articles should reflect journalistic style, being both interesting and objective. Check that students have presented more than one side of the issue they have chosen. Almost any large urban area will have some notable archaeological or paleontological sites, and efforts to excavate the sites are often inconvenient to someone. However, if students need help selecting an area, suggest one of the following locations: New York City, New York; Miami, Florida; Rome, Italy; Athens, Greece; Island of Brac, Croatia; Jerusalem, Israel; Ayodhya, India; Kathmandhu, Nepal; or Yangtze River, China.

People in Science

The Leakey Family

A Family of Fossil Hunters In some families, a business is passed down from one generation to the next. For the Leakey's, the family business is paleoanthropology (PAY lee OH AN thruh PAWL uh jee)—the study of the origin of humans. The first famous Leakey was Dr. Louis Leakey, who was known for his hominid fossil discoveries in Africa in the 1950s. Louis formed many important hypotheses about human evolution. Louis' wife, Mary, made some of the most-important hominid fossil finds of her day.

Louis and Mary's son, Richard, carried on the family tradition of fossil hunting. He found his first fossil, which was of an extinct pig, when he was six years old. As a young man, he went on safari expeditions in which he collected photographs and specimens of African wildlife. Later, he met and married a zoologist named Meave. The photo at right shows Richard (right), Meave (left), and their daughter Louise (middle) Each of the Leakeys has contributed important finds to the study of ancient hominids.

Language Arts Activity

WRITING SKILL Visit the library and look for a book by or about the Leakey family and other scientists who have worked with them. Write a short book review to encourage your classmates to read the book.

To learn more about these Science in Action topics, visit **go.hrw.com** and type in the keyword **HL5HISF.**

Current Science

Check out Current Science® articles related to this chapter by visiting go.hrw.com. Just type in the keyword HL5CS08.

People In Science

Background

In Africa, Richard Leakey may be more widely known for his strong beliefs in political rather than scientific arenas. He has organized and raised money for campaigns against the poaching of wildlife in Kenya. And in 1995, he founded a political party that opposes corruption in the Kenyan government. His political work has been controversial and at times dangerous, as he has received death threats and beatings from opponents.

Activity — GENERAL

Have interested students research and make timelines of major discoveries in hominid evolution. Suggest that they include a map that pinpoints the location of each major discovery. They may wish to present their findings in a poster or Web page.

Answer to Language Arts Activity

Students can find numerous books about members of the Leakey family, books written by some of the Leakeys, and books about their colleagues, such as Dian Fossey, Jane Goodall, Birute Galdikas, and Donald Johansen.

Classification

Chapter Planning Guide

Compression guide:
To shorten instruction because of time limitations, omit the Chapter Lab.

OBJECTIVES	LABS, DEMONSTRATIONS, AND ACTIVITIES	TECHNOLOGY RESOURCES
PACING • 90 min pp. 162–169 **Chapter Opener**	**SE Start-up Activity,** p. 163 GENERAL	**OSP Parent Letter** ■ GENERAL **CD Student Edition on CD-ROM** **CD Guided Reading Audio CD** ■ **TR Chapter Starter Transparency*** **VID Brain Food Video Quiz**
Section 1 Sorting It All Out • Explain how to classify organisms. • List the seven levels of classification. • Explain scientific names. • Describe how dichotomous keys help in identifying organisms.	**TE Demonstration** Classifying Objects, p. 164 GENERAL **SE Quick Lab** A Branching Diagram, p. 165 GENERAL **CRF Datasheet for Quick Lab*** **TE Activity** Branching Diagrams, p. 165 GENERAL **SE Skills Practice Lab** Shape Island, p. 176 GENERAL **CRF Datasheet for Chapter Lab*** **LB EcoLabs & Field Activities** Water Wigglers* GENERAL	**CRF Lesson Plans*** **TR Bellringer Transparency*** **TR** Evolutionary Relationships Among Organisms* **TR** Levels of Classification* **TR** A Dichotomous Key* **SE Internet Activity,** p. 166 GENERAL **VID Lab Videos for Life Science**
PACING • 45 min pp. 170–175 **Section 2 The Six Kingdoms** • Explain how classification schemes for kingdoms developed as greater numbers of different organisms became known. • Describe each of the six kingdoms.	**TE Activity** Grouping Animals, p. 170 GENERAL **TE Connection Activity** Environmental Science, p. 171 GENERAL **TE Connection Activity** Real World, p. 172 ADVANCED **TE Activity** Plant Identification, p. 173 BASIC **SE Connection to Social Studies** Animals That Help, p. 174 GENERAL **LB Long-Term Projects & Research Ideas** The Panda Mystery* ADVANCED **SE Inquiry Lab** Voyage of the USS *Adventure,* p. 190 GENERAL **CRF Datasheet for LabBook***	**CRF Lesson Plans*** **TR Bellringer Transparency*** **TR** *LINK TO EARTH SCIENCE* Intrusive Igneous Rock Formations* **CRF SciLinks Activity*** GENERAL

PACING • 90 min

CHAPTER REVIEW, ASSESSMENT, AND STANDARDIZED TEST PREPARATION

CRF Vocabulary Activity* GENERAL
SE Chapter Review, pp. 178–179 GENERAL
CRF Chapter Review* ■ GENERAL
CRF Chapter Tests A* ■ GENERAL, **B*** ADVANCED, **C*** SPECIAL NEEDS
SE Standardized Test Preparation, pp. 180–181 GENERAL
CRF Standardized Test Preparation* GENERAL
CRF Performance-Based Assessment* GENERAL
OSP Test Generator GENERAL
CRF Test Item Listing* GENERAL

Online and Technology Resources

Visit **go.hrw.com** for a variety of free resources related to this textbook. Enter the keyword **HL5CLS.**

Students can access interactive problem-solving help and active visual concept development with the *Holt Science and Technology* Online Edition available at **www.hrw.com.**

Guided Reading Audio CD

A direct reading of each chapter using instructional visuals as guideposts. For auditory learners, reluctant readers, and Spanish-speaking students. Available in English and Spanish.

KEY

SE Student Edition	CRF	Chapter Resource File	SS	Science Skills Worksheets	*	Also on One-Stop Planner
TE Teacher Edition	OSP	One-Stop Planner	MS	Math Skills for Science Worksheets	◆	Requires advance prep
	LB	Lab Bank	CD	CD or CD-ROM	■	Also available in Spanish
	TR	Transparencies	VID	Classroom Video/DVD		

SKILLS DEVELOPMENT RESOURCES	SECTION REVIEW AND ASSESSMENT	STANDARDS CORRELATIONS
SE Pre-Reading Activity, p. 162 GENERAL **OSP Science Puzzlers, Twisters & Teasers** GENERAL		National Science Education Standards UCP 1
CRF Directed Reading A* ■ BASIC, **B*** SPECIAL NEEDS **CRF Vocabulary and Section Summary*** ■ GENERAL **SE Reading Strategy** Reading Organizer, p. 164 GENERAL **TE Connection to Math,** p. 165 BASIC **TE Inclusion Strategies,** p. 168 **MS Math Skills for Science** A Shortcut for Multiplying Large Numbers* GENERAL	**CRF Critical Thinking** A Breach on Planet Biome* ADVANCED **SE Reading Checks,** pp. 164, 167, 168 GENERAL **TE Reteaching,** p. 168 BASIC **TE Quiz,** p. 168 GENERAL **TE Alternative Assessment,** p. 168 GENERAL **SE Section Review,*** p. 169 ■ GENERAL **CRF Section Quiz*** ■ GENERAL	UCP 1; SAI 2; HNS 1, 2, 3; LS 5a
CRF Directed Reading A* ■ BASIC, **B*** SPECIAL NEEDS **CRF Vocabulary and Section Summary*** ■ GENERAL **SE Reading Strategy** Reading Organizer, p. 170 GENERAL **TE Reading Strategy** Prediction Guide, p. 171 BASIC **SE Math Practice** Ring-Around-the-Sequoia, p. 173 GENERAL **TE Inclusion Strategies,** p. 173 **MS Math Skills for Science** Arithmetic with Decimals* GENERAL **CRF Reinforcement Worksheet** Keys to the Kingdoms* GENERAL	**SE Reading Checks,** pp. 171, 173, 175, GENERAL **TE Homework,** p. 172 ADVANCED **TE Reteaching,** p. 174 BASIC **TE Quiz,** p. 174 GENERAL **TE Alternative Assessment,** p. 174 GENERAL **SE Section Review,*** p. 175 ■ GENERAL **CRF Section Quiz*** ■ GENERAL	UCP 5; SAI 1; HNS 1, 2; LS 1b, 1f, 2a, 2c, 4b, 4c, 4d, 5b; *LabBook:* UCP 1; SAI 1

This convenient CD-ROM includes:
- **Lab Materials QuickList Software**
- **Holt Calendar Planner**
- **Customizable Lesson Plans**
- **Printable Worksheets**
- **ExamView® Test Generator**

cnnstudentnews.com

Find the latest news, lesson plans, and activities related to important scientific events.

www.scilinks.org

Maintained by the **National Science Teachers Association.** See Chapter Enrichment pages for a complete list of topics.

Current Science®

Check out ***Current Science*** articles and activities by visiting the HRW Web site at **go.hrw.com.** Just type in the keyword **HL5CS09T.**

Classroom Videos
- **Lab Videos** demonstrate the chapter lab.
- **Brain Food Video Quizzes** help students review the chapter material.

7 Chapter Resources

Visual Resources

CHAPTER STARTER TRANSPARENCY

Classification — CHAPTER STARTER

This Really Happened!

Skunks have been thrown out of their family. It wasn't their awful smell that got them thrown out, though. It was their DNA.

Skunks were once thought to be most closely related to weasels, ferrets, minks, badgers, and otters. Those furry, short-legged, long-bodied, meat-eating mammals are grouped together in a family called Mustelidae (moo STEL i dee). *Mustelidae* is from the Latin word for "mouse." Skunks were classified along with weasels and ferrets because they all share several physical characteristics with mice, such as short, round ears and short legs.

However, a researcher at the University of New Mexico's Museum of Southwestern Biology discovered that the DNA of skunks is very different from the DNA of the other members of Mustelidae. By comparing the DNA of different species, scientists can tell how closely related the species are. The DNA of two closely related animals—a house cat and a tiger, for example—are more similar than the DNA of two animals that are distantly related—such as a house cat and a chicken.

So where does that leave the little striped stinkers? Right in their own, newly created scientific family—Mephitidae (me FIT i dee). *Mephitid* is from the Latin word that means "bad odor"!

In this chapter you will learn why scientific names are important and how scientists classify organisms. You will also learn about the six major kingdoms into which all organisms are classified.

BELLRINGER TRANSPARENCIES

Classification — BELLRINGER TRANSPARENCY

Section: Sorting It All Out

Think about the different ways humans classify things. List five groups of things that humans classify, such as library books, department store merchandise, and addresses. Is there such a thing as too much classification? What happens when you put something in the wrong group? Can objects or ideas belong in more than one group at the same time?

Record your responses in your **science journal.**

Section: The Six Kingdoms

List seven musical artists, bands, or acts. Categorize the names on your list by style of music. Describe the categories you chose, and explain which bands might fit into more than one category.

Record your responses in your **science journal.**

TEACHING TRANSPARENCIES

Classification — TEACHING TRANSPARENCY

Levels of Classification

Kingdom Animalia	Phylum Chordata	Class Mammalia	Order Carnivora
All animals are in the kingdom Animalia.	All animals in the phylum Chordata have a hollow nerve cord. Most have a backbone.	Animals in the class Mammalia have a backbone. They also nurse their young.	Animals in the order Carnivora have a backbone and nurse their young. They also have special teeth for tearing meat.

Family Felidae	Genus Felis	Species Felis domesticus
Animals in the family Felidae are cats. They have a backbone, nurse their young, have special teeth for tearing meat, and have retractable claws.	Animals in the genus Felis have traits of other animals in the same family. However, these cats cannot roar; they can only purr.	The species Felis domesticus is the common house cat. The house cat shares traits with all of the organisms in the levels above the species level, but it also has unique traits.

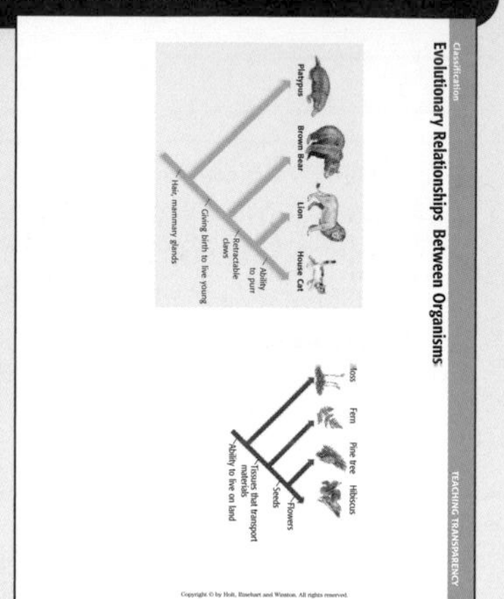

TEACHING TRANSPARENCIES

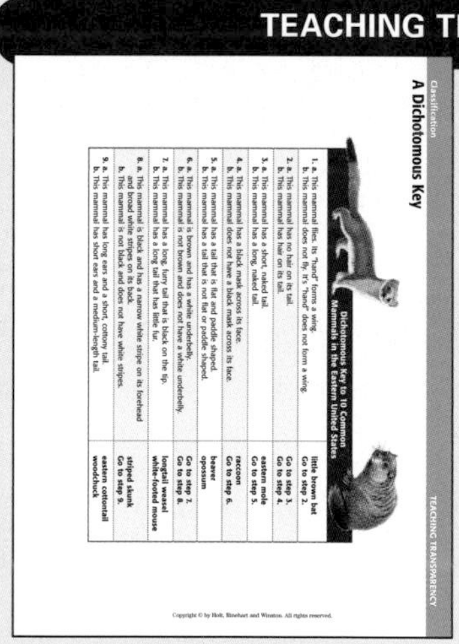

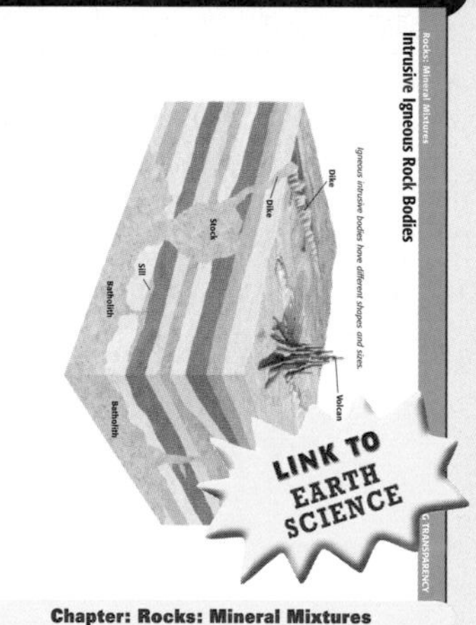

Chapter: Rocks: Mineral Mixtures

CONCEPT MAPPING TRANSPARENCY

The History of Life on Earth — CONCEPT MAPPING

Use the following terms to complete the concept map below:
geologic time scale, mammals, Mesozoic, paleontologists, eukaryotes, relative dating, absolute dating, Paleozoic, dinosaurs, Precambrian

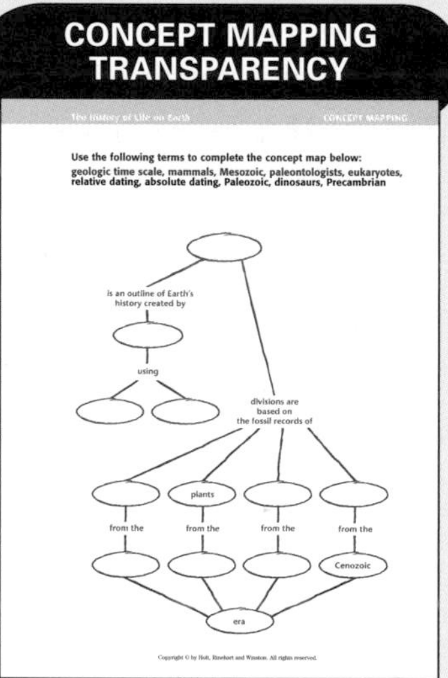

Planning Resources

LESSON PLANS

Lesson Plan SAMPLE

Section: Waves

Pacing

Regular Schedule: with lab(s):2 days without lab(s)2 days
Block Schedule: with lab(s):1 1/2 days without lab(s)1 day

Objectives

1. Relate the seven properties of life to a living organism.
2. Describe seven themes that can help you to organize what you learn about biology.
3. Identify the tiny structures that make up all living organisms.
4. Differentiate between reproduction and heredity and between metabolism and homeostasis.

National Science Education Standards Covered

LSInter6: Cells have particular structures that underlie their functions.

LSMat1: Most cell functions involve chemical reactions.

LSBeh1: Cells store and use information to guide their functions.

UCP1: Cell functions are regulated.

SI1: Cells can differentiate and form complete multicellular organisms.

PS1: Species evolve over time.

ESS1: The great diversity of organisms is the result of more than 3.5 billion years of evolution.

ESS2: Natural selection and its evolutionary consequences provide a scientific explanation for the fossil record of ancient life forms as well as for the striking molecular similarities observed among the diverse species of living organisms.

ST1: The millions of different species of plants, animals, and microorganisms that live on Earth today are related by descent from common ancestors.

ST2: The energy for life primarily comes from the sun.

SPSP1: The complexity and organization of organisms accommodates the need for obtaining, transforming, transporting, releasing, and eliminating the matter and energy used to sustain the organism.

SPSP6: As matter and energy flows through different levels of organization of living systems—cells, organs, communities—and between living systems and the physical environment, chemical elements are recombined in different ways.

HNS1: Organisms have behavioral responses to internal changes and to external stimuli.

PARENT LETTER

SAMPLE

Dear Parent,

Your son's or daughter's science class will soon begin exploring the chapter entitled "The World of Physical Science." In this chapter, students will learn about how the scientific method applies to the world of physical science and the role of physical science in the world. By the end of the chapter, students should demonstrate a clear understanding of the chapter's main ideas and be able to discuss the following topics:

1. physical science as the study of energy and matter (Section 1)
2. the role of physical science in the world around them (Section 1)
3. careers that rely on physical science (Section 1)
4. the steps used in the scientific method (Section 2)
5. examples of technology (Section 2)
6. how the scientific method is used to answer questions and solve problems (Section 2)
7. how our knowledge of science changes over time (Section 2)
8. how models represent real objects or systems (Section 3)
9. examples of different ways models are used in science (Section 3)
10. the importance of the International System of Units (Section 4)
11. the appropriate units to use for particular measurements (Section 4)
12. how area and density are derived quantities (Section 4)

Questions to Ask Along the Way

You can help your son or daughter learn about these topics by asking interesting questions such as the following:

- What are some surprising careers that use physical science?
- What is a characteristic of a good hypothesis?
- When is it a good idea to use a model?
- Why do Americans measure things in terms of inches and yards [...] and meters ?

ALSO IN SPANISH

TEST ITEM LISTING

TEST ITEM LISTING
The World of Science SAMPLE

MULTIPLE CHOICE

1. A limitation of models is that
   a. they are large enough to see.
   b. they do not act exactly like the things that they model.
   c. they are smaller than the things that they model.
   d. they model unfamiliar things.
   Answer: B Difficulty: I Section: 3 Objective: 2
2. The length 10 m is equal to
   a. 100 cm. c. 10,000 mm.
   b. 1,000 cm. d. Both (b) and (c)
   Answer: B Difficulty: I Section: 3 Objective: 2
3. To be valid, a hypothesis must be
   a. testable. c. made into a law.
   b. supported by evidence. d. Both (a) and (b)
   Answer: B Difficulty: I Section: 3 Objective: 2 1
4. The statement "Sheila has a stain on her shirt" is an example of a(n)
   a. law. c. observation.
   b. hypothesis. d. prediction.
   Answer: B Difficulty: I Section: 3 Objective: 2
5. A hypothesis is often developed out of
   a. observations. c. laws.
   b. experiments. d. Both (a) and (b)
   Answer: B Difficulty: I Section: 3 Objective: 2
6. How many milliliters are in 3.5 kL?
   a. 3,500 mL c. 3,500, 000 mL
   b. 0.0035 mL d. 35,000 mL
   Answer: B Difficulty: I Section: 3 Objective: 2
7. A map of Seattle is an example of a
   a. law. c. model.
   b. theory. d. unit.
   Answer: B Difficulty: I Section: 3 Objective: 2
8. A lab has the safety icons shown below. These icons mean that you should wear
   a. only safety goggles. c. safety goggles and a lab apron.
   b. only a lab apron. d. safety goggles, a lab apron, and gloves.
   Answer: B Difficulty: I Section: 3 Objective: 2
9. The law of conservation of mass says the tot al mass before a chemical change is
   a. more than the total mass after the change.
   b. less than the total mass after the change.
   c. the same as the total mass after the change.
   d. not the same as the total mass after the change.
   Answer: B Difficulty: I Section: 3 Objective: 2
10. In which of the following areas might you find a geochemist at work?
    a. studying the chemistry of rocks c. studying fishes
    b. studying forestry d. studying the atmosphere
    Answer: B Difficulty: I Section: 3 Objective: 2

One-Stop Planner® CD-ROM

This CD-ROM includes all of the resources shown here and the following time-saving tools:

- *Lab Materials QuickList Software*
- *Customizable lesson plans*
- *Holt Calendar Planner*
- *The powerful ExamView® Test Generator*

For a preview of available worksheets covering math and science skills, see pages T12–T19. All of these resources are also on the One-Stop Planner®.

Meeting Individual Needs

DIRECTED READING A / DIRECTED READING B

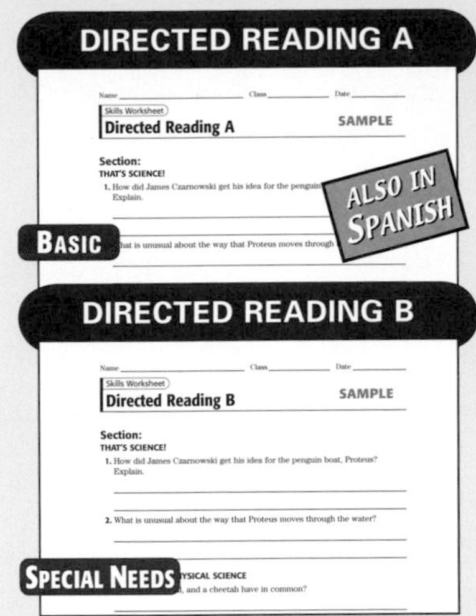

Basic · Also in Spanish · Special Needs

VOCABULARY ACTIVITY / VOCABULARY AND SECTION SUMMARY

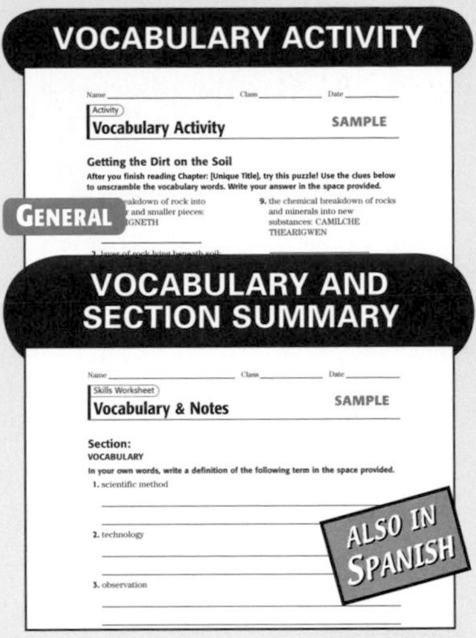

General · Also in Spanish

REINFORCEMENT / CRITICAL THINKING

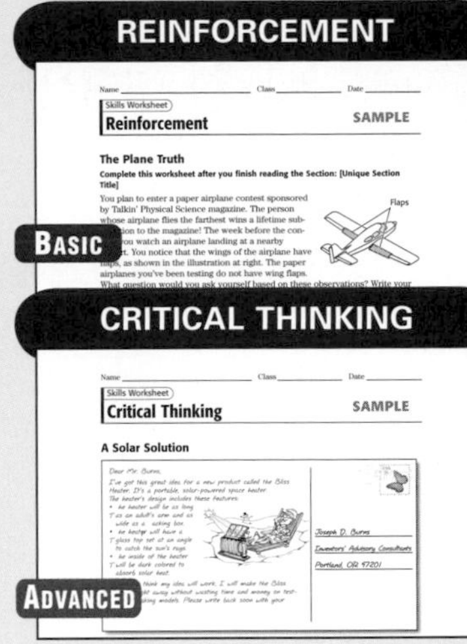

Basic · Advanced

SCILINKS ACTIVITY / SCIENCE PUZZLERS, TWISTERS & TEASERS

General · General

Labs and Activities

ECOLABS & FIELD ACTIVITIES

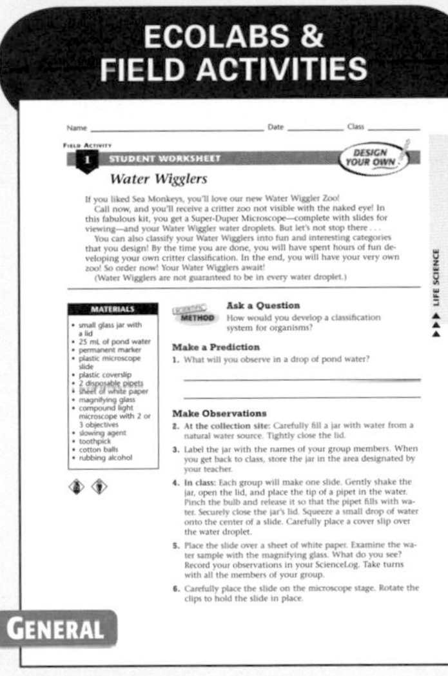

General

LONG-TERM PROJECTS & RESEARCH IDEAS

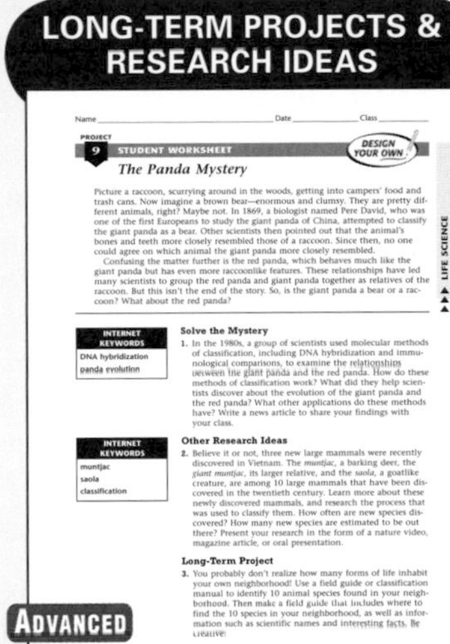

Advanced

DATASHEETS FOR QUICK LABS / DATASHEETS FOR CHAPTER LABS / DATASHEETS FOR LABBOOK

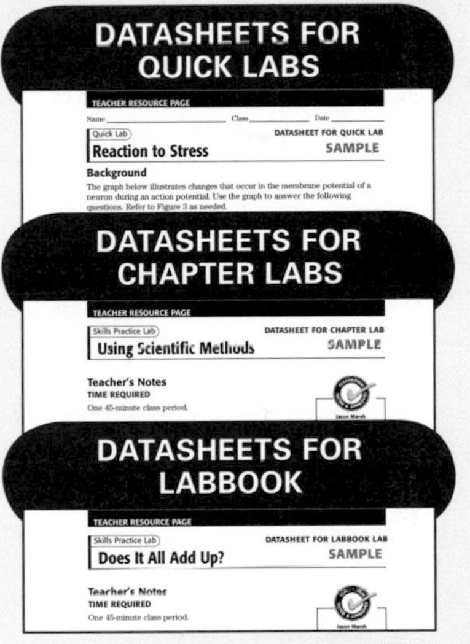

Review and Assessments

SECTION QUIZ / SECTION REVIEW

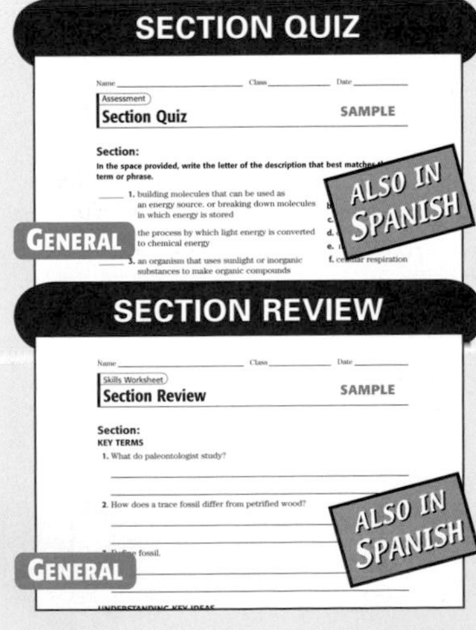

General · Also in Spanish · General · Also in Spanish

CHAPTER REVIEW / CHAPTER TEST A

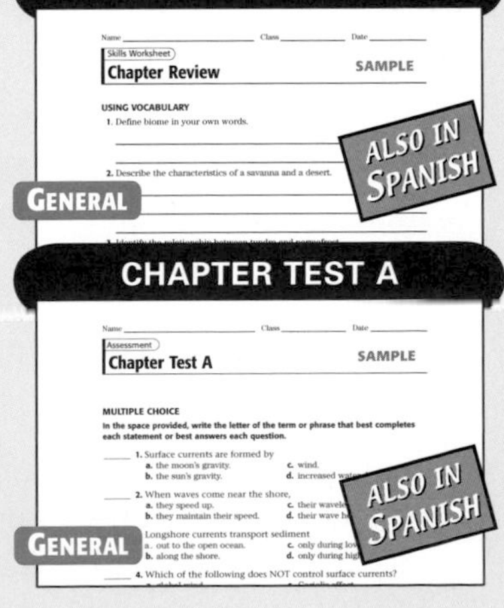

General · Also in Spanish · General · Also in Spanish

CHAPTER TEST B / CHAPTER TEST C

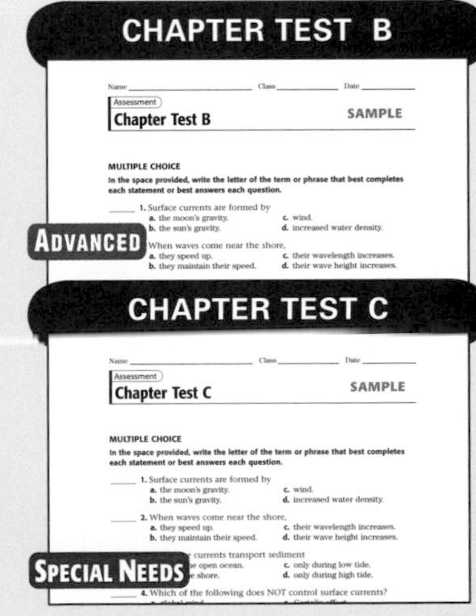

Advanced · Special Needs

STANDARDIZED TEST PREPARATION / PERFORMANCE-BASED ASSESSMENT

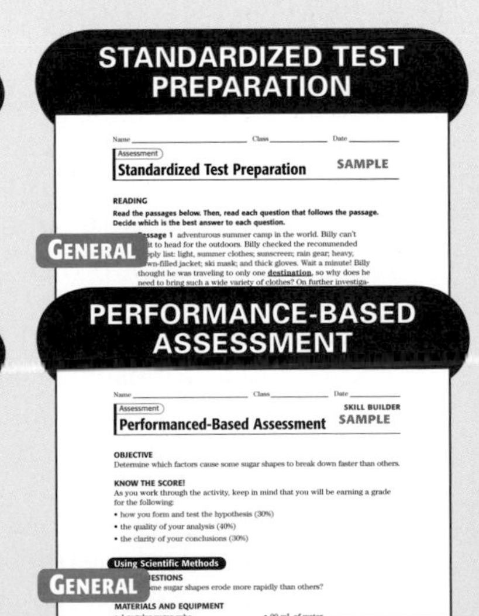

General · General

7 Chapter Enrichment

This Chapter Enrichment provides relevant and interesting information to expand and enhance your presentation of the chapter material.

Section 1

Sorting It All Out

Aristotle's Classification System

- The great Greek philosopher and scientist Aristotle (384–322 BCE) began classifying animals into logical groupings more than 2,000 years ago. Although Aristotle did not view different kinds of organisms as being related by descent, he arranged all living things in an ascending ladder with humans at the top.
- Animals were separated into two major groups—those with red blood and those without red blood—that correspond very closely with our modern classification of vertebrates and invertebrates.
- Animals were further classified according to their way of life, their actions, and their body parts.

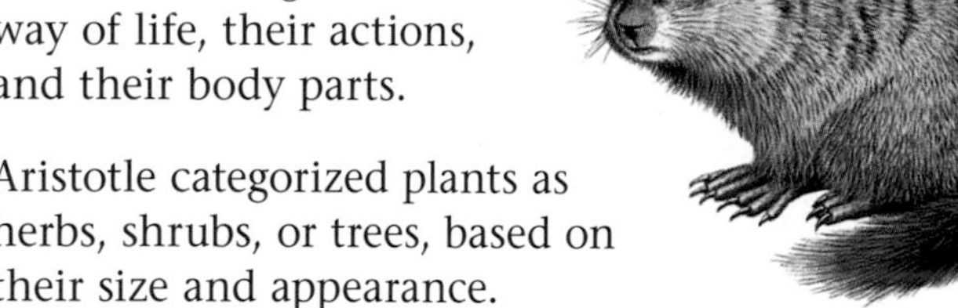

- Aristotle categorized plants as herbs, shrubs, or trees, based on their size and appearance.

Species in Classification

- In the late 1600s, the English scientist John Ray established the species as the basic unit of classification.

Basis for Modern Classification System

- Our modern system of classification was codified by Swedish scientist Carolus Linnaeus. He published a book on plant classification in 1753 and a book on animal classification in 1758.
- Organisms were classified according to their structure. Plants and animals were arranged into the categories of genus and species, and the categories of class and order were introduced.
- Species were given distinctive two-word names. Linnaeus's system is still in use today, although it has gone through many changes.
- "Carolus Linnaeus" is the Latin translation of the Swedish scientist's given name, Carl von Linné.

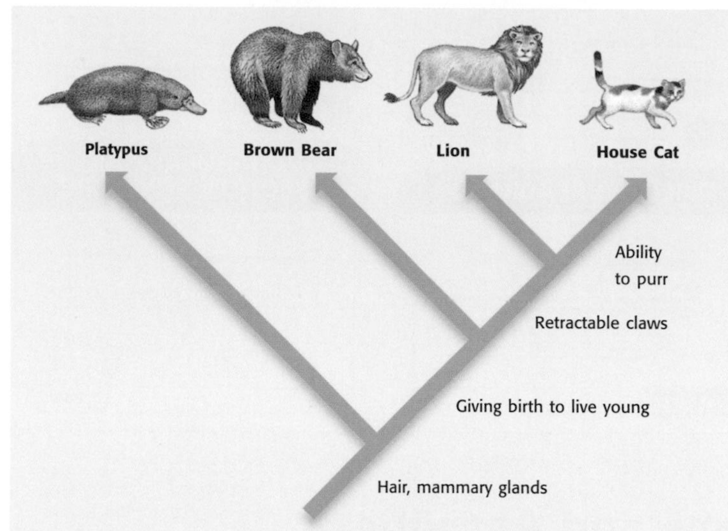

Subgroups in the Animal Kingdom

- Baron Georges Cuvier first divided the animal kingdom into subgroups, such as Vertebrata, Mollusca, Articulata, and Radiata, in 1817.

Section 2

The Six Kingdoms

Variations of the Classification System

- Variations of the five-kingdom classification system introduced by R. H. Whittaker in 1969 are used by some modern scientists. Whittaker's system classifies organisms according to whether they are prokaryotic or eukaryotic, whether they are unicellular or multicellular, and whether they obtain food by photosynthesis, ingestion, or absorption of nutrients from their environment.
- Because studies of DNA indicate significant differences between archaebacteria and eubacteria, many scientists place archaebacteria in a sixth kingdom.

For background information about teaching strategies and issues, refer to the ***Professional Reference for Teachers.***

Different Classification Methods

- Classification methods vary between different fields of biology.
- The classification method microbiologists use is organized by volume, section, family, and genus.
- Some botanists use the term *division* instead of *phylum* for plants and fungi.

Life Within the Planet

- When we organize life on Earth into categories, it is important to remember that organisms are not equally distributed throughout our classification system. We often think of the Earth's living things in terms of plants and animals—organisms that live above the Earth's surface and within its waters. However, the largest kingdom in terms of the number of individuals and total biomass are bacteria. And bacteria's most common home may be deep within the Earth's crust.

- Scientists have known for some time that bacteria exist all around us and that some have the ability to live in extreme environments. Some bacteria live in hot geysers; other bacteria live in water that has such high salt concentrations that no other organisms can survive in the water. Scientists have also known that many archaebacteria can thrive in anaerobic and high-pressure environments, such as those found underground. But only recently have scientists learned just how far underground archaebacteria are found and just how many archaebacteria live there.
- In 1987, scientists were drilling in the rock beneath the Savannah River in South Carolina to investigate the safety of the drinking water. The cores of the rock they investigated harbored bacteria at a depth of 500 m. Other scientists found bacteria in the ocean at a depth 750 m. A South African gold mine yielded other bacteria from as far down as 5 km.
- Once scientists knew to look deep in the Earth for life-forms, they began looking for—and finding—organisms in the sediment under the ocean. Some scientists predict that further exploration will reveal organisms that live as deep as 15 km within the sediment. If that is the case, then the total biomass of these organisms beneath the surface of the Earth may exceed the total biomass of all the living things on the Earth's surface.

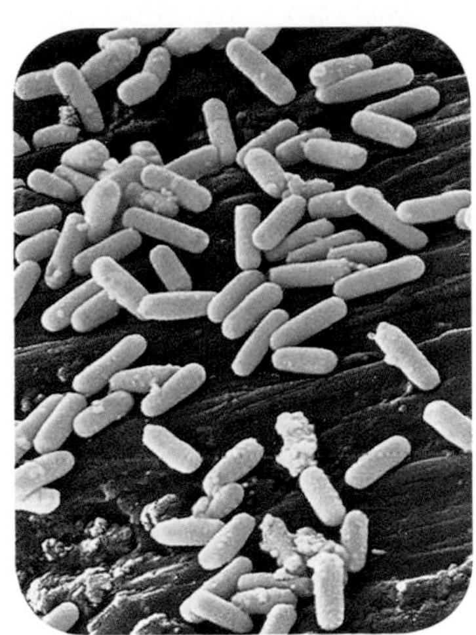

- No one knows exactly how these microorganisms tolerate the tremendous pressures and temperatures of their environment, but scientists have learned that these organisms are meeting their nutritional needs in a variety of ways. Some live on oxidized forms of sulfur; others live on bits of organic matter found in the sediment. Some bacteria have even been found in igneous rocks, where they apparently subsist on the carbon dioxide and hydrogen gas trapped in the rock.

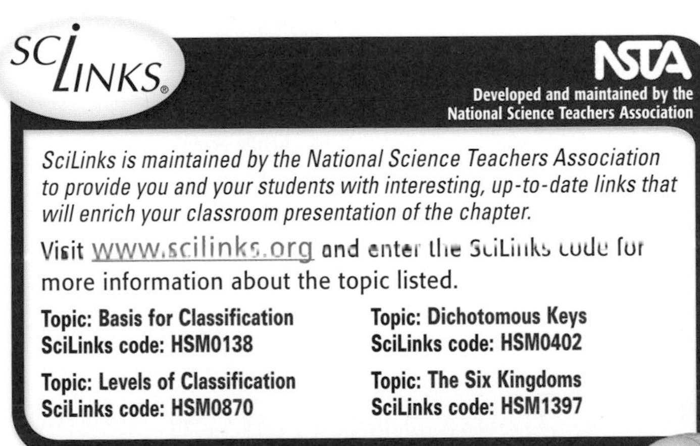

SciLinks is maintained by the National Science Teachers Association to provide you and your students with interesting, up-to-date links that will enrich your classroom presentation of the chapter.

Visit www.scilinks.org and enter the SciLinks code for more information about the topic listed.

Topic: Basis for Classification
SciLinks code: HSM0138

Topic: Dichotomous Keys
SciLinks code: HSM0402

Topic: Levels of Classification
SciLinks code: HSM0870

Topic: The Six Kingdoms
SciLinks code: HSM1397

Overview

Tell students that this chapter will help them learn about classification in life science. The chapter covers methods of classification and the six kingdoms of organisms.

Assessing Prior Knowledge

Students should be familiar with the following topics:

- characteristics of living things
- history of life on Earth

Identifying Misconceptions

As students learn the material in this chapter, some of them may be confused about how scientists classify organisms. Some students categorize by characteristics such as numbers of limbs or the shape of leaves rather than more fundamental distinctions. Furthermore, students often rely on the information found in common names. For example, students may mistakenly categorize a jellyfish as a fish.

7

Classification

About the PHOTO

Look at the katydids, grasshoppers, and mantids in the photo. A scientist is classifying these insects. Every insect has a label describing the insect. These descriptions will be used to help the scientist know if each insect has already been discovered and named. When scientists discover a new insect or other organism, they have to give the organism a name. The name chosen is unique and should help other scientists understand some basic facts about the organism.

PRE-READING ACTIVITY

FOLDNOTES **Booklet** Before you read the chapter, create the FoldNote entitled "Booklet" described in the **Study Skills** section of the Appendix. Label each page of the booklet with a main idea from the chapter. As you read the chapter, write what you learn about each main idea on the appropriate page of the booklet.

Standards Correlations

National Science Education Standards

The following codes indicate the National Science Education Standards that correlate to this chapter. The full text of the standards is at the front of the book.

Chapter Opener
UCP 1

Section 1 Sorting It All Out
UCP 1; SAI 2; HNS 1, 2, 3; LS 5a

Section 2 The Six Kingdoms
UCP 5; SAI 1; HNS 1, 2; LS 1b, 1f, 2a, 2c, 4b, 4c, 4d, 5b; *LabBook:* UCP 1; SAI 1

Chapter Lab
UPC 1; SAI 1

Science in Action
HNS 1, 3; LS 5a

START-UP ACTIVITY

Classifying Shoes

In this group activity, each group will develop a system of classification for shoes.

Procedure

1. Gather **10 shoes.** Number pieces of **masking tape** from 1 to 10. Label the sole of each shoe with a numbered piece of tape.
2. Make a list of shoe features. Make a table that has a column for each feature. Complete the table by describing each shoe.
3. Use the data in the table to make a shoe identification key.
4. The key should be a list of steps. Each step should have two contrasting statements about the shoes. The statements will lead you either to the next step or to a specific shoe.
5. If your shoe is not identified in one step, go on to the next step or steps until the shoe is identified.
6. Trade keys with another group. How did the other group's key help you identify the shoes?

Analysis

1. How was listing the shoe features before making the key helpful?
2. Were you able to identify the shoes using another group's key? Explain.

START-UP ACTIVITY

MATERIALS

FOR EACH GROUP

- marker
- shoes, 10 different kinds (from class members, a secondhand store, or a garage sale)
- tape, masking

Teacher's Notes: Make certain that students understand that the list of shoe characteristics should be unique to a particular set of 10 shoes.

Characteristics of shoes listed should be those that can easily be observed. For example, whether a shoe belongs to a boy or to a girl is not always obvious to an observer.

You can offer the following as a model for the statement for Procedure step 4:

a. This is a red sandal.

b. This is not a red sandal. (Go to step 2.)

Answers

1. Sample answer: Listing the shoes' features helped me find some features that were common and some that were unique.
2. Each student may describe the shoes differently, but the students' descriptions should be clear enough to lead the other students to the same conclusion.

Classification CHAPTER STARTER

This Really Happened!

Skunks have been thrown out of their family. It wasn't their awful smell that got them thrown out, though. It was their DNA.

Skunks were once thought to be most closely related to weasels, ferrets, minks, badgers, and otters. Those furry, short-legged, long-bodied, meat-eating mammals are grouped together in a family called Mustelidae (moo STEL i dee). *Mustelidae* is from the Latin word for "mouse." Skunks were classified along with weasels and ferrets because they all share several physical characteristics with mice, DNA of the other members of Mustelidae. By comparing the DNA of different species, scientists can tell how closely related the species are. The DNA of two closely related animals—a house cat and a tiger, for example—are more similar than the DNA of two animals that are distantly related—such as a house cat and a chicken.

So where does that leave the little striped stinkers? Right in their own, newly created scientific family—Mephitidae (me FIT i dee). *Mephitid* is from the Latin word that means "bad odor"!

Chapter Starter Transparency
Use this transparency to help students begin thinking about classifying organisms.

CHAPTER RESOURCES

Technology

Transparencies READING SKILLS
- Chapter Starter Transparency

Student Edition on CD-ROM

Guided Reading Audio CD
- English or Spanish

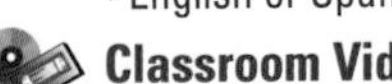

Classroom Videos
- Brain Food Video Quiz

Workbooks

Science Puzzlers, Twisters & Teasers
- Classification GENERAL

SECTION 1

Focus

Overview

In this section, students learn about the modern biological classification system. The section explains how organisms are classified based on their shared characteristics and how their scientific names are determined. Finally, students learn how to identify animals by using a dichotomous key.

Bellringer

Ask students to think about the different ways humans classify things. Ask them to list at least five things that humans classify. You may want to give them examples, such as library books, department-store merchandise, and addresses.

Motivate

Demonstration — GENERAL

Classifying Objects Display a variety of small, solid objects. Ask students for their ideas on ways to put the objects into groups. For each grouping, record the defining characteristic and the objects that belong in the group. Identify objects that fit in more than one grouping. Discuss how putting objects into groups can be helpful. LS Visual English Language Learners

SECTION 1

Sorting It All Out

Imagine that you live in a tropical rain forest and must get your own food, shelter, and clothing from the forest. What do you need to know to survive in the forest?

READING WARM-UP

Objectives

- Explain how to classify organisms.
- List the seven levels of classification.
- Explain scientific names.
- Describe how dichotomous keys help in identifying organisms.

Terms to Learn

classification
taxonomy
dichotomous key

READING STRATEGY

Reading Organizer As you read this section, create an outline of the section. Use the headings from the section in your outline.

To survive in the rain forest, you need to know which plants are safe to eat and which are not. You need to know which animals you can eat and which might eat you. In other words, you need to study the living things around you and organize them into categories, or classify them. **Classification** is putting things into orderly groups based on similar characteristics.

Why Classify?

For thousands of years, humans have classified living things based on usefulness. The Chácabo people of Bolivia know of 360 types of plants that grow in the forest where they live. Of these 360 plant types, 305 are useful to the Chácabo.

Some biologists, such as those shown in **Figure 1,** classify living and extinct organisms. Scientists classify organisms to help make sense and order of the many kinds of living things in the world. Biologists use a system to classify living things. This system groups organisms according to the characteristics they share. The classification of living things makes it easier for biologists to answer many important questions, such as the following:

- How many known species are there?
- What are the defining characteristics of each species?
- What are the relationships between these species?

classification the division of organisms into groups, or classes, based on specific characteristics

Reading Check What are three questions that classifying organisms can help answer? (*See the Appendix for answers to Reading Checks.*)

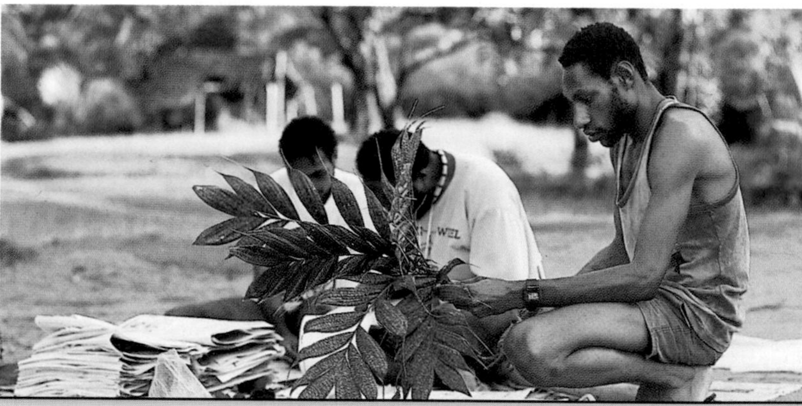

Figure 1 *These biologists are sorting rain-forest plant material.*

CHAPTER RESOURCES

Chapter Resource File

- Lesson Plan
- Directed Reading A BASIC
- Directed Reading B SPECIAL NEEDS

Technology

Transparencies
- Bellringer
- Evolutionary Relationships Among Organisms

Answer to Reading Check

- How many known species are there?
- What are the defining characteristics of each species?
- What are the relationships between these species?

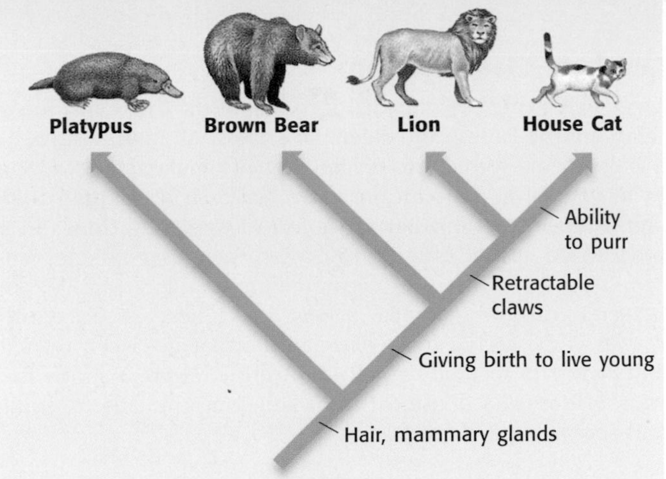

Figure 2 *This branching diagram shows the similarities and differences between four mammals.*

How Do Scientists Classify Organisms?

Before the 1600s, many scientists divided organisms into two groups: plants and animals. But, as more organisms were discovered, some organisms did not fit into either group. In the 1700s, a Swedish scientist named Carolus Linnaeus (KAR uh luhs li NAY uhs) founded modern taxonomy. **Taxonomy** (taks AHN uh mee) is the science of describing, classifying, and naming living things. Linnaeus tried to classify all living things based on their shape and structure. He described a seven-level system of classification, which is still used today.

taxonomy the science of describing, naming, and classifying organisms

Classification Today

Taxonomists use the seven-level system to classify living things based on shared characteristics. Scientists also use shared characteristics to hypothesize how closely related living things are. The more characteristics the organisms share, the more closely related the organisms may be. For example, the platypus, brown bear, lion, and house cat are thought to be related because they share many characteristics. These animals have hair and mammary glands, so they are grouped together as mammals. But they can be further classified into more-specific groups.

Branching Diagrams

Look at the branching diagram in **Figure 2.** Several characteristics are listed along the line that points to the right. Each characteristic is shared by the animals to the right of it. All of the animals shown have hair and mammary glands. But only the bear, lion, and house cat give birth to live young. The lion and the house cat have retractable claws, but the other animals do not. Thus, the lion and the house cat are more closely related to each other than to the other animals.

A Branching Diagram

1. Construct a diagram similar to the one in **Figure 2.**
2. Use a frog, a snake, a kangaroo, and a rabbit in your diagram.
3. Think of one major change that happened before the frog evolved.
4. For the last three organisms, think of a change that happened between one of these organisms and the other two. Write all of these changes in your diagram.

Teach

Activity — General

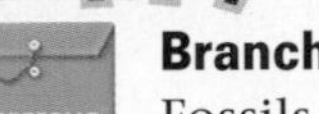

Branching Diagrams Fossils show that one difference between ancestral genera of the modern horse, *Equus,* is the number of toes:

- *Eohippus* (55 mya*)—four toes
- *Mesohippus* (35 mya)—three toes
- *Merychippus* (26 mya)—one large toe, two small toes
- *Pliohippus* (3 mya)—one large toe surrounded by a hoof
- *Equus* (modern)—one large toe, more broad and flat, surrounded by a hoof

(* mya = million years ago)

Have students use this information to construct their own branching diagram. Use the diagram on this page as a model. English Language Learners **LS Logical**

Connection to Math — Basic

Give students the following information:

Suppose there were about 1,260,000 known species of insects in the world. The number of insect species accounts for about 70% of all known species. Have students calculate the approximate total number of Earth's known species. (1,260,000 ÷ 0.7 = 1.8 million species) **LS Logical**

Connection to Environmental Science — General

Rain-Forest Pharmacy Many tropical rain forests are being cut down. Scientists suspect that these forests may be pharmaceutical treasure troves. One-fifth of the world's known plant species live in tropical rain forests. Only a small percentage of the species have been studied. Some unstudied species may be beneficial to humans. Ask students to think of ways that rain forests can be preserved while the needs of humans are also met. **LS Logical/Auditory**

Answer to Quick Lab

Sample answer:

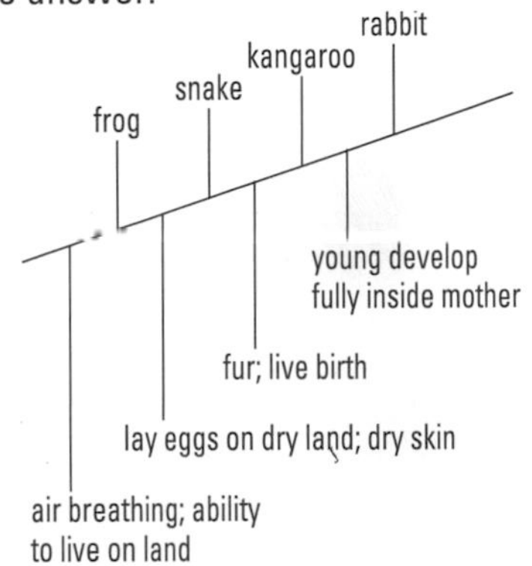

Teach, *continued*

Discussion — Basic

Classification Drill To help students understand what constitutes a species, genus, family, order, class, phylum, and kingdom, ask them the following questions:

- What does a species contain? (organisms that have the same characteristics)
- What does a genus contain? (similar species)
- What does a family contain? (similar genera)
- What does an order contain? (similar families)
- What does a class contain? (similar orders)
- What does a phylum contain? (similar classes)
- What does a kingdom contain? (similar phyla)

LS Logical/Auditory

Using the Figure — General

Classification Refer students to **Figure 3.** Have them answer these questions:

- Which animals are pictured at the kingdom level? (beetle, bird, lion, lynx, bear, human, house cat)
- Which of these pictured animals does not fit the description of a chordate? (the beetle)
- Which of the animals pictured at the chordate level does not fit the description of a mammal? (the bird)

English Language Learners

LS Visual

Internet Activity

For another activity related to this chapter, go to **go.hrw.com** and type in the keyword **HL5CLSW.**

Levels of Classification

Every living thing is classified into one of six kingdoms. Kingdoms are the largest, most general groups. All living things in a kingdom are sorted into several phyla (singular, *phylum*). The members of one phylum are more like each other than they are like members of other phyla. All of the living things in a phylum are further sorted into classes. Each class includes one or more orders. Orders are separated into families. Families are broken into genera (singular, *genus*). And genera are sorted into species. A species is a group of organisms that are closely related and can mate to produce fertile offspring. **Figure 3** shows the classification of a house cat from kingdom Animalia to genus and species, *Felis domesticus*.

Scientific Names

By classifying organisms, biologists are able to give organisms scientific names. A scientific name is always the same for a specific kind of organism no matter how many common names there might be. Before Linnaeus's time, scholars used names that were as long as 12 words to identify species. The names were hard to work with because they were so long. And different scientists named organisms differently, so an organism could have more than one name.

Figure 3 *The seven levels of classification are kingdom, phylum, class, order, family, genus, and species.*

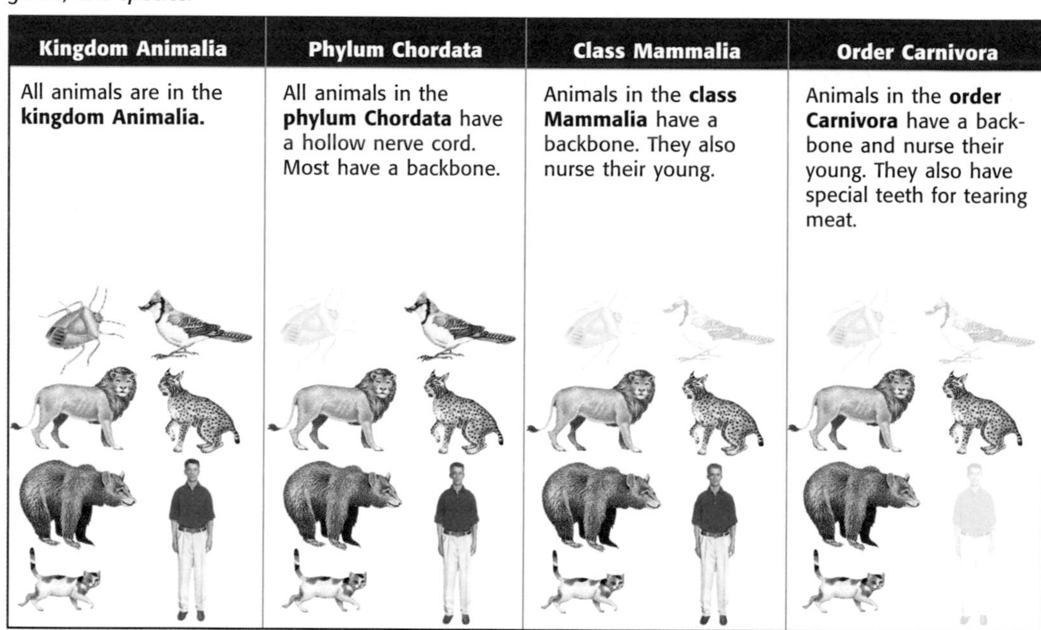

CHAPTER RESOURCES

Technology

Transparencies
- Levels of Classification

Is That a Fact!

The term *dinosaur* wasn't coined until the 19th century. Until then, as dinosaur fossils were uncovered all over the world, the most widely accepted view was that the fossils were the remains of dragons.

Two-Part Names

Linnaeus simplified the naming of living things by giving each species a two-part scientific name. For example, the scientific name for the Asian elephant is *Elephas maximus* (EL uh fuhs MAK suh muhs). The first part of the name, *Elephas,* is the genus name. The second part, *maximus,* is the species name. No other species has both this genus name and this species name. Naming rules help scientists communicate clearly about living things.

All genus names begin with a capital letter. All species names begin with a lowercase letter. Usually, both words are underlined or italicized. But if the surrounding text is italicized, the genus and species names are not italicized, as shown in **Figure 4.** These printing styles show a reader which names are genus and species names.

Scientific names, which are usually in Latin or Greek, contain information about an organism. The name of the animal shown in **Figure 4** is *Tyrannosaurus rex. Tyrannosaurus* is a combination of two Greek words and means "tyrant lizard." The word *rex* is Latin for "king." The name tells you that this animal was probably not a passive grass eater! Sometimes, *Tyrannosaurus rex* is referred to as *T. rex.* The species name is not correct without the genus name or its abbreviation.

Figure 4 *You would never call* Tyrannosaurus rex *just* rex*!*

Reading Check What are the two parts of a scientific name?

Family Felidae	Genus *Felis*	Species *Felis domesticus*
Animals in the **family Felidae** are cats. They have a backbone, nurse their young, have special teeth for tearing meat, and have retractable claws.	Animals in the **genus *Felis*** have traits of other animals in the same family. However, these cats cannot roar; they can only purr.	The **species *Felis domesticus*** is the common house cat. The house cat shares traits with all of the organisms in the levels above the species level, but it also has unique traits.

BRAIN FOOD

Classifying Ideas Have students consider the importance of classification to human thought. Ask students to try to think of something that cannot be classified in some way. Suggest that they test any item or concept they come up with by placing it in the following sentence:

(A) _____ is a type of _____.

For example, if the word is *speech*, the sentence could be filled in as follows:

Speech is a type of communication.

You may wish to hold a contest or have students share their examples in class.

LS Logical/Verbal

MISCONCEPTION ALERT

Scientific Names Though most scientific names have their origins in Latin or Greek, not all scientific names do. Some scientific names indicate the location where the animal was found. And some names include words in languages native to the region where the animal was found.

Is That a Fact!

If you put all of the insects in the world together, they would weigh more than all of the people and the rest of the animals combined.

Answer to Reading Check

genus and species

Close

Reteaching — Basic

Name That Bird Display a picture of a bird whose common name is not well known to your students. Ask students to give the bird a name. List students' answers on the board. Help students understand that scientists would have difficulty sharing information about the bird if they used more than one name for it. **LS Visual**

Quiz — General

1. Why do scientists classify animals? (to make studying them easier)
2. What is the basis of modern classification systems? (shared characteristics)

Alternative Assessment — General

PORTFOLIO **Cartooning** Have students create a cartoon that shows how using different common names for an animal instead of its scientific name creates confusion. Students must include scientific names in their cartoon. English Language Learners **LS Visual**

Answer to Dichotomous Key

Mammal on the top left: 1b, 2b, 4b, 6a, 7a, longtail weasel

Mammal on the top right: 1b, 2b, 4b, 6b, 8b, 9b, woodchuck

Answer to Reading Check

A dichotomous key is an identification aid that helps identify organisms.

Dichotomous Keys

dichotomous key an aid that is used to identify organisms and that consists of the answers to a series of questions

You might someday turn over a rock and find an organism that you don't recognize. How would you identify the organism? Taxonomists have developed special guides to help scientists identify organisms. A **dichotomous key** (die KAHT uh muhs KEE) is an identification aid that uses sequential pairs of descriptive statements. There are only two alternative responses for each statement. From each pair of statements, the person trying to identify the organism chooses the statement that describes the organism. Either the chosen statement identifies the organism or the person is directed to another pair of statements. By working through the statements in the key in order, the person can eventually identify the organism. Using the simple dichotomous key in **Figure 5,** try to identify the two animals shown.

Reading Check What is a dichotomous key?

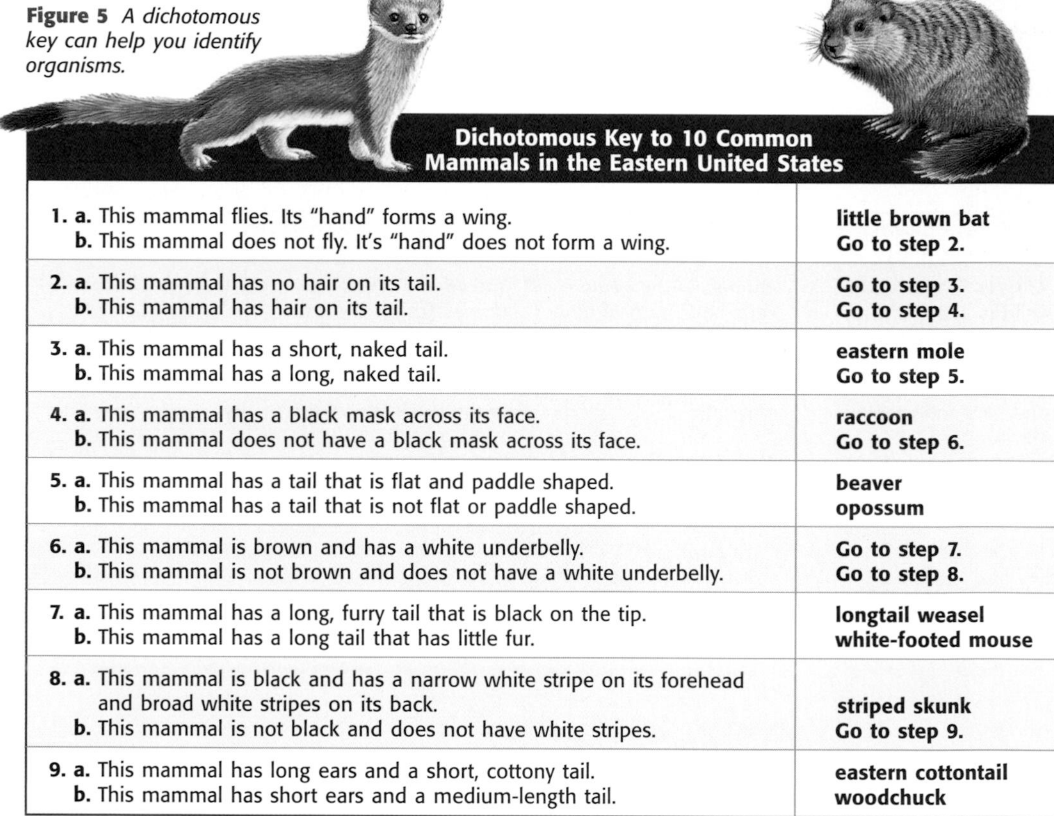

Figure 5 *A dichotomous key can help you identify organisms.*

Dichotomous Key to 10 Common Mammals in the Eastern United States	
1. a. This mammal flies. Its "hand" forms a wing. **b.** This mammal does not fly. It's "hand" does not form a wing.	**little brown bat** **Go to step 2.**
2. a. This mammal has no hair on its tail. **b.** This mammal has hair on its tail.	**Go to step 3.** **Go to step 4.**
3. a. This mammal has a short, naked tail. **b.** This mammal has a long, naked tail.	**eastern mole** **Go to step 5.**
4. a. This mammal has a black mask across its face. **b.** This mammal does not have a black mask across its face.	**raccoon** **Go to step 6.**
5. a. This mammal has a tail that is flat and paddle shaped. **b.** This mammal has a tail that is not flat or paddle shaped.	**beaver** **opossum**
6. a. This mammal is brown and has a white underbelly. **b.** This mammal is not brown and does not have a white underbelly.	**Go to step 7.** **Go to step 8.**
7. a. This mammal has a long, furry tail that is black on the tip. **b.** This mammal has a long tail that has little fur.	**longtail weasel** **white-footed mouse**
8. a. This mammal is black and has a narrow white stripe on its forehead and broad white stripes on its back. **b.** This mammal is not black and does not have white stripes.	**striped skunk** **Go to step 9.**
9. a. This mammal has long ears and a short, cottony tail. **b.** This mammal has short ears and a medium-length tail.	**eastern cottontail** **woodchuck**

INCLUSION *Strategies*

- ***Developmentally Delayed***
- ***Hearing Impaired***

Use this activity to clarify the procedure. Place the following six objects on a table: stapler, marker, zipper bag of ice, book, roll of tape, and piece of wadded-up paper. Use the following questions to identify the items by their physical characteristics.

1. Is it very cold and could melt? If yes, it is ice. If no, go to step 2.
2. Is it made of metal? If yes, it is a stapler. If no, go to step 3.
3. Is it made of paper? If yes, go to step 4. If no, go to step 5.
4. Is it intended to be read? If yes, it is a book. Is it intended to be thrown? If yes, it is a paper wad.
5. Is it used for writing? If yes, it is a marker. Does it have a sticky side? If yes, it is tape. English Language Learners **LS Logical**

A Growing System

You may think that all of the organisms on Earth have already been classified. But people are still discovering and classifying organisms. Some newly discovered organisms fit into existing categories. But sometimes, someone discovers new evidence or an organism that is so different from other organisms that it does not fit existing categories. For example, in 1995, scientists studied an organism named *Symbion pandora* (SIM bee AHN pan DAWR uh). Scientists found *S. pandora* living on lobster lips! Scientists learned that *S. pandora* had some characteristics that no other known organism had. In fact, scientists trying to classify *S. pandora* found that it didn't fit in any existing phylum. So, taxonomists created a new phylum for *S. pandora*.

SECTION Review

Summary

- Classification refers to the arrangement of living things into orderly groups based on their similarities.
- Today's living things are classified by using a seven-level system of organization. The seven levels are kingdom, phylum, class, order, family, genus, and species.
- An organism has only one correct scientific name.
- Dichotomous keys are tools for identifying organisms.

Using Key Terms

1. In your own words, write a definition for each of the following terms: *classification* and *taxonomy*.

Understanding Key Ideas

2. The two parts of a scientific name are the names of the genus and the
   a. species. **c.** family.
   b. phylum. **d.** order.
3. Why do scientists use scientific names for organisms?
4. List the seven levels of classification.
5. Describe how a dichotomous key helps scientists identify organisms.

Critical Thinking

6. **Analyzing Processes** Biologists think that millions of species are not classified yet. Why do you think so many species have not been classified yet?
7. **Applying Concepts** Both dolphins and sharks have a tail and fins. How can you determine if dolphins and sharks are closely related?

Interpreting Graphics

Use the figure below to answer the questions that follow.

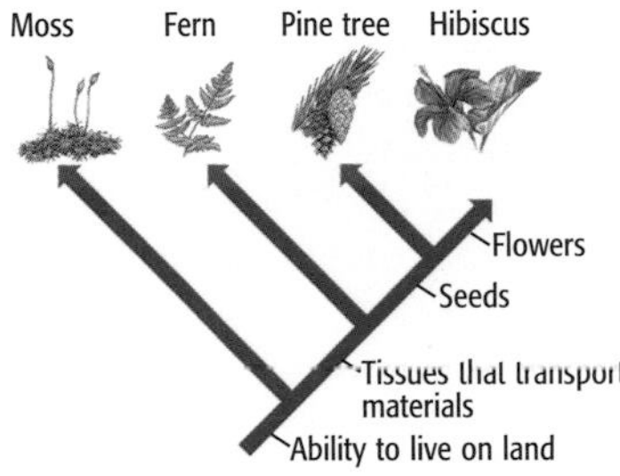

8. Which plant is most similar to the hibiscus?
9. Which plant is least similar to the hibiscus?

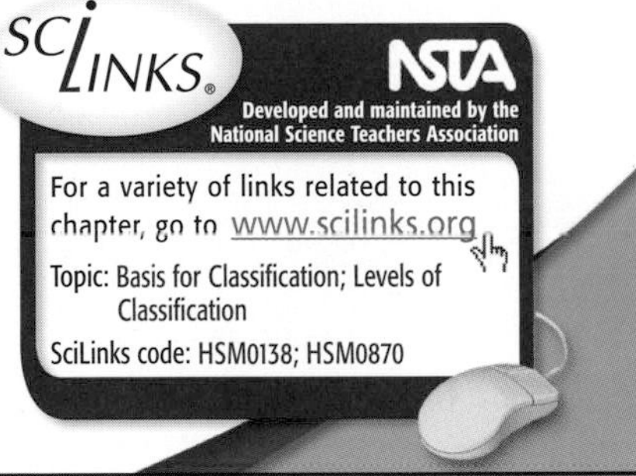

Answers to Section Review

1. Sample answer: Classification is grouping things according to similar characteristics. Taxonomy is the science of classifying and naming organisms.
2. a
3. Sample answer: Scientists use scientific names for organisms to be clear and precise when they refer to a living thing.
4. kingdom, phylum, class, order, family, genus, and species
5. Sample answer: A dichotomous key is organized into a series of pairs of descriptive statements. By matching an organism's characteristics to the statements in the key, one can identify unknown organisms.
6. Sample answer: Before being classified, organisms need to be discovered and then studied. Both discovery and study take time. Scientists are discovering and classifying organisms often, but it is a slow process.
7. Sample answer: I could compare many characteristics to see if the characteristics are shared between dolphins and sharks. For example, I could ask, "Do dolphins and sharks both have gills?" If the answer is no because one animal has gills and the other animal doesn't have gills, then dolphins and sharks would not be very closely related, even though they have fins and similar body shapes.
8. pine tree
9. moss

CHAPTER RESOURCES

Chapter Resource File

- **Section Quiz** GENERAL
- **Section Review** GENERAL
- **Vocabulary and Section Summary** GENERAL
- **Critical Thinking** ADVANCED
- **SciLinks Activity** GENERAL
- **Datasheet for Quick Lab**

Technology

Transparencies
- A Dichotomous Key

SECTION 2

Focus

Overview

This section explains how improved understanding of organisms leads to revisions in our system of biological classification. Students are introduced to the six kingdoms:

Archaebacteria, Eubacteria, Protista, Fungi, Plantae, and Animalia.

They learn how organisms from each kingdom are distinguished.

Bellringer

Have students list seven musical artists, bands, or acts. Then have students categorize the names on their lists by style of music. Ask them to describe in their **science journal** the categories they chose and to explain which bands might fit into more than one category.

Motivate

Activity — GENERAL

Grouping Animals Have students write letters to zoos requesting a copy of their visitors' map. Students could then compare the layouts of many zoos and the ways that animals are grouped. Be sure to have students include a stamped, self-addressed envelope with their letter describing the project.
LS Logical

SECTION 2

The Six Kingdoms

READING WARM-UP

Objectives

- Explain how classification schemes for kingdoms developed as greater numbers of different organisms became known.
- Describe each of the six kingdoms.

Terms to Learn

Archaebacteria	Fungi
Eubacteria	Plantae
Protista	Animalia

READING STRATEGY

Reading Organizer As you read this section, make a table comparing the six classification kingdoms to each other.

What do you call an organism that is green, makes its own food, lives in pond water, and moves? Is it a plant, an animal, or something in between?

For hundreds of years, all living things were classified as either plants or animals. But over time, scientists discovered species that did not fit easily into these two kingdoms. For example, an organism of the genus *Euglena,* such as the one shown in **Figure 1,** has characteristics of both plants and animals. How would you classify such an organism?

What Is It?

Organisms are classified by their characteristics. For example, organisms of the genus *Euglena* are

- single celled and live in pond water
- green and make their own food through photosynthesis

These two characteristics might lead you to conclude that the genus of *Euglena* are plants. However, you should consider the following other characteristics before you form your conclusion:

- Members of the genus *Euglena* move by whipping their "tails," which are called *flagella.*
- *Euglena* can feed on other organisms.

Plants don't move themselves around and usually do not eat other organisms. So, are organisms of the genus *Euglena* animals? As you can see, *Euglena* does not fit into the category of plants or animals. Scientists solved this classification problem by adding another kingdom, the kingdom Protista, to classify organisms such as *Euglena*.

As scientists continued to learn about living things, they added kingdoms that account for the characteristics of different organisms. Currently, most scientists agree that the six-kingdom classification system works best. However, scientists will continue to adjust the system as they learn more.

Figure 1 *How would you classify this organism? This member of the genus* Euglena, *which is shown here highly magnified, has characteristics of both plants and animals.*

CHAPTER RESOURCES

Chapter Resource File

- Lesson Plan
- Directed Reading A BASIC
- Directed Reading B SPECIAL NEEDS

Technology

Transparencies

- Bellringer
- *LINK TO EARTH SCIENCE* Intrusive Igneous Rock Formations

CONNECTION to Earth Science — GENERAL

Rocky Habitats Bacteria have been found living in igneous rocks deep in the Earth's crust. The rocks contain little water and no organic matter. The bacteria subsist on carbon dioxide and hydrogen gas dissolved in the rock and slowly make their own organic compounds. Use the teaching transparency entitled "Intrusive Igneous Rock Formations" to introduce information about igneous rocks that may be unfamiliar to students. **LS Visual**

Figure 2 *The Grand Prismatic Spring in Yellowstone National Park contains water that is about 90°C (194°F). The spring is home to archaebacteria that thrive in its hot water.*

Archaebacteria a kingdom made up of bacteria that live in extreme environments

Eubacteria a kingdom that contains all prokaryotes except archaebacteria

The Two Kingdoms of Bacteria

Bacteria are extremely small, single-celled organisms that differ from all other living things. Bacteria are *prokaryotes* (proh KAYR ee OHTS), organisms that lack nuclei. Many biologists divide bacteria into two kingdoms: Archaebacteria (AHR kee bak TEER ee uh) and Eubacteria (YOO bak TEER ee uh).

Archaebacteria

Prokaryotes that can live in extreme environments are in the kingdom **Archaebacteria.** The prefix *archae-* comes from a Greek word meaning "ancient." Today, most archaebacteria can be found living in places where most organisms could not survive. **Figure 2** shows a hot spring in Yellowstone National Park. The yellow and orange rings around the edge of the hot spring are made up of the billions of archaebacteria that live there.

Eubacteria

Bacteria that are not in the kingdom Archaebacteria are in the kingdom **Eubacteria.** Eubacteria are prokaryotes that live in the soil, in water, and even on and inside the human body! For example, *Escherichia coli,* pictured in **Figure 3,** is present in large numbers in human intestines, where it produces vitamin K. One kind of eubacterium converts milk into yogurt. Another kind of eubacterium causes pneumonia.

Reading Check **Name a type of eubacterium that lives in your body.** (*See the Appendix for answers to Reading Checks.*)

Figure 3 *Specimens of* E. coli *are shown on the point of a pin under a scanning electron microscope. These eubacteria live in the intestines of animals and decompose undigested food.*

Teach

READING STRATEGY — BASIC

Prediction Guide Before students read this section, ask them whether the following statements are true or false. Students will discover the answers as they explore this section.

- All living things were once classified into either the kingdom Plantae or the kingdom Animalia. (true)
- Members of the kingdom Protista are prokaryotes. (false)
- The kingdom Fungi contains multicellular photosynthetic organisms. (false)
- The kingdom Animalia contains multicellular organisms that do not photosynthesize. (true)

LS Verbal/Auditory

Research — ADVANCED

Writing **Sanitation and Hygiene** Have students work in teams to use the Internet to conduct research and write reports on measures that protect humans against harmful bacteria. Have students answer the following questions: "How can bacteria be killed?" "How can bacteria be prevented from growing on living tissues?" "When harmful bacteria get inside the body, how does the body defend itself?" **LS Verbal/Interpersonal** Co-op Learning

Answer to Reading Check

Escherichia coli

CONNECTION ACTIVITY
Environmental Science — GENERAL

Penguin Problem Recently, signs of *Salmonella* infection were found in the droppings of an Antarctic gentoo penguin. The bacteria were most likely introduced from outside the Antarctic. The bacterium, *Salmonella enteritidis,* is not endemic to penguins. Scientists think that sewage dumped from passing ships or visiting albatrosses that feed on waste-contaminated squid in the oceans surrounding South America might be the sources of the bacteria. The bacterium could kill the penguins if it becomes infectious and pathogenic. Have students create a poster that could inform ship owners of the danger human wastes pose to native wildlife. **LS Visual**

Teach, continued

Homework — Advanced

Researching Protists Have students research protists, such as *Paramecium,* slime mold, and giant kelp. Have them write descriptions about each protist, including information about its size, form, method of obtaining nutrients, method of reproduction, and, in the case of the giant kelp and other algae, the commercial uses of the organism. LS **Verbal**

Connection Activity
Real World — Advanced

Exploring Mushrooms Tell students that Pennsylvania, which has many caves, is one of the major mushroom-growing regions of the United States. Caves are ideal places in which to grow some kinds of edible mushrooms. Ask students to research and write a report on mushroom farming in the United States. What kinds of mushrooms are grown commercially, and what special conditions does each species require? Inexpensive kits are available for growing mushrooms, and interested students might enjoy the experience of raising their own. Caution students not to attempt to cultivate or eat mushrooms that they find in the wild. Some toxic species are difficult to distinguish from nontoxic ones. Accidental ingestion of toxic mushrooms can be fatal. LS **Kinesthetic/Verbal**

Figure 4 *This slime mold is a protist.*

Kingdom Protista

Members of the kingdom **Protista** (proh TIST uh), commonly called *protists* (PROH tists), are single-celled or simple multicellular organisms that don't fit into any other kingdom. Unlike bacteria, protists are *eukaryotes,* organisms whose cells have a nucleus and membrane-bound organelles. The kingdom Protista contains all eukaryotes that are not plants, animals, or fungi. Scientists think the first protists evolved from ancient bacteria about 2 billion years ago. Much later, protists gave rise to plants, fungi, and animals as well as to modern protists.

The kingdom Protista contains many kinds of organisms. Animal-like protists are called *protozoans*. Plantlike protists are called *algae*. Slime molds, such as the one shown in **Figure 4,** and water molds are fungus-like protists. Members of *Euglena* are also members of the kingdom Protista.

Kingdom Fungi

Molds and mushrooms are examples of the complex multicellular members of the kingdom Fungi (FUHN JIE). Unlike plants, fungi do not perform photosynthesis, and unlike animals, fungi do not eat food. Instead, members of the kingdom **Fungi** absorb nutrients from substances in their surroundings. Fungi use digestive juices to break down the substances. **Figure 5** shows a very poisonous fungus. Never eat wild fungi.

Protista a kingdom of mostly one-celled eukaryotic organisms that are different from plants, animals, bacteria, and fungi

Fungi a kingdom made up of nongreen, eukaryotic organisms that have no means of movement, reproduce by using spores, and get food by breaking down substances in their surroundings and absorbing the nutrients

Figure 5 *This beautiful fungus of the genus* Amanita *is poisonous.*

Scientists at Odds

To Be or Not to Be a Fungus Is a slime mold a fungus? Slime molds were traditionally classified as fungi because despite other differences, they exhibit a similar life cycle, including the formation of spores on sporangia. But critics point out that some bacteria also exhibit a similar life cycle and those organisms are not reclassified as fungi.

Is That a Fact!

Farmers on the Orkney Islands of Scotland have historically used seaweed as fertilizer and food for their sheep.

Figure 6 *Giant sequoias can measure 30 m around at the base and can grow to more than 91.5 m tall.*

Figure 7 *Plants such as these are common in the Tropics.*

Kingdom Plantae

Although plants vary remarkably in size and form, most people easily recognize the members of the kingdom Plantae. **Plantae** consists of organisms that are eukaryotic, have cell walls, and make food through photosynthesis. For photosynthesis to occur, plants must be exposed to sunlight. Plants can therefore be found on land and in water that light can penetrate.

The food that plants make is important not only for the plants but also for all of the organisms that get nutrients from plants. Most life on Earth is dependent on plants. For example, some members of the kingdoms Fungi, Protista, and Eubacteria consume plants. When these organisms digest the plant material, they get energy and nutrients made by the plants.

Plants also provide habitat for other organisms. The giant sequoias in **Figure 6** and the flowering plants in **Figure 7** provide birds, insects, and other animals with a place to live.

Reading Check How do members of the kingdom Plantae provide energy and nutrients to members of other kingdoms?

Plantae a kingdom made up of complex, multicellular organisms that are usually green, have cell walls made of cellulose, cannot move around, and use the sun's energy to make sugar by photosynthesis

Ring-Around-the-Sequoia

How many students would have to join hands to form a human chain around a giant sequoia that is 30 m in circumference? Assume for this calculation that the average student can extend his or her arms about 1.3 m.

Using the Figure — Basic

Compare Structures Ask students what structures are common to the plants in **Figures 6** and **7.** (stems, leaves, and so on) Ask students to assess differences and similarities between features on various kinds of plants, such as maple leaves and pine needles, or tomato stems and tree trunks. English Language Learners
LS Visual

Plant Identification Have students work together in small groups to find pictures in magazines of ferns and flowering plants that grow in North America. Provide resource books for the students to use to identify the plants. Then, have students mount the plant pictures on poster board and label them. English Language Learners
LS Visual

Answer to Math Practice

24 students (30 ÷ 1.3 = 23.1 students. Round up because you cannot have a fraction of a student)

INCLUSION *Strategies*

- *Learning Disabled*
- *Developmentally Delayed*
- *Hearing Impaired*

Many students have a better time understanding information when the main facts are isolated in a meaningful way. Using two interlocking circles, create a Venn diagram on the board. Ask students to add the similarities and differences between plants and fungi to the diagram. English Language Learners
LS Visual

Cultural Awareness — General

Sequoia The name *sequoia* comes from Sequoyah, the name of a Cherokee who is credited with developing the Cherokee written language during the 1820s.

Answer to Reading Check

Sample answer: Plants make energy through photosynthesis. Some members of the kingdom Fungi, Protista, and Eubacteria consume plants. When these organisms digest plant material, they get nutrients made by the plants.

Close

Reteaching — Basic

New Kingdom Have students describe and illustrate in their **science journal** an organism that might require the formation of a seventh kingdom. Students should explain why they think the organism should be classified in its own kingdom. LS **Visual/Logical**

Quiz — General

1. What causes increases in the number of kingdoms in the modern classification system? (discovery of some organisms that do not fit into established kingdoms)
2. Which of the six kingdoms have prokaryotic organisms, and which have eukaryotic organisms? (prokaryotic: Archaebacteria, Eubacteria, Protista; eukaryotic: Protista, Plantae, Fungi, Animalia)

Alternative Assessment — General

PORTFOLIO **Making a Chart** Have students construct a chart of the six kingdoms. They should list the major characteristics of each kingdom on the chart and include a representative organism for each kingdom. English Language Learners
LS **Visual/Logical**

Animalia a kingdom made up of complex, multicellular organisms that lack cell walls, can usually move around, and quickly respond to their environment

Kingdom Animalia

The kingdom **Animalia** contains complex, multicellular organisms that don't have cell walls, are usually able to move around, and have specialized sense organs. These sense organs help most animals quickly respond to their environment. Organisms in the kingdom Animalia are commonly called *animals*. You probably recognize many of the organisms in the kingdom Animalia. All of the organisms in **Figure 8** are animals.

Animals depend on the organisms from other kingdoms. For example, animals depend on plants for food. Animals also depend on bacteria and fungi to recycle the nutrients found in dead organisms.

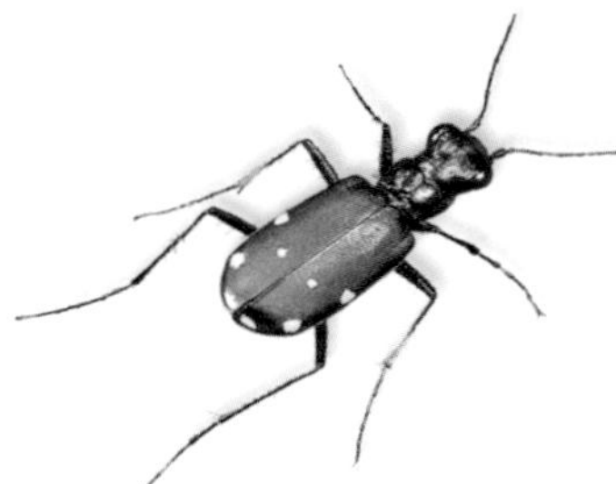

Figure 8 *The kingdom Animalia contains many different organisms, such as eagles, tortoises, and beetles.*

CONNECTION TO Social Studies

WRITING SKILL **Animals That Help** Humans have depended on animals for thousands of years. Many people around the world still use oxen to farm. Camels, horses, donkeys, goats, and llamas are all still used as pack animals. Dogs still help herd sheep, protect property, and help people hunt. Scientists are even discovering new ways that animals can help us. For example, scientists are training bees to help find buried land mines. Using the library or the Internet, research an animal that helps people. Make a poster describing the animal and the animal's scientific name. The poster should show who uses the animal, how the animal is used, and how long people have depended on the animal. Find or draw pictures to put on your poster.

ACTIVITY

MISCONCEPTION ALERT

Misleading Similarities Physical similarities can be misleading indicators of the relatedness of two organisms. For example, a small lizard, such as a skink, may look more like a salamander than like a turtle, but the skink is more closely related to the turtle. Both the lizard and turtle are reptiles, and the salamander is an amphibian.

Simple Animals

When you think of an animal, what do you imagine? You may think of a dog, a cat, or a parrot. All of those organisms are animals. But the animal kingdom also includes some members that might surprise you, such as worms, insects, and corals.

The red cup sponge shown in **Figure 9** is also an animal. Sponges are usually thought of as the simplest animals. They don't have sense organs. Most sponges cannot move. Sponges used to be considered plants. But sponges cannot make food. They must eat other organisms to get nutrients, which is one reason that sponges are classified as animals.

Reading Check Why were sponges once thought to be plants?

Figure 9 *This red cup sponge is a simple animal.*

SECTION Review

Summary

- Most biologists recognize six kingdoms: Archaebacteria, Eubacteria, Protista, Fungi, Plantae, and Animalia.
- Archaebacteria live in extreme environments. Eubacteria live almost everywhere else.
- Plants, animals, fungi, and protists are eukaryotic organisms. Plants perform photosynthesis. Animals eat food and digest it inside their body. Fungi absorb nutrients from material that they break down outside of their body. Protists are organisms that don't fit in other kingdoms.

Using Key Terms

For each pair of terms, explain how the meanings of the terms differ.

1. *Archaebacteria* and *Eubacteria*
2. *Plantae* and *Fungi*

Understanding Key Ideas

3. Biological classification schemes change
   a. as new evidence and more kinds of organisms are discovered.
   b. every 100 years.
   c. when scientists disagree.
   d. only once.
4. Explain the different ways in which plants, fungi, and animals obtain nutrients.
5. Why are protists placed in their own kingdom?
6. Describe the six kingdoms.

Math Skills

7. A certain eubacterium can divide every 30 min. If you begin with 1 eubacterium, when will you have more than 1,000 eubacteria?

Critical Thinking

8. **Identifying Relationships** How are bacteria similar to fungi? How are fungi similar to animals?
9. **Analyzing Methods** Why do you think Linnaeus did not include classification kingdoms for categories of bacteria?
10. **Applying Concepts** The Venus' flytrap does not move around. It can make its own food by using photosynthesis. It can also trap insects and digest the insects to get nutrients. The flytrap also has a cell wall. Into which kingdom would you place the Venus' flytrap? What makes this organism unusual in the kingdom you chose?

For a variety of links related to this chapter, go to www.scilinks.org

Topic: The Six Kingdoms
SciLinks code: HSM1397

Answer to Reading Check

Sponges don't have sense organs, and they usually can't move around.

CHAPTER RESOURCES

Chapter Resource File

- Section Quiz GENERAL
- Section Review GENERAL
- Vocabulary and Section Summary GENERAL
- Reinforcement Worksheet BASIC
- SciLinks Activity GENERAL

Answers to Section Review

1. Sample answer Archaebacteria live in extreme environments. Eubacteria are bacteria that usually live in environments that are not extreme.
2. Sample answer: The kingdom Plantae contains organisms that can make their own food. The kingdom Fungi contains organisms that can't make their own food.
3. a
4. Plants make food using the energy in sunlight. Fungi absorb nutrients from their surroundings after breaking them down with digestive juices. Animals eat plants and other animals to obtain food.
5. The kingdom Protista includes organisms that don't fit in the other kingdoms. Although protists are eukaryotes, they are not plants, animals, or fungi.
6. Sample answer: Archaebacteria includes prokaryotes that live in extreme environments. Eubacteria includes prokaryotes that live in soil, water, and other organisms. Protista includes eukaryotes don't fit into the other kingdoms. Fungi includes eukaryotic organisms that make spores, don't make their own food, and don't move. Plantae includes eukaryotes that use photosynthesis and have cells with cell walls. Animalia includes multicellular organisms that have cells without cell walls, can usually move around, and have specialized sense organs.
7. after 5 h
8. Sample answer: Bacteria and fungi are both decomposers; Fungi and animals are both unable to make their own food.
9. Sample answer: Linnaeus may not have had access to microscopes that could allow him to study bacteria.
10. Sample answer: Plantae; Most plants are not consumers.

Skills Practice Lab

Shape Island

Teacher's Notes

Time Required

One 45-minute class period

Lab Ratings

Teacher Prep 1 flask
Student Set-Up 1 flask
Concept Level 2 flasks
Clean Up 1 flask

Lab Notes

This lab will help students demonstrate an understanding of binomial nomenclature by using a key to assign scientific names to fictional organisms. After completing the lab, students should be able to explain the function of the scientific naming system. This chapter on classification uses the term *two-part scientific name* instead of *binomial nomenclature*. You may wish to introduce the latter here. This activity may be more successful if you review prefixes, suffixes, and root words briefly before beginning. Remind students that the genus name is capitalized but the species name is not and that both words are underlined or italicized.

Skills Practice Lab

Shape Island

OBJECTIVES

Classify organisms.

Name organisms.

You are a biologist exploring uncharted parts of the world to look for new animal species. You sailed for days across the ocean and finally found Shape Island hundreds of miles south of Hawaii. Shape Island has some very unusual organisms. The shape of each organism is a variation of a geometric shape. You have spent more than a year collecting and classifying specimens. You have been able to assign a two-part scientific name to most of the species that you have collected. Now, you must assign a two-part scientific name to each of the last 12 specimens collected before you begin your journey home.

Procedure

1. Draw each of the organisms pictured on the facing page. Beside each organism, draw a line for its name, as shown on the top left of the following page. The first organism pictured has already been named, but you must name the remaining 12. Use the glossary of Greek and Latin prefixes, suffixes, and root words in the table to help you name the organisms.

Greek and Latin roots, prefixes, and suffixes	Meaning
ankylos	angle
antennae	external sense organs
bi-	two
cyclo-	circular
macro-	large
micro-	small
mono-	one
peri-	around
-plast	body
-pod	foot
quad-	four
stoma	mouth
tri-	three
uro-	tail

Analyze Results

1. **Analyzing Results** If you gave species 1 a common name, such as *round-face-no-nose,* would any other scientist know which of the newly discovered organisms you were referring to? Explain. How many others have a round face and no nose?

2. **Organizing Data** Describe two characteristics that are shared by all of your newly discovered specimens.

Maurine Marchani
Raymond Park Middle School
Indianapolis, Indiana

CHAPTER RESOURCES

Chapter Resource File

- Datasheet for Chapter Lab
- Lab Notes and Answers

Technology

Classroom Videos
- Lab Video

LabBook
- The Voyage of the USS *Adventure*

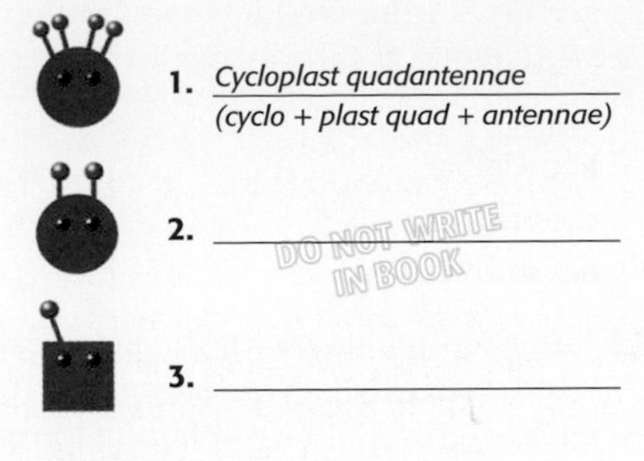

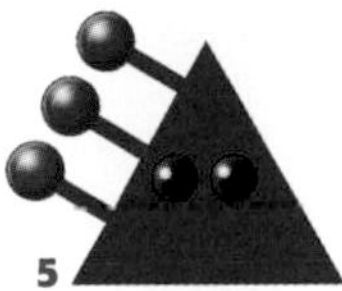

Draw Conclusions

3 **Applying Conclusions** One more organism exists on Shape Island, but you have not been able to capture it. However, your supplies are running out, and you must start sailing for home. You have had a good look at the unusual animal and can draw it in detail. Draw an animal that is different from all of the others, and give it a two-part scientific name.

Applying Your Data

Look up the scientific names *Mertensia virginica* and *Porcellio scaber.* Answer the following questions as they apply to each organism: Is the organism a plant or an animal? How many common names does the organism have? How many scientific names does it have?

Think of the name of your favorite fruit or vegetable. Find out if it has other common names, and find out its two-part scientific name.

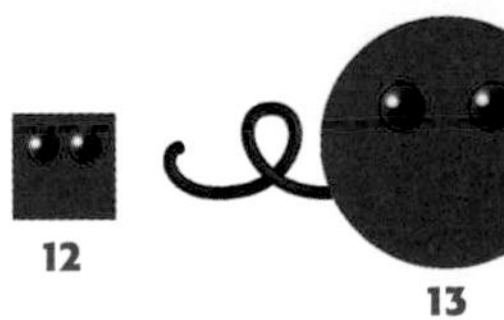

CHAPTER RESOURCES

Workbooks

EcoLabs & Field Activities
- Water Wigglers GENERAL

Long-Term Projects & Research Ideas
- The Panda Mystery ADVANCED

Procedure

1. Students' answers may vary, but students should demonstrate an understanding of the key provided. Each name should consist of two words. The first describes the organism generally, and the second describes it more specifically. Sample answer:
   1. *Cycloplast quadantennae*
   2. *Cycloplast biantennae*
   3. *Quadankylosplast monoantenna*
   4. *Quadankylosplast bipod*
   5. *Triankylosplast triantennae*
   6. *Cycloplast stoma*
   7. *Triankylosplast stoma*
   8. *Quadankylosplast periantennae*
   9. *Cycloplast monopod*
   10. *Triankylosplast uromonopod*
   11. *Triankylos macroplast*
   12. *Quadankylos microplast*
   13. *Cycloplast uro*

Analyze the Results

1. no; Five species have round faces and lack noses.
2. Sample answer: All have geometric shapes and two eyes. All are the same color. All are animals. All are living organisms.

Draw Conclusions

3. Answers may vary. Students should demonstrate an understanding of binomial nomenclature.

Applying Your Data

Mertensia virginica, commonly known as the Virginia bluebell, is a plant; Other common names for this species include Virginia-cowslip, Roanoke-bells, lungwort, and oysterleaf. These wildflowers are found in April and May in shady areas, mostly in moist spots near streams. Flower buds are pink but turn blue when the flower is fully opened. This wildflower is very common in western Kentucky. *Porcellio scaber* is a species of wood louse. Common names for *Porcellio scaber* include dooryard sowbug and common rough woodlouse. Wood lice are crustaceans related to shrimps, crabs, and lobsters, and they belong to a class of arthropods called *Isopoda*.

Chapter Review

Assignment Guide	
Section	**Questions**
1	1, 4, 6–9, 12–13, 15, 17–18, 20–23
2	2–3, 5, 10–11, 14, 16, 19

ANSWERS

Using Key Terms

1. taxonomy
2. Archaebacteria
3. Animalia
4. classification
5. Eubacteria

Understanding Key Ideas

6. a
7. d
8. a
9. b
10. b
11. c

Chapter Review

USING KEY TERMS

Complete each of the following sentences by choosing the correct term from the word bank.

Animalia, Eubacteria, Archaebacteria, taxonomy, Protista, Plantae, classification

1. Linnaeus founded the science of ___.

2. Bacteria that live in extreme environments are in the kingdom ___.

3. Complex multicellular organisms that can usually move around and respond to their environment are in the kingdom ___.

4. A system of ___ can help group animals into categories.

5. Prokaryotes that are not archaebacteria are in the kingdom ___.

UNDERSTANDING KEY IDEAS

Multiple Choice

6. Scientists classify organisms by
   - **a.** arranging the organisms in orderly groups.
   - **b.** giving the organisms many common names.
   - **c.** deciding whether the organisms are useful.
   - **d.** using only existing categories of classification.

7. When the seven levels of classification are listed from broadest to narrowest, which level is fifth in the list?
   - **a.** class
   - **b.** order
   - **c.** genus
   - **d.** family

8. The scientific name for the European white waterlily is *Nymphaea alba*. To which genus does this plant belong?
   - **a.** *Nymphaea*
   - **b.** *alba*
   - **c.** water lily
   - **d.** alba lily

9. *Animalia, Protista, Fungi, Archaebacteria, Eubacteria,* and *Plantae* are the
   - **a.** scientific names of different organisms.
   - **b.** names of kingdoms.
   - **c.** levels of classification.
   - **d.** scientists who organized taxonomy.

10. Bacteria that live in your intestines are classified in the kingdom
    - **a.** Protista.
    - **b.** Eubacteria.
    - **c.** Archaebacteria.
    - **d.** Fungi.

11. What kind of organism thrives in hot springs and other extreme environments?
    - **a.** fungus
    - **b.** eubacterium
    - **c.** archaebacterium
    - **d.** protist

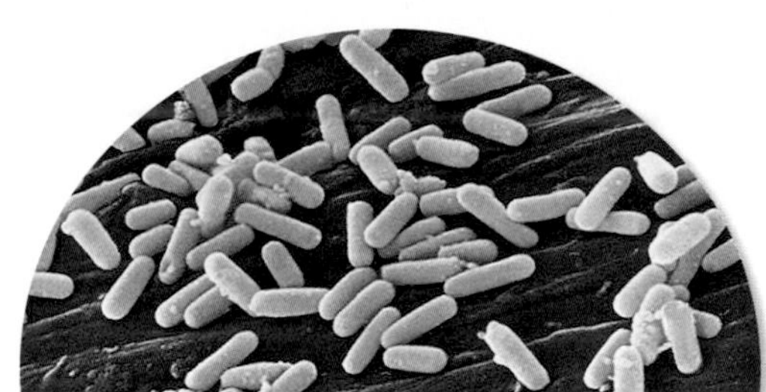

Short Answer

12. Why is the use of scientific names important in biology?

13. What kind of evidence is used by modern taxonomists to classify organisms based on evolutionary relationships?

14. Is a eubacterium a type of eukaryote? Explain your answer.

15. Scientists used to classify organisms as either plants or animals. Why doesn't that classification system work?

CRITICAL THINKING

16. **Concept Mapping** Use the following terms to create a concept map: *kingdom, fern, lizard, Animalia, Fungi, algae, Protista, Plantae,* and *mushroom.*

17. **Analyzing Methods** Explain how the levels of classification depend on the similarities and differences between organisms.

18. **Making Inferences** Explain why two species that belong to the same genus, such as white oak (*Quercus alba*) and cork oak (*Quercus suber*), also belong to the same family.

19. **Identifying Relationships** What characteristic do the members of all six kingdoms have in common?

INTERPRETING GRAPHICS

Use the branching diagram of selected primates below to answer the questions that follow.

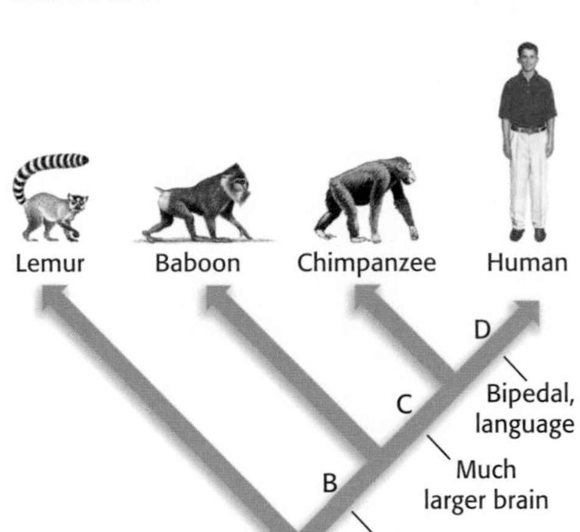

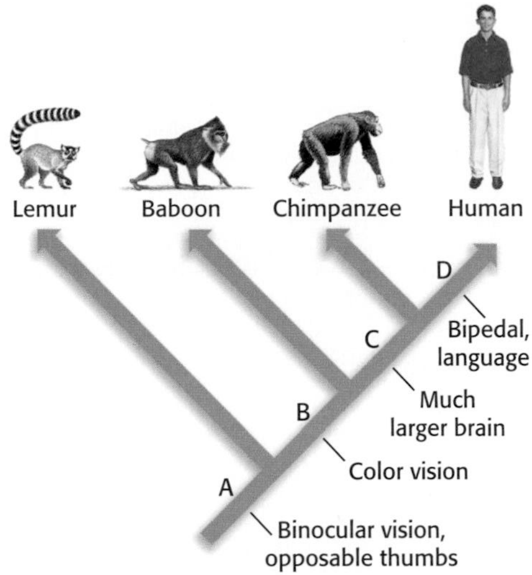

20. Which primate is the closest relative to the common ancestor of all primates?

21. Which primate shares the most traits with humans?

22. Do both lemurs and humans have the characteristics listed at point D? Explain your answer.

23. What characteristic do baboons have that lemurs do not have? Explain your answer.

12. Sample answer: Each species is unique, and scientific names make it possible for scientists to know specifically which organism is being discussed without the confusion of common names.

13. Taxonomists classify organisms based on their shared characteristics.

14. no, A eubacterium is a prokaryote because it does not have a nucleus.

15. Sample answer: Some organisms, such as slime molds and mushrooms, have characteristics that neither plants nor animals have.

Critical Thinking

16. 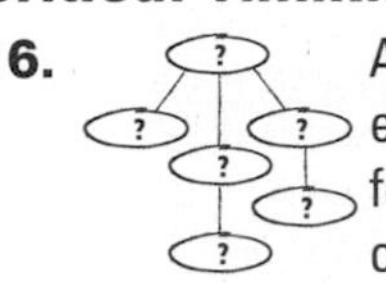 An answer to this exercise can be found at the end of this book.

17. Sample answer: Each level of classification groups organisms according to characteristics they share. At broader levels of classification, such as kingdom and phylum, organisms share fewer characteristics than they do at more specific levels, such as genus and species.

18. Sample answer: The family level of classification contains genera and all the species in those genera. All of the *Quercus* genera are in the same family because of shared characteristics.

19. Sample answer: All members of the six kingdoms are living organisms. They all have DNA.

Interpreting Graphics

20. lemur

21. chimpanzee

22. no; Lemurs branched off between points A and B.

23. Baboons have color vision, but lemurs do not. Color vision appears on the diagram after lemurs branched off and before baboons branched off.

CHAPTER RESOURCES

Chapter Resource File

- Chapter Review GENERAL
- Chapter Test A GENERAL
- Chapter Test B ADVANCED
- Chapter Test C SPECIAL NEEDS
- Vocabulary Activity GENERAL

Workbooks

Study Guide

- Assessment resources are also available in Spanish.

Standardized Test Preparation

Teacher's Note

To provide practice under more realistic testing conditions, give students 20 minutes to answer all of the questions in this Standardized Test Preparation.

MISCONCEPTION ALERT

Answers to the standardized test preparation can help you identify student misconceptions and misunderstandings.

READING

Passage 1

1. A
2. I
3. C

TEST DOCTOR

Question 1: Students selecting an incorrect answer may benefit from a review of how context can help a reader understand new terms. The words *equally* and *categories* offer clues to the reader that the word *distributed* indicates that the organisms are divided into the categories.

Question 3: Answer B is arguably true, but it is not stated in the passage. The correct answer is found in the last sentence of the paragraph.

Standardized Test Preparation

READING

Read each of the passages below. Then, answer the questions that follow each passage.

Passage 1 When organizing life on Earth into categories, we must remember that organisms are not equally distributed throughout the categories of our classification system. We often think of Earth's living things as only the plants and animals that live on Earth's surface. However, the largest kingdoms in terms of the number of individuals and total mass are the kingdoms Archaebacteria and Eubacteria. And a common home of bacteria may be deep within the Earth's crust.

1. In the passage, what does *distributed* mean?

- **A** divided
- **B** important
- **C** visible
- **D** variable

2. According to the passage, what are most of the organisms living on Earth?

- **F** plants
- **G** animals
- **H** fungi
- **I** bacteria

3. Which of the following statements is a fact according to the passage?

- **A** All organisms are equally distributed over Earth's surface.
- **B** Plants are the most important organisms on Earth.
- **C** Many bacteria may live deep within Earth's crust.
- **D** Bacteria are equally distributed over Earth's surface.

Passage 2 When you think of an animal, what do you imagine? You may think of a dog, a cat, or a parrot. All of those organisms are animals. But the animal kingdom also includes some members that might surprise you, such as worms, insects, corals, and sponges.

1. In the passage, what is coral?

- **A** a kind of animal
- **B** a kind of insect
- **C** a color similar to pink
- **D** an organism found in lakes and streams

2. What can you infer from the passage?

- **F** All members of the animal kingdom are visible.
- **G** Parrots make good pets.
- **H** Not all members of the animal kingdom have DNA.
- **I** Members of the animal kingdom come in many shapes and sizes.

3. Which of the following can you infer from the passage?

- **A** Worms and corals make good pets.
- **B** Corals and cats have some traits in common.
- **C** All organisms are animals.
- **D** Worms, corals, insects, and sponges are in the same family.

4. In the passage, what does *members* mean?

- **F** teammates
- **G** limbs
- **H** individuals admitted to a club
- **I** components

Passage 2

1. A
2. I
3. B
4. I

TEST DOCTOR

Question 2: Students may struggle with the task of inferring. None of the answers offered are explicitly stated in the passage. But the fourth sentence links corals, sponges, worms, and insects to dogs, cats, and parrots. Because the passage indicates that both groups are in the animal kingdom, students can conclude that all the organisms mentioned share characteristics.

INTERPRETING GRAPHICS

The Venn diagrams below show two classification systems. Use the diagrams to answer the questions that follow.

Classification system A

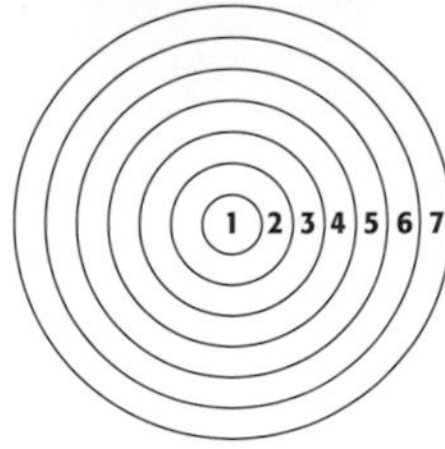

Classification system B

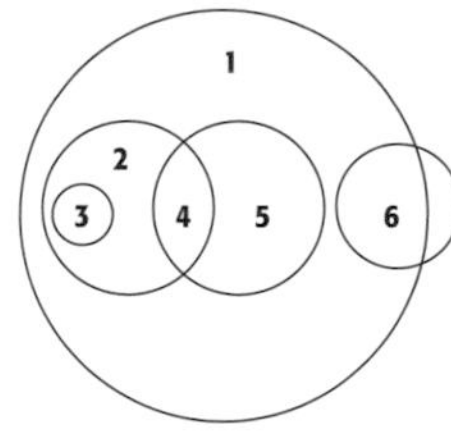

1. For Classification system A, which of the following statements is true?

A All organisms in group 6 are in group 7.
B All organisms in group 5 are in group 4.
C All organisms in group 6 are in group 1.
D All organisms in group 2 are in group 1.

2. For Classification system A, which of the following statements is true?

F All organisms in group 3 are in group 2.
G All organisms in group 3 are in group 4.
H All organisms in group 3 are in group 1.
I All organisms in group 3 are in every other group.

3. For Classification system B, which of the following statements is true?

A All organisms in group 1 are in group 6.
B All organisms in group 6 are in group 1.
C All organisms in group 3 are in group 1.
D All organisms in group 2 are in group 5.

4. For Classification system B, which of the following statements is true?

F All organisms in group 4 are in group 1, 2, and 5.
G All organisms in group 4 are in groups 3 and 5.
H All organisms in group 4 are in groups 5 and 6.
I All organisms in group 4 are in groups 1, 5, and 6.

5. In Classification system B, which group contains organisms that are not in group 1?

A 2
B 4
C 5
D 6

MATH

Read each question below, and choose the best answer.

1. Scientists estimate that millions of species have not yet been discovered and classified. About 1.8 million species have been discovered and classified. If scientists think that this 1.8 million makes up only 10% of the total number of species on Earth, how many species do scientists think exist on Earth?

A 180 million
B 18 million
C 1.8 million
D 180,000

2. Sequoia trees can grow to more than 90 m in height. There are 3.28 feet in 1 meter. How many feet are in 90 m?

F 27.4 ft
G 95.2 ft
H 270 ft
I 295.2 ft

Standardized Test Preparation

INTERPRETING GRAPHICS

1. A
2. G
3. C
4. F
5. D

TEST DOCTOR

Question 1: In Classification system A, the larger number contains all of the organisms in the groups smaller than it. So, the correct answer to this question will have to be the answer that lists a smaller group in a larger group. Answer option A is the only answer with that characteristic.

Question 5: In Classification system B, the only organisms that are not in group 1 are those outside the circle marking group 1. The only organisms outside that circle are in group 6.

MATH

1. B
2. I

Question 1: Students who select incorrect answers here may benefit from a review of how percentages are calculated. Showing students how to transfer the written problem into an equation may help them solve for the correct variable.

CHAPTER RESOURCES

Chapter Resource File

• Standardized Test Preparation GENERAL

State Resources

For specific resources for your state, visit **go.hrw.com** and type in the keyword **HSMSTR.**

Scientific Debate

Background

A 1997 find in Argentina gives some support to the proponents of the birds-from-dinosaurs hypothesis. A 6 ft long fossil found in Argentina shows the most birdlike dinosaur ever discovered. Its skeletal structure indicates it had arms that could flap and fold like wings. It had a birdlike pelvis as well. The sediments in which the dinosaur fossil was found suggest that it is 90 million years old. But this fossil, too, has fueled the debate. Some experts say the dinosaur existed long after the development of modern birds. Birds, they argue, evolved from another line of reptiles.

Scientific Discovery

Background

In basic research, entomologists study insect classification, distribution, and behavior. Entomologists work with farmers and ranchers to help them produce crops or livestock more efficiently. They may also work in forestry to protect trees from insect pests. Forensic entomologists use their knowledge of insect physiology, behavior, and distribution to help law enforcement officials solve crimes or resolve legal issues.

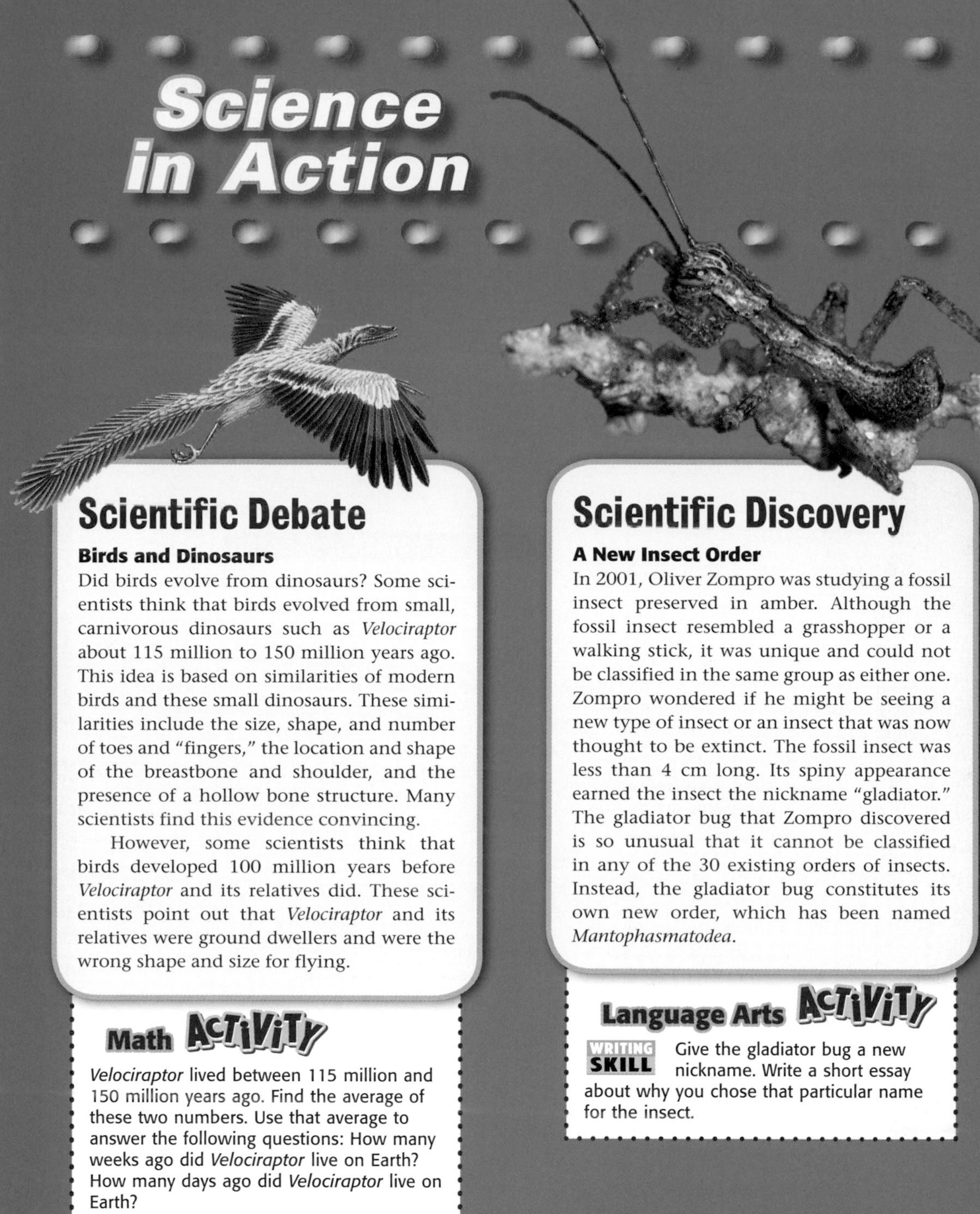

Science in Action

Scientific Debate

Birds and Dinosaurs

Did birds evolve from dinosaurs? Some scientists think that birds evolved from small, carnivorous dinosaurs such as *Velociraptor* about 115 million to 150 million years ago. This idea is based on similarities of modern birds and these small dinosaurs. These similarities include the size, shape, and number of toes and "fingers," the location and shape of the breastbone and shoulder, and the presence of a hollow bone structure. Many scientists find this evidence convincing.

However, some scientists think that birds developed 100 million years before *Velociraptor* and its relatives did. These scientists point out that *Velociraptor* and its relatives were ground dwellers and were the wrong shape and size for flying.

Math ACTIVITY

Velociraptor lived between 115 million and 150 million years ago. Find the average of these two numbers. Use that average to answer the following questions: How many weeks ago did *Velociraptor* live on Earth? How many days ago did *Velociraptor* live on Earth?

Scientific Discovery

A New Insect Order

In 2001, Oliver Zompro was studying a fossil insect preserved in amber. Although the fossil insect resembled a grasshopper or a walking stick, it was unique and could not be classified in the same group as either one. Zompro wondered if he might be seeing a new type of insect or an insect that was now thought to be extinct. The fossil insect was less than 4 cm long. Its spiny appearance earned the insect the nickname "gladiator." The gladiator bug that Zompro discovered is so unusual that it cannot be classified in any of the 30 existing orders of insects. Instead, the gladiator bug constitutes its own new order, which has been named *Mantophasmatodea*.

Language Arts ACTIVITY

WRITING SKILL Give the gladiator bug a new nickname. Write a short essay about why you chose that particular name for the insect.

Answer to Math Activity

115 million years + 150 million years = 265 million years,
265 million years ÷ 2 = 132.5 million years;
132.5 million years × 52 weeks/year = 927.5 million weeks;
927.5 million weeks × 7 days/week = 6.49 billion days

Answer to Language Arts Activity

Nicknames may vary, but essays should all give clear reasons for the name chosen for the insect.

People in Science

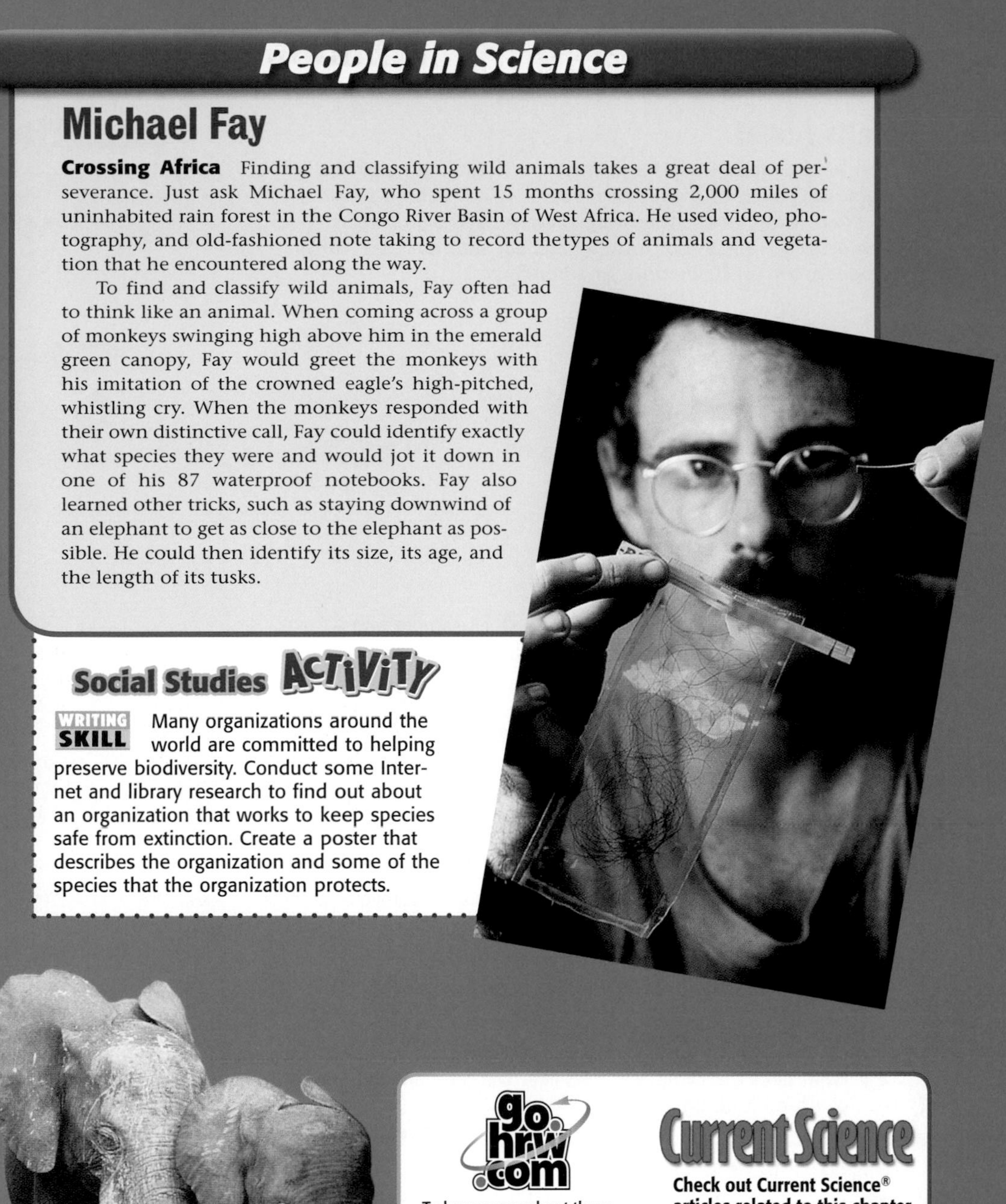

Michael Fay

Crossing Africa Finding and classifying wild animals takes a great deal of perseverance. Just ask Michael Fay, who spent 15 months crossing 2,000 miles of uninhabited rain forest in the Congo River Basin of West Africa. He used video, photography, and old-fashioned note taking to record the types of animals and vegetation that he encountered along the way.

To find and classify wild animals, Fay often had to think like an animal. When coming across a group of monkeys swinging high above him in the emerald green canopy, Fay would greet the monkeys with his imitation of the crowned eagle's high-pitched, whistling cry. When the monkeys responded with their own distinctive call, Fay could identify exactly what species they were and would jot it down in one of his 87 waterproof notebooks. Fay also learned other tricks, such as staying downwind of an elephant to get as close to the elephant as possible. He could then identify its size, its age, and the length of its tusks.

Social Studies ACTIVITY

WRITING SKILL Many organizations around the world are committed to helping preserve biodiversity. Conduct some Internet and library research to find out about an organization that works to keep species safe from extinction. Create a poster that describes the organization and some of the species that the organization protects.

go.hrw.com

To learn more about these Science in Action topics, visit go.hrw.com and type in the keyword **HL5CLSF.**

Current Science

Check out Current Science® articles related to this chapter by visiting go.hrw.com. Just type in the keyword HL5CS09.

People in Science

ACTIVITY — GENERAL

Have students research Michael Fay and "Megatransect," the official name of his exploration, on the Internet. Once students have found a Web site that traces Fay's route, have them trace it along a map or a globe. Then, have students research some of the national parks that have been created since Fay's trip, such as the Wonga-Wongué, the Ogooué Wetlands, and Mont Iboundji. Students could then pick a particular national park and draw their own map of that area, including pictures and information about the plants and animals there.

Answer to Social Studies Activity

Answers may vary. Groups, such as the World Wildlife Fund, the Nature Conservancy, and Conservation International, work around the world to preserve biodiversity. They help protect many animals and habitats and raise public awareness about the importance of preserving biodiversity.

LabBook

Skills Practice Lab

Cells Alive!

Teacher's Notes

Time Required

One 45-minute class period

Lab Ratings

EASY → HARD

Teacher Prep 1

Student Set-Up 1

Concept Level 1

Clean Up 1

MATERIALS

The materials listed on the student page are enough for a group of 3–4 students. Be sure to keep the algae in a warm, damp place out of direct sunlight; a closed plastic bag with water sprayed into it is ideal.

Procedure

4. Chloroplasts are the parts of the cell that are responsible for photosynthesis.
5. The nucleus of a cell controls most of the activities that take place in that cell and contains the hereditary information.
6. The cytoplasm is a clear gel-like substance that fills the cell and surrounds the organelles. The organelles are floating around in the cytoplasm.

Skills Practice Lab

Cells Alive!

You have probably used a microscope to look at single-celled organisms such as those shown below. They can be found in pond water. In the following exercise, you will look at *Protococcus*—algae that form a greenish stain on tree trunks, wooden fences, flowerpots, and buildings.

MATERIALS

- eyedropper
- microscope
- microscope slide and coverslip
- *Protococcus* (or other algae)
- water

SAFETY

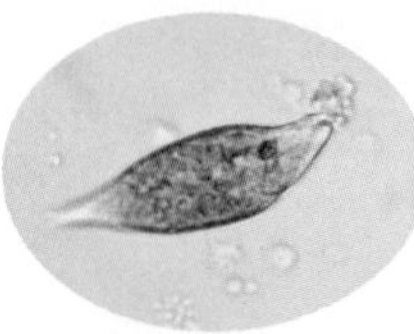

Euglena

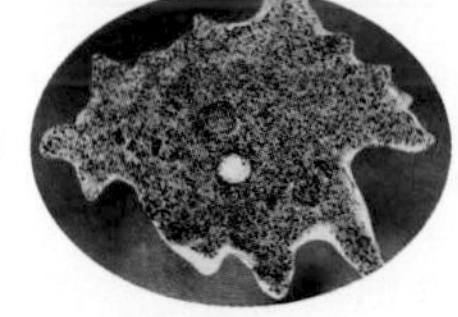

Amoeba

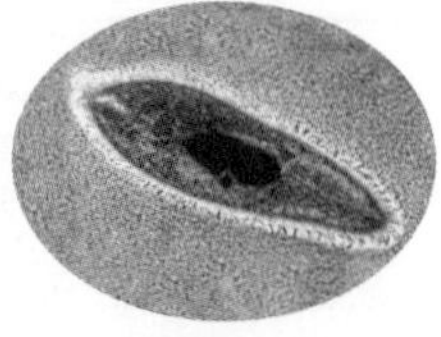

Paramecium

Procedure

1. Locate some *Protococcus*. Scrape a small sample into a container. Bring the sample to the classroom, and make a wet mount of it as directed by your teacher. If you can't find *Protococcus* outdoors, look for algae on the glass in an aquarium. Such algae may not be *Protococcus,* but it will be a very good substitute.
2. Set the microscope on low power to examine the algae. On a separate sheet of paper, draw the cells that you see.
3. Switch to high power to examine a single cell. Draw the cell.
4. You will probably notice that each cell contains several chloroplasts. Label a chloroplast on your drawing. What is the function of the chloroplast?
5. Another structure that should be clearly visible in all the algae cells is the nucleus. Find the nucleus in one of your cells, and label it on your drawing. What is the function of the nucleus?
6. What does the cytoplasm look like? Describe any movement you see inside the cells.

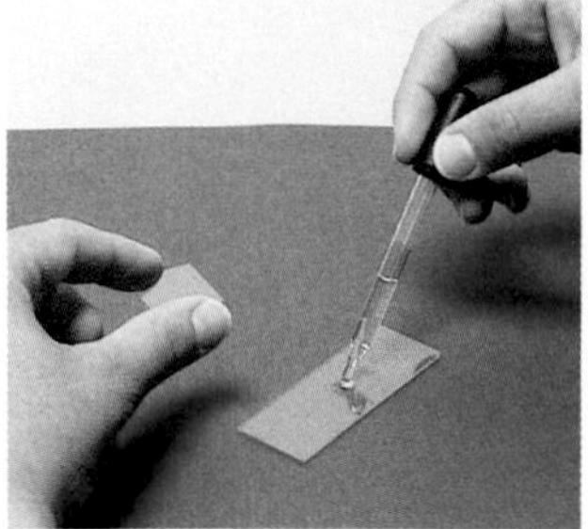

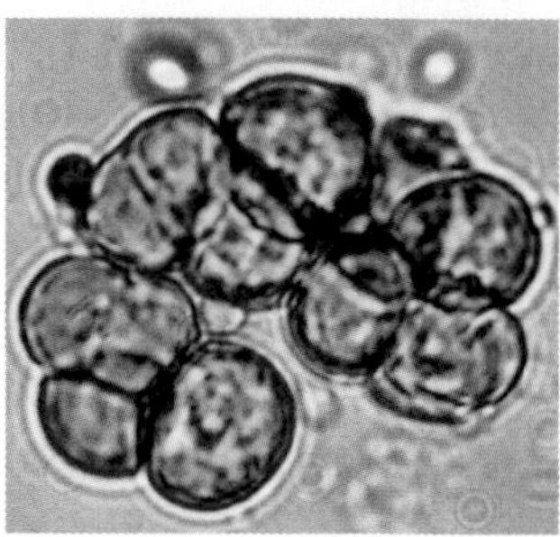

Protococcus

Analyze the Results

1. Are *Protococcus* single-celled organisms or multicellular organisms?
2. How are *Protococcus* different from amoebas?

LabBook

Analyze the Results

1. *Protococcus* is a genus composed of single-celled algae.
2. Many answers are possible, but the following are most likely: *Protococcus* cannot move about as amoebas can; unlike amoebas, they are green and photosynthesize.

CHAPTER RESOURCES

Chapter Resource File

- **Datasheet for LabBook**
- **Lab Notes and Answers**

Terry Rakes
Elmwood Junior High School
Rogers, Arkansas

Skills Practice Lab

Stayin' Alive!

Every second of your life, your body's trillions of cells take in, use, and store energy. They repair themselves, reproduce, and get rid of waste. Together, these processes are called *metabolism.* Your cells use the food that you eat to provide the energy you need to stay alive.

Your Basal Metabolic Rate (BMR) is a measurement of the energy that your body needs to carry out all the basic life processes while you are at rest. These processes include breathing, keeping your heart beating, and keeping your body's temperature stable. Your BMR is influenced by your gender, your age, and many other things. Your BMR may be different from everyone else's, but it is normal for you. In this activity, you will find the amount of energy, measured in Calories, you need every day in order to stay alive.

MATERIALS

- bathroom scale
- tape measure

Procedure

1. Find your weight on a bathroom scale. If the scale measures in pounds, you must convert your weight in pounds to your mass in kilograms. To convert your weight in pounds (lb) to mass in kilograms (kg), multiply the number of pounds by 0.454.

Example: If Carlos weighs 125 lb, his mass in kilograms is:

$$\begin{array}{r} 125 \text{ lb} \\ \times\ \underline{0.454} \\ 56.75 \text{ kg} \end{array}$$

2. Use a tape measure to find your height. If the tape measures in inches, convert your height in inches to height in centimeters. To convert your height in inches (in.) to your height in centimeters (cm), multiply the number of inches by 2.54.

If Carlos is 62 in. tall, his height in centimeters is:

$$\begin{array}{r} 62 \text{ in.} \\ \times\ \underline{2.54} \\ 157.48 \text{ cm} \end{array}$$

LabBook

Skills Practice Lab

Stayin' Alive!

Teacher's Notes

Time Required

One 45-minute class period

Lab Ratings

EASY → HARD

Teacher Prep 1
Student Set-Up 1
Concept Level 1
Clean Up 1

MATERIALS

The materials listed on the student page are enough for each group of 5–6 students. You may wish to have your students use a calculator to complete this activity.

CHAPTER RESOURCES

Chapter Resource File

- Datasheet for LabBook
- Lab Notes and Answers

Kathy LaRoe
East Valley Middle School
East Helena, Montana

Preparation Notes

Some students may consider their height and weight to be personal and won't want to weigh and measure themselves with the others in the class. Give these students the option of using the data of a fictional person, such as one of the following:

Jenny	80 lb	4 ft	age 11
Ben	65 lb	3 ft	age 12
Carlos	110 lb	5 ft 2 in.	age 11
Alexa	120 lb	4 ft 6 in.	age 12
Tasheika	90 lb	4 ft 6 in.	age 13

Lab Notes

Some students will think that their basal metabolic rate, or BMR, is impossibly low. Emphasize that the BMR is the number of Calories a body needs just to keep the heart beating, the lungs breathing, and the cells respiring. The BMR is not the number of Calories a person needs for an active lifestyle.

Of course, a person can consume fewer than that number of Calories for a day, or even for a few days, without dying. Explain that the Calories required to live during starvation conditions are obtained from stored fat. When there is no more fat, then the energy comes from muscle tissue. Under extreme conditions of starvation, the body even begins to shut down some organ functions that use energy but that are not required for survival, such as the uterine cycle in women.

Some students may ask why the BMR numbers are so much higher in males than in females. Explain that before puberty, the numbers are much closer together. But as boys approach puberty, they generally develop a higher muscle-to-fat ratio than girls do. Cellular respiration for muscle tissue requires more energy than for fat tissue.

3. Now that you know your height and mass, use the appropriate formula below to get a close estimate of your BMR. Your answer will give you an estimate of the number of Calories your body needs each day just to stay alive.

Calculating Your BMR	
Females	**Males**
65 + (10 × your mass in kilograms) + (1.8 × your height in centimeters) − (4.7 × your age in years)	66 + (13.5 × your mass in kilograms) + (5 × your height in centimeters) − (6.8 × your age in years)

4. Your metabolism is also influenced by how active you are. Talking, walking, and playing games all take more energy than being at rest. To get an idea of how many Calories your body needs each day to stay healthy, select the lifestyle that best describes yours from the table at right. Then multiply your BMR by the activity factor.

Activity Factors	
Activity lifestyle	**Activity factor**
Moderately inactive (normal, everyday activities)	1.3
Moderately active (exercise 3 to 4 times a week)	1.4
Very active (exercise 4 to 6 times a week)	1.6
Extremely active (exercise 6 to 7 times a week)	1.8

Analyze the Results

1. In what way could you compare your whole body to a single cell? Explain.
2. Does an increase in activity increase your BMR? Does an increase in activity increase your need for Calories? Explain your answers.

Draw Conclusions

3. If you are moderately inactive, how many more Calories would you need if you began to exercise every day?

Applying Your Data

The best energy sources are those that supply the correct amount of Calories for your lifestyle and also provide the nutrients you need. Research in the library or on the Internet to find out which kinds of foods are the best energy sources for you. How does your list of best energy sources compare with your diet?

List everything you eat and drink in 1 day. Find out how many Calories are in each item, and find the total number of Calories you have consumed. How does this number of Calories compare with the number of Calories you need each day for all your activities?

Analyze the Results

1. Sample answer: Just as each cell needs energy on a small scale, your body requires energy on a much larger scale.
2. Sample answer: Technically, the BMR does not change with activity. The BMR is the minimum amount of energy a person needs to stay alive. Activity requires that more energy be added to the BMR, thereby increasing the need for Calories.

Draw Conclusions

3. Students should multiply their own BMR by 1.3 and then multiply their BMR by 1.8. Students should subtract the smaller number from the larger number. This number represents the additional Calories per day the student would expend shifting from a moderately inactive state to an extremely active one.

Inquiry Lab

Tracing Traits

Have you ever wondered about the traits you inherited from your parents? Do you have a trait that neither of your parents has? In this project, you will develop a family tree, or pedigree, similar to the one shown in the diagram below. You will trace an inherited trait through a family to determine how it has passed from generation to generation.

Procedure

1. The diagram at right shows a family history. On a separate piece of paper, draw a similar diagram of the family you have chosen. Include as many family members as possible, such as grandparents, parents, children, and grandchildren. Use circles to represent females and squares to represent males. You may include other information, such as the family member's name, birth date, or picture.
2. Draw a table similar to the one on the next page. Survey each of the family members shown in your family tree. Ask them if they have hair on the middle segment of their fingers. Write each person's name in the appropriate square. Explain to each person that it is normal to have either trait. The presence of hair on the middle segment is the dominant form of this trait.

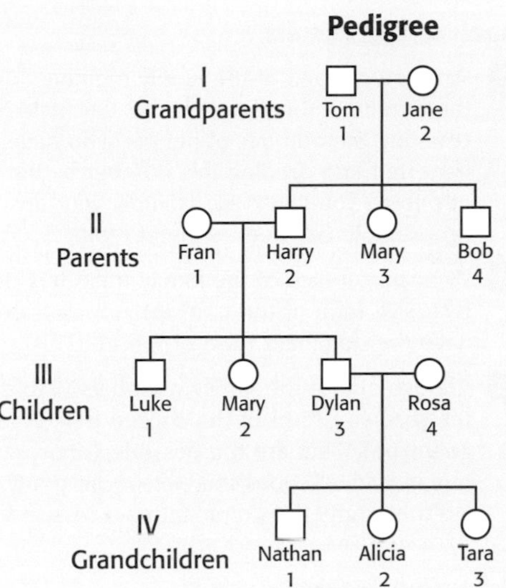

LabBook

CHAPTER RESOURCES

Chapter Resource File

- Datasheet for LabBook
- Lab Notes and Answers

Kerry Johnson
Isbell Middle School
Santa Paula, California

LabBook

Inquiry Lab

Tracing Traits

Teacher's Notes

Time Required

Two 45-minute class periods, separated by several days so students have time to complete their surveys

Lab Ratings

Teacher Prep 1 flask
Student Set-Up 1 flask
Concept Level 2 flasks
Clean Up 1 flask

Lab Notes

Family histories will vary. Encourage students to include at least three generations in their histories.

Survey results will vary. Make sure that students actually surveyed each family member who was available. Responses will vary. You may check family members with shaded symbols against the survey results for accuracy.

Percentages will vary. A family member may receive a recessive allele from the father and a recessive allele from the mother. In such a case, this family member will exhibit the recessive form of the trait rather than the dominant form.

Because so many children are adopted or live in foster homes or group homes, please emphasize to your students that they may choose any family to study.

Analyze the Results

1. Answers may vary.
2. Answers may vary.
3. The genotype of the recessive form of the characteristic must be *hh* (homozygous recessive). Each allele came from one of the individual's parents; Possible genotypes for the parents of the individual expressing the recessive form are *Hh* and *hh;* Does the student know whether either of the parents expresses the recessive form of the trait? Does the student know if the individual chosen has brothers or sisters? Are their genotypes known? If so, have the student decide if each of them has a dominant or recessive genotype. If a dominant genotype is found among the siblings and one of the parents is known to have the recessive form, ask the student what the genotype of the other parent must be (*Hh*).

Draw Conclusions

4. The Punnett square should show *hh* in the bottom right-hand corner. One of the parents must have the genotype *hh.* The other parent must have either *hh* or *Hh.* If any sibling has the dominant trait, the genotype of the other parent must be *Hh.*

Dominant trait	Recessive trait	Family members with the dominant trait	Family members with the recessive trait
Hair present on the middle segment of fingers (*H*)	Hair absent on the middle segment of fingers (*h*)	DO NOT WRITE	IN BOOK

3. Trace this trait throughout the family tree you diagrammed in step 1. Shade or color the symbols of the family members who demonstrate the dominant form of this trait.

Analyze the Results

1. What percentage of the family members demonstrate the dominant form of the trait? Calculate this by counting the number of people who have the dominant trait and dividing this number by the total number of people you surveyed. Multiply your answer by 100. An example has been done at right.

Example: Calculating percentage

$$\frac{10 \text{ people with trait}}{20 \text{ people surveyed}} = \frac{1}{2}$$

$$\frac{1}{2} = 0.50 \times 100 = 50\%$$

2. What percentage of the family members demonstrate the recessive form of the trait? Why doesn't every family member have the dominant form of the trait?
3. Choose one of the family members who demonstrates the recessive form of the chosen trait. What is this person's genotype? What are the possible genotypes for the parents of this individual? Does this person have any brothers or sisters? Do they show the dominant or recessive trait?

Draw Conclusions

4. Draw a Punnett square like the one at right. Use this to determine the genotypes of the parents of the person you chose in step 3. Write this person's genotype in the bottom right-hand corner of your Punnett square. **Hint:** There may be more than one possible genotype for the parents. Don't forget to consider the genotypes of the person's brothers and sisters.

Father

	?	?
Mother ?		
?		

LabBook

Skills Practice Lab

The Half-life of Pennies

Carbon-14 is a special unstable element used in the absolute dating of material that was once alive, such as fossil bones. Every 5,730 years, half of the carbon-14 in a fossil specimen decays or breaks down into a more stable element. In the following experiment you will see how pennies can show the same kind of "decay."

MATERIALS

- container with a cover, large
- pennies (100)

Procedure

1. Place 100 pennies in a large, covered container. Shake the container several times, and remove the cover. Carefully empty the container on a flat surface making sure the pennies don't roll away.
2. Remove all the coins that have the "head" side of the coin turned upward. Record the number of pennies removed and the number of pennies remaining in a data table similar to the one at right.
3. Repeat the process until no pennies are left in the container. Remember to remove only the coins showing "heads."
4. Draw a graph similar to the one at right. Label the *x*-axis "Number of shakes," and label the *y*-axis "Pennies remaining." Using data from your data table, plot the number of coins remaining at each shake on your graph.

Shake number	Number of coins remaining	Number of coins removed
1		
2	DO NOT WRITE IN BOOK	
3		

Half-life of Pennies

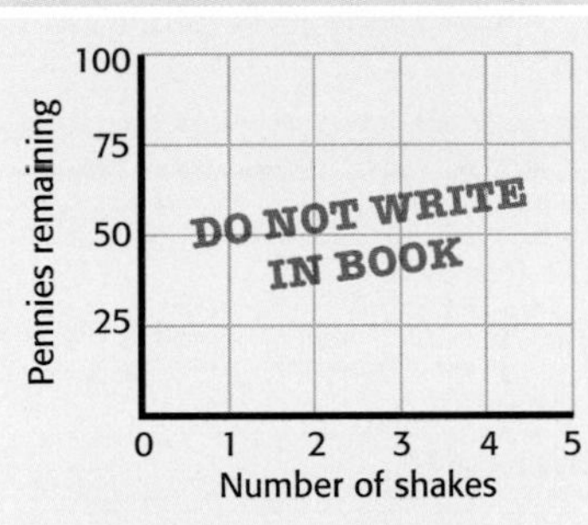

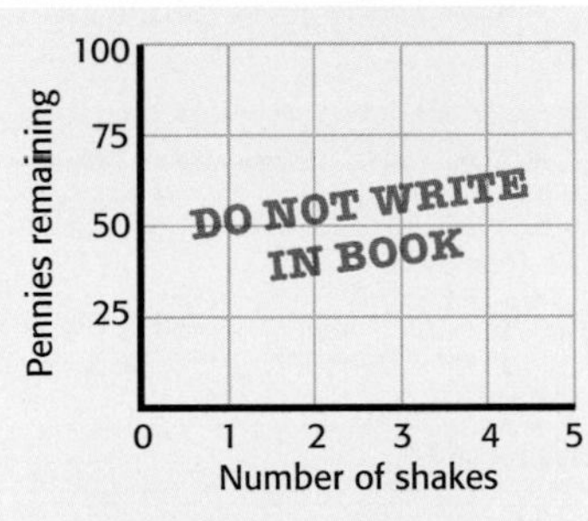

Analyze the Results

1. Examine the Half-life of Carbon-14 graph at right. Compare the graph you have made for pennies with the one for carbon-14. Explain any similarities that you see.
2. Recall that the probability of landing "heads" in a coin toss is 1/2. Use this information to explain why the remaining number of pennies is reduced by about half each time they are shaken and tossed.

Half-life of Carbon-14

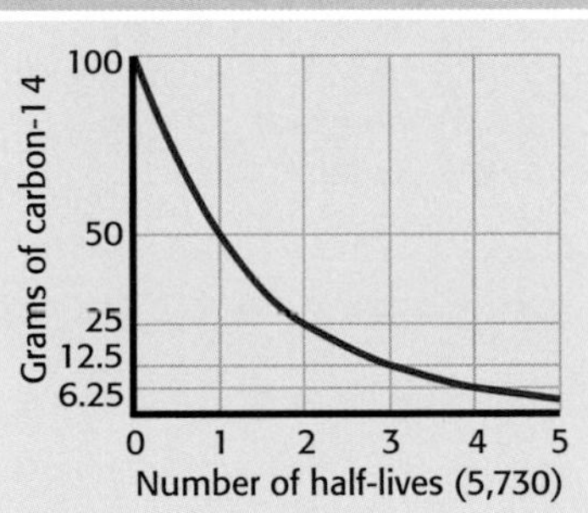

Skills Practice Lab

The Half-life of Pennies

Teacher's Notes

Time Required

One 45-minute class period

Lab Rating

EASY → HARD

Teacher Prep 1

Student Set-Up 1

Concept Level 2

Clean Up 1

Lab Notes

It is useful to use coin tosses to explain half-life because approximately half the coins will land heads and half will land tails. Therefore, about half the entire quantity of coins tossed will be eliminated with each successive toss.

CHAPTER RESOURCES

Chapter Resource File

- Datasheet for LabBook
- Lab Notes and Answers

Karma Houston-Hughes
Kyrene Middle School
Tempe, Arizona

Analyze the Results

1. The graphs should be very similar in shape. With each half-life and each shake, the number remaining will be reduced by half.
2. The remaining number of pennies is reduced by about half each time the pennies are shaken and tossed because there are only two faces on each coin. The rules of probability suggest that half will land heads and half will land tails, and therefore the amount will be reduced by about half with each shake.

LabBook

Skills Practice Lab

Voyage of the USS *Adventure*

Teacher's Notes

Time Required
One 45-minute class period

Lab Ratings
EASY → HARD

Teacher Prep 1
Student Set-Up 1
Concept Level 2
Clean Up 1

Preparation Notes
Some students will find it easier to make the charts on graph paper, so you may wish to supply graph paper to students.

Lab Notes
Students should know that travel outside the solar system is not yet possible. This activity should help students categorize organisms or objects by noticing subtle differences. This activity is a good way to begin a study of classification of animals, rocks, or plants. This lab may be useful before introducing dichotomous keys, for example.

Skills Practice Lab

Voyage of the USS *Adventure*

You are a crew member on the USS *Adventure*. The *Adventure* has been on a 5-year mission to collect life-forms from outside the solar system. On the voyage back to Earth, your ship went through a meteor shower, which ruined several of the compartments containing the extraterrestrial life-forms. Now it is necessary to put more than one life-form in the same compartment.

You have only three undamaged compartments in your starship. You and your crewmates must stay in one compartment, and that compartment should be used for extraterrestrial life-forms only if absolutely necessary. You and your crewmates must decide which of the life-forms could be placed together. It is thought that similar life-forms will have similar needs. You can use only observable characteristics to group the life-forms.

Life-form 2

Life-form 3

Procedure

1. Make a data table similar to the one below. Label each column with as many characteristics of the various life-forms as possible. Leave enough space in each square to write your observations. The life-forms are pictured on this page.

Life-form Characteristics				
	Color	Shape	Legs	Eyes
Life-form 1				
Life-form 2				
Life-form 3				
Life-form 4				

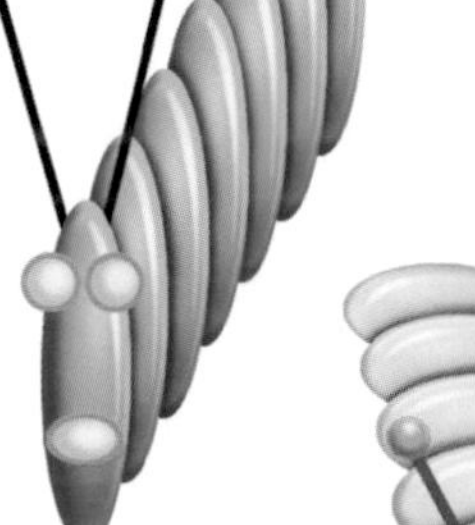

2. Describe each characteristic as completely as you can. Based on your observations, determine which of the life-forms are most alike.

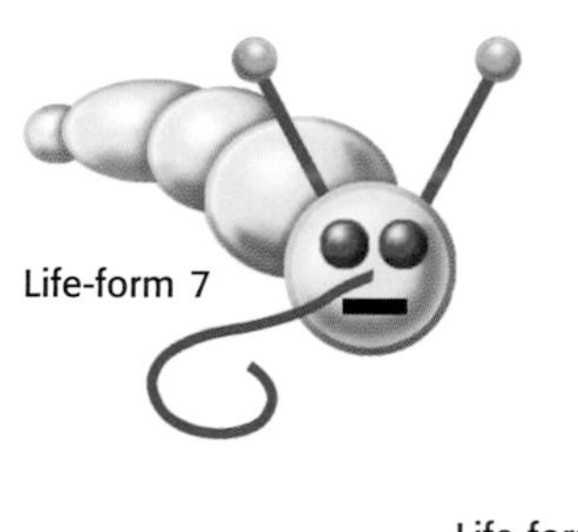

Georgiann Delgadillo
East Valley School District
Continuous Curriculum School
Spokane, Washington

CHAPTER RESOURCES

Chapter Resource File
- Datasheet for LabBook
- Lab Notes and Answers

3. Make a data table like the one below. Fill in the table according to the decisions you made in step 2. State your reasons for the way you have grouped your life-forms.

Life-form Room Assignments		
Compartment	**Life-forms**	**Reasons**
1		
2		
3		

4. The USS *Adventure* has to make one more stop before returning home. On planet X437 you discover the most interesting life-form ever found outside of Earth—the CC9, shown at right. Make a decision, based on your previous grouping of life-forms, about whether you can safely include CC9 in one of the compartments for the trip to Earth.

CC9

Analyze the Results

1. Describe the life-forms in compartment 1. How are they similar? How are they different?
2. Describe the life-forms in compartment 2. How are they similar? How do they differ from the life-forms in compartment 1?
3. Are there any life-forms in compartment 3? If so, describe their similarities. In which compartment will you and your crewmates remain for the journey home?

Draw Conclusions

4. Are you able to transport life-form CC9 safely back to Earth? If so, in which compartment will it be placed? How did you decide?

Applying Your Data

In 1831, Charles Darwin sailed from England on a ship called the HMS *Beagle*. You have studied the finches that Darwin observed on the Galápagos Islands. What were some of the other unusual organisms he found there? For example, find out about the Galápagos tortoise.

Analyze the Results

There are no right or wrong answers in this activity. The objective is to allow students an opportunity to recognize subtle differences and to recognize that organisms may be more alike than they are different. However, you should make sure the students provide good reasons why they grouped certain life-forms together. There are several ways in which these seven organisms are similar. For example, four of them are segmented and have no legs. Three of them are geometrically shaped, and three others have mouths. Have students examine them for less observable characteristics, such as what kind of body plan or symmetry they have, how they might obtain food, or whether they might be land dwelling or aquatic.

Draw Conclusions

4. Answers will depend on how students grouped the life-forms in this lab and which characteristics the students used to classify life-form CC9.

Applying Your Data

The Galápagos tortoise can have a shell length of 1.3 m, have a mass of 180 kg, and live to be 150 years old.

Contents

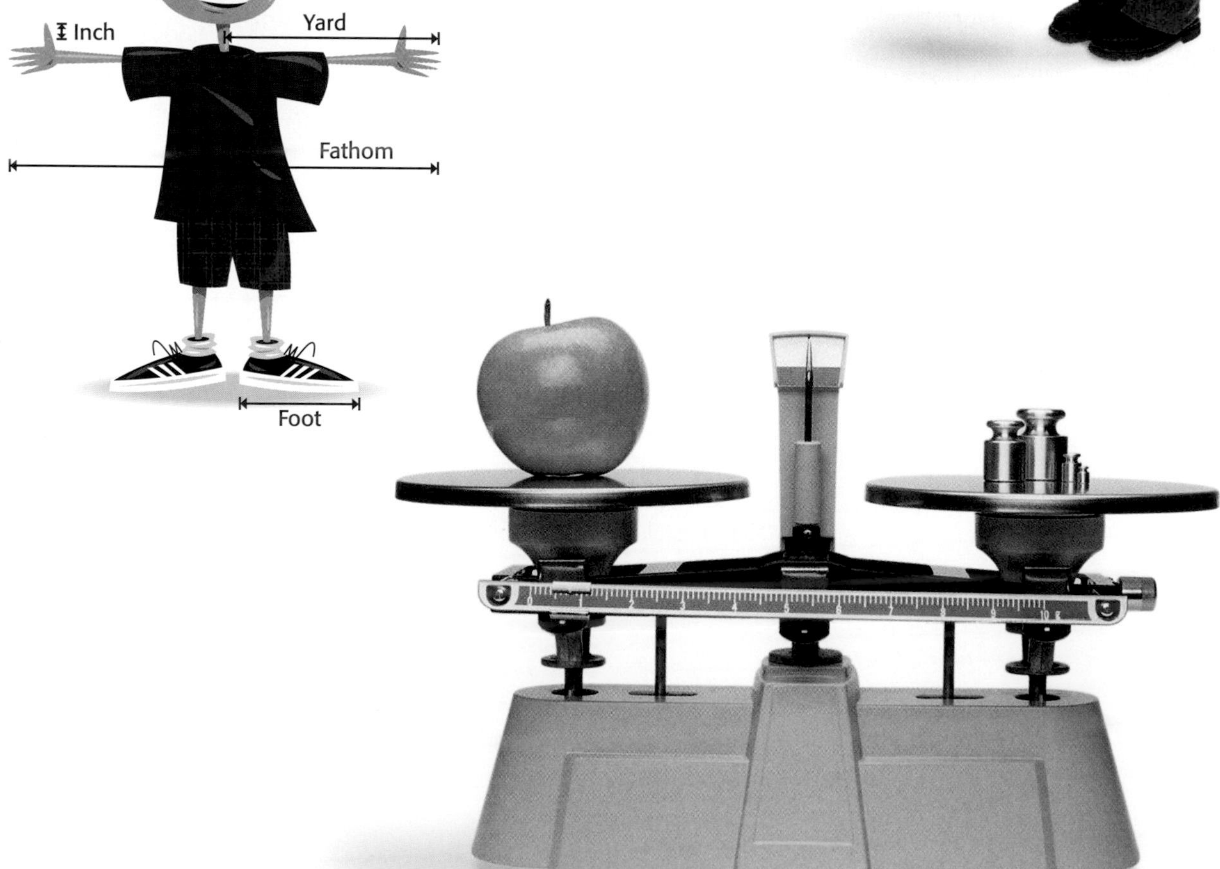

✓Reading Check Answers

Chapter 1 Cells: The Basic Units of Life

Section 1

Page 5: Sample answer: All organisms are made of one or more cells, the cell is the basic unit of all living things, and all cells come from existing cells.

Page 6: If a cell's volume gets too large, the cell's surface area will not be able to take in enough nutrients or get rid of wastes fast enough to keep the cell alive.

Page 7: Organelles are structures within a cell that perform specific functions for the cell.

Page 9: One difference between eubacteria and archaea is that bacterial ribosomes are different from archaebacterial ribosomes.

Page 10: The main difference between prokaryotes and eukaryotes is that eukaryotic cells have a nucleus and membrane-bound organelles and prokaryotic cells do not.

Section 2

Page 12: Plant, algae, and fungi cells have cell walls.

Page 13: A cell membrane encloses the cell and separates and protects the cell's contents from the cell's environment. The cell wall also controls the movement of materials into and out of the cell.

Page 14: The cytoskeleton is a web of proteins in the cytoplasm. It gives the cell support and structure.

Page 16: Most of a cell's ATP is made in the cell's mitochondria.

Page 18: Lysosomes destroy worn-out organelles, attack foreign invaders, and get rid of waste material from inside the cell.

Section 3

Page 20: Sample answer: larger size, longer life, and cell specialization

Page 21: An organ is a structure of two or more tissues working together to perform a specific function in the body.

Page 22: cell, tissue, organ, organ system

Chapter 2 The Cell in Action

Section 1

Page 35: Red cells would burst in pure water because water particles move from outside, where particles were dense, to inside the cell, where particles were less dense. This movement of water would cause red cells to fill up and burst.

Page 37: Exocytosis is the process by which a cell moves large particles to the outside of the cell.

Section 2

Page 39: Cellular respiration is a chemical process by which cells produce energy from food. Breathing supplies the body with the raw materials needed for cellular respiration.

Page 41: the kind that happens in the muscle cells of animals and the kind that occurs in bacteria

Section 3

Page 43: No, the number of chromosomes is not always related to the complexity of organisms.

Page 44: During mitosis in plant cells, a cell plate is formed. During mitosis in animal cells, a cell plate does not form.

Chapter 3 Heredity

Section 1

Page 58: the passing of traits from parents to offspring

Page 61: During his second set of experiments, Mendel allowed the first-generation plants, which resulted from his first set of experiments, to self-pollinate.

Page 62: A ratio is a relationship between two different numbers that is often expressed as a fraction.

Section 2

Page 64: A gene contains the instructions for an inherited trait. The different versions of a gene are called *alleles*.

Page 66: Probability is the mathematical chance that something will happen.

Page 68: In incomplete dominance, one trait is not completely dominant over another.

Section 3

Page 71: 23 chromosomes

Page 72: During meiosis, one parent cell makes four new cells.

Chapter 4 Genes and DNA

Section 1

Page 89: Guanine and cytosine are always found in DNA in equal amounts, as are adenine and thymine.

Page 91: every time a cell divides

Section 2

Page 92: a string of nucleotides that give the cell information about how to make a specific trait

Page 95: They transfer amino acids to the ribosome.

Page 96: a physical or chemical agent that can cause a mutation in DNA

Page 97: Sickle cell anemia is caused by a mutation in a single nucleotide of DNA, which then causes a different amino acid to be assembled in a protein used in blood cells.

Page 98: a near-identical copy of another organism, created with the original organism's genes

Appendix

Chapter 5 The Evolution of Living Things

Section 1

Page 110: if they mate with each other and produce more of the same type of organism

Page 112: by their estimated ages and physical similarities

Page 114: a four-legged land mammal

Page 116: that they have common ancestry

Section 2

Page 119: 965 km (600 mi) west of Ecuador

Page 121: that Earth had been formed by natural processes over a long period of time

Page 122: *On the Origin of Species by Means of Natural Selection*

Section 3

Page 125: because they often produce many offspring and have short generation times

Page 126: Sample answer: A newly formed canyon, mountain range, or lake could divide the members of a population.

Chapter 6 The History of Life on Earth

Section 1

Page 139: absolute dating

Page 141: periods of sudden extinction of many species

Page 142: the idea that the Earth's continents once formed a single landmass surrounded by ocean

Section 2

Page 144: The early Earth was very different from today—there were violent events and a harsh atmosphere.

Page 147: a mass extinction

Page 148: "recent life"

Section 3

Page 151: the hominid family

Page 152: Africa

Page 155: Paleontologists will review their ideas about the evolution of hominids.

Appendix

Chapter 7 Classification

Section 1

Page 166: Sample answer: How many known species are there? What are the defining characteristics of each species? and What are the relationships between these species?

Page 169: genus and species

Page 170: A dichotomous key is an identification aid that uses a series of descriptive statements.

Section 2

Page 173: *Escherichia coli*

Page 175: Sample answer: Plants make energy through photosynthesis. Some members of the kingdoms Fungi, Protista, and Eubacteria consume plants. When these organisms digest the plant material, they get energy and nutrients made by the plants.

Page 177: Sponges don't have sense organs, and they usually can't move around.

Study Skills

FoldNote Instructions

Have you ever tried to study for a test or quiz but didn't know where to start? Or have you read a chapter and found that you can remember only a few ideas? Well, FoldNotes are a fun and exciting way to help you learn and remember the ideas you encounter as you learn science!

FoldNotes are tools that you can use to organize concepts. By focusing on a few main concepts, FoldNotes help you learn and remember how the concepts fit together. They can help you see the "big picture." Below you will find instructions for building 10 different FoldNotes.

Pyramid

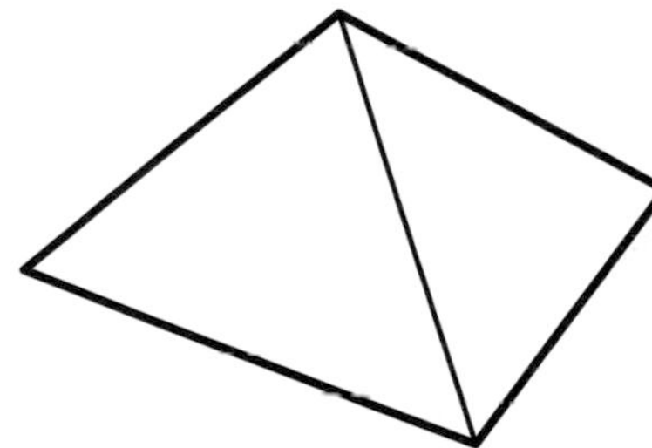

1. Place a sheet of paper in front of you. Fold the lower left-hand corner of the paper diagonally to the opposite edge of the paper.
2. Cut off the tab of paper created by the fold (at the top).
3. Open the paper so that it is a square. Fold the lower right-hand corner of the paper diagonally to the opposite corner to form a triangle.
4. Open the paper. The creases of the two folds will have created an X.
5. Using scissors, cut along one of the creases. Start from any corner, and stop at the center point to create two flaps. Use tape or glue to attach one of the flaps on top of the other flap.

Double Door

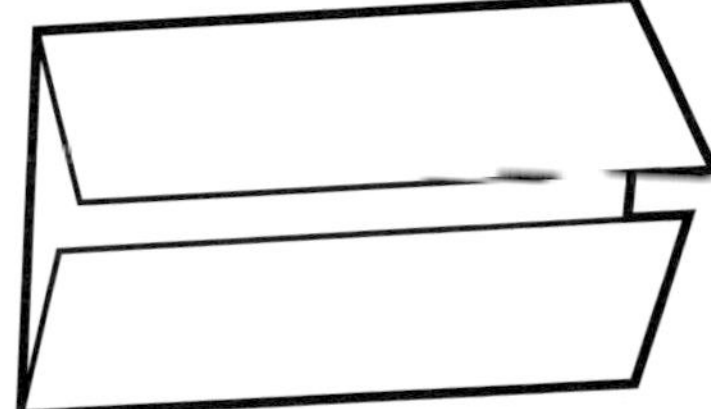

1. Fold a sheet of paper in half from the top to the bottom. Then, unfold the paper.
2. Fold the top and bottom edges of the paper to the crease.

Booklet

1. Fold a sheet of paper in half from left to right. Then, unfold the paper.
2. Fold the sheet of paper in half again from the top to the bottom. Then, unfold the paper.
3. Refold the sheet of paper in half from left to right.
4. Fold the top and bottom edges to the center crease.
5. Completely unfold the paper.
6. Refold the paper from top to bottom.
7. Using scissors, cut a slit along the center crease of the sheet from the folded edge to the creases made in step 4. Do not cut the entire sheet in half.
8. Fold the sheet of paper in half from left to right. While holding the bottom and top edges of the paper, push the bottom and top edges together so that the center collapses at the center slit. Fold the four flaps to form a four-page book.

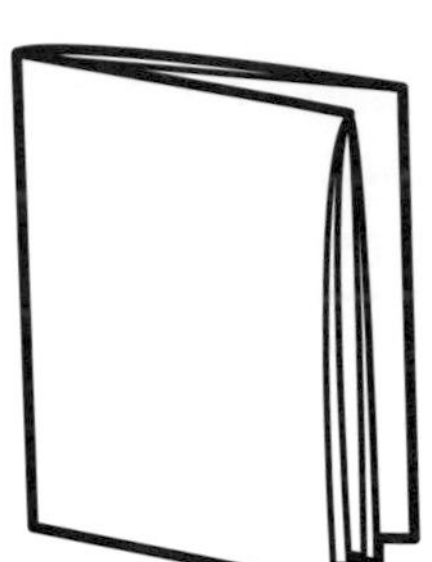

Layered Book

1. Lay one sheet of paper on top of another sheet. Slide the top sheet up so that 2 cm of the bottom sheet is showing.
2. Hold the two sheets together, fold down the top of the two sheets so that you see four 2 cm tabs along the bottom.
3. Using a stapler, staple the top of the FoldNote.

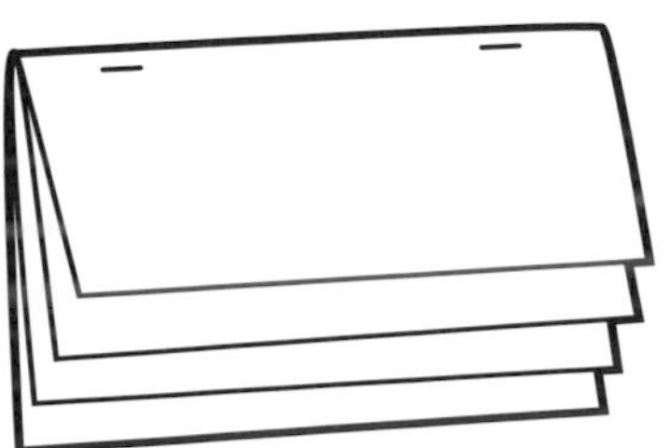

Appendix

Key-Term Fold

1. Fold a sheet of lined notebook paper in half from left to right.
2. Using scissors, cut along every third line from the right edge of the paper to the center fold to make tabs.

Four-Corner Fold

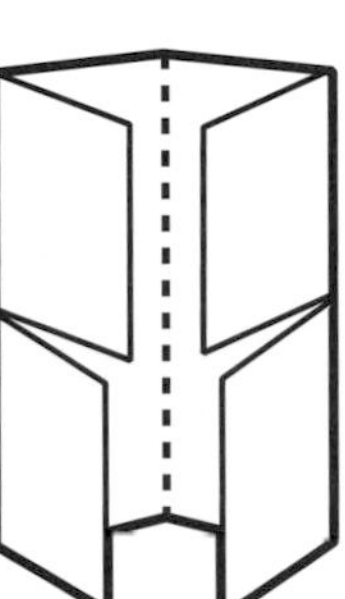

1. Fold a sheet of paper in half from left to right. Then, unfold the paper.
2. Fold each side of the paper to the crease in the center of the paper.
3. Fold the paper in half from the top to the bottom. Then, unfold the paper.
4. Using scissors, cut the top flap creases made in step 3 to form four flaps.

Three-Panel Flip Chart

1. Fold a piece of paper in half from the top to the bottom.
2. Fold the paper in thirds from side to side. Then, unfold the paper so that you can see the three sections.
3. From the top of the paper, cut along each of the vertical fold lines to the fold in the middle of the paper. You will now have three flaps.

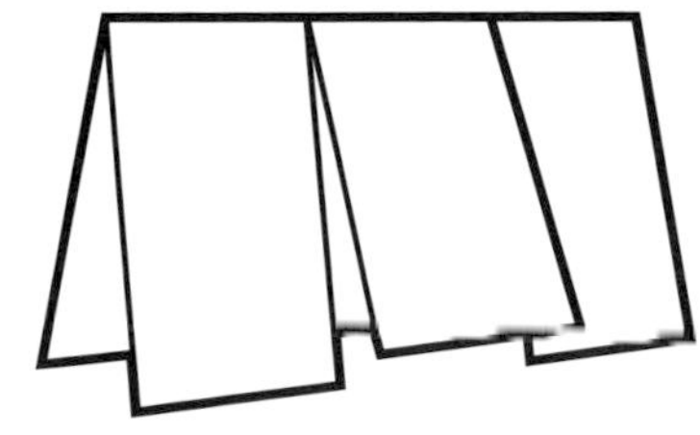

Appendix

Table Fold

1. Fold a piece of paper in half from the top to the bottom. Then, fold the paper in half again.
2. Fold the paper in thirds from side to side.
3. Unfold the paper completely. Carefully trace the fold lines by using a pen or pencil.

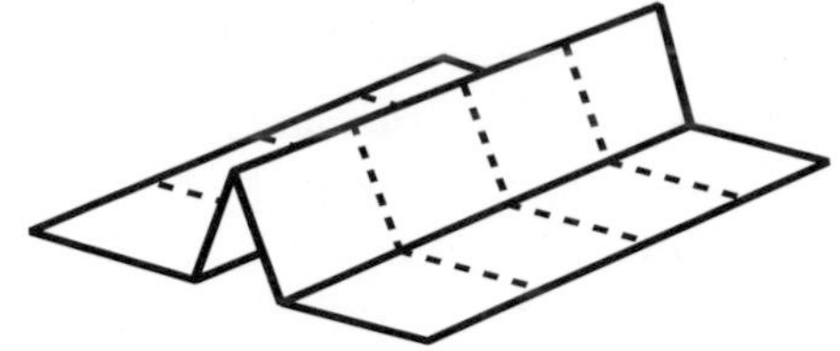

Two-Panel Flip Chart

1. Fold a piece of paper in half from the top to the bottom.
2. Fold the paper in half from side to side. Then, unfold the paper so that you can see the two sections.
3. From the top of the paper, cut along the vertical fold line to the fold in the middle of the paper. You will now have two flaps.

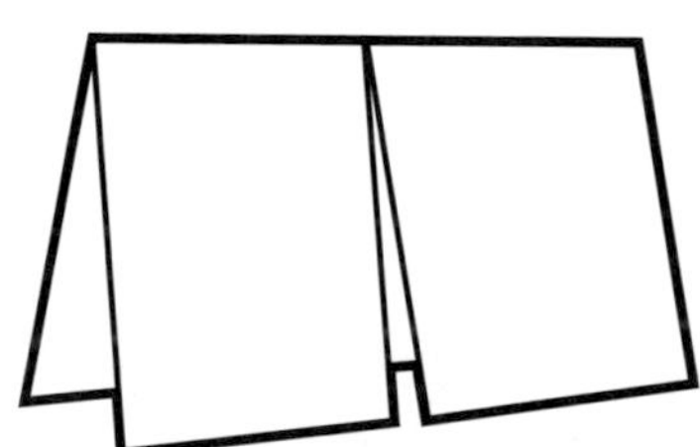

Tri-Fold

1. Fold a piece a paper in thirds from the top to the bottom.
2. Unfold the paper so that you can see the three sections. Then, turn the paper sideways so that the three sections form vertical columns.
3. Trace the fold lines by using a pen or pencil. Label the columns "Know," "Want," and "Learn."

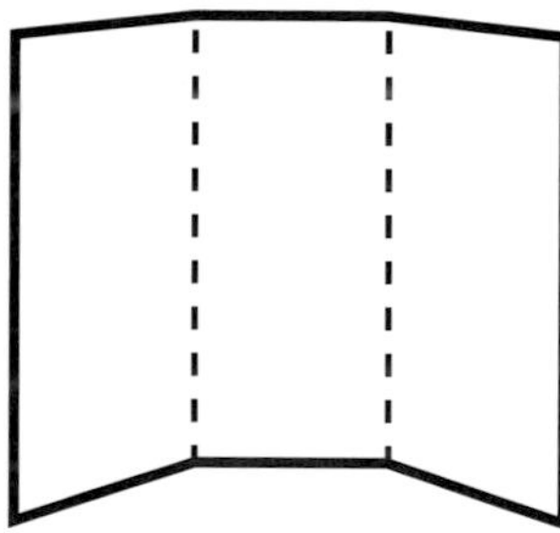

Graphic Organizer Instructions

Have you ever wished that you could "draw out" the many concepts you learn in your science class? Sometimes, being able to *see* how concepts are related really helps you remember what you've learned. Graphic Organizers do just that! They give you a way to draw or map out concepts.

All you need to make a Graphic Organizer is a piece of paper and a pencil. Below you will find instructions for four different Graphic Organizers designed to help you organize the concepts you'll learn in this book.

Spider Map

1. Draw a diagram like the one shown. In the circle, write the main topic.
2. From the circle, draw legs to represent different categories of the main topic. You can have as many categories as you want.
3. From the category legs, draw horizontal lines. As you read the chapter, write details about each category on the horizontal lines.

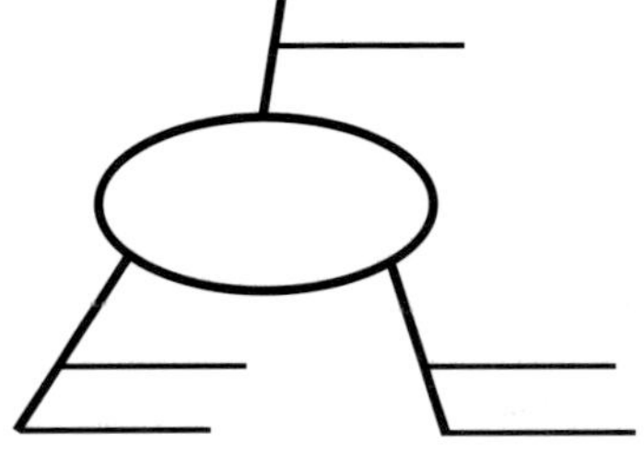

Comparison Table

1. Draw a chart like the one shown. Your chart can have as many columns and rows as you want.
2. In the top row, write the topics that you want to compare.
3. In the left column, write characteristics of the topics that you want to compare. As you read the chapter, fill in the characteristics for each topic in the appropriate boxes.

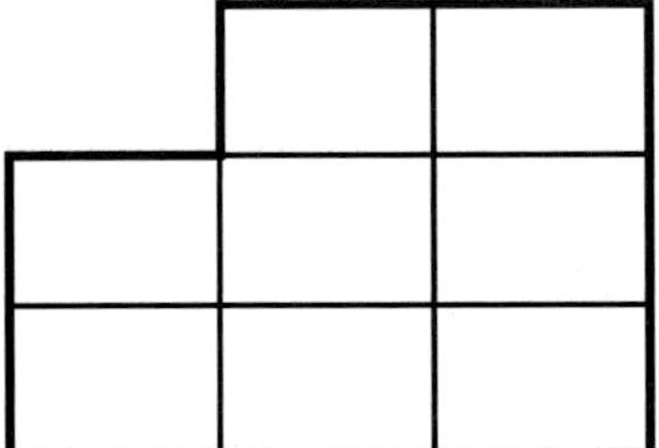

Appendix

Chain-of-Events-Chart

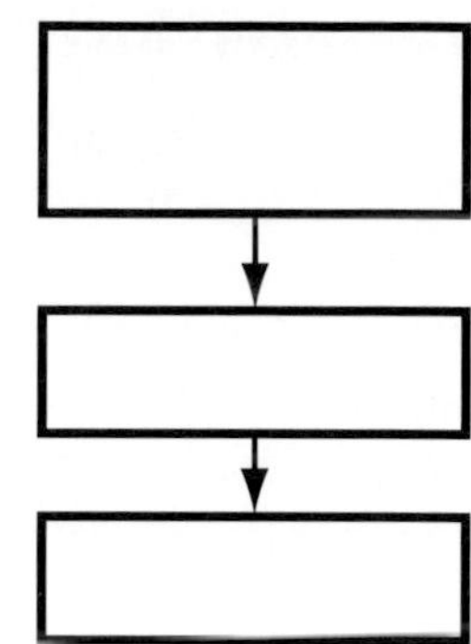

1. Draw a box. In the box, write the first step of a process or the first event of a timeline.
2. Under the box, draw another box, and use an arrow to connect the two boxes. In the second box, write the next step of the process or the next event in the timeline.
3. Continue adding boxes until the process or timeline is finished.

Concept Map

1. Draw a circle in the center of a piece of paper. Write the main idea of the chapter in the center of the circle.
2. From the circle, draw other circles. In those circles, write characteristics of the main idea. Draw arrows from the center circle to the circles that contain the characteristics.
3. From each circle that contains a characteristic, draw other circles. In those circles, write specific details about the characteristic. Draw arrows from each circle that contains a characteristic to the circles that contain specific details. You may draw as many circles as you want.

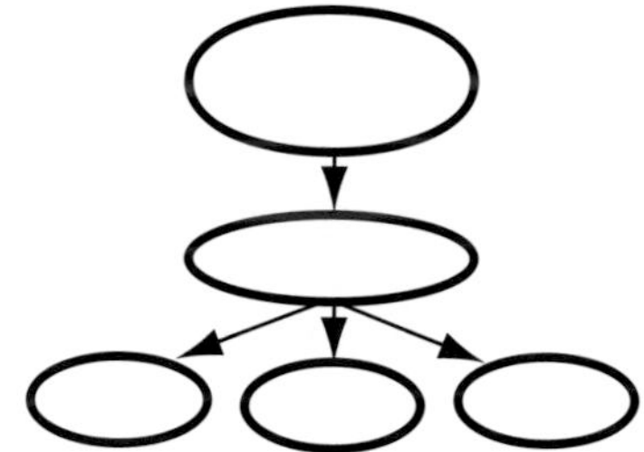

SI Measurement

The International System of Units, or SI, is the standard system of measurement used by many scientists. Using the same standards of measurement makes it easier for scientists to communicate with one another.

SI works by combining prefixes and base units. Each base unit can be used with different prefixes to define smaller and larger quantities. The table below lists common SI prefixes.

SI Prefixes

Prefix	Symbol	Factor	Example
kilo-	k	1,000	kilogram, 1 kg = 1,000 g
hecto-	h	100	hectoliter, 1 hL = 100 L
deka-	da	10	dekameter, 1 dam = 10 m
		1	meter, liter, gram
deci-	d	0.1	decigram, 1 dg = 0.1 g
centi-	c	0.01	centimeter, 1 cm = 0.01 m
milli-	m	0.001	milliliter, 1 mL = 0.001 L
micro-	μ	0.000 001	micrometer, 1 μm = 0.000 001 m

SI Conversion Table

SI units	From SI to English	From English to SI
Length		
kilometer (km) = 1,000 m	1 km = 0.621 mi	1 mi = 1.609 km
meter (m) = 100 cm	1 m = 3.281 ft	1 ft = 0.305 m
centimeter (cm) = 0.01 m	1 cm = 0.394 in.	1 in. = 2.540 cm
millimeter (mm) = 0.001 m	1 mm = 0.039 in.	
micrometer (μm) = 0.000 001 m		
nanometer (nm) = 0.000 000 001 m		
Area		
square kilometer (km^2) = 100 hectares	1 km^2 = 0.386 mi^2	1 mi^2 = 2.590 km^2
hectare (ha) = 10,000 m^2	1 ha = 2.471 acres	1 acre = 0.405 ha
square meter (m^2) = 10,000 cm^2	1 m^2 = 10.764 ft^2	1 ft^2 = 0.093 m^2
square centimeter (cm^2) = 100 mm^2	1 cm^2 = 0.155 $in.^2$	1 $in.^2$ = 6.452 cm^2
Volume		
liter (L) = 1,000 mL = 1 dm^3	1 L = 1.057 fl qt	1 fl qt = 0.946 L
milliliter (mL) = 0.001 L = 1 cm^3	1 mL = 0.034 fl oz	1 fl oz = 29.574 mL
microliter (μL) = 0.000 001 L		
Mass		
kilogram (kg) = 1,000 g	1 kg = 2.205 lb	1 lb = 0.454 kg
gram (g) = 1,000 mg	1 g = 0.035 oz	1 oz = 28.350 g
milligram (mg) = 0.001 g		
microgram (μg) = 0.000 001 g		

Measuring Skills

Using a Graduated Cylinder

When using a graduated cylinder to measure volume, keep the following procedures in mind:

1. Place the cylinder on a flat, level surface before measuring liquid.
2. Move your head so that your eye is level with the surface of the liquid.
3. Read the mark closest to the liquid level. On glass graduated cylinders, read the mark closest to the center of the curve in the liquid's surface.

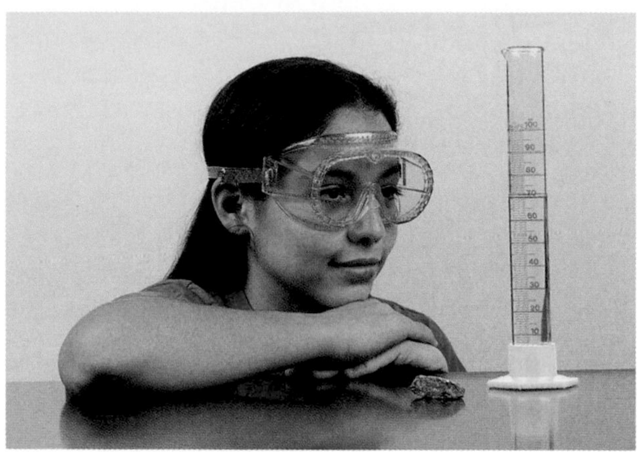

Using a Meterstick or Metric Ruler

When using a meterstick or metric ruler to measure length, keep the following procedures in mind:

1. Place the ruler firmly against the object that you are measuring.
2. Align one edge of the object exactly with the 0 end of the ruler.
3. Look at the other edge of the object to see which of the marks on the ruler is closest to that edge. (Note: Each small slash between the centimeters represents a millimeter, which is one-tenth of a centimeter.)

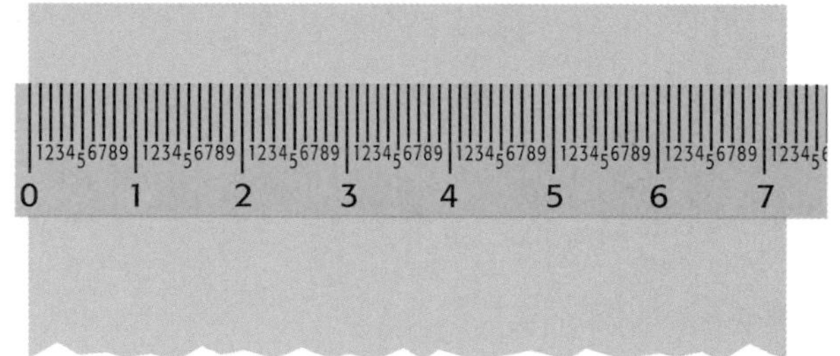

Using a Triple-Beam Balance

When using a triple-beam balance to measure mass, keep the following procedures in mind:

1. Make sure the balance is on a level surface.
2. Place all of the countermasses at 0. Adjust the balancing knob until the pointer rests at 0.
3. Place the object you wish to measure on the pan. **Caution:** Do not place hot objects or chemicals directly on the balance pan.
4. Move the largest countermass along the beam to the right until it is at the last notch that does not tip the balance. Follow the same procedure with the next-largest countermass. Then, move the smallest countermass until the pointer rests at 0.
5. Add the readings from the three beams together to determine the mass of the object.
6. When determining the mass of crystals or powders, first find the mass of a piece of filter paper. Then, add the crystals or powder to the paper, and remeasure. The actual mass of the crystals or powder is the total mass minus the mass of the paper. When finding the mass of liquids, first find the mass of the empty container. Then, find the combined mass of the liquid and container. The mass of the liquid is the total mass minus the mass of the container.

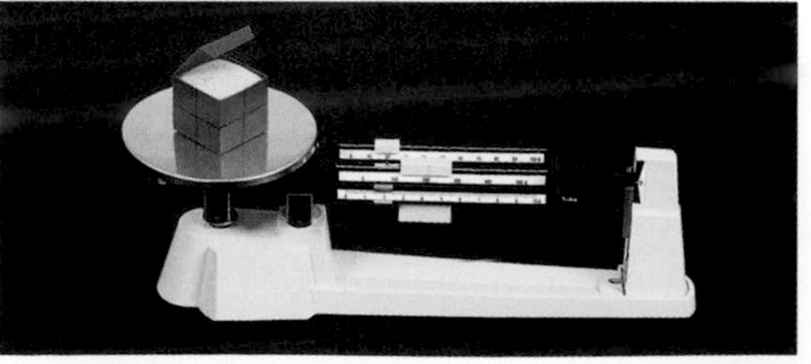

Scientific Methods

The ways in which scientists answer questions and solve problems are called **scientific methods.** The same steps are often used by scientists as they look for answers. However, there is more than one way to use these steps. Scientists may use all of the steps or just some of the steps during an investigation. They may even repeat some of the steps. The goal of using scientific methods is to come up with reliable answers and solutions.

Six Steps of Scientific Methods

1 Ask a Question

Good questions come from careful **observations.** You make observations by using your senses to gather information. Sometimes, you may use instruments, such as microscopes and telescopes, to extend the range of your senses. As you observe the natural world, you will discover that you have many more questions than answers. These questions drive investigations.

Questions beginning with *what, why, how,* and *when* are important in focusing an investigation. Here is an example of a question that could lead to an investigation.

Question: How does acid rain affect plant growth?

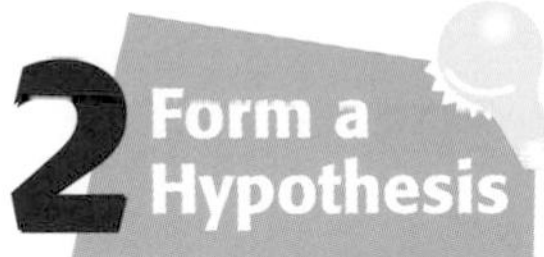

After you ask a question, you need to form a **hypothesis.** A hypothesis is a clear statement of what you expect the answer to your question to be. Your hypothesis will represent your best "educated guess" based on what you have observed and what you already know. A good hypothesis is testable. Otherwise, the investigation can go no further. Here is a hypothesis based on the question, "How does acid rain affect plant growth?"

Hypothesis: Acid rain slows plant growth.

The hypothesis can lead to predictions. A prediction is what you think the outcome of your experiment or data collection will be. Predictions are usually stated in an if-then format. Here is a sample prediction for the hypothesis that acid rain slows plant growth.

Prediction: If a plant is watered with only acid rain (which has a pH of 4), then the plant will grow at half its normal rate.

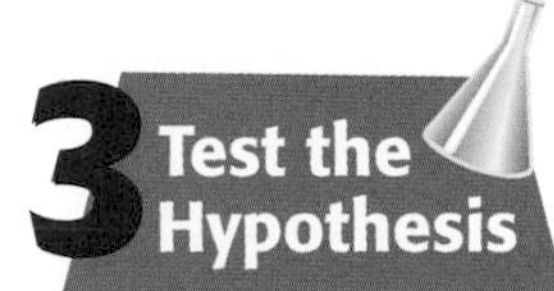

After you have formed a hypothesis and made a prediction, your hypothesis should be tested. One way to test a hypothesis is with a controlled experiment. A **controlled experiment** tests only one factor at a time. In an experiment to test the effect of acid rain on plant growth, the **control group** would be watered with normal rain water. The **experimental group** would be watered with acid rain. All of the plants should receive the same amount of sunlight and water each day. The air temperature should be the same for all groups. However, the acidity of the water will be a variable. In fact, any factor that is different from one group to another is a **variable.** If your hypothesis is correct, then the acidity of the water and plant growth are *dependant variables.* The amount a plant grows is dependent on the acidity of the water. However, the amount of water each plant receives and the amount of sunlight each plant receives are *independent variables.* Either of these factors could change without affecting the other factor.

Sometimes, the nature of an investigation makes a controlled experiment impossible. For example, the Earth's core is surrounded by thousands of meters of rock. Under such circumstances, a hypothesis may be tested by making detailed observations.

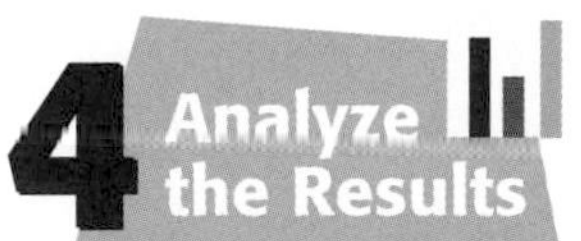

After you have completed your experiments, made your observations, and collected your data, you must analyze all the information you have gathered. Tables and graphs are often used in this step to organize the data.

Appendix

After analyzing your data, you can determine if your results support your hypothesis. If your hypothesis is supported, you (or others) might want to repeat the observations or experiments to verify your results. If your hypothesis is not supported by the data, you may have to check your procedure for errors. You may even have to reject your hypothesis and make a new one. If you cannot draw a conclusion from your results, you may have to try the investigation again or carry out further observations or experiments.

After any scientific investigation, you should report your results. By preparing a written or oral report, you let others know what you have learned. They may repeat your investigation to see if they get the same results. Your report may even lead to another question and then to another investigation.

Scientific Methods in Action

Scientific methods contain loops in which several steps may be repeated over and over again. In some cases, certain steps are unnecessary. Thus, there is not a "straight line" of steps. For example, sometimes scientists find that testing one hypothesis raises new questions and new hypotheses to be tested. And sometimes, testing the hypothesis leads directly to a conclusion. Furthermore, the steps in scientific methods are not always used in the same order. Follow the steps in the diagram, and see how many different directions scientific methods can take you.

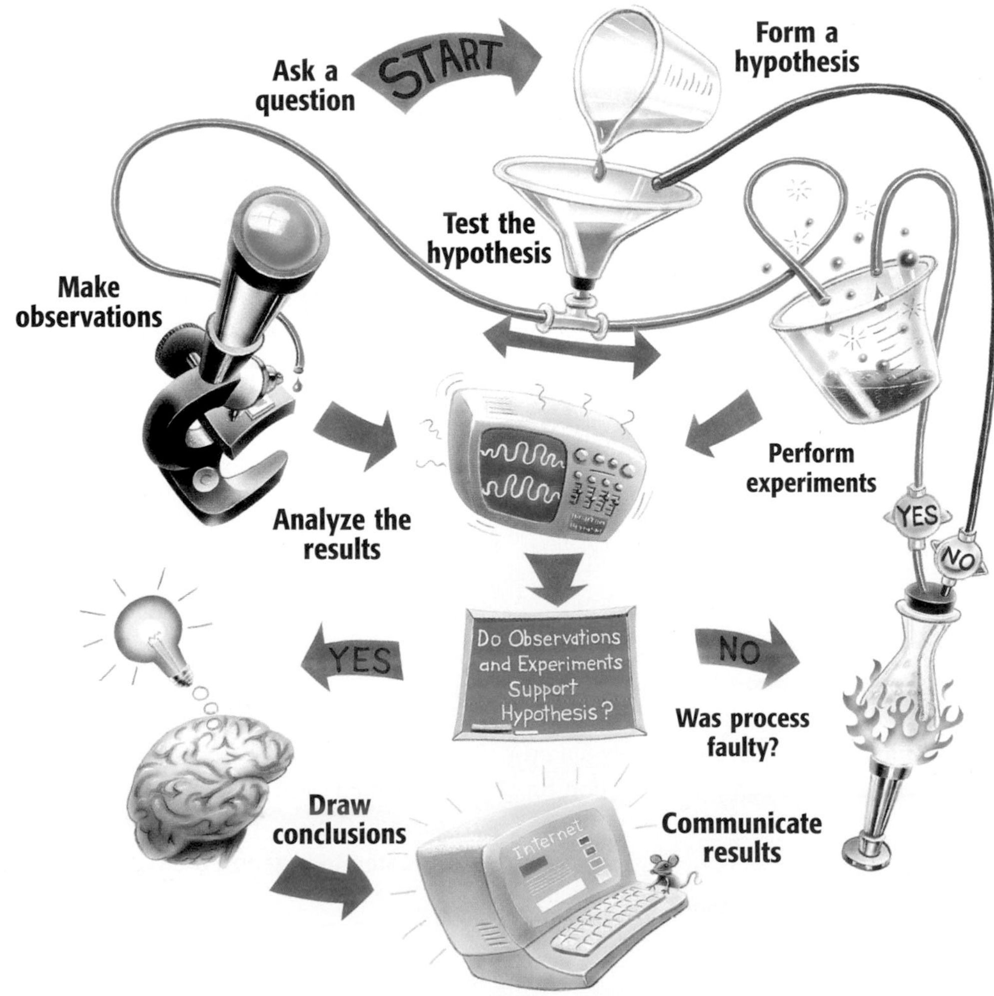

Temperature Scales

Temperature can be expressed by using three different scales: Fahrenheit, Celsius, and Kelvin. The SI unit for temperature is the kelvin (K).

Although 0 K is much colder than 0°C, a change of 1 K is equal to a change of 1°C.

Three Temperature Scales

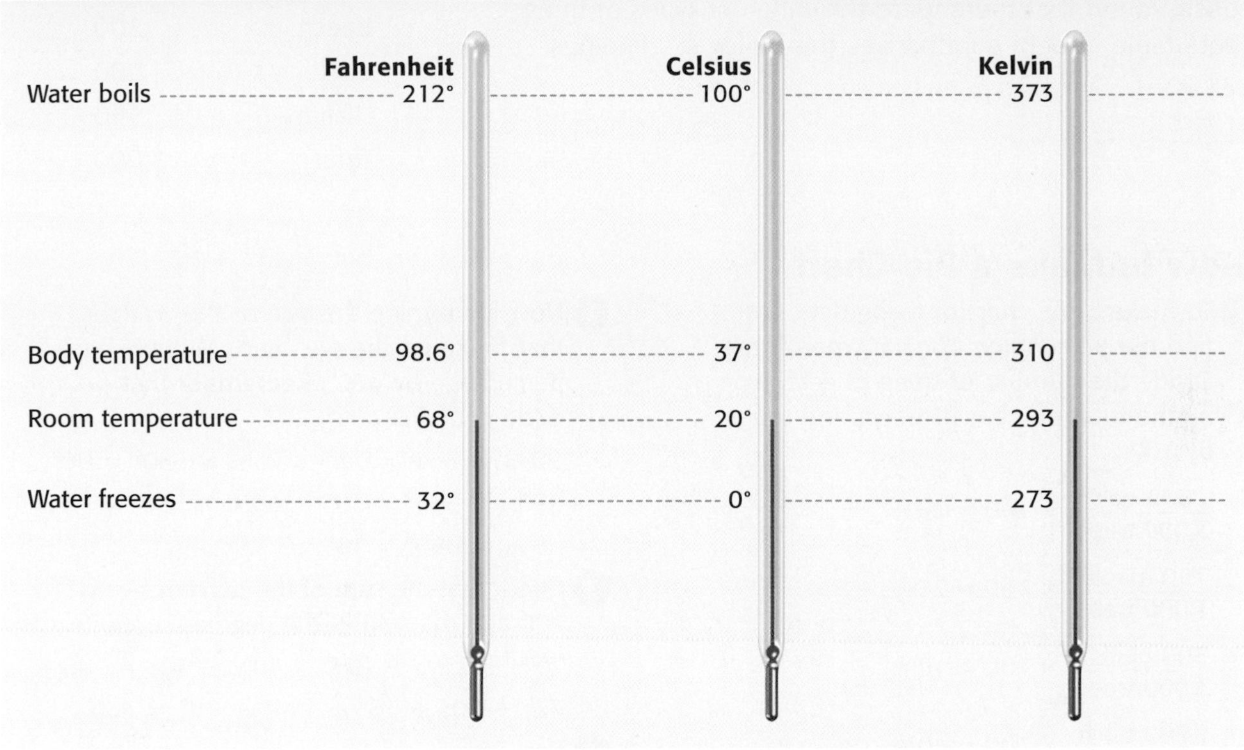

	Fahrenheit	Celsius	Kelvin
Water boils	212°	100°	373
Body temperature	98.6°	37°	310
Room temperature	68°	20°	293
Water freezes	32°	0°	273

Temperature Conversions Table

To convert	Use this equation:	Example
Celsius to Fahrenheit °C → °F	$°F = \left(\frac{9}{5} \times °C\right) + 32$	Convert 45°C to °F. $°F = \left(\frac{9}{5} \times 45°C\right) + 32 = 113°F$
Fahrenheit to Celsius °F → °C	$°C = \frac{5}{9} \times (°F - 32)$	Convert 68°F to °C. $°C = \frac{5}{9} \times (68°F - 32) = 20°C$
Celsius to Kelvin °C → K	$K = °C + 273$	Convert 45°C to K. $K = 45°C + 273 = 318\ K$
Kelvin to Celsius K → °C	$°C = K - 273$	Convert 32 K to °C. $°C = 32K - 273 = -241°C$

Appendix

Making Charts and Graphs

Pie Charts

A pie chart shows how each group of data relates to all of the data. Each part of the circle forming the chart represents a category of the data. The entire circle represents all of the data. For example, a biologist studying a hardwood forest in Wisconsin found that there were five different types of trees. The data table at right summarizes the biologist's findings.

Wisconsin Hardwood Trees	
Type of tree	**Number found**
Oak	600
Maple	750
Beech	300
Birch	1,200
Hickory	150
Total	3,000

How to Make a Pie Chart

1. To make a pie chart of these data, first find the percentage of each type of tree. Divide the number of trees of each type by the total number of trees, and multiply by 100.

$\frac{\text{600 oak}}{\text{3,000 trees}} \times 100 = 20\%$

$\frac{\text{750 maple}}{\text{3,000 trees}} \times 100 = 25\%$

$\frac{\text{300 beech}}{\text{3,000 trees}} \times 100 = 10\%$

$\frac{\text{1,200 birch}}{\text{3,000 trees}} \times 100 = 40\%$

$\frac{\text{150 hickory}}{\text{3,000 trees}} \times 100 = 5\%$

2. Now, determine the size of the wedges that make up the pie chart. Multiply each percentage by 360°. Remember that a circle contains 360°.

$20\% \times 360° = 72°$ $\quad 25\% \times 360° = 90°$

$10\% \times 360° = 36°$ $\quad 40\% \times 360° = 144°$

$5\% \times 360° = 18°$

3. Check that the sum of the percentages is 100 and the sum of the degrees is 360.

$20\% + 25\% + 10\% + 40\% + 5\% = 100\%$

$72° + 90° + 36° + 144° + 18° = 360°$

4. Use a compass to draw a circle and mark the center of the circle.
5. Then, use a protractor to draw angles of 72°, 90°, 36°, 144°, and 18° in the circle.
6. Finally, label each part of the chart, and choose an appropriate title.

A Community of Wisconsin Hardwood Trees

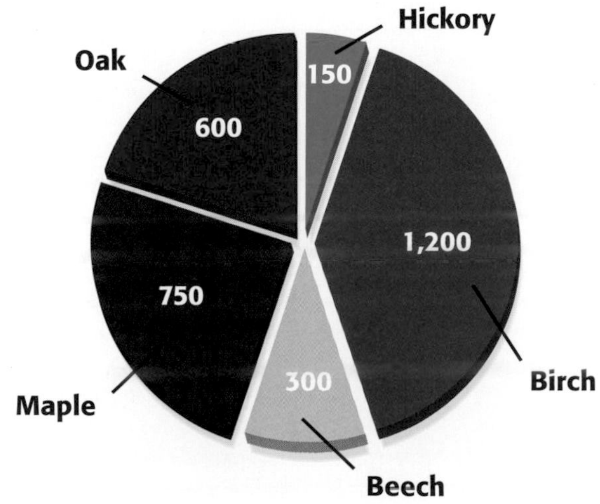

Appendix

Line Graphs

Line graphs are most often used to demonstrate continuous change. For example, Mr. Smith's students analyzed the population records for their hometown, Appleton, between 1900 and 2000. Examine the data at right.

Because the year and the population change, they are the *variables*. The population is determined by, or dependent on, the year. Therefore, the population is called the **dependent variable,** and the year is called the **independent variable.** Each set of data is called a **data pair.** To prepare a line graph, you must first organize data pairs into a table like the one at right.

Population of Appleton, 1900–2000

Year	Population
1900	1,800
1920	2,500
1940	3,200
1960	3,900
1980	4,600
2000	5,300

How to Make a Line Graph

1. Place the independent variable along the horizontal (*x*) axis. Place the dependent variable along the vertical (*y*) axis.
2. Label the *x*-axis "Year" and the *y*-axis "Population." Look at your largest and smallest values for the population. For the *y*-axis, determine a scale that will provide enough space to show these values. You must use the same scale for the entire length of the axis. Next, find an appropriate scale for the *x*-axis.
3. Choose reasonable starting points for each axis.
4. Plot the data pairs as accurately as possible.
5. Choose a title that accurately represents the data.

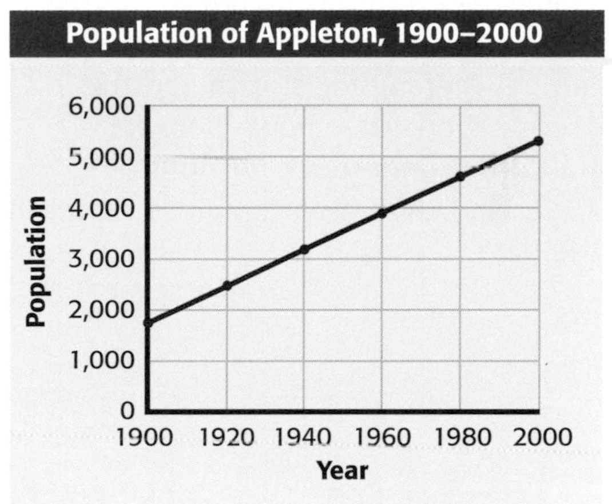

How to Determine Slope

Slope is the ratio of the change in the *y*-value to the change in the *x*-value, or "rise over run."

1. Choose two points on the line graph. For example, the population of Appleton in 2000 was 5,300 people. Therefore, you can define point *a* as (2000, 5,300). In 1900, the population was 1,800 people. You can define point *b* as (1900, 1,800).
2. Find the change in the *y*-value.
   (*y* at point *a*) − (*y* at point *b*) = 5,300 people − 1,800 people = 3,500 people
3. Find the change in the *x*-value.
   (*x* at point *a*) − (*x* at point *b*) = 2000 − 1900 = 100 years
4. Calculate the slope of the graph by dividing the change in *y* by the change in *x*.

$$slope = \frac{change\ in\ y}{change\ in\ x}$$

$$slope = \frac{3{,}500\ \text{people}}{100\ \text{years}}$$

$$slope = 35\ \text{people per year}$$

In this example, the population in Appleton increased by a fixed amount each year. The graph of these data is a straight line. Therefore, the relationship is **linear.** When the graph of a set of data is not a straight line, the relationship is **nonlinear.**

Appendix

Using Algebra to Determine Slope

The equation in step 4 may also be arranged to be

$$y = kx$$

where y represents the change in the y-value, k represents the slope, and x represents the change in the x-value.

$$slope = \frac{change\ in\ y}{change\ in\ x}$$

$$k = \frac{y}{x}$$

$$k \times x = \frac{y \times x}{x}$$

$$kx = y$$

Bar Graphs

Bar graphs are used to demonstrate change that is not continuous. These graphs can be used to indicate trends when the data cover a long period of time. A meteorologist gathered the precipitation data shown here for Hartford, Connecticut, for April 1–15, 1996, and used a bar graph to represent the data.

Precipitation in Hartford, Connecticut April 1–15, 1996

Date	Precipitation (cm)	Date	Precipitation (cm)
April 1	0.5	April 9	0.25
April 2	1.25	April 10	0.0
April 3	0.0	April 11	1.0
April 4	0.0	April 12	0.0
April 5	0.0	April 13	0.25
April 6	0.0	April 14	0.0
April 7	0.0	April 15	6.50
April 8	1.75		

How to Make a Bar Graph

1. Use an appropriate scale and a reasonable starting point for each axis.
2. Label the axes, and plot the data.
3. Choose a title that accurately represents the data.

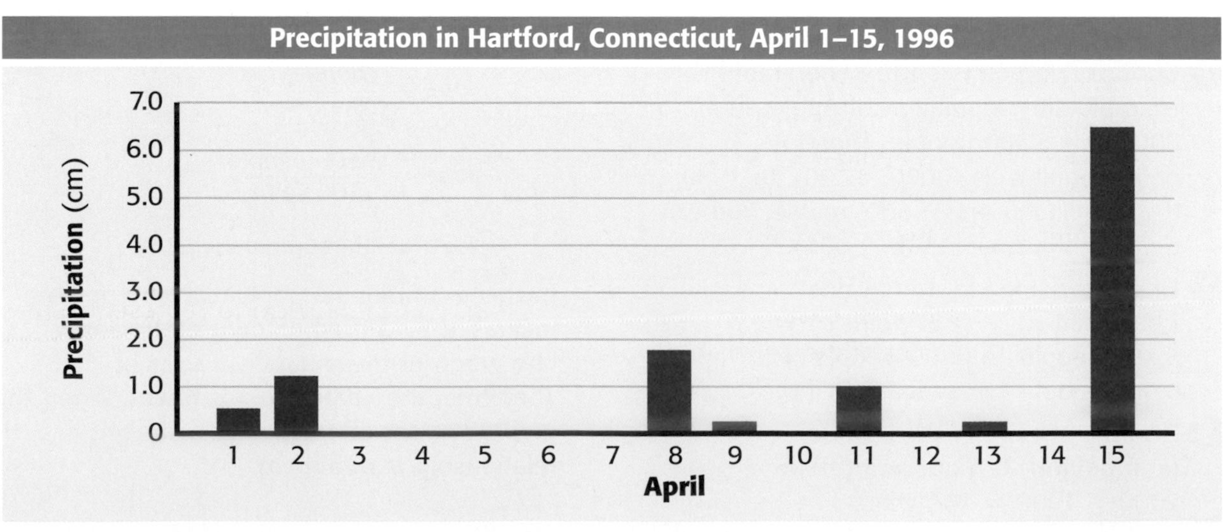

Math Refresher

Science requires an understanding of many math concepts. The following pages will help you review some important math skills.

Averages

An **average,** or **mean,** simplifies a set of numbers into a single number that *approximates* the value of the set.

Example: Find the average of the following set of numbers: 5, 4, 7, and 8.

Step 1: Find the sum.

$5 + 4 + 7 + 8 = 24$

Step 2: Divide the sum by the number of numbers in your set. Because there are four numbers in this example, divide the sum by 4.

$$\frac{24}{4} = 6$$

The average, or mean, is **6.**

Ratios

A **ratio** is a comparison between numbers, and it is usually written as a fraction.

Example: Find the ratio of thermometers to students if you have 36 thermometers and 48 students in your class.

Step 1: Make the ratio.

$$\frac{36 \text{ thermometers}}{48 \text{ students}}$$

Step 2: Reduce the fraction to its simplest form.

$$\frac{36}{48} = \frac{36 \div 12}{48 \div 12} = \frac{3}{4}$$

The ratio of thermometers to students is **3 to 4,** or $\frac{3}{4}$. The ratio may also be written in the form 3:4.

Proportions

A **proportion** is an equation that states that two ratios are equal.

$$\frac{3}{1} = \frac{12}{4}$$

To solve a proportion, first multiply across the equal sign. This is called *cross-multiplication*. If you know three of the quantities in a proportion, you can use cross-multiplication to find the fourth.

Example: Imagine that you are making a scale model of the solar system for your science project. The diameter of Jupiter is 11.2 times the diameter of the Earth. If you are using a plastic-foam ball that has a diameter of 2 cm to represent the Earth, what must the diameter of the ball representing Jupiter be?

$$\frac{11.2}{1} = \frac{x}{2 \text{ cm}}$$

Step 1: Cross-multiply.

$$\frac{11.2}{1} \times \frac{x}{2}$$

$11.2 \times 2 = x \times 1$

Step 2: Multiply.

$22.4 = x \times 1$

Step 3: Isolate the variable by dividing both sides by 1.

$x = \frac{22.4}{1}$

$x = 22.4 \text{ cm}$

You will need to use a ball that has a diameter of **22.4** cm to represent Jupiter.

Percentages

A **percentage** is a ratio of a given number to 100.

Example: What is 85% of 40?

Step 1: Rewrite the percentage by moving the decimal point two places to the left.

0.85

Step 2: Multiply the decimal by the number that you are calculating the percentage of.

$0.85 \times 40 = 34$

85% of 40 is **34.**

Decimals

To **add** or **subtract decimals,** line up the digits vertically so that the decimal points line up. Then, add or subtract the columns from right to left. Carry or borrow numbers as necessary.

Example: Add the following numbers: 3.1415 and 2.96.

Step 1: Line up the digits vertically so that the decimal points line up.

$$\begin{array}{r} 3.1415 \\ +\ 2.96 \\ \hline \end{array}$$

Step 2: Add the columns from right to left, and carry when necessary.

$$\begin{array}{r} {\scriptstyle 1\ 1} \\ 3.1415 \\ +\ 2.96 \\ \hline 6.1015 \end{array}$$

The sum is **6.1015.**

Fractions

Numbers tell you how many; **fractions** tell you *how much of a whole.*

Example: Your class has 24 plants. Your teacher instructs you to put 5 plants in a shady spot. What fraction of the plants in your class will you put in a shady spot?

Step 1: In the denominator, write the total number of parts in the whole.

$$\frac{?}{24}$$

Step 2: In the numerator, write the number of parts of the whole that are being considered.

$$\frac{5}{24}$$

So, $\mathbf{\frac{5}{24}}$ of the plants will be in the shade.

Reducing Fractions

It is usually best to express a fraction in its simplest form. Expressing a fraction in its simplest form is called *reducing* a fraction.

Example: Reduce the fraction $\frac{30}{45}$ to its simplest form.

Step 1: Find the largest whole number that will divide evenly into both the numerator and denominator. This number is called the *greatest common factor* (GCF).

Factors of the numerator 30:

1, 2, 3, 5, 6, 10, **15,** 30

Factors of the denominator 45:

1, 3, 5, 9, **15,** 45

Step 2: Divide both the numerator and the denominator by the GCF, which in this case is 15.

$$\frac{30}{45} = \frac{30 \div 15}{45 \div 15} = \frac{2}{3}$$

Thus, $\frac{30}{45}$ reduced to its simplest form is $\mathbf{\frac{2}{3}}$.

Adding and Subtracting Fractions

To **add** or **subtract fractions** that have the **same denominator,** simply add or subtract the numerators.

Examples:

$$\frac{3}{5}+\frac{1}{5}=? \text{ and } \frac{3}{4}-\frac{1}{4}=?$$

Step 1: Add or subtract the numerators.

$$\frac{3}{5}+\frac{1}{5}=\frac{4}{} \text{ and } \frac{3}{4}-\frac{1}{4}=\frac{2}{}$$

Step 2: Write the sum or difference over the denominator.

$$\frac{3}{5}+\frac{1}{5}=\frac{4}{5} \text{ and } \frac{3}{4}-\frac{1}{4}=\frac{2}{4}$$

Step 3: If necessary, reduce the fraction to its simplest form.

$\mathbf{\frac{4}{5}}$ cannot be reduced, and $\frac{2}{4}=\mathbf{\frac{1}{2}}$.

To **add** or **subtract fractions** that have **different denominators,** first find the least common denominator (LCD).

Examples:

$$\frac{1}{2}+\frac{1}{6}=? \text{ and } \frac{3}{4}-\frac{2}{3}=?$$

Step 1: Write the equivalent fractions that have a common denominator.

$$\frac{3}{6}+\frac{1}{6}=? \text{ and } \frac{9}{12}-\frac{8}{12}=?$$

Step 2: Add or subtract the fractions.

$$\frac{3}{6}+\frac{1}{6}=\frac{4}{6} \text{ and } \frac{9}{12}-\frac{8}{12}=\frac{1}{12}$$

Step 3: If necessary, reduce the fraction to its simplest form.

The fraction $\frac{4}{6}=\mathbf{\frac{2}{3}}$, and $\mathbf{\frac{1}{12}}$ cannot be reduced.

Multiplying Fractions

To **multiply fractions,** multiply the numerators and the denominators together, and then reduce the fraction to its simplest form.

Example:

$$\frac{5}{9}\times\frac{7}{10}=?$$

Step 1: Multiply the numerators and denominators.

$$\frac{5}{9}\times\frac{7}{10}=\frac{5\times 7}{9\times 10}=\frac{35}{90}$$

Step 2: Reduce the fraction.

$$\frac{35}{90}=\frac{35\div 5}{90\div 5}=\mathbf{\frac{7}{18}}$$

Dividing Fractions

To **divide fractions,** first rewrite the divisor (the number you divide by) upside down. This number is called the *reciprocal* of the divisor. Then multiply and reduce if necessary.

Example:

$$\frac{5}{8}\div\frac{3}{2}=?$$

Step 1: Rewrite the divisor as its reciprocal.

$$\frac{3}{2}\rightarrow\frac{2}{3}$$

Step 2: Multiply the fractions.

$$\frac{5}{8}\times\frac{2}{3}=\frac{5\times 2}{8\times 3}=\frac{10}{24}$$

Step 3: Reduce the fraction.

$$\frac{10}{24}=\frac{10\div 2}{24\div 2}=\mathbf{\frac{5}{12}}$$

Scientific Notation

Scientific notation is a short way of representing very large and very small numbers without writing all of the place-holding zeros.

Example: Write 653,000,000 in scientific notation.

Step 1: Write the number without the place-holding zeros.

653

Step 2: Place the decimal point after the first digit.

6.53

Step 3: Find the exponent by counting the number of places that you moved the decimal point.

6.53000000

The decimal point was moved eight places to the left. Therefore, the exponent of 10 is positive 8. If you had moved the decimal point to the right, the exponent would be negative.

Step 4: Write the number in scientific notation.

$\mathbf{6.53 \times 10^{8}}$

Area

Area is the number of square units needed to cover the surface of an object.

Formulas:

area of a square = *side* × *side*

area of a rectangle = *length* × *width*

area of a triangle = $\frac{1}{2}$ × *base* × *height*

Examples: Find the areas.

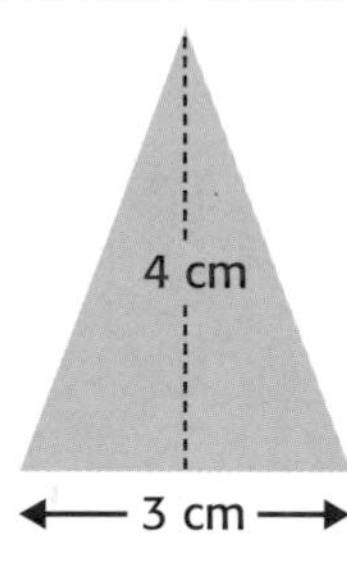

Triangle

area = $\frac{1}{2}$ × *base* × *height*

area = $\frac{1}{2}$ × 3 cm × 4 cm

area = **6 cm²**

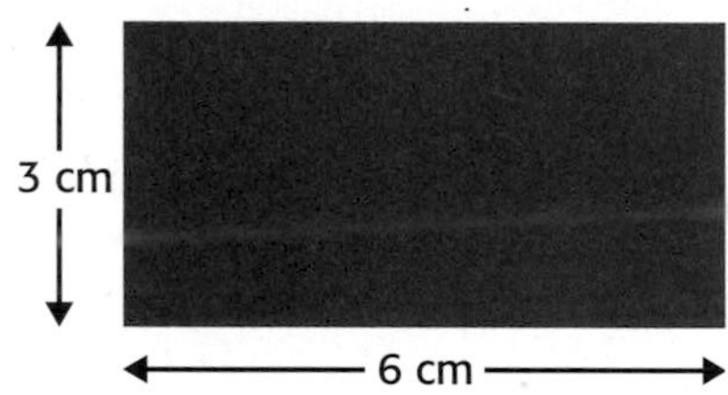

Rectangle

area = *length* × *width*

area = 6 cm × 3 cm

area = $\mathbf{18\ cm^2}$

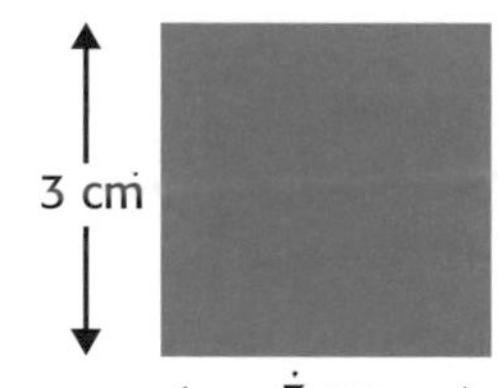

Square

area = *side* × *side*

area = 3 cm × 3 cm

area = $\mathbf{9\ cm^2}$

Volume

Volume is the amount of space that something occupies.

Formulas:

volume of a cube = *side* × *side* × *side*

volume of a prism = *area of base* × *height*

Examples:

Find the volume of the solids.

Cube

volume = *side* × *side* × *side*

volume = 4 cm × 4 cm × 4 cm

volume = $\mathbf{64\ cm^3}$

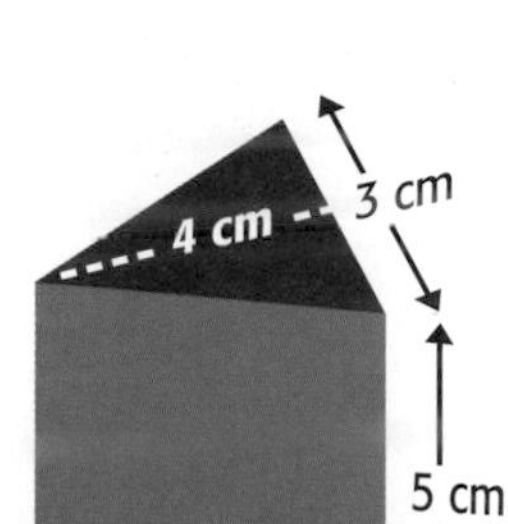
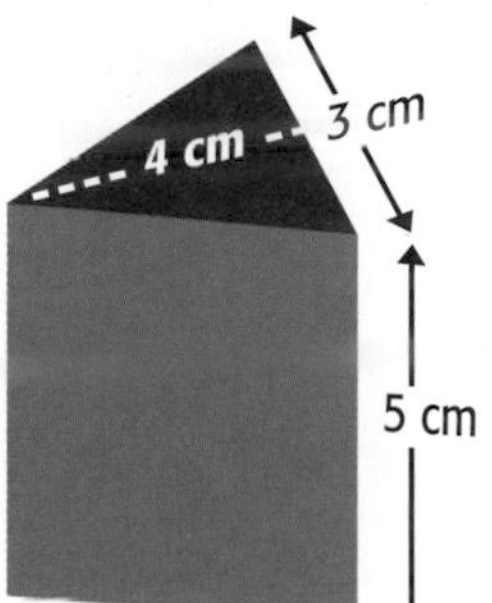

Prism

volume = *area of base* × *height*

volume = (*area of triangle*) × *height*

volume = ($\frac{1}{2}$ × 3 cm × 4 cm) × 5 cm

volume = $6\ cm^2$ × 5 cm

volume = $\mathbf{30\ cm^3}$

Glossary

A

absolute dating any method of measuring the age of an event or object in years (137)

active transport the movement of substances across the cell membrane that requires the cell to use energy (36)

adaptation a characteristic that improves an individual's ability to survive and reproduce in a particular environment (108)

allele (uh LEEL) one of the alternative forms of a gene that governs a characteristic, such as hair color (62)

Animalia a kingdom made up of complex, multicellular organisms that lack cell walls, can usually move around, and quickly respond to their environment (174)

Archaebacteria (AHR kee bak TEER ee uh) a kingdom made up of bacteria that live in extreme environments (171)

C

cell in biology, the smallest unit that can perform all life processes; cells are covered by a membrane and contain DNA and cytoplasm (4)

cell cycle the life cycle of a cell (42)

cell membrane a phospholipid layer that covers a cell's surface and acts as a barrier between the inside of a cell and the cell's environment (7)

cellular respiration the process by which cells use oxygen to produce energy from food (39)

cell wall a rigid structure that surrounds the cell membrane and provides support to the cell (12)

Cenozoic era (SEN uh ZOH ik ER uh) the most recent geologic era, beginning 65 million years ago; also called the *Age of Mammals* (146)

chromosome in a eukaryotic cell, one of the structures in the nucleus that are made up of DNA and protein; in a prokaryotic cell, the main ring of DNA (42)

classification the division of organisms into groups, or classes, based on specific characteristics (164)

cytokinesis the division of the cytoplasm of a cell (44)

D

dichotomous key (die KAHT uh muhs KEE) an aid that is used to identify organisms and that consists of the answers to a series of questions (168)

diffusion (di FYOO zhuhn) the movement of particles from regions of higher density to regions of lower density (34)

DNA **d**eoxyribo**n**ucleic **a**cid, a molecule that is present in all living cells and that contains the information that determines the traits that a living thing inherits and needs to live (86)

dominant trait the trait observed in the first generation when parents that have different traits are bred (59)

E

endocytosis (EN doh sie TOH sis) the process by which a cell membrane surrounds a particle and encloses the particle in a vesicle to bring the particle into the cell (36)

endoplasmic reticulum (EN doh PLAZ mik ri TIK yuh luhm) a system of membranes that is found in a cell's cytoplasm and that assists in the production, processing, and transport of proteins and in the production of lipids (15)

Eubacteria (YOO bak TEER ee uh) a kingdom that contains all prokaryotes except archaebacteria (171)

eukaryote an organism made up of cells that have a nucleus enclosed by a membrane; eukaryotes include animals, plants, and fungi but not archaebacteria or eubacteria (10)

evolution the process in which inherited characteristics within a population change over generations such that new species sometimes arise (109)

exocytosis (EK soh sie TOH sis) the process in which a cell releases a particle by enclosing the particle in a vesicle that then moves to the cell surface and fuses with the cell membrane (37)

extinct describes a species that has died out completely (139)

F

fermentation the breakdown of food without the use of oxygen (39)

fossil the remains or physical evidence of an organism preserved by geological processes (110, 136)

fossil record a historical sequence of life indicated by fossils found in layers of the Earth's crust (110)

function the special, normal, or proper activity of an organ or part (22)

Fungi (FUHN JIE) a kingdom made up of nongreen, eukaryotic organisms that have no means of movement, reproduce by using spores, and get food by breaking down substances in their surroundings and absorbing the nutrients (172)

G

gene one set of instructions for an inherited trait (62)

generation time the period between the birth of one generation and the birth of the next generation (123)

genotype the entire genetic makeup of an organism; also the combination of genes for one or more specific traits (63)

geologic time scale the standard method used to divide the Earth's long natural history into manageable parts (138)

Golgi complex (GOHL jee KAHM PLEKS) cell organelle that helps make and package materials to be transported out of the cell (17)

H

heredity the passing of genetic traits from parent to offspring (56)

hominid a type of primate characterized by bipedalism, relatively long lower limbs, and lack of a tail; examples include humans and their ancestors (149)

homologous chromosomes (hoh MAHL uh guhs KROH muh SOHMZ) chromosomes that have the same sequence of genes and the same structure (43, 68)

Homo sapiens (HOH moh SAY pee UHNZ) the species of hominids that includes modern humans and their closest ancestors and that first appeared about 100,000 to 150,000 years ago (152)

L

lysosome (LIE suh SOHM) a cell organelle that contains digestive enzymes (18)

M

meiosis (mie OH sis) a process in cell division during which the number of chromosomes decreases to half the original number by two divisions of the nucleus, which results in the production of sex cells (gametes or spores) (68)

Mesozoic era (MES oh ZOH ik ER uh) the geologic era that lasted from 248 million to 65 million years ago; also called the *Age of Reptiles* (145)

mitochondrion (MIET oh KAHN dree uhn) in eukaryotic cells, the cell organelle that is surrounded by two membranes and that is the site of cellular respiration (16)

mitosis in eukaryotic cells, a process of cell division that forms two new nuclei, each of which has the same number of chromosomes (43)

mutation a change in the nucleotide-base sequence of a gene or DNA molecule (94)

N

natural selection the process by which individuals that are better adapted to their environment survive and reproduce more successfully than less well adapted individuals do; a theory to explain the mechanism of evolution (120)

nucleotide in a nucleic-acid chain, a subunit that consists of a sugar, a phosphate, and a nitrogenous base (86)

nucleus in a eukaryotic cell, a membrane-bound organelle that contains the cell's DNA and that has a role in processes such as growth, metabolism, and reproduction (7)

O

organ a collection of tissues that carry out a specialized function of the body (21)

organelle one of the small bodies in a cell's cytoplasm that are specialized to perform a specific function (7)

organism a living thing; anything that can carry out life processes independently (22)

organ system a group of organs that work together to perform body functions (22)

osmosis (ahs MOH sis) the diffusion of water through a semipermeable membrane (35)

P

Paleozoic era (PAY lee OH ZOH ik ER uh) the geologic era that followed Precambrian time and that lasted from 543 million to 248 million years ago (244)

passive transport the movement of substances across a cell membrane without the use of energy by the cell (36)

pedigree a diagram that shows the occurrence of a genetic trait in several generations of a family (74)

phenotype (FEE noh TIEP) an organism's appearance or other detectable characteristic (62)

photosynthesis (FOHT oh SIN thuh sis) the process by which plants, algae, and some bacteria use sunlight, carbon dioxide, and water to make food (38)

Plantae a kingdom made up of complex, multicellular organisms that are usually green, have cell walls made of cellulose, cannot move around, and use the sun's energy to make sugar by photosynthesis (173)

plate tectonics the theory that explains how large pieces of the Earth's outermost layer, called *tectonic plates,* move and change shape (140)

Precambrian time (pree KAM bree uhn TIEM) the period in the geologic time scale from the formation of the Earth to the beginning of the Paleozoic era, from about 4.6 billion to 543 million years ago (142)

primate a type of mammal characterized by opposable thumbs and binocular vision (148)

probability the likelihood that a possible future event will occur in any given instance of the event (64)

prokaryote (pro KAR ee OHT) an organism that consists of a single cell that does not have a nucleus (8)

Protista (proh TIST uh) a kingdom of mostly one-celled eukaryotic organisms that are different from plants, animals, bacteria, and fungi (172)

R

recessive trait a trait that is apparent only when two recessive alleles for the same characteristic are inherited (59)

relative dating any method of determining whether an event or object is older or younger than other events or objects (137)

ribosome a cell organelle composed of RNA and protein; the site of protein synthesis (15, 93)

RNA **r**ibo**n**ucleic **a**cid, a molecule that is present in all living cells and that plays a role in protein production (92)

S

selective breeding the human practice of breeding animals or plants that have certain desired characteristics (118)

sex chromosome one of the pair of chromosomes that determine the sex of an individual (73)

speciation (SPEE shee AY shuhn) the formation of new species as a result of evolution (124)

species a group of organisms that are closely related and can mate to produce fertile offspring (108)

structure the arrangement of parts in an organism (22)

T

taxonomy (taks AHN uh mee) the science of describing, naming, and classifying organisms (165)

tissue a group of similar cells that perform a common function (21)

trait a genetically determined characteristic (118)

V

vesicle (VES i kuhl) a small cavity or sac that contains materials in a eukaryotic cell; forms when part of the cell membrane surrounds the materials to be taken into the cell or transported within the cell (17)

Spanish Glossary

A

absolute dating/datación absoluta cualquier método que sirve para determinar la edad de un suceso u objeto en años (137)

active transport/transporte activo el movimiento de substancias a través de la membrana celular que requiere que la célula gaste energía (36)

adaptation/adaptación una característica que mejora la capacidad de un individuo para sobrevivir y reproducirse en un determinado ambiente (108)

allele/alelo una de las formas alternativas de un gene que rige un carácter, como por ejemplo, el color del cabello (62)

Animalia/Animalia un reino formado por organismos pluricelulares complejos que no tienen pared celular, normalmente son capaces de moverse y reaccionan rápidamente a su ambiente (174)

Archaebacteria/arqueobacteria un reino formado por bacterias que viven en ambientes extremos (171)

C

cell/célula en biología, la unidad más pequeña que puede realizar todos los procesos vitales; las células están cubiertas por una membrana y tienen ADN y citoplasma (4)

cell cycle/ciclo celular el ciclo de vida de una célula (42)

cell membrane/membrana celular una capa de fosfolípidos que cubre la superficie de la célula y funciona como una barrera entre el interior de la célula y el ambiente de la célula (7)

cellular respiration/respiración celular el proceso por medio del cual las células utilizan oxígeno para producir energía a partir de los alimentos (39)

cell wall/pared celular una estructura rígida que rodea la membrana celular y le brinda soporte a la célula (12)

Cenozoic era/era Cenozoica la era geológica más reciente, que comenzó hace 65 millones de años; también llamada *Edad de los Mamíferos* (146)

chromosome/cromosoma en una célula eucariótica, una de las estructuras del núcleo que está hecha de ADN y proteína; en una célula procariótica, el anillo principal de ADN (42)

classification/clasificación la división de organismos en grupos, o clases, en función de características específicas (164)

cytokinesis/citoquinesis la división del citoplasma de una célula (44)

D

dichotomous key/clave dicotómica una ayuda para identificar organismos, que consiste en las respuestas a una serie de preguntas (168)

diffusion/difusión el movimiento de partículas de regiones de mayor densidad a regiones de menor densidad (34)

DNA/ADN **á**cido **d**esoxirribo**n**ucleico, una molécula que está presente en todas las células vivas y que contiene la información que determina los caracteres que un ser vivo hereda y necesita para vivir (86)

dominant trait/carácter dominante el carácter que se observa en la primera generación cuando se cruzan progenitores que tienen caracteres diferentes (59)

E

endocytosis/endocitosis el proceso por medio del cual la membrana celular rodea una partícula y la encierra en una vesícula para llevarla al interior de la célula (36)

endoplasmic reticulum/retículo endoplásmico un sistema de membranas que se encuentra en el citoplasma de la célula y que tiene una función en la producción, procesamiento y transporte de proteínas y en la producción de lípidos (15)

Eubacteria/Eubacteria un reino que agrupa a todos los procariotes, excepto a las arqueobacterias ni las eubacterias (171)

Spanish Glossary

eukaryote/eucariote un organismo cuyas células tienen un núcleo rodeado por una membrana; entre los eucariotes se encuentran los animales, las plantas y los hongos, pero no las arqueobacterias (10)

evolution/evolución el proceso por medio del cual las características heredadas dentro de una población cambian con el transcurso de las generaciones de manera tal que a veces surgen nuevas especies (109)

exocytosis/exocitosis el proceso por medio del cual una célula libera una partícula encerrándola en una vesícula que luego se traslada a la superficie de la célula y se fusiona con la membrana celular (37)

extinct/extinto término que describe a una especie que ha desaparecido por completo (139)

F

fermentation/fermentación la descomposición de los alimentos sin utilizar oxígeno (39)

fossil/fósil los restos o las pruebas físicas de un organismo preservados por los procesos geológicos (110, 136)

fossil record/registro fósil una secuencia histórica de la vida indicada por fósiles que se han encontrado en las capas de la corteza terrestre (110)

function/función la actividad especial, normal o adecuada de un órgano o parte (22)

Fungi/Fungi un reino formado por organismos eucarióticos no verdes que no tienen capacidad de movimiento, se reproducen por esporas y obtienen alimento al descomponer substancias de su entorno y absorber los nutrientes (172)

G

gene/gene un conjunto de instrucciones para un carácter heredado (62)

generation time/tiempo de generación el período entre el nacimiento de una generación y el nacimiento de la siguiente generación (123)

genotype/genotipo la constitución genética completa de un organismo; *también* la combinación genes para uno o más caracteres específicos (63)

geologic time scale/escala de tiempo geológico el método estándar que se usa para dividir la larga historia natural de la Tierra en partes razonables (138)

Golgi complex/aparato de Golgi un organelo celular que ayuda a hacer y a empacar los materiales que serán transportados al exterior de la célula (17)

H

heredity/herencia la transmisión de caracteres genéticos de padres a hijos (56)

hominid/homínido un tipo de primate caracterizado por ser bípedo, tener extremidades inferiores relativamente largas y no tener cola; incluye a los seres humanos y sus ancestros (149)

homologous chromosomes/cromosomas homólogos cromosomas con la misma secuencia de genes y la misma estructura (43, 68)

Homo sapiens/Homo sapiens la especie de homínidos que incluye a los seres humanos modernos y a sus ancestros más cercanos; apareció hace entre 100,000 y 150,000 años (152)

L

lysosome/lisosoma un organelo celular que contiene enzimas digestivas (18)

M

meiosis/meiosis un proceso de división celular durante el cual el número de cromosomas disminuye a la mitad del número original por medio de dos divisiones del núcleo, lo cual resulta en la producción de células sexuales (gametos o esporas) (68)

Mesozoic era/era Mesozoica la era geológica que comenzó hace 248 millones de años y terminó hace 65 millones de años; también llamada *Edad de los Reptiles* (145)

mitochondrion/mitocondria en las células eucarióticas, el organelo celular rodeado por dos membranas que es el lugar donde se lleva a cabo la respiración celular (16)

mitosis/mitosis en las células eucarióticas, un proceso de división celular que forma dos núcleos nuevos, cada uno de los cuales posee el mismo número de cromosomas (43)

mutation/mutación un cambio en la secuencia de la base de nucleótidos de un gene o de una molécula de ADN (94)

N

natural selection/selección natural el proceso por medio del cual los individuos que están mejor adaptados a su ambiente sobreviven y se reproducen con más éxito que los individuos menos adaptados; una teoría que explica el mecanismo de la evolución (120)

nucleotide/nucleótido en una cadena de ácidos nucleicos, una subunidad formada por un azúcar, un fosfato y una base nitrogenada (86)

nucleus/núcleo en una célula eucariótica, un organelo cubierto por una membrana, el cual contiene el ADN de la célula y participa en procesos tales como el crecimiento, metabolismo y reproducción (7)

O

organ/órgano un conjunto de tejidos que desempeñan una función especializada en el cuerpo (21)

organelle/organelo uno de los cuerpos pequeños del citoplasma de una célula que están especializados para llevar a cabo una función específica (7)

organism/organismo un ser vivo; cualquier cosa que pueda llevar a cabo procesos vitales independientemente (22)

organ system/aparato (o sistema) de órganos un grupo de órganos que trabajan en conjunto para desempeñar funciones corporales (22)

osmosis/ósmosis la difusión del agua a través de una membrana semipermeable (35)

P

Paleozoic era/era Paleozoica la era geológica que vino después del período Precámbrico; comenzó hace 543 millones de años y terminó hace 248 millones de años (244)

passive transport/transporte pasivo el movimiento de substancias a través de una membrana celular sin que la célula tenga que usar energía (36)

pedigree/pedigrí un diagrama que muestra la incidencia de un carácter genético en varias generaciones de una familia (74)

phenotype/fenotipo la apariencia de un organismo u otra característica perceptible (62)

photosynthesis/fotosíntesis el proceso por medio del cual las plantas, las algas y algunas bacterias utilizan la luz solar, el dióxido de carbono y el agua para producir alimento (38)

Plantae/Plantae un reino formado por organismos pluricelulares complejos que normalmente son verdes, tienen una pared celular de celulosa, no tienen capacidad de movimiento y utilizan la energía del Sol para producir azúcar mediante la fotosíntesis (173)

plate tectonics/tectónica de placas la teoría que explica cómo se mueven y cambian de forma las placas tectónicas, que son grandes porciones de la capa más externa de la Tierra (140)

Precambrian time/tiempo Precámbrico el período en la escala de tiempo geológico que abarca desde la formación de la Tierra hasta el comienzo de la era Paleozoica; comenzó hace aproximadamente 4.6 mil millones de años y terminó hace 543 millones de años (142)

primate/primate un tipo de mamífero caracterizado por tener pulgares oponibles y visión binocular (148)

probability/probabilidad la probabilidad de que ocurra un posible suceso futuro en cualquier caso dado del suceso (64)

prokaryote/procariote un organismo que está formado por una sola célula y que no tiene núcleo (8)

Protista/Protista un reino compuesto principalmente por organismo eucarióticos unicelulares que son diferentes de las plantas, animales, bacterias y hongos (172)

R

recessive trait/carácter recesivo un carácter que se hace aparente sólo cuando se heredan dos alelos recesivos de la misma característica (59)

relative dating/datación relativa cualquier método que se utiliza para determinar si un acontecimiento u objeto es más viejo o más joven que otros acontecimientos u objetos (137)

ribosome/ribosoma un organelo celular compuesto de ARN y proteína; el sitio donde ocurre la síntesis de proteínas (15, 93)

RNA/ARN **á**cido **r**ibo**n**ucleico, una molécula que está presente en todas las células vivas y que juega un papel en la producción de proteínas (92)

S

selective breeding/reproducción selectiva la práctica humana de cruzar animales o plantas que tienen ciertas características deseadas (118)

sex chromosome/cromosoma sexual uno de los dos cromosomas que determinan el sexo de un individuo (73)

speciation/especiación la formación de especies nuevas como resultado de la evolución (124)

species/especie un grupo de organismos que tienen un parentesco cercano y que pueden aparearse para producir descendencia fértil (108)

structure/estructura el orden y distribución de las partes de un organismo (22)

T

taxonomy/taxonomía la ciencia de describir, nombrar y clasificar organismos (165)

tissue/tejido un grupo de células similares que llevan a cabo una función común (21)

trait/carácter una característica determinada genéticamente (118)

V

vesicle/vesícula una cavidad o bolsa pequeña que contiene materiales en una célula eucariótica; se forma cuando parte de la membrana celular rodea los materiales que van a ser llevados al interior la célula o transportados dentro de ella (17)

Index

Boldface page numbers refer to illustrative material, such as figures, tables, margin elements, photographs, and illustrations.

A

B

C

D

E

Index

I

J

K

L

M

Credits

Abbreviations used: (t) top, (c) center, (b) bottom, (l) left, (r) right, (bkgd) background

PHOTOGRAPHY

Front Cover Dennis Kunkel/Phototake

Skills Practice Lab Teens Sam Dudgeon/HRW

Connection to Astrology Corbis Images; **Connection to Biology** David M. Phillips/Visuals Unlimited; **Connection to Chemistry** Digital Image copyright © 2005 PhotoDisc; **Connection to Environment** Digital Image copyright © 2005 PhotoDisc; **Connection to Geology** Letraset Phototone; **Connection to Language Arts** Digital Image copyright © 2005 PhotoDisc; **Connection to Meteorology** Digital Image copyright © 2005 PhotoDisc; **Connection to Oceanography** © ICONOTEC; **Connection to Physics** Digital Image copyright © 2005 PhotoDisc

Table of Contents iv (tr), James Beveridge/Visuals Unlimited; iv (bl), ©National Geographic Image Collection/Ned M. Seidler; v (tr), Sam Dudgeon/HRW; x (bl), Sam Dudgeon/HRW; xi (tl), John Langford/HRW; xi (b), Sam Dudgeon/HRW; xii (tl), Victoria Smith/HRW; xii (bl), Stephanie Morris/HRW; xii (br), Sam Dudgeon/HRW; xiii (tl), Patti Murray/Animals, Animals; xiii (tr), Jana Birchum/HRW; xiii (b), Peter Van Steen/HRW

Chapter One 2–3, Dennis Kunkel/Phototake; 4 (l), Visuals Unlimited/Kevin Collins; 4 (r), Leonard Lessin/Peter Arnold; 5 (r), T.E. Adams/Visuals Unlimited; 5 (cl), Roland Birke/Peter Arnold, Inc.; 5 (bkgd), Jerome Wexler/Photo Researchers, Inc.; 5 (cr), Biophoto Associates/Photo Researchers, Inc.; 5 (l), M.I. Walker/Photo Researchers, Inc.; 6 Photodisc, Inc.; 7 (t), William Dentler/BPS/Stone; 7 (b), Dr. Gopal Murti/Science Photo Library/Photo Researchers, Inc.; 9 Wolfgang Baumeister/Science Photo Library/Photo Researchers, Inc.; 10 (l), Biophoto Associates/Photo Researchers, Inc.; 14 (bl), Don Fawcett/Visuals Unlimited; 14 (t), Dr. Peter Dawson/Science Photo Library/Photo Researchers, Inc.; 15 (r), R. Bolender–D. Fawcett/Visuals Unlimited; 16 (cl), Don Fawcett/Visuals Unlimited; 16 (bl), Newcomb & Wergin/BPS/Tony Stone Images; 17 (br), Garry T Cole/BPS/Stone; 18 (tl), Dr. Gopal Murti/Science Photo Library/Photo Researchers, Inc.; 18 (cl), Dr. Jeremy Burgess/Science PhotoLibrary/Science Source/Photo Researchers; 76 Quest/Science Photo Library/Photo Researchers, Inc.; 77 Manfred Kage/Peter Arnold, Inc. ; 24 (b), Sam Dudgeon/HRW; 30 (r), Photo Researchers, Inc.; 30 (l), Science Photo Library/Photo Researchers, Inc.; 31 (b), Digital Image copyright © 2005 Artville; 31 (t), Courtesy Caroline Schooley

Chapter Two 32–33 © Michael & Patricia Fogden/CORBIS; 34 Sam Dudgeon/HRW; 36 (br), Photo Researchers; 37 (tr), Birgit H. Satir; 38 (l), Runk/Schoenberger/Grant Heilman; 39 (r), John Langford/HRW Photo; 41 Corbis Images; 42 CNRI/Science Photo Library/Photo Researchers, Inc ; 43 (t), L. Willatt, East Anglian Regional Genetics Service/Science Photo Library/Photo Researchers, Inc ; 43 (b), Biophoto Associates/Photo Researchers; 44 (b), Visuals Unlimited/R. Calentine; 44 (cl), Ed Reschke/Peter Arnold, Inc.; 44 (c), Ed Reschke/Peter Arnold, Inc.; 44 (cr), Ed Reschke/Peter Arnold, Inc.; 45 (cl), Ed Reschke/Peter Arnold, Inc.; 45 (c), Biology Media/Photo Researchers, Inc.; 45 (cr), Biology Media/Photo Researchers, Inc.; 46 Sam Dudgeon/HRW; 47 Sam Dudgeon/HRW; 48 (l), Runk/Schoenberger/Grant Heilman; 49 (cl), Biophoto Associates/Science Source/Photo Researchers; 49 (cr), Biophoto Associates/Science Source/Photo Researchers; 49 (br), John Langford/HRW Photo; 52 (l), Lee D. Simons/Science Souce/Photo Researchers; 53 (tr), Courtesy Dr. Jarrel Yakel; 53 (tr), David McCarthy/SPL/Photo Researchers, Inc.

Chapter Three 54–55 © Maximilian Weinzierl/Alamy Photos; 56 Ned M. Seidler/National Geographic Society Image Collection; 61 © Andrew Brookes/CORBIS; 62 © Joe McDonald/Visuals Unlimited; 63 (b), Sam Dudgeon/HRW; 65 Digital Image copyright © 2005 PhotoDisc; 66 (b), © Mervyn Rees/Alamy Photos; 67 (b), Image Copyright ©2001 Photodisc, Inc.; 67 (tl), Sam Dudgeon/HRW; 67 (tr), Sam Dudgeon/HRW; 68 (br), Biophoto Associates/Photo Researchers, Inc.; 68 (b), Phototake/CNRI/Phototake NYC; 73 (b), © Rob vanNostrand; 74 (b), © ImageState; 75 © ImageState; 76 (b), Sam Dudgeon/HRW; 77 (b), Sam Dudgeon/HRW; 79 (r), © Mervyn Rees/Alamy Photos; 79 (l), © Rob vanNostrand; 82 (c), Dr. F. R. Turner, Biology Dept., Indiana University; 82 (r), Dr. F. R. Turner, Biology Dept., Indiana University; 82 (l), Hank Morgan/Rainbow; 83, Courtesy of Stacey Wong

Chapter Four 84–85 US Department of Energy/Science Photo Library/Photo Researchers, Inc.; 87 (r), Science Photo Library/Photo Researchers, Inc.; 87 (l), Hulton Archive/Getty Images; 90 (l), Sam Dudgeon/HRW; 90 (l), Sam Dudgeon/HRW; 91 (bl), David M. Phillips/Visuals Unlimited; 91 (cl), J.R. Paulson & U.K. Laemmli/University of Geneva; 95 (br), Jackie Lewin/Royal Free Hospital/Science Photo Library/Photo Researchers, Inc.; 95 (tr), Jackie Lewin/Royal Free Hospital/Science Photo Library/Photo Researchers, Inc.; 96 (t), Visuals Unlimited/Science Visuals Unlimited/Keith Wood ; 96 (b), Volker Steger/Peter Arnold; 97 Sam Dudgeon/HRW; 99 Victoria Smith/HRW; 104 (l), Robert Brook/Science Photo Library/Photo Researchers, Inc.; 105 (r), Photo courtesy of the Whitehead Institute for Biomedical Research at MIT; 105 (l), Garry Watson/Science Photo Library/Photo Researchers, Inc.106–107 (t), © Stuart Westmorland/CORBIS; 108 (bl), James Beveridge/Visuals Unlimited; 108 (tc), © Gail Shumway/Getty Images/FPG International; 108 (br), Doug Wechsler/Animals Animals

Chapter Five 110 (l), Ken Lucas; 110 (r), John Cancalosi/Tom Stack & Associates; 111 (cl), © SuperStock; 111 (bl), © Martin Ruegner/Alamy Photos; 111 (br), © James D. Watt/Stephen Frink Collection/Alamy Photos; 111 (tl), © Ron Kimball/Ron Kimball Stock; 111 (tr), © Carl & Ann Purcell/CORBIS; 111 (cr), © Martin B. Withers; Frank Lane Picture Agency/CORBIS; 112 (tr), Illustration by Carl Buell, and taken from http://www.neoucom.edu/Depts/Anat/Pakicetid.html. ; 112 (tl), Courtesy of Research Casting International and Dr. J. G. M. Thewissen; 112 (b), Courtesy of Research Casting International and Dr. J. G. M. Thewissen; 113 (t), © 1998 Philip Gingerich/Courtesy of the Museum of Paleontology, The University of Michigan; 113 (c), Courtesy of Betsy Webb, Pratt Museum, Homer, Alaska; 113 (b), Courtesy of Betsy Webb, Pratt Museum, Homer, Alaska; 115 (b), Visuals Unlimited/H.W. Robison; 115 (t), James Beveridge/Visuals Unlimited; 116 (l), Christopher Ralling; 116 (r), © William E. Ferguson; 118 (b), Carolyn A. McKeone/Photo Researchers, Inc. ; 122 (b), Getty Images/Stone; 125 (r), Zig Leszczynski/Animals Animals/Earth Scenes; 125 (l), Gary Mezaros/Visuals Unlimited; 127 Victoria Smith/HRW; 128 (t), James Beveridge/Visuals Unlimited; 129 Courtesy of Betsy Webb, Pratt Museum, Homer, Alaska; 132 (l), Doug Wilson/Westlight; 133 (r), Wally Emerson/Courtesy of Raymond Pierotti; 133 (l), George D. Lepp/Photo Researchers, Inc.; 134–135 © Reuters NewMedia Inc./CORBIS

Chapter Six 148 (r), © Daniel J. Cox/Getty Images/Stone; 150 (b), John Reader/Science Photo Library/Photo Researchers; 150 (t), © Partick Robert/CORBIS; 151 John Gurche; 152 (tr), John Reader/Science Photo Library/Photo Researchers; 152 (tl & tc), E.R. Degginger/Bruce Coleman; 152 (bl), Neanderthal Museum; 152 (br), Volker Steger/Nordstar–4 Million Years, of Man/Science Photo Library/Photo Researchers, Inc.; 160 (r), Rikki Larma/AP/Wide World Photos; 160 (l), Copyright M. Ponce de León and Christoph Zollikofer, Zurich; 161 (r), © Robert Campbell/CORBIS; 161 (l), John Reader/Science Photo Library/Photo Researchers, Inc.; 162–163 Frans Lanting/Minden Pictures

Chapter Seven 164 Gerry Ellis/ENP Images; 170 Biophoto Associates/Photo Researchers; 171 (t), Sherrie Jones/Photo Researchers; 171 (br), Dr. Tony Brian & David Parker/Science Photo Library/Photo Researchers; 171 (inset), Dr. Tony Brian & David Parker/Science Photo Library/Photo Researchers; 172 (t), Visuals Unlimited/Stanley Flegler; 172 (b), © Phil A. Dotson/Photo Researchers, Inc.; 173 (l), © Royalty–free/CORBIS; 173 (r), David R. Frazier/Photo Researchers, Inc.; 174 (tl), SuperStock; 174 (tr), © G. Randall/Getty Images/FPG International; 174 (bl), SuperStock; 175, Corbis Images; 175 Dr. Tony Brian & David Parker/Science Photo Library/Photo Researchers; 178 (b), Dr. Tony Brian & David Parker/Science Photo Library/Photo Researchers; 179 (b), SuperStock; 182 (r), Piotr Naskrecki/Conservation International; 183 (r), © National Geographic Image Collection/Michael Nichols; 183 (l), © National Geographic Image Collection/Michael Fay

Lab Book/Appendix "LabBook Header", "L", Corbis Images; "a", Letraset Phototone; "b", and "B", HRW; "o", and "k", images ©2006 PhotoDisc/HRW; 184 (tl), Runk/Schoenberger/Grant Heilman; 184 (tc), Runk/Schoenberger/Grant Heilman; 184 (tr), Michael Abbey/Photo Researchers, Inc.; 184 (tr), Sam Dudgeon/HRW; 184 (br), Runk/Schoenberger/Grant Heilman; 185 (b), Sam Dudgeon/HRW; 185 (c), Sam Dudgeon/HRW; 186 Sam Dudgeon/HRW; 189 (all), Sam Dudgeon/HRW; 198 Sam Dudgeon/HRW; 199 Sam Dudgeon/HRW; 204 (t), Peter Van Steen/HRW; 204 (b), Sam Dudgeon/HRW

TEACHER EDITION CREDITS

1E (tl), Visuals Unlimited/Kevin Collins; 1F (l), Quest/Science Photo Library/Photo Researchers, Inc.; 31E (t), Photo Researchers, Inc.; 31F (tl), L. Willatt, East Anglian Regional Genetics Service/Science Photo Library/Photo Researchers, Inc.; 31F (r), Ed Reschke/Peter Arnold, Inc.; 53E (l), Ned M. Seidler/National Geographic Society Image Collection; 83E (l), Hulton Archive/Getty Images; 83F (l), Visuals Unlimited/Science Visuals Unlimited/Keith Wood; 83F (r), Volker Steger/Peter Arnold; 105E (l), Courtesy of Betsy Webb, Pratt Museum, Homer, Alaska; 133F (r), Neanderthal Museum; 161F (tl), Biophoto Associates/Photo Researchers , Inc.; 161F (bl), Sherrie Jones/Photo Researchers, Inc.; 161F (r), Dr. Tony Brian & David Parker/Science Photo Library/ Photo Researchers, Inc.

Answers to Concept Mapping Questions

The following pages contain sample answers to all of the concept mapping questions that appear in the Chapter Reviews. Because there is more than one way to do a concept map, your students' answers may vary.

CHAPTER 1 Cells: The Basic Units of Life

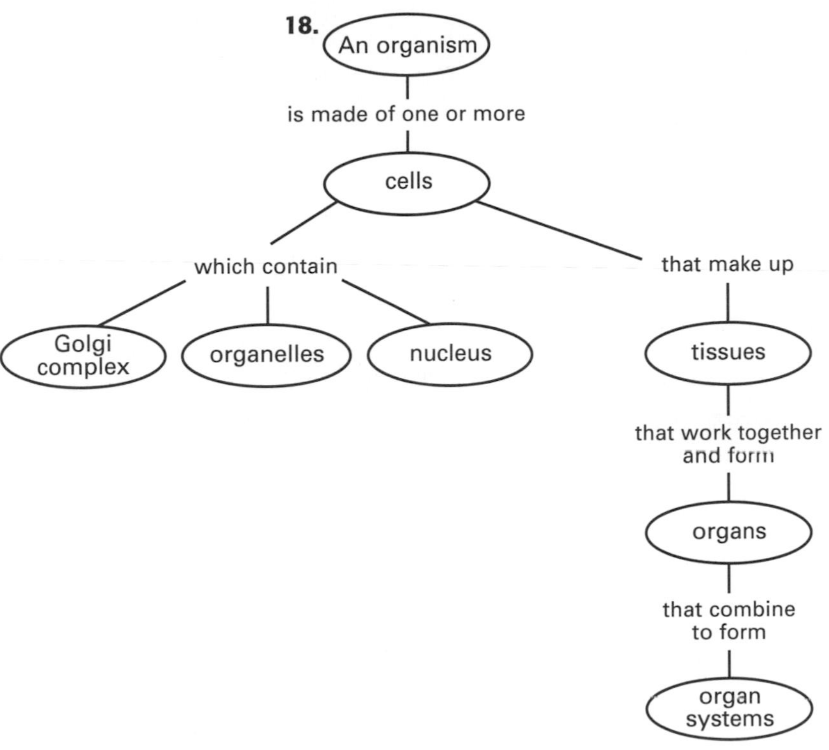

CHAPTER 2 The Cell in Action

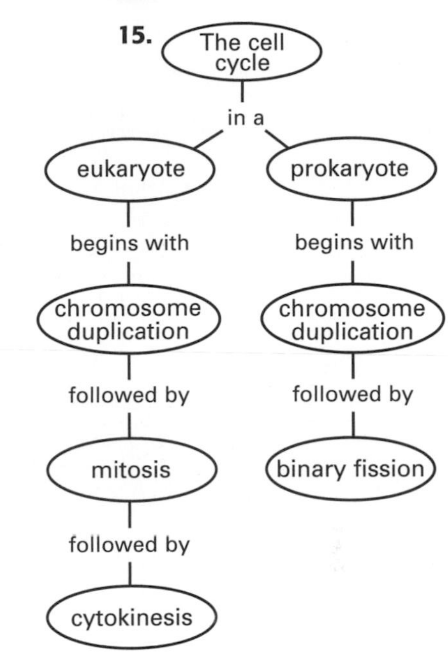

CHAPTER 3 Heredity

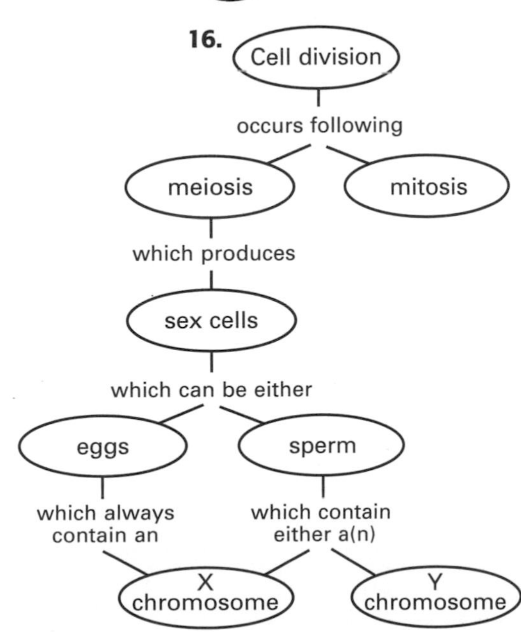

CHAPTER 4 Genes and DNA

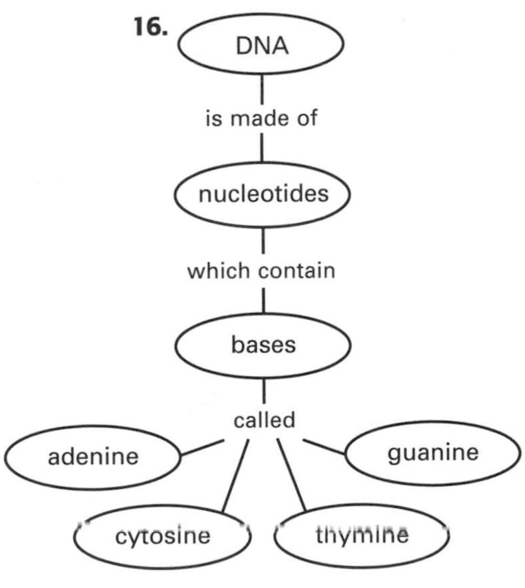

CHAPTER 5 The Evolution of Living Things

17.

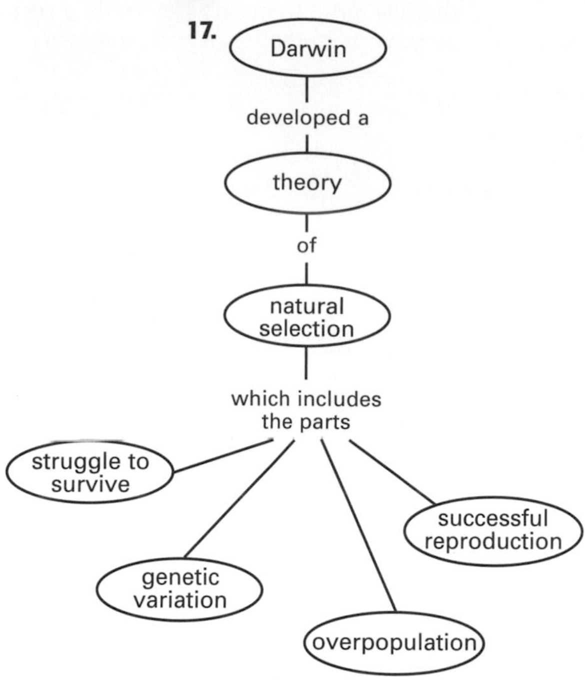

CHAPTER 6 The History of Life on Earth

19.

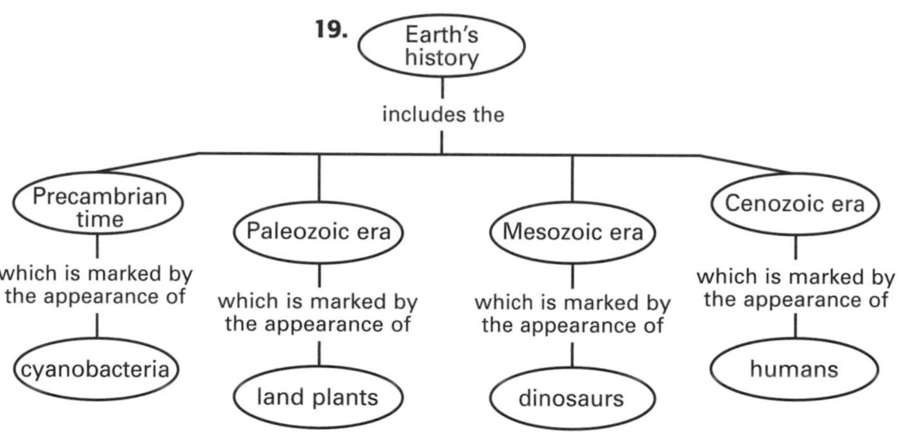

CHAPTER 7 Classification

16.

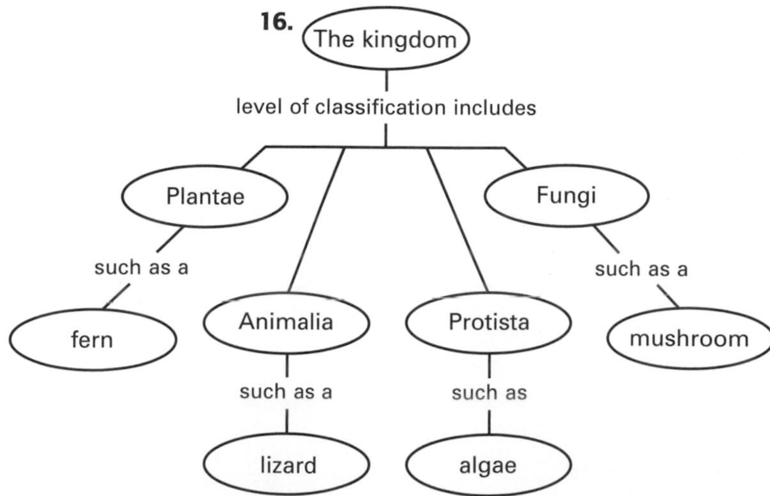